Biology of Plant Volatiles

Biology of Plant Volatiles

Second Edition

Edited by
Eran Pichersky
Natalia Dudareva

CRC Press
Taylor & Francis Group
Boca Raton London New York

CRC Press is an imprint of the
Taylor & Francis Group, an **informa** business

Cover photo: Male thynnine wasps are lured to the flowers of the Australian orchid *Caladenia plicata*, because the flowers emit two volatiles, (S)-β-citronellol and 2-hydroxy-6-methylacetophenone, that also serve as the sex pheromone of female thynnine wasps. The flowers also morphologically resemble the female wasps, and upon landing on a flower, the male wasp tries to copulate with it, in the process picking up pollen on its body and also depositing on the stigma of the present flower pollen that was picked up during earlier visits to other flowers. This form of "pollination by sexual deception" is wide spread among orchids in multiple continents. The picture of this *Caladenia plicata* flower was taken in Western Australia by Liza Pichersky.

CRC Press
Taylor & Francis Group
6000 Broken Sound Parkway NW, Suite 300
Boca Raton, FL 33487-2742

First issued in paperback 2022

© 2020 by Taylor & Francis Group, LLC

CRC Press is an imprint of Taylor & Francis Group, an Informa business

No claim to original U.S. Government works

ISBN-13: 978-1-138-31649-2 (hbk)
ISBN-13: 978-1-03-233614-5 (pbk)
DOI: 10.1201/9780429455612

Publisher's Note

The publisher has gone to great lengths to ensure the quality of this reprint but points out that some imperfections in the original copies may be apparent.

Library of Congress Control Number: 2020938232

Typeset in Times
by Lumina Datamatics Limited

Visit the Taylor & Francis Web site at
http://www.taylorandfrancis.com

and the CRC Press Web site at
http://www.crcpress.com

Contents

SECTION I   Chemistry of Plant Volatiles

SECTION II   Biochemistry, Molecular Biology, and Evolution of Plant Volatiles

SECTION III   Volatiles in Plant–Plant, Plant–Insect and Plant–Microbial Interactions

SECTION IV *Genetic Improvements of Plant Volatiles*

Preface

Volatiles – chemicals with high vapor pressure that thus vaporize well below their boiling point – are common in plants. The scents of flowers immediately come to our mind, as do the smells we detect from the crushed leaves of plants we often find ourselves stepping on, brushing against, or adding to our foods. Indeed, volatiles emanating from plants serve as signals that can be detected by another living organism – be it an animal, a microorganism or even another plant. Such signals have been found to benefit the plant from which the volatiles are released. Floral volatiles, for example, attract pollinators to the flower. But, as detailed in this book, volatile compounds have a multitude of other functions beyond pollinator attraction. For example, volatiles emitted from a plant under herbivoral attack may attract an enemy of the herbivore, and many volatiles are toxic to organisms that harm plants in one way or another, thus protecting plants from pathogenes, parasites, and herbivores. Potential plant enemies often learn to avoid plants containing such compounds, and therefore these volatiles can be said to also serve as repellents and deterrents.

Some nonvolatile plant compounds also serve as toxic, defensive compounds, while other nonvolatile compounds can function as long-distance signals – for example, pigments that give flowers specific colors and patterns. Therefore, why treat volatile compounds as a separate class and devote a whole book to this topic? The simple answer is that investigating the presence, biosynthesis, and function of volatiles in plants requires a unique set of experimental methods, techniques, and instrumentations. Plant scents are not the kind of materials that we can observe and measure with our eyes or ears. Even our noses are not up to the task – there is too much variation among humans in the quantitative and qualitative detection of specific volatiles. Fortunately, the last half century has seen a significant improvement in the type of instrumentation available for studying volatiles. This in turn has led to a vast increase in the numbers of researchers investigating volatiles in plants as well as in the amount and quality of research in the field of plant volatile biology.

In 2006, we edited a book, appropriately titled *The Biology of Floral Scent*, that provided authoritative reviews covering the most up-to-date knowledge of the chemistry, biochemistry, physiology, and ecology of floral volatiles. With the significant progress in the study of volatiles produced in the rest of the plant body, we feel it is now an opportune time to assemble another compendium of authoritative reviews on all aspects of plant volatile biology. Floral scent is still extensively covered, and a few of the chapters in this new book are updated versions of chapters appearing in *Biology of Floral Scent*. But the rest are completely new chapters covering new topics in floral scent as well as plant volatiles found in leaves, stems, roots and fruits. We hope that this book, too, will serve to stimulate additional research to advance our understanding of all aspects of plant volatile biology.

Editors

Eran Pichersky is the Michael M. Martin Collegiate Professor in the Department of Molecular, Cellular, and Developmental Biology (MCDB) at the University of Michigan. He earned his B.Sc. degree from the University of California, Berkeley in 1980, and a Ph.D. from the University of California, Davis in 1984. After doing research as a postdoctoral fellow at the Rockefeller University in New York, he has been on the faculty of the University of Michigan since 1986, serving as the first Chair of the newly created MCDB Department from 2001 to 2003. His awards include a Fulbright fellowship and an Alexander von Humboldt fellowship, both received in 2000, and a Guggenheim fellowship in 2015. He was elected a fellow by the American Association for the Advancement of Science (AAAS) in 2012 and by the American Society of Plant Biologists in 2017. Dr. Pichersky has served on the editorial boards of several major scientific journals that cover plant research and had previously edited (together with Dr. Natalia Dudareva) a book on the *Biology of Floral Scent* published by CRC Press.

Dr. Pichersky's research has concentrated on identifying the myriad compounds that are found uniquely in plants, many of which are extensively used by people, with emphasis on those that impart scent and flavor. His group further elucidates how plants synthesize these compounds and how this information can be used to enhance the production by plants of such valuable chemicals. Over the years Dr. Pichersky's research group has collaborated with many other research groups around the world, and Dr. Pichersky himself has spent extensive time as a visiting scholar doing research at scientific institutes around the world, including the United States, Germany, Israel and Australia. Dr. Pichersky has authored more than 250 reports, reviews, letters and editorials in scientific publications, and is a recipient of several patents.

Natalia Dudareva is a distinguished professor in the Department of Biochemistry at Purdue University. She earned her M.S. in biology and biochemistry from Novosibirsk State University, Russia and her Ph.D. at the Institute of Biochemistry, Kiev, Ukraine and the University of Louis Pasteur, Strasbourg, France. After a postdoctoral training at the University of Michigan, Dr. Dudareva joined the Department of Horticulture and Landscape Architecture, Purdue University in 1997 and received a distinguished professorship in 2010. She joined the Department of Biochemistry in 2013. Research in the Dudareva laboratory focuses on understanding biochemical and molecular mechanisms controlling the formation of primary and secondary (phenylpropanoid and terpenoid) metabolites in plants using the power of genetic and biochemical approaches combined with metabolic flux analysis and modeling. With initial focus on plant volatiles, her group has not only identified many genes involved in the final steps of their biosynthesis but also discovered the primary metabolic networks that supply precursors for volatile formation. Her group also showed that active biological mechanisms are involved in transporting volatile compounds from plant cells to the atmosphere.

Dr. Dudareva has published 120 papers, 25 book chapters and 2 books, and has given more than 200 invited lectures at conferences and other universities. She has received recognition for her research as Purdue University faculty scholar and the Wickersham chair of excellence in Agricultural Research. She has been awarded the Purdue University Agricultural Research Award, Sigma Xi Faculty Research Award, the Alexander von Humboldt Research Award (Germany) and 2018 Herbert Newby McCoy Award, Purdue's most prestigious award given for outstanding work in the natural sciences. In 2010 she was elected a Fellow of the American Association for the Advancement of Science.

Contributors

Sylvie Baudino
University of Lyon, UJM-Saint-Etienne, CNRS
Saint-Etienne, France

Björn Bohman
School of Molecular Sciences
The University of Western Australia
Crawley, Western Australia, Australia

Anna-Karin Borg-Karlson
Department of Chemistry
School of Engineering Sciences in Chemistry,
    Biotechnology and Health
Royal Institute of Technology
Stockholm, Sweden

Harro Bouwmeester
Plant Hormone Biology Group
Swammerdam Institute for Life Sciences
University of Amsterdam
Amsterdam, The Netherlands

Jean-Claude Caissard
University of Lyon, UJM-Saint-Etienne, CNRS
Saint-Etienne, France

Feng Chen
Department of Plant Sciences
University of Tennessee
Knoxville, Tennessee

and

College of Horticulture
Nanjing Agricultural University
Nanjing, China

Marcel Dicke
Laboratory of Entomology
Wageningen University
Wageningen, The Netherlands

Lemeng Dong
Plant Hormone Biology Group
Swammerdam Institute for Life Sciences
University of Amsterdam
Amsterdam, The Netherlands

Natalia Dudareva
Department of Biochemistry
Purdue University
West Lafayette, Indiana

Franziska Eberl
Department of Biochemistry
Max Planck Institute for Chemical Ecology
Jena, Germany

Gavin R. Flematti
School of Molecular Sciences
The University of Western Australia
Crawley, Western Australia, Australia

Jonathan Gershenzon
Department of Biochemistry
Max Planck Institute for Chemical Ecology
Jena, Germany

Antonio Granell
Plant Genomics and Biotechnology Lab.
    IBMCP (CSIC-UPV) Ingeniero Fausto
Valencia, Spain

Michel A. Haring
Department of Plant Physiology
Swammerdam Institute for Life Sciences
University of Amsterdam
Amsterdam, The Netherlands

Philippe Hugueney
Université de Strasbourg
INRAE, SVQV UMR-A 1131
Colmar, France

Mwafaq Ibdah
Newe Ya'ar Research Center
Agriculture Research Organization
Ramat Yishay, Israel

Yoko Iijima
Department of Nutrition and Life Science
Kanagawa Institute of Technology
Atsugi, Kanagawa, Japan

Yifan Jiang
College of Horticulture
Nanjing Agricultural University
Nanjing, China

Jette T. Knudsen
Nattaro Labs AB
Lund, Sweden

Takao Koeduka
Division of Agricultural Sciences
Graduate School of Sciences and Technology
    for Innovation
Yamaguchi University
Yamaguchi, Japan

Pan Liao
Department of Biochemistry
Purdue University
West Lafayette, Indiana

Bernd Markus Lange
M.J. Murdock Metabolomics Laboratory
Institute of Biological Chemistry
Washington State University
Pullman, Washington

Efraim Lewinsohn
Newe Ya'ar Research Center
Agriculture Research Organization
Ramat Yishay, Israel

Dani Lucas-Barbosa
Laboratory of Entomology
Wageningen University
Wageningen, The Netherlands

and

Bio-Communication & Ecology, ETH Zürich
Zürich, Switzerland

Joseph H. Lynch
Department of Biochemistry
Purdue University
West Lafayette, Indiana

Itay Maoz
Department of Biochemistry
Purdue University
West Lafayette, Indiana

Kenji Matsui
Division of Agricultural Sciences
Graduate School of Sciences and Technology
    for Innovation
Yamaguchi University
Yamaguchi, Japan

Bhagwat Nawade
Newe Ya'ar Research Center
Agriculture Research Organization
Ramat Yishay, Israel

Naomi Okubo
Institute of Vegetable and Floriculture Science
Tsukuba, Japan

Rod Peakall
Ecology and Evolution, Research School of
    Biology
The Australian National University
Canberra, Australian Capital Territory,
    Australia

Erica Perreca
Department of Biochemistry
Max Planck Institute for Chemical Ecology
Jena, Germany

Eran Pichersky
Department of Molecular, Cellular, and
    Developmental Biology
University of Michigan
Ann Arbor, Michigan

Milan Plasmeijer
Department of Plant Physiology
Swammerdam Institute for Life Sciences,
    Green Life Sciences Cluster
University of Amsterdam
and
Department of Green Biotechnology
Domain for Agri, Food and Life Sciences
Inholland University of Applied Sciences
Amsterdam, The Netherlands

Robert A. Raguso
Department of Neurobiology and Behavior
Cornell University
Ithaca, New York

José L. Rambla
Plant Genomics and Biotechnology Lab.
  IBMCP (CSIC-UPV) Ingeniero Fausto Elio
Valencia, Spain

and

Department of Agricultural and Environmental
  Sciences
Universitat Jaume I
Castellón de la Plana, Spain

Ursula S. R. Röse
Department of Biology
University of New England
Biddeford, Maine

Robert C. Schuurink
Department of Plant Physiology
Swammerdam Institute for Life Sciences
University of Amsterdam
Amsterdam, The Netherlands

Koichi Sugimoto
Science Research Center
Yamaguchi University
Yamaguchi, Japan

Pulu Sun
Department of Plant Physiology
Swammerdam Institute for Life Sciences
University of Amsterdam
Amsterdam, The Netherlands

Fukuyo Tanaka
Central Region Agricultural Research Center
National Agriculture and Food Research
  Organization (NARO)
Tsukuba, Japan

Dorothea Tholl
Department of Biological Sciences
Virginia Polytechnic Institute and State
  University
Blacksburg, Virginia

Alexander Weinhold
Molecular Interaction Ecology
German Centre for Integrative Biodiversity
  Research (iDiv) Halle-Jena-Leipzig
Leipzig, Germany

and

Molecular Interaction Ecology
Institute of Biodiversity
Friedrich Schiller University Jena
Jena, Germany

Darren C. J. Wong
Ecology and Evolution
Research School of Biology
The Australian National University
Canberra, Australian Capital Territory,
  Australia

Mossab Yahyaa
Newe Ya'ar Research Center
Agriculture Research Organization
Ramat Yishay, Israel

Section I

Chemistry of Plant Volatiles

1 Practical Approaches to Plant Volatile Collection and Analysis

Dorothea Tholl, Alexander Weinhold, and Ursula S. R. Röse

CONTENTS

1.1 INTRODUCTION

Any treatment of the subject of plant volatile organic compounds (VOCs) must begin with a description of how these compounds are detected and identified. The analysis of plant volatiles was developed more than 40 years ago as so-called "headspace" analysis, a term derived from the beer industry, where the identification of volatile compounds in the "head" of the beer was first established. Since then, methods to sample plant VOCs in different environments have substantially advanced and analytical techniques from gas chromatography-mass spectrometry (GC-MS) to VOC sensors and online instrumentation have become sensitive enough to monitor VOC emissions at low concentrations. Several previous reviews have discussed and compared different techniques in plant

VOC analysis (Knudsen et al., 1993; Millar and Sims, 1998; Raguso and Pellmyr, 1998; Tholl et al., 2006; Penuelas et al., 2014; Materic et al., 2015; van Dam et al., 2016). This chapter will provide an overview of the practical approaches to plant VOC analysis and discuss their advantages and limitations as well as future developments.

The first step in choosing a sample technique should always focus on the biology of the plant system and the purpose of the VOC analysis. While most studies in the past have addressed functions of VOCs emitted from flowers and leaves, increased interest in recent years has been placed on understanding the role of plant VOCs in the soil (van Dam and Bouwmeester, 2016; van Dam et al., 2016) or even the aquatic environment (Fink, 2007; Pohnert et al., 2007; Giordano et al., 2017). Depending on the overall complexity of VOCs in the background of these environments and the presence or absence of water, suitable instrumentation and devices need to be selected. Furthermore, it has to be decided whether VOCs need to be collected in the field or if the compounds can be sampled in the laboratory without affecting the composition of the blend. While some collection methods are transportable and can easily be taken to the field, other more sophisticated methods may require a complicated setup suitable only in the laboratory. If VOCs from roots are collected, the researcher needs to decide whether to sample the compounds *in situ* or first measure them from detached root tissue.

Another consideration should focus on whether the VOCs of the investigated plant are already known and identified and only need to be confirmed (e.g., repetitive insect behavioral experiments), or whether the volatile blend is unknown and needs to be identified. Characterization of a volatile blend often requires additional analytical steps and more source material for compound identification. Also, if several species are screened for the presence of only one or two compounds, the most appropriate technique may differ from the one necessary for identification of a complex volatile blend. Depending on the expected detectability of the volatile mixture, one may have to choose different types of collection methods that vary in their sensitivity.

Another important consideration is the developmental stage of the plant and the timing of the volatile collection. For example, flowers of the orchid *Ophrys sphegodes* are known to change their odor emission after pollination has occurred (Schiestl and Ayasse, 2001). The volatile profile emitted by flowers may also vary depending on the time of day (e.g., Effmert et al., 2005) as some flowers are mainly pollinated by moth and emit volatiles at night to attract their pollinators while others are pollinated by insects that are mainly active during the day (Raguso and Pichersky, 1995). VOC emissions from photosynthetic tissues are also controlled by diurnal cycles and circadian rhythms (e.g., Joo et al., 2019). Moreover, emissions of root derived VOCs may be affected by the type of aboveground stress a plant is exposed to or vice versa (Kaplan et al., 2008; Tytgat et al., 2013). Depending on the time intervals of volatile sampling, some methods allow for an easy automated setup for 24 h collection, while some are most appropriate for taking a "snapshot" of the current volatile release. Latest developments in precision agriculture and phenotyping require fast measurements of such snapshots or fingerprints for early monitoring of disease outbreaks (Jansen et al., 2011; Aksenov et al., 2014; Martinelli et al., 2015).

In the following sections we present collection methods that range from low-tech, inexpensive, quick sampling methods to automated high-tech methods that collect samples in short time intervals over a period of several days. We also present an overview of detection and identification methods for volatiles, including GC-MS and multidimensional GC, and discuss the latest developments in ultrafast volatile analysis techniques such as proton-transfer reaction mass spectrometry (PTR-MS) and portable sensing platforms.

1.2   PLANT VOLATILE SAMPLING TECHNIQUES

To construct a chamber for headspace collection, all materials that may retain volatiles or cause bleeding of compounds that contaminate the system need to be avoided. Good choices for materials include glass, Teflon, and metal, which are easy to clean and do not show bleeding, whereas materials such as rubber, plastic, glues, adhesives, and wood should be avoided. Details on the materials for the construction of such chambers are discussed by Millar and Sims (1998).

1.2.1 STATIC HEADSPACE OR *IN SITU* SAMPLING TECHNIQUES

Sampling of VOCs from a static headspace of whole plants or plant tissues is most often performed in closed tubes or chambers. This method enriches emitted compounds and reduces background impurities resulting from a continuous air stream which is an advantage when collecting from low-emitting plants like *Arabidopsis thaliana*. However, for longer sampling times in a static airspace, humidity and a lack of gas exchange may interfere with normal physiological processes and affect the emission of VOCs. If VOCs are sampled in the presence of additional light, a temperature increase in the chamber may affect the emission of compounds. For time course experiments with an expected change in volatile emission this technique is not suitable, because not all of the emitted VOCs are removed at one sampling time and changes in emission will be difficult to determine. In addition to sampling of VOCs from plants enclosed in chambers, the devices applied to static headspace sampling can also be used *in situ* for nonenclosed environments such as canopies or soil as described below.

1.2.1.1 Solid Phase Microextraction

Solid phase microextraction (SPME) is a very fast, effective and simple method to collect volatiles. The method is based on an adsorption-desorption technique using an inert fiber coated with different types of adsorbents that can vary in polarity and thickness and can be selected according to different applications (Zhu et al., 2013). The adsorbent-coated fiber is mounted to a modified syringe and can be extended out of a needle by pushing the plunger and exposing the fiber to volatiles. SPME devices are available from Supelco/Millipore Sigma (St. Louis, MO, USA). To collect from static headspace of a tissue sample that is enclosed in a glass container sealed with a septum, the needle of the SPME-holder is pierced through the septum and the fiber subsequently extended into the headspace (Figure 1.1a). VOCs are then adsorbed to the exposed fiber for several minutes

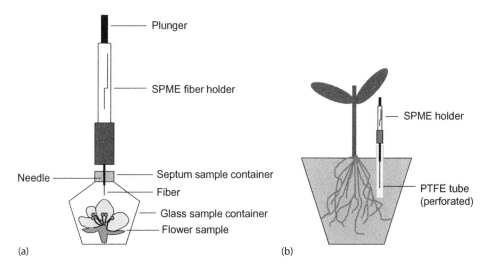

(a)   (b)

FIGURE 1.1 SPME device to extract VOCs from (a) the headspace of a flower enclosed in a glass sample container. The adsorbent-coated fiber is mounted to an SPME fiber holder, similar to a modified syringe that is injected through the septum of the sample container. By pushing the plunger of the SPME fiber holder, the fiber can be extended out of the needle, exposing the fiber to volatiles. After collection, the fiber is retracted into the needle and the SPME is removed for GC analysis. (b) Example of static belowground sampling. A probe is introduced into the soil and a SPME fiber inserted. The SPME fiber can be also replaced by PDMS-based silicone tubing. (Adapted from Deasy, W. et al., *Phytochem. Anal.*, 27, 343–353, 2016a.).

up to an hour until equilibrium is reached. After volatile collection, the fiber is retracted into the needle which is transferred to a GC injector where the fiber is exposed to thermal desorption of the compounds. The direct desorption of VOCs from the fiber into a GC injector eliminates the need for solvent-mediated desorption and thereby reduces background of solvent contaminants in the analysis that may obscure some volatile compound peaks. Two important limitations of SPME sampling are that only one injection per sample can be done (no repeated injections) and that the amount of material obtained from sampling by SPME is generally enough for GC analysis but not sufficient for structure elucidation of unknown compounds.

The amount of compounds adsorbed by the SPME fiber depends on the thickness of the fiber and the distribution constant of the analyte that generally increases with molecular weight and boiling point. For most VOCs a thick coating is recommended to optimally retain compounds until thermal desorption, whereas semivolatile compounds may be better detected with a thin coating. Thicker coatings desorb analytes more slowly which increases the risk of carry-overs. In general, fibers should be cleaned carefully by heating before reusing them. The adsorption of VOCs further depends on the varying polarity and porosity of the surface area. Nonpolar VOCs and nonpolar semivolatiles are effectively extracted with nonpolar fiber coatings such as polydimethylsiloxane (PDMS). Polar volatiles can be extracted with PDMS/Divinylbenzene fibers, and trace level volatiles with a PDMS/Carboxen™ fiber.

The effectiveness of SPME extraction is influenced by the concentration of the VOC. At low concentrations, changes in headspace sample volume do not affect responses, because equilibrium is concentration-dependent. However, at higher concentrations the sample volume has a strong affect. The amount of analyte removed from the sample is not sufficient to change the concentration in a volume of >5 mL. The response throughout a calibration curve is mostly exponential and is linear only for low concentrations (50 ppb). For collection of volatiles from mature leaves or flowers, a small volume of 5 mL or less is often not practical. An external or internal calibration for some compounds may be possible but is often difficult when samples contain a wide range of compounds of different concentrations. Detailed information on theory, optimization and different types of fiber adsorbents is available from Supelco. The company also offers a portable field sampler that has a sealing mechanism to allow storing of samples for later analysis in the laboratory.

The detection of volatiles by SPME has been applied in a large number of studies as summarized by Zhu et al. (2013). Many analyses have focused on VOCs from fruits, flowers, and leaves, while fewer applications have been used for the detection of VOCs from stems, seeds, and roots. To analyze root volatiles or collect volatiles *in situ* directly from the soil, additional pre-collection considerations are required to protect the fragile fiber from breaking or being contaminated with soil (Figure 1.1b). Rasmann et al. (2005) used SPME to determine the diffusion of (*E*)-beta-caryophyllene in sand. SPME has also been applied in the greenhouse and the field by directly measuring VOCs released by *Brassica* roots after infection by the root feeding larvae of *Delia radicum* (Deasy et al., 2016a, b).

1.2.1.2 PDMS-Based Silicone Tubing

In recent years, static headspace sampling with silicone-based tubing has become increasingly popular as an alternative to the easy to use but rather expensive SPME method. The tubing is made of polydimethylsiloxane (PDMS), the same material that is used for SPME. Instead of applying it in the shape of fragile and expensive SPME fibers, the material comes in the form of much cheaper silicone laboratory tubing. The tubing is relatively easy to prepare (Kallenbach et al., 2015) and allows simpler collections with more replicates in the field. This method has been used to sample volatiles from wild tobacco plants (Kallenbach et al., 2014) or poplar trees in the greenhouse (Fabisch et al., 2019). Limitations and drawbacks of using PDMS tubing are similar to those of SPME. An advantage is that large numbers of tubes can be collected and stored in a freezer until later analysis. Analysis of captured VOCs is normally done by thermal desorption coupled with GC. Since silicone tubing is not primarily manufactured for thermal desorption an additional desorption of silicone compounds may occur. Therefore, regular blanks and cleaning of the equipment (e.g., TD sampler tubes) is necessary. This technique has been used successfully to capture VOCs from the rhizosphere of dandelion in a mesocosm experiment (Eilers et al., 2015).

1.2.1.3 Use of Stir-Bar Sorptive Extraction (SBSE) in Headspace Sampling

Another alternative to SPME is the use of stir bars under the brand name Twister®, which are coated with PDMS adsorbent. Similar to PDMS tubing, Twisters® have a larger volume for the adsorption of compounds. Desorption of compounds occurs in a thermal desorption unit coupled with GC (see Section 1.3). The method was developed mostly for the extraction of organic compounds in aqueous samples (see Section 1.9). However, SBSE has also been applied to measure VOCs on the surface of leaves (Kfoury et al., 2017) and to monitor changes in volatile blends induced by disease outbreaks in the field. For example, Twisters® were positioned in stainless steel tea strainers in the canopy of citrus trees to determine VOC profile changes associated with citrus greening disease and tristeza virus infection (Aksenov et al., 2014; Cheung et al., 2015).

1.2.2 Dynamic Headspace Sampling Techniques

Sampling of volatiles from a dynamic headspace eliminates some of the problems that are connected to sampling from static headspace. In general, larger amounts of volatiles can be collected over longer time periods by adsorption (see Section 1.3) in a continuous air stream, allowing not only subsequent detection but also structure elucidation of compounds. In addition, systems with a continuous incoming air stream provide a sufficient temperature and gas exchange and avoid accumulation of compounds in the headspace that may affect the volatile release (pull– and push–pull systems). Relative humidity can be adjusted in a push–pull system to a desired percentage by adding a humidified airstream to the incoming air and mixing it with dry air. However, care needs to be taken to avoid background impurities by cleaning the incoming air carefully with filters containing for example activated charcoal. The problem of background contaminants resulting from continuous incoming air is reduced in closed-loop stripping systems where a limited air volume is sampled repeatedly. All of the above-mentioned considerations hold true for analysis of belowground volatiles. However, the different nature of the matrix has to be taken into account (see Section 1.7).

1.2.2.1 Closed-Loop Stripping

In closed-loop stripping systems volatiles are collected inside closed devices by circulating headspace adsorption. Boland (1984) developed a system, in which plants or detached plant parts are placed in small glass containers like 1 or 3 l desiccators. The top of the container is connected to the odor-collecting device consisting of a circulation pump, stainless steel tubes and a stainless steel housing containing the volatile trap (Figure 1.2a). Headspace air is circulated through the container and the connected trap at flow rates of approximately 3 l per min, allowing a continuous quantitative collection of emitted volatiles. The system has been applied for the analysis of herbivore induced volatiles but also for trapping floral volatiles, including collections from flowers with very low emission rates such as those from *Arabidopsis thaliana* (Engelberth et al., 2001; Chen et al., 2003; Dudareva et al., 2005; Tholl et al., 2006; Huang et al., 2010). Floral volatile profiles and compositions analyzed with this method were comparable to those obtained by semi-open headspace volatile trapping, but showed a clearly improved signal to noise ratio, thus allowing a detailed analysis of minor components of the complex terpene volatile mixture.

Closed-loop stripping systems do not require space confiscating tubing and equipment and can easily be set up in controlled climate chambers. They allow volatile collections from several plants at the same time which makes them suitable for the analysis of several replicates and screening purposes. A disadvantage of the closed-loop stripping procedure is the strongly reduced air exchange between the inside and outside of the chamber that affects the gas exchange of the plant during longer sampling times. In addition, transpiration can enhance relative humidity during the collection requiring intermediate venting of the system between trapping cycles. It is, therefore, recommended to combine closed-loop stripping analyses with other open headspace sampling techniques for additional result verification and exclusion of potential artifacts.

1.2.2.2 Pull Systems

A simple form of a pull system is an adsorbent trap connected to a vacuum pump that is directly positioned next to a plant organ (Kaiser and Kraft, 2001). This may work well for tissues that emit high amounts of VOCs, however the risk of trapping ambient air that contains impurities unrelated to the plant source is high and could obscure minor sample compounds during GC analysis. An enclosure of the entire plant, a plant part or an entire branch in a glass container or a polyacetate cooking bag that releases very little VOCs may reduce the amount of impurities from ambient air (Figure 1.2b and c).

For the collection of floral scent, systems have been described in which air enters the container through a purifying filter and is drawn from the chamber by pulling a defined volume of air through an adsorbent trap that can be extracted for further analysis (Raguso and Pellmyr, 1998). Although

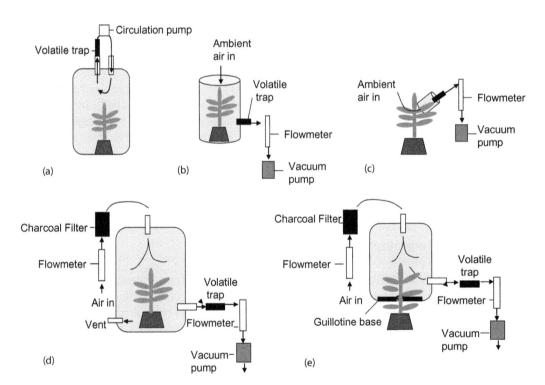

FIGURE 1.2 Examples of dynamic headspace collection systems to sample VOCs from aboveground plant tissues. (a) Volatile collection by the closed-loop stripping procedure. Air is circulated by a pump in and out of a chamber containing the plant sample. VOCs are collected on an adsorbent trap housed in a stainless steel tube attached to the pump. (b) Sampling of VOCs in a simple "pull" headspace open-top chamber. An air stream, regulated by a flow meter, is pulled over the plant through a VOC collecting adsorbent trap. (c) A "pull" headspace open-top chamber for collecting VOCs from a single leaf. (d) VOC sampling in a "push–pull" headspace collection system. Pressurized, purified air, regulated by a flow meter, enters the top of the collection chamber. The incoming air passes over the plant sample and is pulled through an adsorbent volatile trap at the lower side of the chamber with a defined rate controlled by a second flow meter. A vent on the lower side of the chamber allows excess air to escape. (e) A modified "push–pull" collection system to sample VOCs from upper parts of a plant. The base of the chamber around the stem of the plant is closed by Teflon-coated guillotine-like blades. (Modified from Tholl, D. et al., *Plant J.*, 45, 540–560, 2006.)

this method is very easy to set up and portable to the field, it has some drawbacks. Temperature can increase when exposed to direct sunlight and relative humidity in the chamber can increase to nearly 100% within a short time and may lead to condensation on the chamber walls. If a bag is used, it may collapse and damage plant tissue, which may alter VOC emission. To circumvent some of these problems, Kessler and colleagues developed a field VOC collection systems based on 400-mL polystyrene chambers fitted with two openings (Kessler and Baldwin, 2001; Kessler et al., 2004). VOCs were collected for 6 h from leaves of *Nicotiana attenuata* plants by pulling air with a vacuum pump through the chamber and subsequently through a charcoal air-sampling trap (Kessler et al., 2004). It has to be considered though that openings in the chamber may provide additional sources of nonpurified air entering the system.

For the analysis of belowground volatiles, an extra enclosure might not be necessary. The surrounding soil already provides a closed environment. However, some limitations (e.g., humidity, soil type, see Section 1.7) have to be taken into account. A pull system can be as simple as a steel probe buried in the soil (Figure 1.3a) with a tubing connection to a vacuum pump. VOCs can be collected onto suitable adsorbents. Such a system was used to identify volatiles that are important in the interaction of root knot nematodes with *Capsicum annum* (Kihika et al., 2017). Root volatiles of barley have also been collected by pulling air through a glass reactor filled with artificial soil (Figure 1.3b) (Delory et al., 2016a).

1.2.2.3 Push–Pull Systems

In push–pull systems, purified and humidified air is pushed into a chamber containing the plant at a controlled flow rate regulated by a flow meter. A defined portion of this pushed air is pulled through a collector trap by a vacuum pump regulated by a second flow meter (Figure 1.2d). Thus a known percentage of the VOCs emitted by the plant tissues are collected. To avoid a vacuum or

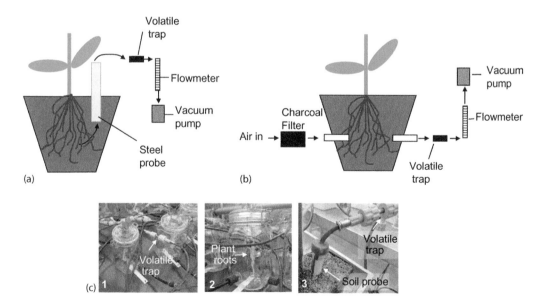

FIGURE 1.3 Volatile collection of belowground volatiles: (a) Example of a pull collection system. (Adapted from Kihika, R. et al., *Sci. Rep.*, 7, 2903, 2017.). Air is pulled out of the belowground compartment. The air flow is regulated by a flowmeter and the volatiles are captured on a suitable adsorbent. (b) Example of belowground volatile pull collection system using an extra air inlet. (Adapted from Delory, B.M. et al., *Plant Physiol. Biochem.*, 104, 134–145, 2016a.). (C1 and 2) Dynamic headspace sampling with pull system on excavated tomato roots. (C3) Dynamic headspace sampling with pull system using a self-made soil probe. Pictures: Alexander Weinhold.

overpressuring of the system a vent is included in the chamber. Positive-pressure venting prevents ambient air from contaminating the volatile collection. This type of system has been used to collect VOCs from a variety of different plants under constitutive or biotic stress conditions (e.g., Chen et al., 2003; Kunert et al., 2005; Kigathi et al., 2009). The positive-pressure venting is also employed in large glass collection chambers on top of a multiport guillotine base. The guillotine base contains concentric gas-sampling ports and two Teflon-coated removable blades that close the bottom of the chamber around the stem of the plant, leaving an opening for the stem where the blades fit together (Figure 1.2e). This collection system allows for the sampling of VOCs from parts of intact growing plants while completely isolating the lower section of the plant including soil and pots (Röse et al., 1996). Various applications of the system have been described such as time course measurements of floral or herbivore induced emissions (Degenhardt and Gershenzon, 2000; Effmert et al., 2005). To collect VOCs from larger plants such as tree saplings, glass collection chambers might not be suitable. In this case the foliage can be enclosed in polyethylene terephthalate (PET) bags connected to an air in- and outlet. The method has been applied to collect VOCs from herbivore-damaged or fungus-infected leaves of poplar (McCormick et al., 2014; Eberl et al., 2018). Despite the common use of push–pull systems for the collection of VOCs from aboveground plant parts, there is, to our knowledge, no study that has used a push–pull system for the analysis of belowground volatiles, which may be due to the constraints of the soil matrix (see Section 1.7).

1.3 ADSORPTION AND DESORPTION OF VOLATILES

Evaluations regarding the choice and application of volatile adsorbents have been published previously (Millar and Sims, 1998; Raguso and Pellmyr, 1998; Tholl et al., 2006; Niinemets et al., 2011), therefore, this chapter will primarily summarize the most important practical aspects in the use of different adsorbent materials for dynamic volatile collection.

For dynamic headspace volatile collections, adsorbents are usually packed in beds of approximately 2–50 mm inside glass or metal tubes between glass wool plugs or metal grids. The amount of adsorbent material to use depends on the chemical properties of the compounds to be trapped, the adsorbing capacity of the matrix, the sampling volume and the flow rate of the collection system. Particularly in case of collecting VOCs with low emissions, the amount of the trapping matrix should be kept low to reduce the volume of solvent required for subsequent compound elution. Smaller amounts of adsorbent also minimize potential artifacts and the resistance to the air flow. However, in case of using insufficient amounts of trapping media, volatile compounds might break through the trap as a consequence of adsorbent surface saturation. Breakthrough sampling volumes (per gram adsorbent) are specified by the supplier or need to be determined by using a series of two traps for collection.

A variety of different adsorbents have been applied for trapping VOCs from plants. The most common media are the polymer-based Porapak Q® (80–100 mesh size, Supelco) and its refined version Super Q®, Tenax GC® and its cleaner version Tenax TA® (60–80 mesh, Supelco) and activated charcoal. Other carbon-based adsorbents are carbon molecular sieves (Carboxen™, Carbosieve™, Supelco) and graphitized carbon blacks (Carbotrap®, Supelco). Both porous polymers, Tenax® (2,6-diphenyl-p-phenylene oxide) and Porapak® (ethylvinylbenzene-divinylbenzene) have a high affinity for lipophilic to medium polarity organic compounds of intermediate molecular weight and low affinity for polar and/or low molecular weight compounds such as ethanol and water.

Trapped volatiles are usually eluted from the adsorbents into glass vials with pure or mixtures of low-boiling-point organic solvents such as pentane, hexane, ether, acetone or dichloromethane, the latter being preferably used with Porapak®. A defined amount of one or two standard compounds (e.g., 1-bromodecane, n-octane, nonyl acetate) is generally added to the sample for quantitative analysis. Volatile extracts can be further concentrated by solvent evaporation at ambient temperature or under a nitrogen stream before they are stored at freezing temperatures. Under field conditions, extracts are ideally stored in flame sealed glass ampoules to exclude sample evaporation (Kaiser, 1991). After compound elution, adsorbents are reconditioned by rinsing with clean solvent and dried at room temperature or by flushing with clean nitrogen.

A disadvantage of eluting compounds with solvents can be a reduced sensitivity due to dilution. An alternative method is thermal desorption, where VOCs are collected on a trap that is directly inserted into a small oven (thermal desorber) placed on top of a GC, thereby transferring the volatiles to the analytical column. Despite its lower capacity for small molecules compared to Porapak®, Tenax® is particularly suited for thermal desorption of VOCs because of its higher thermal stability (350°C). Thermal desorption in combination with cryofocusing (see also Section 1.4) allows the analysis of the total sample and therefore can enhance the detection limit compared to the analysis of aliquots from solvent eluted samples. Limitations of thermal desorption include the impossibility of repeated sample injections, although newer thermal desorption devices can re-collect some of the desorbed volatiles. Other limitations include incomplete desorption of high molecular weight compounds, degradation of thermally unstable compounds, especially at very high desorption temperatures, and artifacts produced from the trapping media (Vercammen et al., 2000; McLeod, 2005; Tholl et al., 2006). Such artifacts do not only occur with thermal desorption but can in general be the result of reactions of the adsorbent material itself or reactions of the adsorbed compounds on the surface of the adsorbing polymers.

Activated charcoal is known as a cheap alternative adsorbent with high adsorbing capacity. It can be used in small commercially available traps that are eluted with small volumes of organic solvent such as dichloromethane and reconditioned by extensive rinsing with solvents of different polarity. Artifacts such as the oxidation of terpenes on the active surface of charcoal have been described (Surburg et al., 1993). Moreover, charcoal has been reported to be less efficient than Tenax® in trapping aromatic aldehydes (Raguso and Pellmyr, 1998) leading to the combined application of both adsorbents (Williams and Whitten, 1983). In addition, a combination of Tenax® and Carbopack B (Markes Intl.) was used to identify differences in the volatile blends of cabbage after herbivore attack (van Dam et al., 2010). Furthermore, Kännaste et al. (2014) used multi-bed cartridges filled with Carbotrap C, Carbopack B, and Carbotrap X for the simultaneous analysis of volatile products of the lipoxygenase pathway, isoprene, and other terpenes. The adsorbents were arranged in order of increasing adsorbent strength, from sample inlet to sample outlet.

1.4   GAS CHROMATOGRAPHIC SEPARATION OF VOLATILES

Gas chromatography (GC) is known as the most efficient chromatographic technique for the separation, identification and quantification of volatile organic compounds including plant volatiles. Hence, numerous research and review articles have been published describing developments and advances in GC analysis technology (e.g., Lockwood, 2001; Dewulf and Van Langenhove, 2002; Eiceman et al., 2006; Waseem and Low, 2015; Lebanov et al., 2019; Wong et al., 2019).

Plant VOCs are usually trapped and pre-concentrated on adsorbent matrices prior to GC analysis (see Section 1.3). Samples eluted from adsorbents with organic solvents are typically injected into the column in a split or splitless mode. The splitless mode is preferred for high-sensitivity analysis of samples with low concentrations. The temperature of the injection liner is typically adjusted to 230°C–250°C to ensure complete vaporization of all sample components. However, adjustments to lower temperatures have to be considered in case of compound decomposition. For example, a conversion of the sesquiterpene germacrene A to β-elemene can be eliminated at an injection temperature around 150°C (de Kraker et al., 2001).

When samples are thermally desorbed from adsorbents such as Tenax®, the solid material is directly placed in a thermal desorption tube that is heated to 250°C–300°C. In a two-stage thermal desorber, the thermally released volatiles are then transported with the carrier gas to a cold or cryo-trap for pre-concentration prior to their injection into the GC column. Thermal desorption units are available from different suppliers (Gerstel, Markes, Perkin-Elmer, Shimadzu) and online systems combining volatile trapping with automated thermal desorption have been developed (see Section 1.6).

The separation of volatiles in GC analysis is most frequently achieved by the use of fused-silica capillary columns. Columns are usually 30 m long and have a stationary phase film thickness of 0.2–0.3 μm and an internal diameter of 0.25 mm. The most common stationary phases are the nonpolar dimethyl polysiloxanes, including DB-1, DB-5, CPSil 5, SE-30 and OV-1, and the more polar polyethylene glycol polymers, including Carbowax™ 20M, DB-Wax, and HP-20M. For different stationary phases, retention index systems have been developed, using reference compounds, to facilitate compound characterization and identification. Such retention indices have been determined and summarized for several hundreds of volatile compounds (Jennings and Shibamoto, 1980; Davies, 1990; Adams, 2017).

1.5   VOLATILE DETECTION AND IDENTIFICATION

For the detection of volatile compounds separated by gas chromatography, principally two different types of detectors can be applied. The first type, for example a flame ionization detector (FID), provides only information on retention times, while detectors of the second type such as mass spectrometry (MS) and Fourier transform infrared (FT-IR) spectroscopy detectors allow additional structure evaluation. FID detectors are primarily employed in quantitative analysis due to their wide linear dynamic range (10^6–10^8), a very stable response and high sensitivity with detection limits in the order of 0.05–0.5 ng per compound. Since isomers with the same molecular formula and carbon content (e.g., sesquiterpene hydrocarbons) principally generate the same FID signal response, relative response factors can be calculated for compounds that are not available in pure form for calibration. Other detectors are the thermal conductivity detector (no sample destruction but moderate sensitivity), the nitrogen phosphorus detector with very high sensitivity for nitrogen or phosphorus containing compounds (Knudsen et al., 2004) and the photoionization detector that has been employed in monitoring plant isoprene and monoterpene emissions (Loreto et al., 2000).

Besides flame ionization detection, mass spectrometry is one of the most widely used detection techniques in GC analysis. The most common configuration of bench top GC-MS systems comprises a GC with a single capillary column directly coupled to a quadrupole mass spectrometer with ionization by electron impact (EI) or by chemical ionization that causes less massive fragmentation. Mass spectrometry is a highly sensitive detection method with a minimum detectable quantity in the range of 0.1–1 ng per compound. The sensitivity can be increased approximately 100- to 1000-fold in the selected ion monitoring (SIM) mode in which only selected ions representing particular compounds are scanned.

Due to the high popularity of EI-MS for routine analysis of volatile compounds, several comprehensive mass spectral libraries (NIST 17 and Wiley Libraries) have been established that are used in EI-mass spectra searches for compound identification. Another database for mass spectral comparison of volatile compounds was developed by Adams (2017) and the MassFinder software (https://massfinder.com/wiki/Main_Page) was originally established by König, Hochmuth, and coworkers. MassFinder allows the comparison of GC-MS data with those of the provided mass spectral library as well as retention indices obtained under identical instrumental and experimental conditions. The software is bundled with the "Terpenoids Library", which contains approximately 2000 spectra of monoterpene, sesquiterpene, diterpene, aliphatic and aromatic plant volatiles. MS data of newly identified compounds can be incorporated into the library. Shimadzu also offers the FFNSC2 MS Library developed by Mondello (https://www.ssi.shimadzu.com/sites/ssi.shimadzu.

com/files/Products/literature/gc/c146-e197.pdf). This library contains more than 3,000 spectra of natural and synthetic flavor and fragrance compounds.

Tandem mass spectrometry systems have been established to allow separate analyses of single compounds of complex GC peaks. For example, GC-tandem mass spectrometry was applied to determine the floral scent composition of *Cucurbita pepo* flowers and VOCs in rhizomes of *Rhodiola* species (Granero et al., 2004; Wang and Guo, 2004).

GC-MS is not sufficient for structure elucidation of unknown volatile compounds and usually multiple analytical steps need to be considered. Most often sufficient amounts of single compounds need to be isolated for 1- and 2-dimensional NMR spectroscopic techniques (¹H NMR, ¹³C NMR, ¹H¹H-COSY, HMQC, HMBC, NOESY). If compounds have to be obtained directly from plant material, preparative isolation can be accomplished by multiple chromatographic steps including preparative packed GC columns and thick-film capillary columns with highly selective cyclodextrin matrices. To reduce the amount of compound required for NMR analysis, a simple efficient NMR sample preparation technique for volatile chemicals has been described using a micropreparative GC system (Nojima et al., 2011). Further details regarding structure elucidation are given elsewhere (König and Hochmuth, 2004) and are not within the scope of this chapter.

GC-MS analysis can also be complemented by capillary GC-FT-IR, for example for the differentiation of closely related isomers with very similar EI-mass spectra (Visser, 2002). Since FT-IR provides information on the intact molecular structure, unique spectra even for similar isomers can be obtained. The spectroscopic method has been applied, for example, for the identification of essential oil components (Michelina et al., 2019). Drawbacks of GC-FT-IR are the lack of sensitivity, difficulties in quantification and time-consuming data interpretation due to the absence of reference databases.

1.5.1 ENANTIOSELECTIVE GC, MULTIDIMENSIONAL GC

The chirality of plant derived VOCs can be crucial for the olfactory response of pollinators and herbivores. Hence, determining the enantiomeric composition of VOCs is critical in understanding plant-animal interactions. Enantioselective capillary columns with chiral phases such as different hydrophobic cyclodextrin derivatives have been developed for enantiomer resolution of a variety of chiral volatile compounds, primarily from essential oils (König and Hochmuth, 2004). As a general rule, polar compounds are better resolved on acylated cyclodextrin derivatives, while nonpolar analytes are better separated on pre-alkylated cyclodextrin derivatives (König and Hochmuth, 2004). König and coworkers assembled an enormous amount of data for the identification and enantiomeric recognition of hundreds of sesquiterpene hydrocarbons (König, 1993a–c; Joulain and König, 1998). Other examples for the application of cyclodextrin derivatives in flavor and fragrance analysis were documented by Schreier et al. (1995).

In situations where complex volatile mixtures cannot be sufficiently separated on a single chiral column, comprehensive two dimensional gas chromatography (GC × GC) can be employed. In this system two columns with different polarity are directly coupled by a cryogenic modulator. The method increases peak resolution and improves quantification or identification of volatile components, which makes it particularly useful for the analysis of essential oils (Prebihalo et al., 2018; Lebanov et al., 2019). To separate enantiomers, chiral columns can be used as the second dimension (e.g., Krupcik et al., 2016). Multidimensional GC has been further expanded by the use of time of flight (TOF) mass spectrometers. In TOF mass spectrometers, ions move through a flight tube to reach the detector and the time required to hit the detector is directly related to the mass of the ion. Advantages of TOF-MS are a higher mass accuracy and resolution paired with enhanced sensitivity and faster acquisition. Mateus et al. (2010) used (GC × GC)-TOF-MS in combination with automatic peak detection and deconvolution algorithms to determine more than 200 volatile compounds in needles of different pine species. Because of its resolution power, applications of (GC x GC)-TOF-MS have increased in the field of metabolomics and analysis of volatile fingerprints from different organisms (Prebihalo et al., 2018).

1.6   VOLATILE ANALYSIS TECHNIQUES WITH HIGH TIME RESOLUTION

A detailed understanding of regulatory mechanisms governing plant VOC biosynthesis and emissions such as circadian or diurnal control regimes require analysis techniques that monitor volatile emission changes with appropriate time resolution. Computer assisted and online dynamic headspace trapping systems are capable of collecting volatiles in hourly or shorter time intervals. In a commercially available online thermal desorption system (Gerstel Online-TDS-G, Gerstel, Germany) volatile samples are drawn automatically and, following a cryofocusing step, compounds are flash heated and directly injected on the column. The time resolution depends on the time necessary to collect sufficient amounts of material from the emitting plant tissue for analysis, the compounds themselves, and the time for chromatographic separation on the GC-MS. The TDS-G system can be connected to collecting containers with plant samples as described in detail by Vercammen et al. (2001). This method allows for a high time resolution, depending on the compounds and a high sensitivity, but requires an extensive laboratory setup. Moreover, analysis by GC presents a time limiting factor. Described below are alternative analytical systems that have been developed to allow faster and sensitive volatile analyses and hence represent promising tools for continuous monitoring of plant VOC emissions.

1.6.1   TRANSPORTABLE GC

To allow for fast real-time analysis of plant VOCs in the laboratory and the field, portable GC instruments have been developed (Yashin and Yashin, 2001). An example of a portable GC is the zNose™ (Electronic Sensor Technology, Newbury Park, CA, USA), which has been applied for the analysis of floral scent and induced VOC emissions from herbivore-damaged plants (Kunert et al., 2002; Oh, 2018). The zNose™ separates compounds by fast gas chromatography and operates with a highly sensitive surface acoustic wave (SAW) quartz microbalance detector, which drastically reduces the volatile sampling and pre-concentration time of the system. As a drawback, the SAW detector does not allow structure elucidation. Therefore, volatiles need to be analyzed by GC-MS prior to calibration of the system with authentic standards. Moreover, the short gas chromatography column reduces the compound resolution. Thus, monitoring changes of volatiles with similar elution profiles is limited.

As another example for a GC system used in the field, Wong et al. (2019) combined SPME with a portable GC containing a low-thermal mass (LTM) column in combination with a miniature ion trap mass spectrometer. LTM columns allow efficient heating and cooling within a short analytical time. This system as well as the zNose™ are applicable for screenings of natural variants or mutant populations and suitable for monitoring kinetics of volatile emissions from floral and vegetative tissues under different endogenous and environmental conditions.

1.6.2   PROTON-TRANSFER REACTION MASS SPECTROMETRY (PTR-MS)

The PTR-MS analysis technology was developed ca. 20 years ago at the University of Innsbruck by Lindinger and coworkers (Lindinger et al., 1998). PTR-MS systems operate independently of gas chromatographic separation and allow online measurements of VOCs with concentrations in the pptv range. Originally developed for monitoring changes of VOCs in the atmosphere, in food control and medical analyses, PTR-MS has been applied for real-time analysis of volatile emissions from plants above- and belowground. PTR-MS instruments are still relatively expensive and their operation requires extended training by experienced researchers.

For detection by PTR-MS, volatiles undergo a chemical ionization by proton-transfer reactions with H_3O^+ ions. Differences in proton affinities allow a proton transfer from H_3O^+ ions to a large number of organic volatiles but prevent a reaction of H_3O^+ ions with the main constituents of the air. The proton-transfer reaction takes place under defined conditions in a homogeneous electric field applied to a drift tube (Figure 1.4). Ions exiting the tube are finally mass analyzed by a quadrupole or TOF mass spectrometer (Graus et al., 2010). The soft ionization of compounds by protonation causes only low fragmentation, hence mainly one product ion species occurs for each reactant. The extremely fast time response of the instrument results from less than one second that volatiles spend in the drift tube.

The analysis of volatile mixtures by PTR-MS is limited by the ability to determine only the molecular mass of products. Compounds of the same nominal mass cannot be identified separately using a quadrupole MS. Therefore, systems with additional analysis by GC either in parallel or by coupling to the PTR-MS have been developed or instruments using triple-quad MS analysis have been applied to distinguish between isomeric species (Graus et al., 2010; Materic et al., 2015). Moreover, the development of PTR-TOF-MS has allowed the separation of isobaric VOCs in complex mixtures (Graus et al., 2010).

The PTR-MS technique has been applied in several studies to measure fluctuations of volatile emissions from various plants. Usually whole plants or plant parts are enclosed in glass containers, inert bags or dynamic cuvette systems (Tholl et al., 2006) with a continuous air stream and controlled temperature, humidity and light conditions, and aliquots of the exiting air are analyzed by PTR-MS. Emissions of volatiles including isoprene and monoterpenes from trees and other plants have been monitored under laboratory and field conditions at different developmental

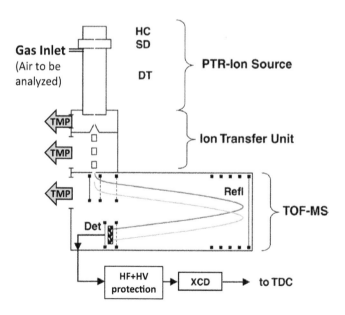

FIGURE 1.4 Schematic representation of a PTR-TOF-MS instrument according to Graus et al., (2010). Chemical ionization of VOCs by proton-transfer reaction occurs in the drift tube (DT). In the ToF mass spectrometer ions are pulsed orthogonally toward the reflectron (Refl) and are refocused into the detector (Det) plane. An amplifier/discriminator unit (XCD) preprocesses the signal to generate individual ion counting events, which are further processed by a time to digital-converter (TDC). HC, hollow cathode; SD, source drift region; TMP, turbo molecular pump. (Modified from Graus, M. et al., *J. Am. Soc. Mass Spectrom.*, 21, 1037–1044, 2010. With permission.)

stages and in response to changes of abiotic factors like light and temperature or biotic factors such as herbivory (e.g., Loreto et al., 2006; Brilli et al., 2011; Giacomuzzi et al., 2016; Mozaffar et al., 2018).

Besides measurements of VOCs from aboveground plant tissues, PTR-MS has been used successfully for noninvasive analysis of VOCs from roots. Danner et al. (2012, 2015) measured sulfur containing compounds emitted by *Brassica* roots after herbivory by positioning a two-part cuvette on top of the soil at the stem-root interface. A constant flow of purified air was applied to flush the root-cuvettes. The instrument was used in SIM mode optimized for measuring sulfur compounds. In a different system developed by Acton et al. (2018) a root glass chamber was designed, in which VOC-free air enters the vessel at its base through a ring of tubing and exits the chamber at a port located above the soil level. Here, a PTR-TOF-MS with selective reagent ionization (SRI-TOF-MS) was used to detect VOCs with high time and mass resolution and identify distinct functional groups of compounds.

1.6.3 Automated Systems for Volatilome Analysis

The emission of VOCs can be considered a phenotypic trait that provides information about the developmental and physiological state of an organism and its exposure to biotic and abiotic stress. Therefore, new systems have been developed to monitor so-called plant "volatilomes" representing the entirety of VOCs emitted by a plant including CO_2 and water vapor (Jud et al., 2018). The VOC-SCREEN platform established by Schnitzler and coworkers allows the simultaneous, noninvasive online analysis of VOCs emitted from the foliage of 24 plants (Figure 1.5). In a test example, whole potted barley plants were enclosed in flow-through cuvettes. VOCs were sampled

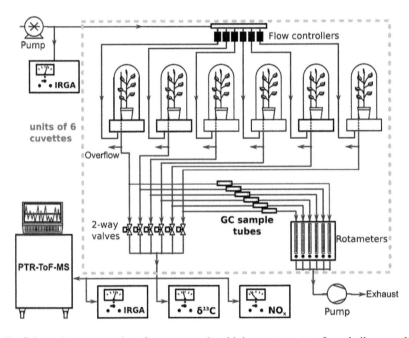

FIGURE 1.5 Schematic representation of an automated multiple cuvette system for volatilome analysis according to Jud et al. (2018). Air from the phytotron chamber is pumped into cuvettes containing individual plants (blue lines). Air exiting the cuvettes (red lines) undergoes analysis by PTR-ToF-MS and other gas analyzers. Sampling occurs from one cuvette at a time. A portion of the air can be sampled for VOCs for GC-MS analysis. IRGA, infrared gas analyzer. (Reprinted from Jud, W. et al., *Plant Methods*, 14, 109, 2018. With permission.)

every 2 h for 5 min from each cuvette and measured quantitatively and in a nontargeted way by PTR-TOF-MS with a high mass resolution and detection limits in the low ppt range.

A portion of the VOCs is trapped on adsorbent tubes to allow for GC-MS analysis of different isomers. The cuvette based system has the advantage that CO_2 assimilation and transpiration rates can be measured simultaneously with the analysis of VOCs and thereby allow correlations of VOC release with the physiological condition of the plant. The screening platform can be installed in controlled environmental chambers and is suitable for monitoring VOC profiles under transient conditions or long term scenarios.

1.6.4 SENSORS FOR VOC PROFILING IN THE FIELD

Developments in smart farming and precision agriculture are based on obtaining accurate information on crop plants, soil, weather and environmental conditions. Various sensor technologies have been developed to monitor these conditions. Among these, there is growing interest in obtaining information about the physiological state of plants by sensing VOC emissions. VOC sensing is considered a diagnostic tool in determining plant health by recognizing infested or infected plant material at an early stage of outbreak. Targeted VOCs include compounds involved in plant defense as well as VOCs released from pathogens or pests (Cellini et al., 2017).

Electronic noses have been developed for monitoring mixtures of VOCs that are released from different materials. An e-nose operates with an array of sensors, whose surfaces interact with the gas phase and have different sensibilities to volatile molecules. Based on electric signals from the sensors, patterns or profiles corresponding to specific VOC compositions are generated (Wilson, 2013). While e-noses do not identify individual VOC compounds they can be applied for the non-destructive quick monitoring or recognition of VOC blends. Several applications of commercially available e-nose models in plant disease and pest diagnosis have been summarized by Cellini et al. (2017). Parallel testing of an e-nose with GC-MS analysis provides more precise information on the sensor response to specific compounds and their discrimination between different odor sources (Long et al., 2019).

A challenge of using e-noses is their limited sensitivity, but cartridges with VOC sorbent materials such as Tenax have been coupled with electronic noses (Cellini et al., 2017). In these systems VOCs are adsorbed and thermally desorbed in a small air volume prior to their detection by the sensor. The adsorbent can provide some selectivity toward VOCs that are of specific interest concerning distinct disease profiles. Another challenge in the application of e-noses is that VOC recognition can be compromised by background noise, natural variation, or variability of VOC blends due to environmental conditions other than those targeted in the analysis. Therefore, it has been suggested that e-nose technology might best be combined with other phenotyping and molecular methods to optimize plant disease diagnostics.

In addition to e-noses, differential mobility spectrometry (DMS) coupled with gas chromatography (GC-DMS) is emerging as a promising tool for VOC based disease diagnostics. DMS is a form of field asymmetric waveform ion mobility spectrometry (FAIMS), that identifies molecules according to their difference in ion behavior under alternating low and high electric fields (Anderson et al., 2008). Combining DMS with GC allows for the profiling of complex VOC blends with high sensitivity and an analysis time faster than that of a standard GC. The system is portable and, in contrast to the e-nose, can build libraries based on chemical standards (Anderson et al., 2008).

Davis and coworkers applied GC-DMS to determine VOC fingerprints characteristic of the early onset of citrus greening disease (Aksenov et al., 2014). The method was found to be about 90% accurate at early stages of infection before trees do show visual symptoms and PCR based detection methods are applied. The measurements were paired with SPME-GC-MS analysis (using Twisters®) to determine the identities of the most discriminating compounds between healthy and infected trees (Figure 1.6).

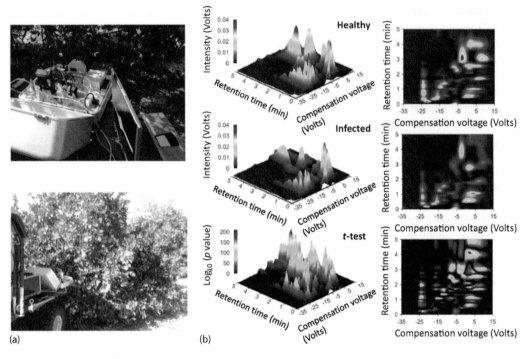

FIGURE 1.6 VOC monitoring by GC-DMS from the canopy of citrus trees to diagnose citrus greening disease. (a) Field sampling with a GC-DMS device. (b) GC/DMS plots representing VOC signatures for healthy (noninoculated) (top) and inoculated (middle) trees, which were asymptomatic for the disease. VOC signatures show differential abundances of compounds in inoculated and noninoculated trees. The differences are plotted on the Student's t-test plot ($p<0.05$) (bottom). (Reprinted from Aksenov, A.A. et al., *Anal. Chem.*, 86, 2481–2488, 2014. With permission.)

1.7 SPECIAL CONSIDERATIONS FOR BELOWGROUND VOLATILE COLLECTION

In recent years, more and more focus has been on the belowground compartment of plant systems. The understanding of plant root interactions with soil-dwelling organisms like microbes, fungi, or herbivores, are crucial to understand plant resistance. Many of these interactions are mediated by soil-borne volatiles (for extensive review see Delory et al., 2016b; van Dam et al., 2016). These soil volatile compounds can originate from the plant root itself or from other soil living organism. Therefore, it is important to conduct volatile sampling in a way that allows distinguishing between the different volatile sources. A simple solution is to remove roots from the soil and collect root volatiles in glass vessels (Figure 1.3c) or to harvest roots and analyze volatile profiles of ground tissue, but this analysis may not reflect real biological conditions. Volatile profiles collected can be altered by damage inflicted during harvest. Furthermore, measurements of ground tissue will rather give information about the total amount of volatiles in the roots than about volatiles that are emitted. Hence, only *in vivo* collections of belowground plant scents will provide ecological meaningful insight. As for all chemical analyses, *in vivo* belowground volatile collection should be done with adequate control samples, e.g., sterile soil to exclude microbes, or trapping volatiles solely from soil without plants or roots. Soil is less homogenous and therefore another point of consideration in the belowground compartment is that the adsorption capacity of the soil, grain size, and the water binding capacity of the soil will influence distribution and diffusion of volatiles (Hiltpold and Turlings, 2008). To achieve comparable results, soil should always have the same composition and humidity. In field experiments, this can hardly be controlled. Therefore, one can expect a higher variation in amounts of collected volatiles. All of the above-mentioned constraints in the belowground environment restrict the applicability of some volatile collection methods. For instance, push–pull and closed-loop systems are hard to realize because of inconsistent airflow through the nonhomogeneous soil matrix. Delory et al. (2016a) for instance, used artificial soil comprised of sand and glass beads to correlate emission patterns of volatile aldehydes of barley roots with plant age. Nevertheless, belowground volatile analysis is gaining increased attention. When conducted in a meaningful way it will enhance the understanding of plant–plant, plant–herbivore and plant-microbe interactions that are mediated by soil-borne scents.

1.8   SPECIFIC CONSIDERATIONS IN AQUATIC SYSTEMS

Volatiles also have important ecological roles in aquatic systems (Fink, 2007). They can be released as defensive compounds by algae (Van Alstyne, 2008) in response to wounding (Wiesemeier et al., 2007), function as sex pheromones in macroalgae and diatoms (Pohnert and Boland, 2002), or aid in the location of food or habitat (Gabric et al., 2001; Nevitt et al., 2004; Juettner et al., 2010). Additionally, halogenated VOCs in aquatic systems may further act as inorganic antioxidants in algae (Kuepper et al., 2008).

Compounds that are detected by marine organisms are not limited to water-soluble compounds. Low molecular weight terpenes that are volatile in air, but not water-soluble can be detected by aquatic organisms and convey information about food palatability (Giordano et al., 2017). To collect VOCs from aquatic organisms, it is often possible to remove the organism from the aquatic system and employ one of the terrestrial volatile collection techniques discussed above. For example, volatiles from algae and seagrass were collected from the headspace by SPME (Jerkovic et al., 2018). If this is not possible, there are a number of techniques that can be employed to allow detection of VOCs from the aquatic phase.

The detection and identification of volatiles in aquatic systems is well explored in the context of water quality and volatiles in drinking water. Several techniques are listed as standardized procedures in EPA manuals for evaluation of contaminants. Semivolatile compounds can be extracted from drinking water by a liquid-solid extraction technique (EPA 525) (Munch, 1195) that can partition compounds onto a C18 organic phase bonded to a solid disk or cartridge. Water samples and a standard are passed through the C18 column, dried with nitrogen gas and subsequently eluted from the column with ethyl acetate followed by methylene chloride. The elution time can be reduced by applying a partial vacuum to the cartridge. Samples are subsequently concentrated through evaporation of part of the solvent under a stream of nitrogen gas to prevent oxidation of compounds and analyzed by GC-FID and GC-MS. More recent methods have improved the efficiency and accuracy for the extraction of semivolatile organic compounds by automating the extraction (Horizon Technology, Inc., Salem, NH).

A modification of this method was developed (Cirri et al., 2016) that facilitates a disruption-free solid phase extraction of surface metabolites from macroalgae. This method applies a C18 solid phase powder directly to the surface of organisms from which volatiles are to be extracted. The powder is subsequently washed of the organism and trapped in a polypropylene column by a polyethylene frit. Compounds are eluted from the C18 phase by a suitable solvent and analyzed by GC-MS or LC-MS.

Solid phase microextraction (SPME), as discussed earlier, can also be used for the collection of volatiles from water samples. It has an advantage over liquid-solid extraction when humic acids and suspended mineral oxides are a concern that showed no effect on the SPME analysis of VOC (Nilsson et al., 1998). Additionally, SPME collection avoids salts in the water that could affect analysis by GC-MS. Volatiles can be collected from the headspace above the water sample or by immersion into the aqueous solution (Figure 1.7a and b). SPME fibers that are suitable for collection of volatile and semivolatile compounds by immersion into water are for example polydimethylsiloxane (PDMS) and Carboxen/polydimethylsiloxane fibers, depending on the molecular weights and polarity of the analytes. Polar semivolatiles from water are best extracted with polyacrylate film fibers (Supelco, application bibliography for fiber thickness and exposure times available).

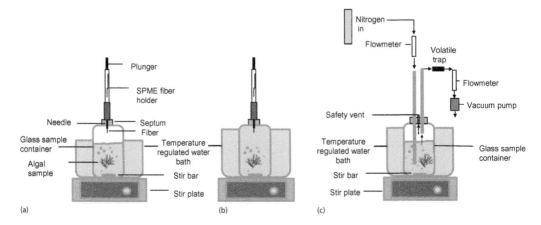

FIGURE 1.7 Collection of VOCs from aquatic systems. Volatile collection by static headspace sampling with a SPME fiber from above the aquatic phase (a) or by immersion of the fiber into the aquatic phase (b). The sample is stirred with a stir bar and the temperature is regulated by a water bath to obtain reproducible results and to increase volatility of the sample. (c) Dynamic headspace sampling above the water sample by pushing a stream of nitrogen gas (to avoid oxidation) or air through the water sample, controlled by a flowmeter. Volatiles are collected onto an adsorbent trap that is connected to a flowmeter attached to a vacuum pump. The glass sample container is placed into a temperature controlled water bath on a stir plate with a stir bar to increase volatilization of compounds into the headspace. A safety vent is inserted into the lid to avoid over pressuring if connections are airtight and the apparatus is placed under a fume hood if nitrogen is used.

If volatiles are collected by immersion of the SPME fiber into the sample, extraction efficiency can be increased by agitating the sample and by warming it on a hot plate or in a water bath (Figure 1.7b). For static volatile collections from headspace above the aquatic phase, sample vials needs to be small enough so that 2/3 of the vial are filled with the sample to avoid dilution of volatile compounds in the headspace (Figure 1.7a). For reproducibility, the time of exposure of the fiber to volatiles should to be consistent and the temperature of the aquatic phase needs to stabilize before exposing the needle if the sample is warmed. A number of official methods and standards of SPME for the analysis of water samples are summarized by Piri-Moghadam et al. (2016). While SPME fibers are a very suitable way to extract volatiles from aquatic samples, fibers are very fragile and need to be handled carefully to avoid damage. More recently, metal wires have been used as a supporting structure for different sorbent materials (Feng et al., 2013) increasing the lifetime of the fiber dramatically and PDMS coating to steel wire has been successfully used (Zali et al., 2016). Other developments include thin-film microextraction (TFME) that consists of a flat, thin sheet providing a higher surface area-to-volume ratio thus increasing its sensitivity for trace level analytes (Jiang and Pawliszyn, 2012). After volatile collection, the thin film can be thermally desorbed in a special GC-TD large volume inlet or by extraction into a solvent. The extraction efficiency of compounds from water samples can be increased by rotation of the TFME (Qin et al., 2008). If the TFME unit is enclosed in metal mesh attached to a rotating device, for example an electronic drill, it can be further used for active water sampling in the field (Jiang and Pawliszyn, 2012; Piri-Moghadam et al., 2016). This technique is also more sensitive compared to stir-bar sorptive extraction (SBSE) that uses a PDMS adsorbent-coated magnetic stir bar (Twisters®), but has a thicker coating, resulting in longer extraction times (Qin et al., 2008). Recent attempts have been made to improve SPME fibers by applying the sorption phase to a cylindrical steel rod that runs through the entire fiber thereby increasing the fiber robustness and the amount of sorption phase volume. In this Prep And Load SPME (PAL SPME Arrow) termed extraction method, the tip of the fiber contains a little arrow that serves to facilitate injection through the septum with less force and simultaneously caps the fiber when it is retracted into the needle preventing contamination during the transport prior to analysis (Kremser et al., 2016).

Dynamic headspace collection described above can also be applied to aquatic samples by sampling above the water sample. A stream of air or nitrogen (to prevent oxidation of compounds) is pushed through the water sample, controlled by a flowmeter, and volatiles are collected onto an adsorbent trap that is connected to a flowmeter attached to a vacuum pump (Figure 1.7c). The glass sample container can be placed into a temperature controlled water bath that is placed onto a stir plate to increase volatilization of compounds into the headspace. A safety vent should be inserted into the lid to avoid over pressuring if connections are airtight and the apparatus should be placed under a fume hood if nitrogen gas is used. This system can be adapted for small samples in the mL range by using a glass vial with a Teflon cap that can be pierced to insert a nitrogen stream while simultaneously collecting volatiles onto an adsorbent filter inserted into the cap. To increase volatility, the vial can be placed into a heat block. Different types of dynamic headspace collections have been successfully used to collect volatile compounds from algae (Beauchene et al., 2000; Le Pape et al., 2004) and marine phospholipids (Lu et al., 2013). Finally, closed loop stripping in aquatic environments has been applied to investigate volatile compounds in algae (Bruchet, 2006) or fish tissue (Osemwengie and Steinberg, 2003) with detailed schematic drawings of the closed-loop-stripping apparatus used.

1.9 CONCLUSIONS

Since the earlier edition of this chapter was published more than 10 years ago, several advances have been made in the field of VOC collection and analysis. While many of the here described basic methods in VOC trapping and GC-MS based analysis have not changed significantly, important steps have been made to explore plant VOC dynamics in belowground and even aquatic environments to gain a better understanding of these metabolites in the interaction of plants with soil-borne or aquatic organisms (see also Chapter 18). Especially, SPME-based applications have substantially increased in the past decade due to the simplicity of the method in VOC trapping and desorption. On the other hand, online monitoring of plant volatilomes as phenotypic markers has advanced thanks in large to further improvements in PTR-MS technology. A promising area in VOC measurements is the monitoring of VOC profiles or fingerprints in crop pest and disease diagnostics by the application of electronic noses and other equivalent systems. This is a transdisciplinary field requiring combined knowledge of plant biologists, pathologists, entomologists, chemists, and engineers. With the development of smaller, yet sensitive and accurate sensors and portable devices, VOC monitoring may become an important component in smart farming and precision agriculture. These advanced technologies may also be applicable, in combination with traditional methods, to measure VOC emissions from plants in natural environments and hence further improve our understanding of plant volatile biology.

REFERENCES

Acton, W.J.F., Jud, W., Ghirardo, A., et al. 2018. The effect of ozone fumigation on the biogenic volatile organic compounds (BVOCs) emitted from *Brassica napus* above- and below-ground. *PLoS One* 13: e0208825.

Adams, R.P. 2017. *Identification of Essential Oil Components by Gas Chromatography/Mass Spectrometry*. Gruver, Texas: Texensis Publishing.

Aksenov, A.A., Pasamontes, A., Peirano, D.J., et al. 2014. Detection of huanglongbing disease using differential mobility spectrometry. *Anal. Chem.* 86: 2481–2488.

Anderson, A.G., Markoski, K.A., Shi, Q., et al. 2008. DMS-IMS2, GC-DMS, DMS-MS: DMS hybrid devices combining orthogonal principles of separation for challenging applications - art. no. 69540H. In *Chemical, Biological, Radiological, Nuclear, and Explosives, Proceedings of the Society of Photo-Optical Instrumentation Engineers* (SPIE), ed. A. W. Fountain Iii, and P. J. Gardner, H9540–H9540, Bellingham, Washington: SPIE.

Beauchene, D., Grua-Priol, J., Lamer, T., Demaimay, M., and Quemeneur, F. 2000. Concentration by pervaporation of aroma compounds from *Fucus serratus*. *J. Chem. Technol. Biotechnol.* 75: 451–458.

Boland, W. 1984. A "closed-loop-stripping" technique as a versatile tool for metabolic studies of volatiles. In *Analysis of Volatiles*, ed. P. Schreier, 371. New York: Walter de Gruyter.

Brilli, F., Ruuskanen, T.M., Schnitzhofer, R., et al. 2011. Detection of plant volatiles after leaf wounding and darkening by Proton Transfer Reaction "Time-of-Flight" Mass Spectrometry (PTR-TOF). *PLoS One* 6: e20419.

Bruchet, A. 2006. State of the art analytical methods for solving taste and odour episodes. In *5th World Water Congress: Drinking Water Treatment Processes, Water Science and Technology: Water Supply*, ed. H. Kroiss, 157–165, London, UK: IWA Publishing.

Cellini, A., Blasioli, S., Biondi, E., et al. 2017. Potential applications and limitations of electronic nose devices for plant disease diagnosis. *Sensors* 17: 2596.

Chen, F., Tholl, D., D'Auria, J.C., et al. 2003. Biosynthesis and emission of terpenoid volatiles from *Arabidopsis* flowers. *Plant Cell* 15: 481–494.

Cheung, W.H.K., Pasamontes, A., Peirano, D.J., et al. 2015. Volatile organic compound (VOC) profiling of *Citrus tristeza* virus infection in sweet orange citrus varietals using thermal desorption gas chromatography time of flight mass spectrometry (TD-GC/TOF-MS). *Metabolomics* 11: 1514–1525.

Cirri, E., Grosser, K., and Pohnert, G. 2016. A solid phase extraction based non-disruptive sampling technique to investigate the surface chemistry of macroalgae. *Biofouling* 32: 145–153.

Danner, H., Brown, P., Cator, E.A., et al. 2015. Aboveground and belowground herbivores synergistically induce volatile organic sulfur compound emissions from shoots but not from roots. *J. Chem. Ecol.* 41: 631–640.

Danner, H., Samudrala, D., Cristescu, S.M., and Van Dam, N.M. 2012. Tracing hidden herbivores: Time-resolved non-invasive analysis of belowground volatiles by proton-transfer-reaction mass spectrometry (PTR-MS). *J. Chem. Ecol.* 38: 785–794.

Davies, N.W. 1990. Gas-chromatographic retention indexes of monoterpenes and sesquiterpenes on methyl silicone and Carbowax 20M phases. *J. Chromatogr.* 503: 1–24.

de Kraker, J.W., Franssen, M.C.R., de Groot, A., Shibata, T., and Bouwmeester, H.J. 2001. Germacrenes from fresh costus roots. *Phytochemistry* 58: 481–487.

Deasy, W., Shepherd, T., Alexander, C.J., Birch, A.N., and Evans, K.A. 2016a. Field-based evaluation of a novel SPME-GC-MS method for investigation of below-ground interaction between *Brassica* roots and larvae of cabbage root fly, *Delia radicum* L. *Phytochem. Anal.* 27: 343–353.

Deasy, W., Shepherd, T., Alexander, C.J., Birch, A.N.E., and Evans, K.A. 2016b. Development and validation of a SPME-GC-MS method for in situ passive sampling of root volatiles from glasshouse-grown broccoli plants undergoing below-ground herbivory by larvae of cabbage root fly, *Delia radicum* L. *Phytochem. Anal.* 27: 375–393.

Degenhardt, J. and Gershenzon, J. 2000. Demonstration and characterization of (*E*)-nerolidol synthase from maize: A herbivore-inducible terpene synthase participating in (3*E*)-4,8-dimethyl-1,3,7-nonatriene biosynthesis. *Planta* 210: 815–822.

Delory, B.M., Delaplace, P., du Jardin, P., and Fauconnier, M.-L. 2016a. Barley (*Hordeum distichon* L.) roots synthesise volatile aldehydes with a strong age-dependent pattern and release (*E*)-non-2-enal and (*E,Z*)-nona-2,6-dienal after mechanical injury. *Plant Physiol. Biochem.* 104: 134–145.

Delory, B.M., Delaplace, P., Fauconnier, M.-L., and du Jardin, P. 2016b. Root-emitted volatile organic compounds: can they mediate belowground plant-plant interactions? *Plant Soil* 402: 1–26.

Dewulf, J. and Van Langenhove, H. 2002. Analysis of volatile organic compounds using gas chromatography. *Trac-Trends Anal. Chem.* 21: 637–646.

Dudareva, N., Andersson, S., Orlova, I., et al. 2005. The nonmevalonate pathway supports both monoterpene and sesquiterpene formation in snapdragon flowers. *Proc. Natl. Acad. Sci. USA* 102: 933–938.

Eberl, F., Perreca, E., Vogel, H., et al. 2018. Rust infection of black poplar trees reduces photosynthesis but does not affect isoprene biosynthesis or emission. *Front. Plant Sci.* 9: 1733.

Effmert, U., Grosse, J., Röse, U.S.R., et al. 2005. Volatile composition, emission pattern, and localization of floral scent emission in *Mirabilis jalapa* (Nyctaginaceae). *Am. J. Bot.* 92: 2–12.

Eiceman, G.A., Gardea-Torresdey, J., Dorman, F., et al. 2006. Gas chromatography. *Anal. Chem.* 78: 3985–3996.

Eilers, E.J., Pauls, G., Rillig, M.C., et al. 2015. Novel set-up for low-disturbance sampling of volatile and non-volatile compounds from plant roots. *J. Chem. Ecol.* 41: 253–266.

Engelberth, J., Koch, T., Schüler, G., et al. 2001. Ion channel-forming alamethicin is a potent elicitor of volatile biosynthesis and tendril coiling. Cross talk between jasmonate and salicylate signaling in lima bean. *Plant Physiol.* 125: 369–377.

Fabisch, T., Gershenzon, J., and Unsicker, S.B. 2019. Specificity of herbivore defense responses in a woody plant, black poplar (*Populus nigra*). *J. Chem. Ecol.* 45: 162–177.

Feng, J., Qiu, H., Liu, X., and Jiang, S. 2013. The development of solid-phase microextraction fibers with metal wires as supporting substrates. *Trac-Trends Anal. Chem.* 46: 44–58.

Fink, P. 2007. Ecological functions of volatile organic compounds in aquatic systems. *Mar. Freshwat. Behav. Physiol.* 40: 155–168.

Gabric, A., Gregg, W., Najjar, R., Erickson, D., and Matrai, P. 2001. Modeling the biogeochemical cycle of dimethylsulfide in the upper ocean: A review. *Chemosphere Global Change Sci.* 3: 377–392.

Giacomuzzi, V., Cappellin, L., Khomenko, I., et al. 2016. Emission of volatile compounds from apple plants infested with *Pandemis heparana* larvae, antennal response of conspecific adults, and preliminary field trial. *J. Chem. Ecol.* 42: 1265–1280.

Giordano, G., Carbone, M., Ciavatta, M.L., et al. 2017. Volatile secondary metabolites as aposematic olfactory signals and defensive weapons in aquatic environments. *Proc. Natl. Acad. Sci. USA* 114: 3451–3456.

Granero, A.M., Gonzalez, F.J.E., Frenich, A.G., Sanz, J.M.G., and Vidal, J.L.M. 2004. Single step determination of fragrances in *Cucurbita* flowers by coupling headspace solid-phase microextraction low-pressure gas chromatography-tandem mass spectrometry. *J. Chromatogr. A* 1045: 173–179.

Graus, M., Muller, M., and Hansel, A. 2010. High resolution PTR-TOF: Quantification and formula confirmation of VOC in real time. *J. Am. Soc. Mass Spectrom.* 21: 1037–1044.

Heath, R.R. and Manukian, A. 1992. Development and evaluation of systems to collect volatile semiochemicals from insects and plants using a charcoal-infused medium for air purification. *J. Chem. Ecol.* 18: 1209–1226.

Hiltpold, I. and Turlings, T.C.J. 2008. Belowground chemical signaling in maize: When simplicity rhymes with efficiency. *J. Chem. Ecol.* 34: 628–635.

Huang, M.S., Abel, C., Sohrabi, R., et al. 2010. Variation of herbivore-induced volatile terpenes among *Arabidopsis* ecotypes depends on allelic differences and subcellular targeting of two terpene synthases, TPS02 and TPS03. *Plant Physiol.* 153: 1293–1310.

Jansen, R.M.C., Wildt, J., Kappers, I.F., et al. 2011. Detection of diseased plants by analysis of volatile organic compound emission. In *Annual Review of Phytopathology*. ed. N. K. VanAlfen, G. Bruening, and J. E. Leach, 157–174, Palo Alto, CA: Annual Reviews.

Jennings, W. and Shibamoto, T. 1980. *Qualitative Analysis of Flavor and Fragrance Volatiles by Glass Capillary Gas Chromatography*. New York: Academic Press.

Jerkovic, I., Marijanovic, Z., Roje, M., et al. 2018. Phytochemical study of the headspace volatile organic compounds of fresh algae and seagrass from the Adriatic Sea (single point collection). *PLoS One* 13: e0196462.

Jiang, R. and Pawliszyn, J. 2012. Thin-film microextraction offers another geometry for solid-phase microextraction. *Trac-Trends Anal. Chem.* 39: 245–253.

Joo, Y., Schuman, M.C., Goldberg, J.K., et al. 2019. Herbivory elicits changes in green leaf volatile production via jasmonate signaling and the circadian clock. *Plant Cell Environ.* 42: 972–982.

Joulain, D. and König, W.A. 1998. *The Atlas of Spectral Data of Sesquiterpene Hydrocarbons*. Hamburg, Germany: E.B.-Verlag.

Jud, W., Winkler, J.B., Niederbacher, B., Niederbacher, S., and Schnitzler, J.P. 2018. Volatilomics: A non-invasive technique for screening plant phenotypic traits. *Plant Methods* 14: 109.

Juettner, F., Messina, P., Patalano, C., and Zupo, V. 2010. Odour compounds of the diatom *Cocconeis scutellum*: Effects on benthic herbivores living on *Posidonia oceanica*. *Mar. Ecol. Prog. Ser.* 400: 63–73.

Kaiser, R. 1991. Trapping, investigation and reconstitution of flower scents. In *Perfumes: Art, Science, Technology*, ed. P. M. Mueller, and D. Lamparsky, 213–250. London, UK: Elsevier Applied Science.

Kaiser, R. and Kraft, P. 2001. Surprising aromatic experiences—New and unusual fascinating floral scents from nature. *Chemie in Unserer Zeit* 35: 8–22.

Kallenbach, M., Oh, Y., Eilers, E.J., et al. 2014. A robust, simple, high-throughput technique for time-resolved plant volatile analysis in field experiments. *Plant J.* 78: 1060–1072.

Kallenbach, M., Veit, D., Eilers, E.J., and Schuman, M.C. 2015. Application of silicone tubing for robust, simple, high-throughput, and time-resolved analysis of plant volatiles in field experiments. *Bio Protoc.* 5: 1391.

Kännaste, A., Copolovici, L., and Niinemets, U. 2014. Gas chromatography-mass spectrometry method for determination of biogenic volatile organic compounds emitted by plants. In *Plant Isoprenoids: Methods and Protocols*, Methods in Molecular Biology, ed. M. Rodriguez Concepcion, 161–169. Totowa, NJ: Humana Press.

Kaplan, I., Halitschke, R., Kessler, A., Sardanelli, S., and Denno, R.F. 2008. Constitutive and induced defenses to herbivory in above- and belowground plant tissues. *Ecology* 89: 392–406.

Kessler, A. and Baldwin, I.T. 2001. Defensive function of herbivore-induced plant volatile emissions in nature. *Science* 291: 2141–2144.

Kessler, A., Halitschke, R., and Baldwin, I.T. 2004. Silencing the jasmonate cascade: Induced plant defenses and insect populations. *Science* 305: 665–668.

Kfoury, N., Scott, E., Orians, C., and Robbat, A. 2017. Direct contact sorptive extraction: A robust method for sampling plant volatiles in the field. *J. Agric. Food Chem.* 65: 8501–8509.

Kigathi, R.N., Unsicker, S.B., Reichelt, M., et al. 2009. Emission of volatile organic compounds after herbivory from *Trifolium pratense* (L.) under laboratory and field conditions. *J. Chem. Ecol.* 35: 1335–1348.

Kihika, R., Murungi, L.K., Coyne, D., et al. 2017. Parasitic nematode *Meloidogyne incognita* interactions with different *Capsicum annum* cultivars reveal the chemical constituents modulating root herbivory. *Sci. Rep.* 7: 2903.

Knudsen, J.T., Tollsten, L., and Bergstrom, L.G. 1993. Floral scents—A checklist of volatile compounds isolated by headspace techniques. *Phytochemistry* 33: 253–280.

Knudsen, J.T., Tollsten, L., Groth, I., Bergstrom, G., and Raguso, R.A. 2004. Trends in floral scent chemistry in pollination syndromes: Floral scent composition in hummingbird-pollinated taxa. *Bot. J. Linn. Soc.* 146: 191–199.

König, W.A. 1993a. Collection of enantiomer separation factors obtained by capillary gas chromatography on chiral stationary phases. *HRC-J. High. Res. Chrom.* 16: 569.

König, W.A. 1993b. Collection of enantiomer separation factors obtained by capillary gas chromatography on chiral stationary phases. *HRC-J. High. Res. Chrom.* 16: 312.

König, W.A. 1993c. Collection of enantiomer separation factors obtained by capillary gas chromatography on chiral stationary phases. *HRC-J. High. Res. Chrom.* 16: 338.

König, W.A. and Hochmuth, D.H. 2004. Enantioselective gas chromatography in flavor and fragrance analysis: Strategies for the identification of known and unknown plant volatiles. *J. Chromatogr. Sci.* 42: 423–439.

Kremser, A., Jochmann, M.A., and Schmidt, T.C. 2016. PAL SPME Arrow-evaluation of a novel solid-phase microextraction device for freely dissolved PAHs in water. *Anal. Bioanal. Chem.* 408: 943–952.

Krupcik, J., Gorovenko, R., Spanik, I., Armstrong, D.W., and Sandra, P. 2016. Enantioselective comprehensive two-dimensional gas chromatography of lavender essential oil. *J. Sep. Sci.* 39: 4765–4772.

Kuepper, F.C., Carpenter, L.J., McFiggans, G.B., et al. 2008. Iodide accumulation provides kelp with an inorganic antioxidant impacting atmospheric chemistry. *Proc. Natl. Acad. Sci. USA* 105: 6954–6958.

Kunert, G., Otto, S., Röse, U.S.R., Gershenzon, J., and Weisser, W.W. 2005. Alarm pheromone mediates production of winged dispersal morphs in aphids. *Ecol. Lett.* 8: 596–603.

Kunert, M., Biedermann, A., Koch, T., and Boland, W. 2002. Ultrafast sampling and analysis of plant volatiles by a hand-held miniaturised GC with pre-concentration unit: Kinetic and quantitative aspects of plant volatile production. *J. Sep. Sci.* 25: 677–684.

Le Pape, M.A., Grua-Priol, J., Prost, C., and Demaimay, M. 2004. Optimization of dynamic headspace extraction of the edible red algae *Palmaria palmata* and identification of the volatile components. *J. Agric. Food Chem.* 52: 550–556.

Lebanov, L., Tedone, L., Kaykhaii, M., Linford, M.R., and Paull, B. 2019. Multidimensional gas chromatography in essential oil analysis. Part 2: Application to characterisation and identification. *Chromatographia* 82: 399–414.

Lindinger, W., Hansel, A., and Jordan, A. 1998. On-line monitoring of volatile organic compounds at pptv levels by means of proton-transfer-reaction mass spectrometry (PTR-MS)—Medical applications, food control and environmental research. *Int. J. Mass Spectrom.* 173: 191–241.

Lockwood, G.B. 2001. Techniques for gas chromatography of volatile terpenoids from a range of matrices. *J. Chromatogr. A* 936: 23–31.

Long, Q., Li, Z., Han, B., et al. 2019. Discrimination of two cultivars of *Alpinia officinarum* Hance using an electronic nose and gas chromatography-mass spectrometry coupled with chemometrics. *Sensors* 19: 572.

Loreto, F., Barta, C., Brilli, F., and Nogues, I. 2006. On the induction of volatile organic compound emissions by plants as consequence of wounding or fluctuations of light and temperature. *Plant Cell Environ.* 29: 1820–1828.

Loreto, F., Nascetti, P., Graverini, A., and Mannozzi, M. 2000. Emission and content of monoterpenes in intact and wounded needles of the Mediterranean Pine, *Pinus pinea*. *Funct. Ecol.* 14: 589–595.

Lu, F.S.H., Nielsen, N.S., and Jacobsen, C. 2013. Comparison of two methods for extraction of volatiles from marine PL emulsions. *Eur. J. Lipid Sci. Technol.* 115: 246–251.

Martinelli, F., Scalenghe, R., Davino, S., et al. 2015. Advanced methods of plant disease detection. A review. *Agron. Sustain. Dev.* 35: 1–25.

Materic, D., Bruhn, D., Turner, C., et al. 2015. Methods in plant foliar volatile organic compounds research. *Appl. Plant Sci.* 3:150044.

Mateus, E., Barata, R.C., Zrostlikova, J., da Silva, M., and Paiva, M.R. 2010. Characterization of the volatile fraction emitted by *Pinus* spp. by one- and two-dimensional chromatographic techniques with mass spectrometric detection. *J. Chromatogr. A* 1217: 1845–1855.

McCormick, A.C., Irmisch, S., Reinecke, A., et al. 2014. Herbivore-induced volatile emission in black poplar: Regulation and role in attracting herbivore enemies. *Plant Cell Environ.* 37: 1909–1923.

McLeod, P. 2005. Evaluation of insecticides for control of harlequin bugs on turnip. *Arthropod Manage. Tests* 30: E96.

Michelina, C., Naviglio, D., Gallo, M., and Severina, P. 2019. FT-IR and GC-MS analyses of an antioxidant leaf essential oil from sage plants cultivated as an alternative to tobacco production. *J. Essent. Oil Res.* 31: 138–144.

Millar, J.G. and Sims, J.J. 1998. Preparation, cleanup, and preliminary fractionation of extracts. In *Methods in Chemical Ecology*, eds. J. G. Millar, and K. F. Haynes, 1–37. Boston, MA: Kluwer Academic Publishers.

Mozaffar, A., Schoon, N., Bachy, A., et al. 2018. Biogenic volatile organic compound emissions from senescent maize leaves Chock for and a comparison with other leaf developmental stages. *Atmos. Environ.* 176: 71–81.

Munch, J.W. 1195. Method 525.2, Revision 2.0. Determination of organic compounds in drinking water by liquid-solid extraction and capillary column gas chromatography/mass spectrometry. National Exposure Research Laboratory, Office of Research and Development, U.S. Environmental Protection Agency.

Nevitt, G., Reid, K., and Trathan, P. 2004. Testing olfactory foraging strategies in an Antarctic seabird assemblage. *J. Exp. Biol.* 207: 3537–3544.

Niinemets, U., Kuhn, U., Harley, P.C., et al. 2011. Estimations of isoprenoid emission capacity from enclosure studies: Measurements, data processing, quality and standardized measurement protocols. *Biogeosciences* 8: 2209–2246.

Nilsson, T., Montanarella, L., Baglio, D., et al. 1998. Analysis of volatile organic compounds in environmental water samples and soil gas by solid-phase microextraction. *Int. J. Environ. Anal. Chem.* 69: 217–226.

Nojima, S., Kiemle, D.J., Webster, F.X., Apperson, C.S., and Schal, C. 2011. Nanogram-scale preparation and NMR analysis for mass-limited small volatile compounds. *PLoS One* 6: e18178.

Oh, S.Y. 2018. Rapid monitoring of pharmacological volatiles of night-flowering evening-primrose according to flower opening or closing by fast gas chromatography/surface acoustic wave sensor (electronic znose). *Phytochem. Anal.* 29: 275–283.

Osemwengie, L.I. and Steinberg, S. 2003. Closed-loop stripping analysis of synthetic musk compounds from fish tissues with measurement by gas chromatography-mass spectrometry with selected-ion monitoring. *J. Chromatogr. A* 993: 1–15.

Penuelas, J., Asensio, D., Tholl, D., et al. 2014. Biogenic volatile emissions from the soil. *Plant Cell Environ.* 37: 1866–1891.

Piri-Moghadam, H., Ahmadi, F., and Pawliszyn, J. 2016. A critical review of solid phase microextraction for analysis of water samples. *Trac-Trends Anal. Chem.* 85: 133–143.

Pohnert, G. and Boland, W. 2002. The oxylipin chemistry of attraction and defense in brown algae and diatoms. *Nat. Prod. Rep.* 19: 108–122.

Pohnert, G., Steinke, M., and Tollrian, R. 2007. Chemical cues, defence metabolites and the shaping of pelagic interspecific interactions. *Trends Ecol. Evol.* 22: 198–204.

Prebihalo, S.E., Berrier, K.L., Freye, C.E., et al. 2018. Multidimensional gas chromatography: Advances in instrumentation, chemometrics, and applications. *Anal. Chem.* 90: 505–532.

Qin, Z., Bragg, L., Ouyang, G., and Pawliszyn, J. 2008. Comparison of thin-film microextraction and stir bar sorptive extraction for the analysis of polycyclic aromatic hydrocarbons in aqueous samples with controlled agitation conditions. *J. Chromatogr. A* 1196: 89–95.

Raguso, R.A. and Pellmyr, O. 1998. Dynamic headspace analysis of floral volatiles: A comparison of methods. *Oikos* 81: 238–254.

Raguso, R.A. and Pichersky, E. 1995. Floral volatiles from *Clarkia breweri* and *C. concinna* (Onagraceae) - Recent evolution of floral scent and moth pollination. *Plant Syst. Evol.* 194: 55–67.

Rasmann, S., Kollner, T.G., Degenhardt, J., et al. 2005. Recruitment of entomopathogenic nematodes by insect-damaged maize roots. *Nature* 434: 732–737.

Röse, U.S.R., Manukian, A., Heath, R.R., and Tumlinson, J.H. 1996. Volatile semiochemicals released from undamaged cotton leaves—A systemic response of living plants to caterpillar damage. *Plant Physiol.* 111: 487–495.

Schiestl, F.P. and Ayasse, M. 2001. Post-pollination emission of a repellent compound in a sexually deceptive orchid: A new mechanism for maximising reproductive success? *Oecologia* 126: 531–534.

Schreier, P., Bernreuther, A., and Huffner, A. 1995. *Analysis of Chiral Organic Molecules, Methodology and Applications.* Berlin, Germany: de Gruyter.

Surburg, H., Guentert, M., and Harder, H. 1993. Volatile compounds from flowers. Analytical and olfactory aspects. In *Bioactive Volatile Compounds from Plants*, ACS Symposium Series, eds. R. Teranishi, R. G. Buttery, and H. Sugisawa, 168–186. Washington, DC: American Chemical Society.

Tholl, D., Boland, W., Hansel, A., et al. 2006. Practical approaches to plant volatile analysis. *Plant J.* 45: 540–560.

Tytgat, T.O.G., Verhoeven, K.J.F., Jansen, J.J., et al. 2013. Plants know where it hurts: Root and shoot jasmonic acid induction elicit differential responses in *Brassica olracea*. *PLoS One* 8: e65502.

Van Alstyne, K.L. 2008. The distribution of DMSP in green macroalgae from northern New Zealand, eastern Australia and southern Tasmania. *J. Mar. Biol. Assoc. U.K.* 88: 799–805.

van Dam, N.M. and Bouwmeester, H.J. 2016. Metabolomics in the rhizosphere: Tapping into belowground chemical communication. *Trends Plant Sci.* 21: 256–265.

van Dam, N.M., Qiu, B.L., Hordijk, C.A., Vet, L.E.M., and Jansen, J.J. 2010. Identification of biologically relevant compounds in aboveground and belowground induced volatile blends. *J. Chem. Ecol.* 36: 1006–1016.

van Dam, N.M., Weinhold, A., and Garbeva, P. 2016. Calling in the dark: The role of volatiles for communication in the rhizosphere. In *Deciphering Chemical Language of Plant Communication*, ed. J. D. Blande, and R. Glinwood, 175–210. Cham, Switzerland: Springer International Publishing.

Vercammen, J., Pham-Tuan, H., and Sandra, P. 2001. Automated dynamic sampling system for the on-line monitoring of biogenic emissions from living organisms. *J. Chromatogr. A* 930: 39–51.

Vercammen, J., Sandra, P., Baltussen, E., Sandra, T., and David, F. 2000. Considerations on static and dynamic sorptive and adsorptive sampling to monitor volatiles emitted by living plants. *HRC-J. High Res. Chrom.* 23: 547–553.

Visser, T. 2002. FT-IR detection in gas chromatography. *Trac-Trends Anal. Chem.* 21: 627–636.

Wang, H.Y. and Guo, Y.L. 2004. Rapid analysis of the volatile compounds in the rhizomes of *Rhodiold sachalinensis* and *Rhodiola sacra* by static headspace-gas chromatography-tandem mass spectrometry. *Anal. Lett.* 37: 2151–2161.

Waseem, R. and Low, K.H. 2015. Advanced analytical techniques for the extraction and characterization of plant-derived essential oils by gas chromatography with mass spectrometry. *J. Sep. Sci.* 38: 483–501.

Wiesemeier, T., Hay, M.E., and Pohnert, G. 2007. The potential role of wound-activated volatile release in the chemical defence of the brown alga *Dictyota dichotoma*: Blend recognition by marine herbivores. *Aquat. Sci.* 69: 403–412.

Williams, N. and Whitten, W.M. 1983. Orchid floral fragrances and male euglossine bees: Methods and advances in the last sesquidecade. *Biol. Bull.* 164: 355–395.

Wilson, A.D. 2013. Diverse applications of electronic-nose technologies in agriculture and forestry. *Sensors* 13: 2295–2348.

Wong, Y.F., Yan, D.D., Shellie, R.A., Sciarrone, D., and Marriott, P.J. 2019. Rapid plant volatiles screening using headspace SPME and person-portable gas chromatography-mass spectrometry. *Chromatographia* 82: 297–305.

Yashin, Y.I. and Yashin, A.Y. 2001. Miniaturization of gas-chromatographic instruments. *J. Anal. Chem.* 56: 794–805.

Zali, S., Jalali, F., Es-haghi, A., and Shamsipur, M. 2016. New nanostructure of polydimethylsiloxane coating as a solid-phase microextraction fiber: Application to analysis of BTEX in aquatic environmental samples. *J. Chromatogr. B* 1033: 287–295.

Zhu, F., Xu, J.Q., Ke, Y.Y., et al. 2013. Applications of in vivo and in vitro solid-phase microextraction techniques in plant analysis: A review. *Anal. Chim. Acta* 794: 1–14.

2 Analysis of Internal Pools of Plant Volatiles

Yoko Iijima, Naomi Okubo, and Fukuyo Tanaka

CONTENTS

2.1 INTRODUCTION

Plant volatiles can be roughly categorized on spatio-temporal considerations into three groups. The first group are volatiles stored in plant tissues and are often found in secretary cells or in specific areas in organs such as leaves, fruit skin, petals, roots and rhizomes, and in tree barks (see Chapter 13). These volatiles are secreted and collected in sub-dermal intercellular compartments such as glandular cavities after biosynthesis, and are easily released following physical damage to the cells. The second group contains volatiles released during physiological events in plants, such as flowering and the ripening of fruit. The expression of genes and enzymes regulating the biosynthesis of these volatiles is influenced by such physiological changes, and the produced volatiles are transported and released into the atmosphere. The third group contains volatiles sequentially produced after wounding. Cells damaged by wounding induce secondary enzymatic reactions with pooled precursors. Green leaf volatiles (GLVs), isothiocyanates, and sulfides are well known examples belonging to this class of compounds.

The precise monitoring and analysis of volatiles from plants requires knowledge of their production, pooling, and liberation processes. Each organic volatile shows different chemical properties, such as volatility and logKow (octanol-water partition coefficient). Volatiles easily evaporate, thus changing their concentration in the plant and making their analysis challenging compared to nonvolatile metabolites. Changes in the concentration and composition of volatiles are affected by variables such as the cultivar, the physiological condition of the plant, and time period, even in a single plant. Therefore, analytical conditions appropriate for each variable must be chosen to avoid misinterpretation of the results. In this chapter, we introduce recent advances in the analysis of plant volatiles and their applications, focusing on the determination of internal composition and concentration of plant volatiles.

2.2 ANALYSIS OF POOLED VOLATILES

2.2.1 EXTRACTION OF VOLATILES

Various extraction procedures were developed for plant volatiles (Ormeño et al. 2011), (see also Chapter 1). Many volatiles in plants are biosynthesized in specialized tissues and stored in sub-dermal storage cavities. Such secretory cavities are found in the glandular trichomes of plants such as Lamiaceae and Solanaceae (Wagner 1991, Turner et al. 2000, Werker 2000, Schilmiller et al. 2009, Tissier et al. 2017), and are attached to epithelial cells on the peel of Citrus fruits (Voo et al. 2012). These cavities can store high concentrations of volatiles and other hydrophobic specialized metabolites. Secretory cells and cavities are thus "efficient factories" producing essential oils in aromatic plants. Many trichome cavities on the leaves of plants are covered with a relatively thin cuticle, allowing easy disruption of the cavities by physical or chemical stimulation and herbivore attack to release stored volatiles (Peiffer et al. 2009, Widhalm et al. 2015). Volatiles in glands are typically analyzed following extraction using organic solvents such as diethyl ether, methyl t-butyl ether (MTBE) or dichloromethane, because these solvents can easily penetrate and break the cuticle. For example, simple dipping of leaf samples in MTBE for 1 min allowed the analysis of volatiles in tomato (*Solanum lycopersicum*) glandular trichomes (Schilmiller et al. 2009). Although severe, this extraction procedure is effective for rapid analysis of a large number of samples. HS (Head Space)-SPME/GC-MS analysis after the disruption of cells is a common and facile analytical procedure (Ormeño et al. 2011) that does not require solvent for extraction.

However, the above extraction procedures provide volatiles from both trichomes and other tissues and thus volatiles specifically stored in gland cavities cannot be identified. Direct extraction using micro-capillary insertion into gland cavities addresses this drawback and has been successfully applied to sage (*Salvia officinalis*) (Grassi et al. 2004), oregano (*Origanum vulgare*) (Johnson et al. 2004) and grapefruit peel (Voo et al. 2012). Recently, we collected droplets from leaf glands of Japanese pepper (*Zanthoxylum piperitum*) using a micro-syringe and confirmed that volatile terpenes were concentrated in these glands (Fujita et al. 2017).

2.2.2 DETECTION OF VOLATILES BY IMAGING MASS SPECTROMETRY (IMS)

Mass spectrometry-based imaging technologies have recently been developed. These approaches do not require the extraction of metabolites and allow visualization of the distribution of specific metabolites in various tissues and organelles (Boughton et al. 2016). Matrix-assisted laser desorption/ionization (MALDI)-MS, atmospheric pressure ion-source chamber for laser desorption/ionization (AP-LDI)-MS, and nano-particle laser desorption/ionization (nano-PALDI)-MS are frequently used for such analyses and have allowed the visualization of metabolites *in situ* for various plant species and tissues (Holscher et al. 2009, Bjarnholt et al. 2014, Qin et al. 2018). For example, acyl sugars, sesquiterpene acids, and flavonoids were detected in the trichomes of wild tomato leaf (*Solanum habrochaites*) using AP-LDI-MS (Li et al. 2014) and flavonoids and sesquiterpene lactones were detected by LDI and MALDI-MS in the trichomes of sunflower leaf (Silva et al. 2017). However, several technical problems remain. For example, samples must be handled carefully to prevent loss of volatiles prior to analysis. Furthermore, identifying the volatiles can be challenging because monoterpenes and sesquiterpenes generate similar, overlapping molecular ions and fragment ions, complicating their visualization. Consequently, there are relatively few examples of the visualization of volatiles compared to nonvolatile metabolites. One example of the former is the use of AP-LDI-MS microscopy to map the distribution of monoterpene, diterpene and 6-gingerol MS ions in ginger (*Zingiber officinale*) rhizome tissue (Harada et al. 2009).

Despite recent advances in understanding the biosynthesis of plant volatiles, little is known regarding the mechanisms of their distribution and transportation, i.e., how volatiles are secreted from glandular cells to storage cavities or released to the atmosphere (Linka and Theodoulou 2013, Adebesin et al. 2017, Tissier et al. 2017) (see also Chapter 14). Improved sequential mass images of volatiles in tissues will help elucidate the movement of volatiles in plant tissues.

2.2.3 Availability of Stable Isotope Labeling Analyses of Biosynthesis and the Quantification of Volatile Metabolites

Isotope labeling of trace metabolites is a classic and powerful technique for studying the biosynthetic pathway and metabolic flows of various metabolites because isotope labeled and non-labeled endogenous metabolites have the same chemical properties. Therefore, after a labeled compound is incorporated into a plant, it is metabolized identically to the endogenous compound. Trace amounts of radio-isotope labeled compounds are detectable and their enzymatic reactions and distribution in organs can be measured using a radio-isotope detector. On the other hand, stable isotope labeling is safe and the isotopes are detectable using mass spectrometry. Deuterium, ^{14}C, ^{15}N, and ^{34}S are common isotopes for the trace analysis of metabolites. The parent and fragment ions from the chemical building block containing the heavy isotope are detected at higher m/z values than the non-labeled ions, and show the same fragment patterns and similar retention times on GC or HPLC. Therefore, labeled metabolites can be distinguished from non-labeled metabolites (Creek et al. 2012).

We synthesized deuterium labeled geraniol (D_2-geraniol) and incorporated it in ginger rhizome (*Zingiber officinale*). Gereniol-related volatiles such as geraniol, citral (geranial and neral), geranyl acetate and citronellol significantly contribute to the aroma of fresh ginger rhizomes. We clarified the biotransformations of these volatiles by extracting them from D_2-geraniol-containing and control rhizomes. GC-MS detected the parent ion of D_2-geraniol at m/z 156 and endogenous geraniol ion at m/z 154. Other geraniol-related compounds such as D_2-geranyl acetate, D_1-geranial, D_1-geraniol, and D_1- and D_2-citronellol were also detected and suggested that geraniol is divergently converted to other compounds as shown in Figure 2.1 via the indicated enzymatic reactions. In particular, the

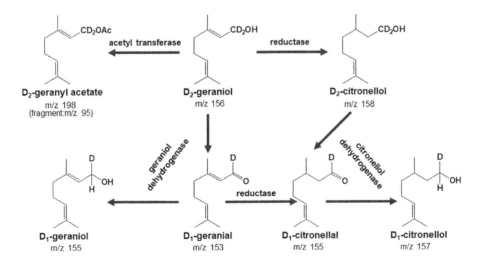

FIGURE 2.1 Probable enzymatic reactions based on detected deuterium-labeled geraniol relatives in ginger rhizome after supplying D_2-geraniol. Volatiles in ginger rhizome were extracted using diethyl ether and analyzed by GC-MS.

detection of both D_1- and D_2-geraniol demonstrated that incorporated D_2-geraniol is oxidized to D_1-geranial but some is reduced to D_1-geraniol. These results led to the characterization of geraniol dehydrogenase in ginger rhizome (Iijima et al. 2014).

Stable isotope labeling of volatiles is also effective for quantification of the corresponding endogenous metabolite. Most volatiles easily diffuse during sample preparation, making their accurate quantification difficult. As stated above, labeled and non-labeled compounds show the same chemical properties. Therefore, spiking known amounts of labeled standards as internal standards in samples prior to extraction allows accurate quantification of the endogenous corresponding compound. This quantification method called stable isotope dilution analysis has been widely used to quantify various metabolites such as plant hormones (Balcke et al. 2012), volatiles (Schieberle and Grosch 1987, Pollnitz et al. 2004) and other nonvolatile metabolites using mass spectrometers (Stokvis et al. 2005).

2.3 COMPOSITIONAL CHANGES IN POOLED AND RELEASED PLANT VOLATILES

2.3.1 DIFFERENCE IN POOLED AND RELEASED VOLATILES FROM FRUITS

Climacteric fruits such as apple (*Malus pumila*), pear (*Pyrus communis*) and cantaloupe (*Cucumis melo*) produce and release characteristic fruity aromas as the fruits ripen, and the profiles of these aroma volatiles influence food quality. Most of these pleasant aromas are due to aliphatic esters and lactones, but whole fruits and fruit flesh can have different aromas. We compared the volatile profiles of whole ripe intact peach (*Prunus persica*) fruit and homogenate peach juice (Figure 2.2). Emitted volatiles from whole fruit were extracted from static headspace gas using a MonoTrap (GL Science, Tokyo, Japan). Volatiles in homogenate juice were extracted by immersing a MonoTrap in the juice. Both extracted volatile samples were introduced into a GC-MS instrument by thermal desorption. The most common compounds in both samples were lactones such as γ-decalactone, whereas linalool and several esters were detected exclusively in the headspace samples. Benzaldehyde and jasmin lactone were characteristic compounds in juice, together with aldehydes and straight chain

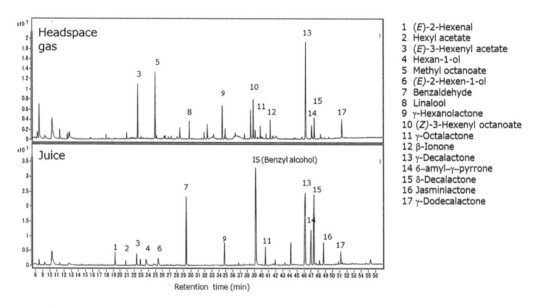

FIGURE 2.2 Comparison of GC-MS chromatograms of emitted volatiles and volatiles in homogenized peach juice. Emitted volatiles were extracted using a MonoTrap from the headspace gas in a chamber containing a whole peach fruit. The volatiles from homogenate juice were extracted with a MonoTrap by immersion in the juice.

alcohols. Consequently, the compositions of the emitted and endogenous volatiles in the two samples are different and thus the extraction method for volatiles should be carefully chosen according to the purpose of the research.

2.3.2 DUAL ANALYSIS OF EMITTED AND INTERNAL VOLATILES IN FLOWERS

Floral scents guide pollinators to the reproductive organs of plants and thus flowers emit floral scent compounds to coincide with the activity of the pollinator (e.g., bees at noon, moths at night). Time-course analysis is important for clarifying plant endogenous volatiles whose amounts in scents change drastically, such as in flowers. Many petunia cultivars planted in flower beds lack in fragrance, but *Petunia axillaris*, the parental species of horticultural petunia cultivars, emits a strong sweet scent at night and its pollinator is the hawk moth (Ando et al. 2001). The floral scent compounds of *P. axillaris* are all benzenoids, with iso-eugenol and methyl benzoate being major components (Oyama-Okubo et al. 2005). Emitted volatiles of *P. axillaris* were collected every 4 h after flowering using the dynamic-headspace method (Figure 2.3); quantitative analysis showed diurnal oscillations, with maxima at night and minima at noon. Simultaneously, one *P. axillaris* was harvested every 6 h after flowering, then frozen. After frozen flowers were ground, endogenous volatiles were collected by extraction with pentane in a microwave oven (700 W) for 20 s in the presence of ethyl decanoate as an internal standard. After concentration, the extract was analyzed by GC-MS. The endogenous concentrations of all volatiles studied correlated well with the oscillating emission patterns (Figure 2.4), showing that analysis of endogenous flower volatiles must take into account the time period during which floral scent is emitted and to harvest flowers when the largest amount of volatiles is being emitted.

2.3.3 REAL-TIME MONITORING OF THE COMPOSITION OF VOLATILES IN WOUNDED PLANTS

GLVs from various plant species, isothiocyanates from Brassicaceae and sulfides from Allium, are characteristic plant volatiles almost absent in intact plant although. Their precursors and synthetic enzymes are present in different compartments in intact cells. Following cell destruction by, for example, wounding or grinding, these volatiles are enzymatically produced as secondary reaction using precursors, resulting in time-dependent variations in the amounts and compositions of the volatiles. These sequential variations can be observed using real-time monitoring analysis methods such as proton-transfer reaction (PTR)-MS. PTR-MS ionizes target compounds at atmospheric pressure using proton transfer reagent ions such as H_3O^+. For example, differences in the emission patterns of three GLVs, (Z)-3-hexenal, (Z)-3-hexen-1-ol and (Z)-3-hexen-1-yl acetate after wounding *Arabidopsis* leaves were monitored by PTR-MS (D'Auria et al. 2007). Direct analysis by real-time

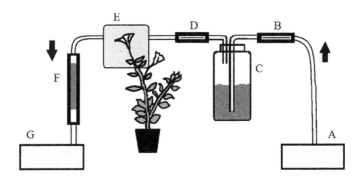

FIGURE 2.3 Schematic of the dynamic-headspace method. A, air-supply pump; B, flow meter (50 L min^{-1}); C, activated charcoal filter; D, flow meter (500 mL min^{-1}); E, tedlar bag (500 mL); F, Tenax-TA trap; G, vacuum pump.

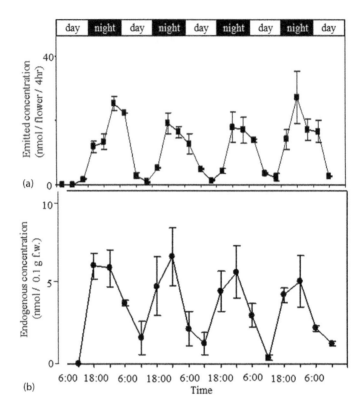

FIGURE 2.4 Time-course change in the concentration of iso-eugenol from *P. axillaris* flowers during a 4-day period after flower opening. (a) emitted volatiles. Measurements were conducted every 4 h; (b) endogenous volatiles. Measurements were conducted every 6 h. Mean values ±SE of three or more repetitions were plotted against the midpoints of these intervals.

ion-source mass spectrometry (DART-MS) can also be used to ionize metabolites in gas at atmospheric pressure (Cody 2008, Musah et al. 2012) and allowed monitoring of sequential changes in characteristic sulfur metabolites after crushing garlic (*Allium sativum*) (Block et al. 2010).

2.4 APPLICATIONS OF VOLATILE PROFILES TO METABOLOMICS STUDIES

2.4.1 COMPREHENSIVE ANALYSIS OF PLANT VOLATILES

Metabolomics is effective for elucidating various biochemical events, the dynamics of biological systems, and the biosynthesis of specific metabolites, based on the comprehensive analysis of metabolites (Fiehn 2002, Sumner et al. 2003). Typically, samples from numerous species or of a single species under various physiological conditions are systematically collected, and their multiple metabolic profiles are used for metabolome analysis. Metabolic profiles are obtained comprehensively and simultaneously using one or more analytical instruments, such as GC-MS (Fiehn et al. 2000), capillary electrophoresis-mass spectrometry (CE-MS) (Soga et al. 2003), liquid chromatography-mass spectrometry (LC-MS) (Yoshida et al. 2007), Fourier transform ion cyclotron resonance mass spectrometry (Aharoni et al. 2002) and nuclear magnetic resonance spectrometry (Reo 2002).

Volatile metabolomics data are frequently obtained by GC-MS and used for multi-sample data analysis. When a single GC-column cannot separate each volatile, separation is achieved by connecting two columns in series (GC×GC). Metabolomics analysis requires minimizing variations in extraction efficiency and analytical conditions between samples. Therefore, simple and quick extraction methods such as HS-SPME and SBSE (stir-bar sorptive extraction) are preferred. Comparative analysis of multi-sample data helps clarify differences between samples and identify the metabolites contributing to these differences. Multi-sample data analysis of volatiles is improved by using the software package for batch analysis, supplied by each instrument manufacturer. We recently reported a procedure for comparative volatile analysis of multiple samples using two free programs, MetAlign (Lommen 2009) for peak detection and peak alignment, and AIoutput (Tsugawa et al. 2011) for peak identification (Iijima et al. 2016) (Figure 2.5).

2.4.2 Application of Volatile Metabolomics to Classification and Quantitative Trait Loci (QTL) Mapping

Volatile profiling is used for chemotaxonomic analysis, to classify and characterize organisms according to their chemical compositions, and to evaluate their biological diversity and evolution. For example, the profiles of essential oils were determined to allow chemotaxonomic study of medicinal and aromatic plants (Barra 2009, Grayer et al. 1996). The relationships between the chemotypes of essential oils and their genetic diversity were investigated in *Salvia fruticosa* (Skoula et al. 1999), *Rosmarinus officinalis* (Angioni et al. 2004), and *Ocimum gratissimum* (Vieira et al. 2001). Recently, quantitative trait loci (QTL) mapping of various metabolites was carried out mainly in model plants such as *Arabidopsis thaliana* and tomato (*S. lycopersicum*) (Fernie and Klee 2011) to identify QTL loci for biosynthesis and regulation of specialized metabolites. QTL mapping of volatiles for apple (*Malus* sp., "Discovery" × "Prima") (Dunemann et al. 2009), tomato (*S. lycopersicum* × *S. pennellii*) (Mathieu et al. 2009, Zanor et al. 2009), strawberry (*Fragaria* × *ananassa*) (Zorrilla-Fontanesi et al. 2012) and petunia (*Petunia axillaris* × *Petunia exserta*) (Klahre et al. 2011) have been reported.

2.4.3 Relationships between Volatiles and Nonvolatiles

Most precursors of volatiles are nonvolatile metabolites. Floral volatile precursors include neutral carbohydrates, phosphorylated saccharides, organic acids, and amino acids. Here, we describe the metabolomics of floral volatiles and their precursors as determined using several instruments.

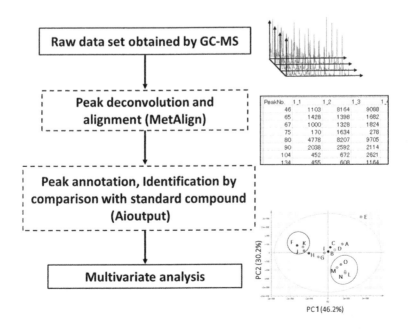

FIGURE 2.5 Typical comprehensive analytical scheme of volatiles for metabolome analysis.

Benzenoid/phenylpropanoid compounds are mainly responsible for the floral scent of wild *Petunia axillaris*, with different lines emitting different concentrations (Kondo et al. 2006). Metabolomics analysis of corolla metabolites in *P. axillaris* that emit scent either strongly or weakly helped clarify the biosynthesis steps controlling the nocturnal rhythm of volatiles and the intensity of the fragrance (Oyama-Okubo et al. 2013). Volatiles and carbohydrates were extracted from the corollas of strongly and weakly scented lines and analyzed using GC-MS and HPLC, respectively (Ichimura et al. 2000). Polar metabolites were analyzed by CE-MS and 23 compounds associated with the biosynthesis of volatiles were selected for further analysis. In the strongly scented line, 6-phosphogluconate (6PG), sedoheptulose-7-phosphate (S7P) and ribose-5-phosphate (R5P) in the pentose phosphate cycle, as well as other downstream volatiles, showed nocturnal concentration changes similar to that of floral volatiles (Figure 2.6). In contrast, the concentrations of several metabolites, including 6PG and downstream metabolites of shikimic acid were remarkably lower in the weakly scented line, while glucose and glucose 6-phosphate (G6P) showed nocturnal changes. These results suggested that the nocturnal change in the concentration of floral volatiles in *P. axillaris* occurred in relation to sugar primary metabolism (e.g., glycolysis and the pentose phosphate cycle).

Several glyco-conjugates of alcoholic and phenolic volatiles are precursors of aroma in rose petals, tea leaves, wine grapes, tomatoes, and other plants (Sarry and Günata 2004). Therefore, comparative investigations of glyco-conjugates and the corresponding aglycone volatiles are valuable for understanding the pleasant aromas of plants. A nontargeted fusion metabolome analysis using GC-MS for volatiles and LC-MS for nonvolatiles was performed for different tomato (*S. lycopersicum*) genotypes. Phenolic aroma compounds emitted from tomato fruit were generated from precursor sugar-conjugates, and cleavage of the disaccharide occurred upon cell disruption during tomato ripening. Furthermore, low phenolic aroma genotypes contained more complicated conjugation forms such as trisaccharides and malonylated disaccharides (Tikunov et al. 2010). These results indicate that multiplatform metabolomics analyses have strong potential for elucidating the synthesis and regulation of volatiles and their effects on other physiological and ecological functions.

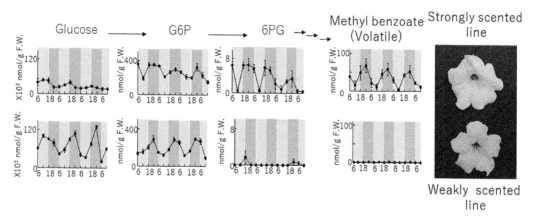

FIGURE 2.6 Metabolome analysis of a strongly scented line and a weakly scented line of *P. axillaris*. Carbohydrates, volatiles and other metabolites were extracted from corolla tissues harvested every 6 h starting at 06:00 on the second day after flower opening until 12:00 of the fifth day. Plants were grown under a 12/12 h (06:00–18:00 light (light gray background)/18:00–06:00 h dark (dark gray background)) photoperiod. Carbohydrates were analyzed by HPLC, volatiles were analyzed by GC-MS and other metabolites were analyzed by CE-MS. Each point is the average of three independent experiments. Standard errors are indicated by vertical bars. n.d. signifies not detected.

2.4.4 Volatile Metabolomics Involved in Other Physiological Changes: Examples from Apple Fruits

Physiological changes and differences in plants induce metabolic changes which influence the nutritional and flavor qualities of edible plants. Here, we focused on the flavor of apples and investigated the influence of flesh browning and watercore on their volatile profiles.

Flesh browning due to long-term storage, even in intact apple fruit, is undesirable. We noninvasively predicted browning apples by screening volatiles as a biomarker using metabolomics. "Fuji" apples were stored under a low concentration of ethylene by 1-MCP treatment or in controlled atmosphere storage, then the severity of flesh browning was assessed. Volatiles were trapped from the static head space of whole apple fruit using a MonoTrap and analyzed by thermal desorption GC-MS. The methyl esters correlated with the browning area at the development stage, indicating that emitted volatiles can reflect physiological changes occurring in internal apple tissues (Tanaka et al. 2018).

The characteristic flavor of watercored apples was investigated by integrated analysis of sensory and metabolic profiles. Two juices from watercored and nonwatercored Fuji apples, were sensory evaluated with and without the use of nose clips to survey the role of volatiles. When the taste was compared with nose clips on the sensory panel participants, the intensity was not significantly different between watercored and nonwatercored apples. However, in the absence of nose clips, the perceptions of floral, sweet and fruity attributes were enhanced in the watercored samples. The concentrations of ethyl 2-methylbutanoate, ethyl hexanoate, and several other ethyl esters present in fruity sweet aromas were significantly higher in juice from watercored apples, showing that ethyl esters in watercored apples enhance sweet and fruity flavors (Tanaka et al. 2016).

2.5 CONCLUSION

This chapter described current analytical technologies for volatiles and the role of volatiles in plants. Distinctive volatiles are pooled and released from particular tissues and cells or under specific conditions. Biochemical studies on the biosynthesis of plant volatiles have clarified where and when they are produced at the molecular level, but the behavior of volatiles following synthesis remains poorly understood. Progress in micro-technologies such as imaging MS will address this issue. Broad surveys using metabolomics, involving data mining from vast volatile data collections obtained under various conditions, can help us understand events occurring in plants. Basic analytical technologies for volatiles are well established but further chemical applications, involving biochemical, physiological, and ecological experiments, will further aid research on plant volatiles.

REFERENCES

Adebesin, F., J.R. Widhalm, B. Boachon, F. Lefevre, B. Pierman, J.H. Lynch, I. Alam, B. Junqueira, et al. 2017. Emission of volatile organic compounds from petunia flowers is facilitated by an ABC transporter. *Science* 356:1386–1388.

Aharoni, A., C.H. Ric de Vos, H.A. Verhoeven, C.A. Maliepaard, G. Kruppa, R. Bino, and D.B. Goodenowe. 2002. Nontargeted metabolome analysis by use of Fourier Transform Ion Cyclotron Mass Spectrometry. *OMICS* 6:217–234.

Ando, T., M. Nomura, J. Tsukahara, H. Watanabe, H. Kokubun, T. Tsukamoto, G. Hashimoto, E. Marchesi, and I.J. Kitching. 2001. Reproductive isolation in a native population of *Petunia sensu* Jussieu (Solanaceae). *Ann Bot-London* 88:403–413.

Angioni, A., A. Barra, E. Cereti, D. Barile, J.D. Coïsson, M.Arlorio, S. Dessi, V. Coroneo, and P. Cabras. 2004. Chemical composition, plant genetic differences, antimicrobial and antifungal activity investigation of the essential oil of *Rosmarinus officinalis* L. *J Agr Food Chem* 52:3530–3535.

Balcke, G.U., V. Handrick, N. Bergau, M. Fichtner, A. Henning, H. Stellmach, A. Tissier, B. Hause, and A. Frolov. 2012. An UPLC-MS/MS method for highly sensitive high-throughput analysis of phytohormones in plant tissues. *Plant Methods* 8:47. doi:10.1186/1746-4811-8-47.

Barra, A. 2009. Factors affecting chemical variability of essential oils: A review of recent developments. *Nat Prod Commun* 4:1147–1154.

Bjarnholt, N., B. Li, J. D'Alvise, and C. Janfelt. 2014. Mass spectrometry imaging of plant metabolites—Principles and possibilities. *Nat Prod Rep* 31:818–837.

Block, E., A.J. Dane, S. Thomas, and R.B. Cody. 2010. Applications of direct analysis in real time mass spectrometry (dart-ms) in Allium chemistry. 2-propenesulfenic and 2-propenesulfinic acids, diallyl tri-sulfane s-oxide, and other reactive sulfur compounds from crushed garlic and other Alliums. *J Agric Food Chem* 58:4617–4625.

Boughton, B.A., D. Thinagaran, D. Sarabia, A. Bacic, and U. Roessner. 2016. Mass spectrometry imaging for plant biology: A review. *Phytochem Rev* 15:445–488.

Cody, Robert B. 2008. Observation of molecular ions and analysis of nonpolar compounds with the direct analysis in real time ion source. *Anal Chem* 81:1101–1107.

Creek, D.J., A. Chokkathukalam, A. Jankevics, K.E. V. Burgess, R. Breitling, and M.P. Barrett. 2012. Stable isotope-assisted metabolomics for network-wide metabolic pathway elucidation. *Anal Chem* 84:8442–8447.

D'Auria, J.C., E. Pichersky, A. Schaub, A. Hansel, and J. Gershenzon. 2007. Characterization of a BAHD acyltransferase responsible for producing the green leaf volatile (Z)-3-hexen-1-yl acetate in *Arabidopsis thaliana*. *Plant J* 49:194–207.

Dunemann, F., D. Ulrich, A. Boudichevskaia, C. Grafe, and W.E. Weber. 2009. QTL mapping of aroma compounds analysed by headspace solid-phase microextraction gas chromatography in the apple progeny "Discovery" x "Prima". *Mol Breeding* 23:501–521.

Fernie, A.R., and H.J. Klee. 2011. The use of natural genetic diversity in the understanding of metabolic organization and regulation. *Front Plant Sci* 2:59. https://www.frontiersin.org/articles/10.3389/fpls.2011.00059/full.

Fiehn, O., J. Kopka, P. Dormann, T. Altmann, R.N. Trethewey, and L. Willmitzer. 2000. Metabolite profiling for plant functional genomics. *Nature Biotechnol* 18:1157–1161.

Fiehn , O. 2002. Metabolomics—the link between genotypes and phenotypes. *Plant Mol Biol* 48:155–171.

Fujita, Y., T. Koeduka, M. Aida, H. Suzuki, Y. Iijima, and K. Matsui. 2017. Biosynthesis of volatile terpenes that accumulate in the secretory cavities of young leaves of Japanese pepper (*Zanthoxylum piperitum*): Isolation and functional characterization of monoterpene and sesquiterpene synthase genes. *Plant Biotechnol* 34:17–28.

Grassi, P., J. Novak, H. Steinlesberger, and C. Franz. 2004. A direct liquid, non-equilibrium solid-phase micro-extraction application for analysing chemical variation of single peltate trichomes on leaves of *Salvia officinalis*. *Phytochem Anal* 15:198–203.

Grayer, R.J., G.C. Kite, F.J. Goldstone, S.E. Bryan, A. Paton, and E. Putievsky. 1996. Infraspecific taxonomy and essential oil chemotypes in sweet basil, *Ocimum basilicum*. *Phytochemistry* 43:1033–1039.

Harada, T., A. Yuba-Kubo, Y. Sugiura, N. Zaima, T. Hayasaka, N. Goto-Inoue, M. Wakui, M. Suematsu, K. Takeshita, K. Ogawa, Y. Yoshida, and M. Setou. 2009. Visualization of volatile substances in different organelles with an atmospheric-pressure mass microscope. *Anal Chem* 81:9153–9157.

Holscher, D., R. Shroff, K. Knop, M. Gottschaldt, A. Crecelius, B. Schneider, D.G. Heckel, U.S. Schubert, and A. Svatos. 2009. Matrix-free UV-laser desorption/ionization (LDI) mass spectrometric imaging at the single-cell level: Distribution of secondary metabolites of *Arabidopsis thaliana* and *Hypericum* species. *Plant J* 60:907–918.

Ichimura, K., K. Kohata, and R. Goto. 2000. Soluble carbohydrates in *Delphinium* and their influence on sepal abscission in cut flowers. *Physiol Plant* 108:307–313.

Iijima, Y., T. Koeduka, H. Suzuki, and K. Kubota. 2014. Biosynthesis of geranial, a potent aroma compound in ginger rhizome (*Zingiber officinale*): Molecular cloning and characterization of geraniol dehydrogenase. *Plant Biotechnol* 31:525–534.

Iijima, Y., Y. Iwasaki, Y. Otagiri, H. Tsugawa, T. Sato, H. Otomo, Y. Sekine, and A. Obata. 2016. Flavor characteristics of the juices from fresh market tomatoes differentiated from those from processing tomatoes by combined analysis of volatile profiles with sensory evaluation. *Biosci Biotech Bioch* 80:2401–2411.

Johnson, C.B., A. Kazantzis, M. Skoula, U. Mitteregger, and J. Novak. 2004. Seasonal, populational and onto-genic variation in the volatile oil content and composition of individuals of *Origanum vulgare* subsp. *hirtum*, assessed by GC headspace analysis and by SPME sampling of individual oil glands. *Phytochem Anal* 15:286–292.

Klahre, U., A. Gurba, K. Hermann, M. Saxenhofer, E. Bossolini, P.M. Guerin, and C. Kuhlemeier. 2011. Pollinator choice in Petunia depends on two major genetic loci for floral scent production. *Curr Biol* 21:730–739.

Kondo, M., N. Oyama-Okubo, T. Ando, E. Marchesi, and M. Nakayama. 2006. Floral scent diversity is differently expressed in emitted and endogenous components in Petunia axillaris lines. *Ann Bot-London* 98:1253–1259.

Li, C., Z.Z. Wang, and A.D. Jones. 2014. Chemical imaging of trichome specialized metabolites using contact printing and laser desorption/ionization mass spectrometry. *Anal Bioanal Chem* 406:171–182.

Linka, N., and F.L. Theodoulou. 2013. Metabolite transporters of the plant peroxisomal membrane: Known and unknown. *Subcell Biochem* 69:169–94.

Lommen, A. 2009. MetAlign: Interface-driven, versatile metabolomics tool for hyphenated full-scan mass spectrometry data preprocessing. *Anal Chem* 81:3079–3086.

Mathieu, S., V.D. Cin, Z. Fei, H. Li, P. Bliss, M.G. Taylor, H.J. Klee, and D.M. Tieman. 2009. Flavour compounds in tomato fruits: Identification of loci and potential pathways affecting volatile composition. *J Exp Bot* 60:325–337.

Musah, R.A., M.A. Domin, M.A. Walling, and J.R.E. Shepard. 2012. Rapid identification of synthetic cannabinoids in herbal samples via direct analysis in real time mass spectrometry. *Rapid Commun Mass Sp* 26:1109–1114.

Ormeño, E., A. Goldstein, and Ü. Niinemets. 2011. Extracting and trapping biogenic volatile organic compounds stored in plant species. *TrAC Trends Anal Chem* 30:978–989.

Oyama-Okubo, N., T. Ando, N. Watanabe, E. Marchesi, K. Uchida, and M. Nakayama. 2005. Emission mechanism of floral scent in *Petunia axillaris*. *Biosci Biotech Bioch* 69:773–777.

Oyama-Okubo, N., T. Sakai, T. Ando, M. Nakayama, and T. Soga. 2013. Metabolome profiling of floral scent production in *Petunia axillaris*. *Phytochemistry* 90:37–42.

Peiffer, M., J.F. Tooker, D.S. Luthe, and G.W. Felton. 2009. Plants on early alert: Glandular trichomes as sensors for insect herbivores. *New Phytologist* 184:644–656.

Pollnitz, A.P., K.H. Pardon, M. Sykes, and M.A. Sefton. 2004. The effects of sample preparation and gas chromatograph injection techniques on the accuracy of measuring guaiacol, 4-methylguaiacol and other volatile oak compounds in oak extracts by stable isotope dilution analyses. *J Agr Food Chem* 52:3244–3252.

Qin, L., Y.W. Zhang, Y.Q. Liu, H.X. He, M.M. Han, Y.Y. Li, M.M. Zeng, and X.D. Wang. 2018. Recent advances in matrix-assisted laser desorption/ionisation mass spectrometry imaging (MALDI-MSI) for in situ analysis of endogenous molecules in plants. *Phytochemical Analysis* 29:351–364.

Reo, N.V. 2002. NMR-based metabolomics. *Drug Chem Toxicol* 25:375–382.

Sarry, J.E., and Z. Günata. 2004. Plant and microbial glycoside hydrolases: Volatile release from glycosidic aroma precursors. *Food Chem* 87 (4):509–521.

Schieberle, P., and W. Grosch. 1987. Quantitative analysis of aroma compounds in wheat and rye bread crusts using a stable isotope dilution assay. *J Agric Food Chem* 35:252–257.

Schilmiller, A.L., I. Schauvinhold, M. Larson, R. Xu, A.L. Charbonneau, A. Schmidt, C. Wilkerson, R.L. Last, and E. Pichersky. 2009. Monoterpenes in the glandular trichomes of tomato are synthesized from a neryl diphosphate precursor rather than geranyl diphosphate. *Proc Natl Acad Sci USA* 106:10865–10870.

Silva, D.B., A.K. Aschenbrenner, N.P. Lopes, and O. Spring. 2017. Direct analyses of secondary metabolites by mass spectrometry imaging (MSI) from Sunflower (*Helianthus annuus* L.) trichomes. *Molecules* 22. https://www.mdpi.com/1420-3049/22/5/774/pdf/1.

Skoula, M., I.E.l. Hilali, and A.M. Makris. 1999. Evaluation of the genetic diversity of *Salvia fruticosa* Mill. clones using RAPD markers and comparison with the essential oil profiles. *Biochem Syst Ecol* 27:559–568.

Soga, T., Y. Ohashi, Y. Ueno, H. Naraoka, M. Tomita, and T. Nishioka. 2003. Quantitative metabolome analysis using capillary electrophoresis mass spectrometry. *J Proteome Res* 2:488–494.

Stokvis, E., H. Rosing, and J.H. Beijnen. 2005. Stable isotopically labeled internal standards in quantitative bioanalysis using liquid chromatography/mass spectrometry: Necessity or not? *Rapid Commun Mass Spectrom.* 19:401–407.

Sumner, L.W., P. Mendes, and R.A. Dixon. 2003. Plant metabolomics: Large-scale phytochemistry in the functional genomics era. *Phytochemistry* 62:817–836.

Tanaka, F., K. Okazaki, T. Kashimura, Y. Ohwaki, M. Tatsuki, A. Sawada, T. Ito, and T. Miyazawa. 2016. Profiles and physiological mechanisms of sensory attributes and flavor components in watercored apple. *J Jpn Soc Food Sci* 63:101–116.

Tanaka, F., M. Tatsuki, K. Matsubara, K. Okazaki, M. Yoshimura, and S. Kasai. 2018. Methyl ester generation associated with flesh browning in "Fuji" apples after long storage under repressed ethylene function. *Postharvest Biol Tech* 145:53–60.

Tikunov, Y.M., R.C. H. de Vos, A.M. G. Paramas, R.D. Hall, and A.G. Bovy. 2010. A Role for differential glycoconjugation in the emission of phenylpropanoid volatiles from tomato fruit discovered using a metabolic data fusion approach. *Plant Physiol* 152:55–70.

Tissier, A., J.A. Morgan, and N. Dudareva. 2017. Plant volatiles: Going "In" but not "Out" of trichome cavities. *Trends Plant Sci* 22:930–938.

Tsugawa, H., Y. Tsujimoto, M. Arita, T. Bamba, and E. Fukusaki. 2011. GC/MS based metabolomics: Development of a data mining system for metabolite identification by using soft independent modeling of class analogy (SIMCA). *BMC Bioinformatics* 12:131. https://bmcbioinformatics.biomedcentral.com/articles/10.1186/1471-2105-12-131.

Turner, G.W., J. Gershenzon, and R.B. Croteau. 2000. Development of peltate glandular trichomes of peppermint. *Plant Physiol* 124:665–679.

Vieira, Roberto F., Renée J. Grayer, Alan Paton, and James E. Simon. 2001. Genetic diversity of *Ocimum gratissimum* L. based on volatile oil constituents, flavonoids and RAPD markers. *Biochem Syst Ecol* 29:287–304.

Voo, S.S., H.D. Grimes, and B.M. Lange. 2012. Assessing the biosynthetic capabilities of secretory glands in Citrus peel. *Plant Physiol* 159:81–94.

Wagner, G.J. 1991. Secreting glandular trichomes—More than just hairs. *Plant Physiol* 96:675–679.

Werker, E. 2000. Trichome diversity and development. *Advances in Botanical Research* 31:1–35.

Widhalm, J.R., R. Jaini, J.A. Morgan, and N. Dudareva. 2015. Rethinking how volatiles are released from plant cells. *Trends Plant Sci* 20:545–550.

Yoshida, H., T. Mizukoshi, K. Hirayama, and H. Miyano. 2007. Comprehensive analytical method for the determination of hydrophilic metabolites by high-performance liquid chromatography and mass spectrometry. *J Agr Food Chem* 55 (3):551–560.

Zanor, M.I., J.L. Rambla, J. Chaib, A. Steppa, A. Medina, A. Granell, A.R. Fernie, and M. Causse. 2009. Metabolic characterization of loci affecting sensory attributes in tomato allows an assessment of the influence of the levels of primary metabolites and volatile organic contents. *J Exp Bot* 60:2139–2154.

Zorrilla-Fontanesi, Y., J.L. Rambla, A. Cabeza, J.J. Medina, J.F. Sanchez-Sevilla, V. Valpuesta, M.A. Botella, A. Granell, and I. Amaya. 2012. Genetic analysis of strawberry fruit aroma and identification of *O*-methyltransferase FaOMT as the locus controlling natural variation in mesifurane content. *Plant Physiol* 159:851–870.

3 Bioassay-Guided Semiochemical Discovery in Volatile-Mediated Specialized Plant–Pollinator Interactions with a Practical Guide to Fast-Track Progress

Björn Bohman, Anna-Karin Borg-Karlson, and Rod Peakall

CONTENTS

3.1   INTRODUCTION

3.1.1   OVERVIEW

More than 85% of all flowering plants depend on animals for pollination (Ollerton et al., 2011), with floral olfactory and visual cues typically responsible for the long-distance attraction of potential pollinators (Pichersky and Gershenzon, 2002). At the flower, other factors such as the presence or absence of reward, type of reward, flower size and morphology often act as filters promoting some degree of pollinator specialization (Armbruster, 2017).

While specialized pollination is less common than more generalized pollination strategies, thousands of plant species are nonetheless pollinated by just one or a few species. Examples involving rewards for pollinators include: euglossine bee pollination (Milet-Pinheiro et al., 2015) and pollination by oil-collecting species (Schaffler et al., 2015). Many other examples involve deception and/or mimicry, including: food-deceptive floral mimicry, oviposition mimicry, brood-site mimicry and sexual mimicry (Johnson and Schiestl, 2016). Deceptive, and often highly specialized pollination strategies dominate the Orchidaceae, involving an estimated 1/3 of the family (~10,000 species) (Jersáková et al., 2006). Yet, examples of pollination by deception and mimicry extend beyond orchids. For example, oviposition mimicry using sulfurous pollinator attractants, is found across many plant families (Johnson and Schiestl, 2016).

While floral volatiles are expected to play key roles in securing specific pollinator attraction in many (but not all) cases of pollination by deception and mimicry, surprisingly the chemistry of these interactions is still poorly understood. Indeed, there are remarkably few studies that fulfill all (or even most) of the following criteria: (1) Experimental confirmation that volatiles play a key role in pollinator attraction. (2) Experimental confirmation of the source(s) and tissue specificity of volatile emission. (3) Isolation and identification of candidate compounds. (4) Confirmation of biological activity in bioassays with reference compounds. (5) Extended bioassays to investigate the role of variation in blend constituents and ratios, and the biological activity of structural analogs. Satisfying even a subset of these is no easy task, particularly when the chemistry involves compounds not commercially available or even new to science.

Some recent pollination related case studies, where many of the above criteria have been met, include: *Orchis* pollination by bumblebees (Valterová et al., 2007); the obligate interactions between *Breynia* and *Epicephala* moths (Svensson et al., 2010); green-leaf volatiles mimicry in *Epipactis* orchids (Brodmann et al., 2008); honey bee alarm pheromone mimicry in *Ceropegia* (Heiduk et al., 2016); aphid alarm pheromone mimicry in *Epipactis* (Stökl et al., 2011); olfactory mimicry of yeast targeting *Drosophila* in *Arum* (Stökl et al., 2010); oviposition site mimicry in *Dracula* orchids (Policha et al., 2016), among others.

In addition to these examples, the specific chemistry of orchid pollination by sexual deception has now been elucidated for several species of European and Australian orchids (Ayasse et al., 2011, Bohman et al., 2016a; Chapter 15). These studies provide compelling examples of the use of bioassay-guided approaches to aid chemical discovery.

3.1.2 WHAT IS BIOASSAY-GUIDED CHEMICAL DISCOVERY?

The idea of bioassay-guided chemical discovery has a surprisingly long history. For example, Figure 3.1 is adapted from Ställberg-Stenhagen et al. (1973). With the terminology updated, we use this figure as a helpful framework in this chapter. The first step commences with knowledge based on field observations. Field bioassays are then used to determine the active tissue to target for the isolation of candidate compound(s). Thereafter, the experimental approach involves the iterative use of field- and/or laboratory experiments with the pollinator across the entire process of chemical investigation (Figure 3.1). In practice, few cases have fully applied this framework in a single study. More often, the incremental knowledge gains are made in a series of studies.

Here, we use "bioassay" to mean any experimental procedure for determining the response of pollinators to: flowers or dissected floral parts, isolated floral volatiles, and reference compounds. The objective is to determine biological activity and ultimately identify the semiochemicals, i.e., the chemical cues being used by the plant to communicate with its pollinator. While often field-based, bioassays may also be conducted in a glasshouse or laboratory. Included within our definition are electrophysiology experiments, such as standalone electroantennography (EAG), or coupled with GC: GC-Electroantennography Detection (GC-EAD) and GC-Single Sensillum Recording (GC-SSR) (Schiestl and Marion-Poll, 2002).

To illustrate how applying a bioassay-guided framework can fast-track semiochemical discovery, we first draw on the history and progress in chemical ecology research on sexually deceptive orchids. These plants secure pollination by the chemical and physical mimicry of the female of their specific male insect pollinator (Peakall, 1990). The semiochemicals involved in the sexual attraction of the pollinator are now known to include unusual combinations of common compounds, rare plant volatiles, and many new-to-science compounds (Ayasse et al., 2011, Bohman et al., 2016a, Chapter 15). Despite the focus on pollination by sexual deception, the general principles are broadly relevant to any plant–insect pollination system where floral volatiles mediate specific pollinator attraction, irrespective of the pollination strategy and the target pollinator species. Next, we outline some important considerations: the prior biological knowledge, recommendations on the handling and treatment of samples in the field and laboratory, and general advice on isolation of semiochemicals (see also Chapter 1 on this topic). Finally, we briefly discuss some of the major biological and chemical challenges that can slow progress in semiochemical discovery, and how they might be overcome.

3.2 A BRIEF HISTORY OF CHEMICAL DISCOVERY IN SEXUALLY DECEPTIVE EUROPEAN AND AUSTRALIAN ORCHIDS

3.2.1 THE DISCOVERY OF SEXUALLY DECEPTION

The initially controversial French publication on pollination by "pseudocopulation" in *Ophrys* by Correvon and Pouyanne (1916) was brought to light in the English literature by Godfery (1925), whose independent work reconfirmed the conclusion of Pouyanne. In parallel, Coleman published a series of studies on the pollination of Australian *Cryptostylis* orchids. Based on observations and simple experiments she concluded pollination was also achieved by pseudocopulation in this genus with "scent and mimicry undoubtedly involved" (Coleman, 1927, 1938).

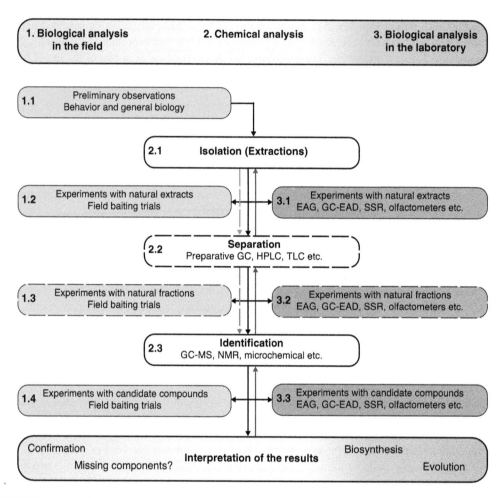

FIGURE 3.1 A schematic of the steps in bioassay-guided semiochemical discovery. The process begins with preliminary field observations and background biological knowledge (1.1). After each of the three chemical analysis steps of isolation (2.1), separation (2.2) and identification (2.3), field- (1.2 to 1.4) and/or laboratory bioassays (3.1 to 3.3) are conducted to confirm biological activity, and inform the next step in the process. This may include returning to a previous step(s), as shown with the grey arrows, or perhaps even returning to the start following the interpretation of the results. Note that the identification step (2.3), will include obtaining or synthesizing the candidate compound(s). Dashed lines indicate optional steps (1.3 & 3.2). Based on Ställberg-Stenhagen et al. (1973) with updated terminology.

3.2.2   European Sexually Deceptive Orchids

3.2.2.1   Field Experiments in *Ophrys* Confirm a Key Role for Volatiles

Beginning in 1948, Kullenberg started a multi-decade study of *Ophrys*, where he demonstrated in field experiments that floral volatiles held the key to pollinator attraction (Kullenberg, 1956, 1961). Remarkably, even before gas chromatography was invented, he also identified putative chemical components of *Ophrys*, including farnesol and nerolidol, by using his nose. Reference compounds, isolated from Cabreuva oil, were then tested in field bioassays, revealing farnesol repeatedly secured nonsexual attraction of the pollinator of *O. insectifera* (Kullenberg, 1956).

3.2.2.2 Field Experiments with Natural Extracts and Fractions

In 1973, Kullenberg published a summary of the outcomes of field experiments conducted in Sweden involving both solvent extracts and thin-layer chromatography (TLC) fractions. These experiments showed that hexane extracts of *O. tenthredinifera* labella elicited the full repertoire of sexual behavior from male *Eucera* bees, including attempted copulation. Other solvent extracts and fractions of *Ophrys* were less attractive (Kullenberg, 1973). Later Borg-Karlson (1979) used TLC and preparative gas chromatography (GC) to fractionate extracts prepared in solvents of different polarity. Field bioassays revealed that fractions from *O. bombyliflora* containing alcohols and carboxylic acids induced low rates of attempted copulation in the *Eucera* bee pollinator. This finding was consistent with the best TLC fraction found by Kullenberg (1973). Borg-Karlson and Tengö (1986) also reported striking similarities in the chemical composition of *Andrena* bee mandibular gland secretions and *O. lutea* labella. In field tests, a blend of (*E, E*)-farnesol, geranial and neral elicited strong attraction by male bees, including attempted copulation.

Following the methods of Bergström et al. (1980), Borg-Karlson and co-authors later comprehensively compared solvent extraction and headspace sampling methods (Borg-Karlson et al., 1985, 1987). These studies showed that no one method recovered the full set of compounds found in the *Ophrys* flower, a finding that is still highly relevant. Thus, to maximize successful downstream chemical elucidation, additional field- or laboratory bioassays will always be required to test which isolation method yields the extracts eliciting the strongest pollinator responses for any given study system.

3.2.2.3 Electroantennography as a New Tool

Electroantennography (EAG) was developed in 1957 and soon became a powerful tool (Schneider, 1957). Priesner (1973) was the first to apply EAG in *Ophrys*, screening extracts from multiple orchids against their bee pollinators to identify electrophysiologically active compounds, including multicyclic sesquiterpenes. Ågren and Borg-Karlson (1984) showed that the male *Argogorytes* wasp pollinator, unlike the conspecific female, responded to alkanes, alkenes and terpenes in extracts of *O. insectifera* flowers. The coupling of EAG with gas chromatography (GC-EAD) developed by Arn et al. (1975), would later help unlock the chemistry of attraction in sexually deceptive orchids (see Section 3.2.2.5, Figure 3.2).

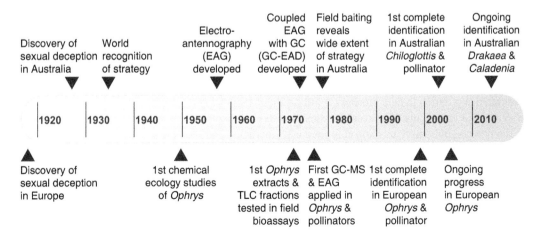

FIGURE 3.2 A timeline showing key points in the research and method development on the biology and chemistry of sexually deceptive orchids in Europe and Australia.

3.2.2.4   Development of Bioassay Protocols for *Ophrys* Studies

In tests with *Ophrys* pollinators, both bees and wasps, Kullenberg quickly abandoned early trials with multi-choice bioassays to avoid "the risk of confusion of the dummies by the insects." Instead, he employed dual-choice bioassays consisting of a scented velvet test "dummy" and a scentless down-wind control (~20 cm away), for a duration of 5 min. These experiments were conducted within areas where the target insects were known to be patrolling for mates (Kullenberg, 1973). Later, odorless dead bees were used as dummies to test floral extracts, extracts of the females and synthetic blends. These dummies were rendered odorless via Soxhlet-extraction, a method of continuous extraction with refluxing solvent. Similar dummies have continued to be used in more recent *Ophrys* studies. Kullenberg's 4-stage classification of bee and wasp pollinator behavior into increasing levels of sexual intensity: quick inspection, long-lasting inspection, and alighting with or without copulation, has also been widely used in other *Ophrys* studies (e.g., Borg-Karlson and Tengö, 1986, Borg-Karlson, 1990, Schiestl et al., 2000), or simplified further into approach, touch and pseudocopulation (e.g., Cuervo et al., 2017), or just contact versus attempted copulation (Ayasse et al., 2003).

3.2.2.5   Discovery and Confirmation of the Chemistry of Sexual Attraction in *Ophrys*

Schiestl et al. (1999) published in *Nature* their discovery of the chemistry of sexual attraction in *O. sphegodes*, with further details in Schiestl et al. (2000). A key to this breakthrough was the use of GC-EAD, which revealed a shared set of 14 electrophysiologically active compounds between the orchid and the female of its bee pollinator, *Andrena nigroaenea*. These compounds were then identified by GC-MS, including micro-derivatization experiments, as a series of alkanes and alkenes. In the critical final step, field bioassays with synthetic reference compounds at blend ratios matching the orchids and females, respectively, elicited sexual behavior including attempted copulation.

In *O. speculum*, which is pollinated by male *Dasyscolia ciliata* scoliid wasps (formerly *Campsoscolia ciliata*), Ayasse et al. (2003) combined GC-EAD, chiral-phase GC, GC-MS, synthesis and field bioassays to show that a mixture of just three hydroxy- and keto acids, in specific enantiomeric ratios, was sufficient to induce attempted copulation on dummies.

In *O. exaltata*, a species pollinated by male *Colletes cunicularius* bees, Mant et al. (2005) unraveled the pollination chemistry in a series of steps that nicely illustrate the bioassay-guided approach (see Figure 3.1). First, hexane extracts from the head and cuticles of virgin females were compared in field bioassays, revealing that the cuticular extracts elicited the strongest sexual behavior. Second, GC-EAD analysis detected 24 electrophysiologically active compounds in the cuticular extracts and the orchid labellum. Third, field bioassays with 12 synthetic candidate alkanes and alkenes, showed strong sexual attraction. Finally, in a series of subtractive bioassays, it was confirmed that a mixture of just three Z-7 alkenes was sufficient to induce attempted copulation.

3.2.2.6   Progress on the Chemistry of Sexual Attraction in Other *Ophrys*

Cuervo et al. (2017), working with *O. leochroma*, pollinated by *Eucera* bees, secured low rates of attempted copulation in field bioassays with synthetic blends of GC-EAD active alkanes, aldehydes, alcohols and fatty acids. However, the degree of sexual response was weaker than solvent extracts of the flowers, indicating missing compounds. Nonetheless, straight chain aldehydes (C7-C11, C13, C16) from polar fractions were shown to be involved in pollinator attraction. This finding matches earlier studies using TLC separations (Kullenberg, 1973, Borg-Karlson, 1979), providing a salient reminder that work in the early literature may provide vital clues to advance current progress.

3.2.3   AUSTRALIAN SEXUALLY DECEPTIVE ORCHIDS

3.2.3.1   Baiting Reveals the Wide Extent and Diversity of Sexual Deception

Stoutamire (1974, 1983) used baiting with picked orchid flowers to discover that thynnine wasp pollination by pseudocopulation is a widespread phenomenon among Australian *Arthrochilus*,

Caladenia, Chiloglottis, Drakaea, Paracaleana and *Spiculaea* orchid genera. Simple floral dissection experiments also revealed that the source of chemical attraction varied among species/genera (e.g., calli on the labellum in *Chiloglottis*; sepal tips in some *Caladenia*).

3.2.3.2 Development of Experimental Pollinator Baiting Protocols for Australian Orchids

Based on Stoutamire's method, Peakall (1990) developed experimental protocols that have formed the basis of subsequent pollinator experiments in Australian research on sexually deceptive orchids, involving flowers, floral dissections or semiochemical bioassays (e.g., Peakall and Handel, 1993, Bower, 1996, Schiestl et al., 2003, Peakall et al., 2010, de Jager and Peakall, 2019). A hierarchical behavioral sequence of: (1) approach, (2) land, (3) attempted copulation, is scored in these experiments (e.g., Peakall et al., 2010, Xu et al., 2017). While initially optimized for male thynnine wasps, the methods have been successfully extended to male ichneumonid wasps (Bohman et al., 2019) and male fungus gnat pollinators (Phillips et al., 2014, Reiter et al., 2019).

A key feature of the experimental baiting protocol is the short presentation time at a single location (often just 3 min, but as short as 1 min), with subsequent relocation (~10 to 100 m), to secure a renewed response. The protocol is based on experiments showing that peak male thynnine wasp responses occurred within 1 min, with mark-recapture revealing that the sharp decline can be attributed to a learned short-term site-specific response (Peakall, 1990, Whitehead and Peakall, 2013).

3.2.3.3 Other Insights from Baiting Experiments with Thynnine Wasp Pollinators

Bower (1996, 2006) demonstrated that multi-choice tests involving up to four orchid species at a time were informative for thynnine wasp pollinators. He also introduced a sequential baiting test to control for variation in the attractiveness of flowers (usually different species). In phase 1, a single test flower was presented for 3 min. Phase 2 followed as a dual-choice test, with the addition of a second control flower for a further 3 min. The control flower was known to be attractive to pollinators at the study site in a pre-screening baiting test. This, and later work, was pivotal to enabling the semiochemical discoveries in *Chiloglottis* that followed (3.2.3.4).

3.2.3.4 Discovery and Confirmation of the Chemistry of Sexual Attraction in *Chiloglottis*

Soon after the key chemical discoveries in *Ophrys*, Schiestl et al. (2003) published in *Science* the details of the chemistry of sexual attraction in the Australian sexually deceptive *Chiloglottis trapeziformis*. They showed that a single compound, 2-ethyl-5-propylcyclohexan-1,3-dione (chiloglottone 1), was both the semiochemical responsible for pollinator attraction and the female sex pheromone of the thynnine wasp pollinator, *Neozeleboria cryptoides*. The application of GC-EAD was again pivotal to success, with GC-high-resolution MS (GC-HRMS) and GC coupled with infrared spectroscopy (GC-FTIR) used as analytical tools to identify the new semiochemical. Field bioassays with synthetic chiloglottone 1 elicited the full sexual behavior of the male pollinator. These bioassays included a direct comparison with the orchid flower (unlike work in *Ophrys*). By achieving similar attempted copulation rates across orchid flowers, solvent extracts of flowers, and the reference compound, the authors demonstrated the paramount importance of chemistry in this system.

Building on this initial advance, and the background knowledge of pollinator specificity, rapid progress in the discovery of additional semiochemicals related to chiloglottone 1 (chiloglottone 2-6) was made across *Chiloglottis* (Franke et al., 2009, Poldy et al., 2009, Peakall et al., 2010, Peakall and Whitehead, 2014). To this day, *Chiloglottis* remains the most well characterized sexually deceptive genus, with bioassays confirming the chemical basis of pollinator attraction and specificity across more than ten species (~1/3 of the genus – see Chapter 15 for full details).

3.2.3.5 The Chemistry of Sexual Attraction in *Drakaea*

Although the genus *Drakaea* is closely related to *Chiloglottis* and both exploit male thynnine wasps, Bohman et al. (2014) revealed that pyrazines, not chiloglottones, are active semiochemicals in *D. glyptodon*. As the crucial first steps in this discovery, comparisons of the electrophysiologically active compounds between solvent extracts of active floral tissue (labellum) and female wasp, revealed four compounds in common. Once identified and synthesized, a series of sequential field bioassays were employed to confirm the optimal blend as an alkylpyrazine and a hydroxymethyl-pyrazine at a 3:1 ratio (see Section 3.2.4.2).

3.2.3.6   The Chemistry of Sexual Attraction in *Caladenia*

In *Caladenia plicata*, an unusual combination of (*S*)-β-citronellol and a new floral volatile com-pound, 2-hydroxy-6-methylacetophenone, secure pollination by sexual deception. This discovery was achieved by the application of GC-EAD and GC-MS. GC-MS/EAD (simultaneous detection by MS and EAD) was also used to circumvent the complexity of comparing GC-Flame Ionization (GC-FID) and GC-MS traces run on separate instruments. Comparisons of solvent extracts of active and nonactive floral tissues, whose activity was previously determined by choice tests, were also helpful (Xu et al., 2017).

In *C. crebra*, pollinated by male *Campylothynnus flavopictus* thynnine wasps, the application of GC-EAD failed. Instead, following field bioassay confirmation that sepal tips are the active tissue, a set of semiochemical candidates was defined by their joint presence in this tissue and the female wasps. GC-HRMS analysis of these candidates indicated a series of phenols with sulfur-containing substituents, one of which had a match in the NIST mass spectral library. Using this clue as start-ing point, a series of four (methylthio)phenols (three new to science) were identified by GC-MS and synthesis. Consecutive single bioassays (see Section 3.2.4.2), in additive and subtractive combina-tion trials of the four compounds, revealed that only two were essential to achieve strong sexual attraction, including frequent attempted copulation (Bohman et al., 2017).

3.2.3.7   Progress toward the Chemistry of Sexual Attraction in *Cryptostylis*

We highlight this study of *Cryptostylis ovata* and its ichneumonid *Lissopimpla excelsa* wasp pol-linator, because it aptly illustrates the iterative process of bioassay-guided chemical discovery. Following the experimental re-confirmation that pollinator attraction was chemically mediated, Bohman et al. (2019) tested floral extracts of different polarity in the field. While no extract elic-ited attempted copulation, polar extracts were more attractive than nonpolar hexane extracts. Next, bioassay-guided fractionation was conducted using two separate methods in parallel: solid phase extraction (SPE, C-18, from floral extract in water) and semi-preparative GC (from floral extract in methanol). Both methods used the outcomes of field bioassays to select the most active frac-tion, independently leading to the isolation of a single fraction containing a previously unknown semiochemical: (*S*)-2-(tetrahydrofuran-2-yl)-acetic acid. When presented in field bioassays, this compound was strongly attractive to male wasps, but did not elicit attempted copulation.

In future research, the next step would be to return to the bioassay-guided process (perhaps restarting near the top of scheme in Figure 3.1) to broaden the search for an extraction methodology that isolates a fraction which elicits stronger sexual attraction. In parallel, it would be ideal to opti-mize the visual and tactile cues required for maximum sexual response in the bioassays involving this kind of wasp (Section 3.3.1).

3.2.4   EXPERIMENTAL BIOASSAY DESIGNS FOR SEXUALLY DECEPTIVE ORCHIDS

3.2.4.1   Experimental Semiochemical Bioassay Protocols for *Ophrys* Studies

The earlier work of Kullenberg set the stage for the two complementary bioassay designs used in more recent studies of *Ophrys*: (1) dual-choice bioassays, and (2) consecutive single bioassays including controls.

In a dual-choice field bioassay design comparing control and spiked flowers, Schiestl and Ayasse (2001) revealed the repellent effect of farnesyl hexanoate, which naturally increases in amount in pollinated flowers. Soon after, Ayasse et al. (2003) compared the attractiveness of frozen virgin wasp females and *O. speculum* flowers in dual-choice tests. Furthermore, in an innovative extension, they drew the chemical headspace from either a hidden female wasp or fresh hidden flower, and then blew it across dead scentless wasps as visible dummies.

Although not called as such, "consecutive single bioassays including the control" have been the main bioassay method use in the studies of *Ophrys* (e.g., Schiestl et al., 1999, Ayasse et al., 2003, Mant et al., 2005, Cuervo et al., 2017). In this method, consecutive short 3-min presentations of either a control, or one of several test blends, are moved to different locations within the mate search area. To account for variation in the availability of patrolling males between days and sites, some studies report standardized responses (e.g., Ayasse et al., 2003, Cuervo et al., 2017). Here the raw number of male responses is standardized as a percentage by dividing the observed number of test reactions by the mean male activity and multiplying by 100. Estimates of male activity is based on counts of the number of males crossing a 1 m wire every 30 min (Ayasse et al., 2003, Cuervo et al., 2017). While multiple tests are usually reported, it is not always clear whether a single test is replicated in a consecutive set, or whether multiple tests including the control are run in random order as in Cuervo et al. (2017). The latter randomized approach would seem to be the optimal experimental design.

3.2.4.2 Experimental Semiochemical Bioassay Protocols for Australian Studies

Three complementary bioassay designs have been routinely employed in studies of the Australian thynnine pollinators of the orchid genera *Chiloglottis*, *Drakaea* and *Caladenia*: (1) multi-choice bioassays, (2) sequential bioassays, (3) consecutive single bioassays.

Since the female thynnine wasps only appear briefly above ground to mate and feed, they are difficult to find (Peakall, 1990). Furthermore, the extreme sexual dimorphism (small wingless females versus large winged males) means that the more accessible males are not suitable as dummies. Therefore, "female dummies" are made with dressmaking pins with a black plastic pin head (usually 4 mm in diameter), onto which the test blends are presented. In turn, the pins are attached to bamboo skewers at the relevant height of the associated orchid flowers (range 5–20 cm). Two other common features across the three bioassay designs are the short duration (typically 3 min, range from 1 to 10 min) and the relocation between trials to other locations within the patrolling zones of the male pollinators (usually 10–100 m apart).

Multi-choice bioassays are ideally suited for tests involving two pure synthetic compounds and a specific test blend of the two compounds (Figure 3.3). Each experiment consists of a minimum of four "baiting" trials at which three test blends and control are simultaneously presented, 15–50 cm apart (depending on the study system), and perpendicular to the prevailing wind direction (if any).

The sequential bioassay design comprises two phases: phase 1 involves the presentation of a single test blend for 3 min. Phase 2 follows for a further 3 min with the addition of a control, providing a dual-choice test. Phase 2 also provides a check for pollinator availability in the event that the phase 1 treatment is not attractive. Therefore, any trials failing to secure a response in phase 2 are discarded. The control may be an orchid flower (e.g., Bohman et al., 2014) or a previously known optimal semiochemical blend (e.g., Bohman et al., 2018). The rationale is that in a choice test with a very strong stimulus (such as an orchid flower), against a weaker stimulus (moderately active blend), the weak stimulus may appear to be entirely unattractive. Whereas when presented on its own first, it may prove to be partially attractive, a crucial finding in the hunt for bioactive compounds (Figure 3.4).

When all test blends are already known to be at least partially active, consecutive single bioassays may be the best strategy to determine an optimal blend (Bohman et al., 2016b; Figure 3.5). Before the experiment, a set of two to four different test blends are prepared and loaded onto dummies before storage in an airtight container. Thereafter, one experimental trial consists of the consecutive presentation of the test blend set, one at a time, and in random order, with each bioassay performed at a new baiting location. A total of four or more trials are run per experiment, to ensure an adequate number of responses. Each replicated experiment involves a new test blend set and runs for the same number of trials as the first experiment. Any trials that do not secure a response are aborted.

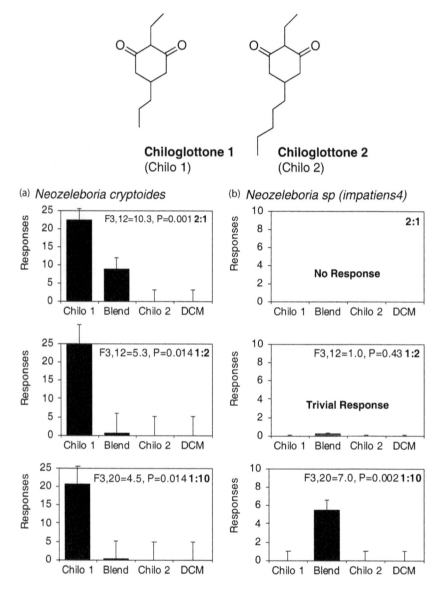

FIGURE 3.3 Outcomes of multi-choice bioassays testing the responses of male thynnine wasp pollinators of *Chiloglottis* orchids to alternative chiloglottones and blends. Mean responses of two *Neozeleboria* thynnine wasp species to choice tests offering chiloglottone 1 (Chilo 1), chiloglottone 2 (Chilo 2), a blend of the two compounds at the respective ratios of 2:1, 1:2 and 1:10 (Blend), and the control solvent dichloromethane (DCM). (a) Male *N. cryptoides* the pollinator of the orchid *Chiloglottis trapeziformis*, (b) Male *N.* sp. (impatiens4) the pollinator of the orchid *C.* aff. *valida*. Error bars represent standard errors. The results show that *N. cryptoides* did not respond to chiloglottone 2 on its own. Furthermore, this species responded weakly to a 2:1 blend, barely to a 1:2 blend, and not at all to a 1:10 blend. By contrast, *N.* sp. (impatiens4) showed no response to chiloglottones 1 or 2 on their own, instead it responded only with a blend ratio of 1:10, or higher. These results were consistent with the findings that *C. trapeziformis* flowers only contain chiloglottones 1, while *C.* aff. *valida* flowers contain both chiloglottones 1 and 2 at an average ratio of 1:14. Based on a subset of the results originally presented in full in Figure 4 of Peakall, R. et al., *New Phytol.*, 188:437–450, 2010.

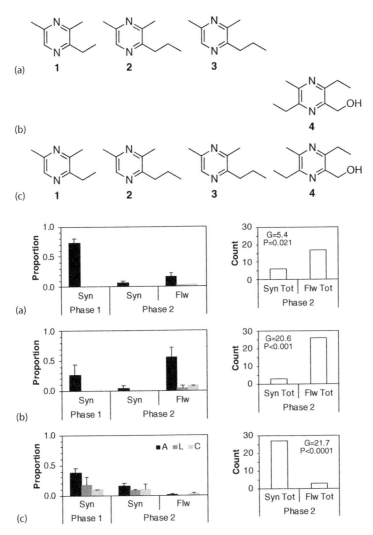

FIGURE 3.4 Outcomes of sequential bioassays testing the response of males of the thynnine wasp *Zaspilothynnus trilobatus* to three different blends of synthetic sex pheromone constituents. Responses to three test blends of synthetic pyrazines identified as sex pheromone constituents of *Zaspilothynnus trilobatus* wasp females, and semiochemicals of *Drakaea glyptodon* orchid flowers pollinated by male *Z. trilobatus*. (a) The three alkylpyrazines **1–3** only, (b) The hydroxymethylpyrazine **4** only, (c). All four compounds **1–4**. Phase 1 consisted of the presentation of synthetic compound(s) for 3 min (Syn). This was followed by phase 2, where both synthetic compound(s) (Syn) and orchid flower (Flw) were presented simultaneously for a further 3 min. In the left panel, male pollinator responses are shown as mean proportions of the total (± se), further partitioned into approaches (A, black bars), lands (L, dark grey bars) and attempted copulations (C, light grey bars) (left panel). In the right panel, the total count across experiments is shown separately for Phase 2. All bioassays were performed at the base ratio of 1:1:300:100 **1–4**. Error bars represent standard errors. The results show that in phase 1 neither the alkylpyrazines on their own, nor the hydroxymethylpyrazines secured lands or attempted copulation. In phase 2, both combinations were barely attractive in the choice test against the flower (a). By contrast, the combination of a specific blend of the four compounds was strongly attractive in phase 1, and remained attractive against the control flower in phase 2. Furthermore, during both phases the test blend elicited the full repertoire of sexual behavior despite the use of a 4 mm bead as a simplistic dummy female/flower (b). Outcomes with just the hydroxymethylpyrazine and one alkylpyrazine were similar to the full blend of 4 compounds (c). Based on a subset of the results originally presented in full in Figure 4 of Bohman B. et al., *New Phytol.*, 203, 939–952, 2014.

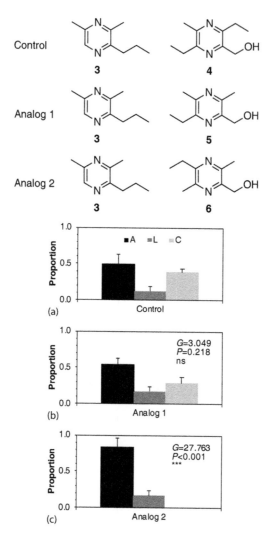

FIGURE 3.5 Outcomes of consecutive single bioassays testing the response of males of the thynnine wasp *Zaspilothynnus trilobatus* to analogs of the sex pheromone of its female. Responses across consecutive single bioassays to three different test blends of synthetic pyrazines. (a) Control – compounds **3 + 4**, $n = 40$, (b) Analog 1- compounds (**3 + 5**), $n = 51$, and (c) Analog 2 - compounds (**3 + 6**), $n = 36$. Pollinator responses are shown as mean proportions of the total ($\pm$se), further partitioned into approaches (A, black bars), lands (L, dark grey bars) and attempted copulations (C, light grey bars). All bioassays were performed using individual blends of the alkyl-pyrazine in a ratio of 3:1 with one of three hydroxymethylpyrazines. *G*-test results shown are relative to the control for each experiment, which is shown in the top graph. Error bars represent standard errors. The results show that across the three test blends, all previously known to be partially attractive, the Analog 1 blend (**3 + 5**) secured a similar sexual response to the control blend. Thus, precise semiochemical matching between orchid and pollinator may not be necessary to achieve pollination in all cases of sexual deception. Based on a subset of the results originally presented in full in Figure 2 of Bohman B. et al., *J. Chem. Ecol.*, 42, 17–23, 2016b.

Consecutive single bioassays can be more informative than choice bioassays in cases where test blends show only subtle differences in activity, and the prime focus is on differences in the degree of sexual attraction (as measured by the proportion of approaches, lands and attempted copulations). The protocol is also particularly useful when pollinator availability is limited. In such cases, sequential bioassays may fail due to insufficient pollinators to secure a sustained response across both phases.

The consecutive single bioassay design closely matches the common bioassay protocol in *Ophrys* (see Section 3.2.4.1), but with the absence of the direct inclusion of an odorless control. This is because, unlike pollinators of *Ophrys* (Tengö, 1979), male thynnine wasps have never been reported to "pounce" on control odorless dummies. Further, the hierarchical responses of pollinators, relative to an optimal test blend or flower, is of most interest. Any standardization for variation in male availability is avoided, and likely difficult to apply, given thynnine wasp species are often indistinguishable on the wing.

3.2.4.3 Potential Future Laboratory-Based Bioassay Opportunities

Wind tunnels, dual-choice- and more sophisticated olfactometers are widely used for sex phero-mone investigations, as well as in some other pollination studies where reared insects are available. Yet, such procedures have not been employed for semiochemical discovery in sexually deceptive orchids. This is likely in part due to a lack of methods for rearing the solitary bee and wasp pol-linators. Nonetheless, the recent isolation of a female-emitted sex pheromone component of the fungus gnat *Lycoriella ingenua* illustrates bioassay-guided laboratory methods relevant to studies of orchids pollinated by male fungus gnats.

Using reared insects, Andreadis et al. (2015) employed a Y-tube olfactometer to confirm that males were most strongly attracted to unmated females. Subsequently, GC-EAD analysis of extracts from virgin females against male antennae revealed a single electrophysiologically active peak. Next, GC coupled with behavioral bioassays (GC-BB) was employed. Here, the airflow normally diverted across the antennae in GC-EAD, was instead delivered into a small cylindrical glass vial containing the gnats. Strong sexual behavior was only observed on the elution of the candidate active peak. This promising bioassay method should be applicable to any insect species where males can be induced to sexually respond to females, or female extracts, in vials.

3.3 KEY CONSIDERATIONS FOR SUCCESSFUL CHEMICAL DISCOVERY

3.3.1 BIOLOGY OF PLANTS AND POLLINATORS

As recognized in the first step of Figure 3.1, an understanding of the basic ecology of the plant and pollinator is fundamental for effective chemical ecology investigations. Furthermore, knowledge of the distribution, abundance and seasonality of the plant and pollinators is vital for planning an effective bioassay-guided approach. Knowledge of the source of active compounds is also crucial, with prior studies on related species potentially offering strong clues. If unknown, pollinator choice experiments using dissected flower parts should be performed.

3.3.2 TREATMENT OF BIOLOGICAL MATERIAL FOR CHEMICAL ANALYSIS AND BIOASSAYS

Careful treatment of samples is critical for reliable outcomes of bioassays and chemical analysis. The safest approach is to process samples as soon as possible after their collection in the field. Solvent extractions can be conducted in the field, with a kit of solvents, syringes and vials, an insu-lated storage box with ice or a portable refrigerator. If processing in the field is not possible, it is important to keep the plant or insect material fresh by storing below 10°C and packed in such a way as to minimize contamination risk.

Irrespective of where the sampling is conducted, utmost care needs to be taken if re-using syringes or other means of sample transfer while handling solvent extracts. Multiple rinses with two or more solvents of contrasting polarity is recommended to avoid cross-contamination between samples. Using containers of glass or other inert material with appropriate lids will also minimize the background of extractable degradation products which can originate from plastic containers and rubber seals. Chemical analysis should be conducted as soon as possible, as volatile com-pounds easily evaporate or may degrade over time, with subsequent chemical profile changes as samples age. If this is not possible, storage in suitable vials at −80°C is recommended.

3.3.3 Isolation of Semiochemicals

Once the active tissue is identified, it is crucial to design the isolation protocol so that it fits the purpose of the study. Chapter 1 provides a comprehensive review of the methods available for volatile sampling and analysis. Here we highlight the importance of following a bioassay-guided approach, particularly when the target compounds are unknown. Implementing bioassays to test which isolation method(s) retain maximum activity, are recommended (Figure 3.1). To minimize variability and reduce the number of bioassays, we suggest, where possible, to work with bulk multi-flower/multi-insect extracts.

3.4 BIOLOGICAL CHALLENGES

3.4.1 Plant and Pollinator Availability

In many of the most interesting cases of chemically mediated specialized pollination (such as sexual deception), both the plant and the pollinator may be uncommon. Furthermore, flowering and pollinator flight times may be seasonally restricted, and pollinator activity constrained by suitable weather conditions. Consequently, it is rarely possible to sample from a large number of flowers or pollinators, which is required for NMR spectroscopy. Therefore, GC-MS is generally the go-to method for the initial analysis of floral and pollinator extracts.

Not to be neglected as a potentially valuable resource for plant material are botanic gardens and private collections. For example, endangered orchid species have been successfully cultivated at botanical gardens and used for biological experiments in the field (Reiter et al., 2018). Similarly, exploring options for the laboratory breeding of pollinators, and downstream laboratory experiments, although challenging, could considerably aid bioassay-guided chemical discoveries in future studies.

3.5 CHEMICAL CHALLENGES

3.5.1 Optimal Solvents/Extraction Methods

The key to successful bioassay-guided chemical discovery is to ensure that during the fractionation process all fractions are adequately tested and confirmed to have retained the biological activity before further fractionations proceed. In a subtractive manner, the number of potential compounds of interest will be reduced in each step, considerably aiding the final isolation and identification of the active compound(s). Further, it is important to evaluate possible combinations of fractions, if several active components are present. Wherever possible, we strongly recommend that extracts of different polarity and different extraction methods are tested for attraction (e.g., solvent extracts, static versus dynamic headspace sampling etc.).

3.5.2 EAG and GC-EAD-Difficulties

As seen above, electrophysiological methods can be very helpful to detect chemical candidates. Nonetheless, these methods suffer from several issues. First, behavioral bioassays still need to be used to confirm the function of any electrophysiologically active compounds, be they identified by EAG, GC-EAD or GC-SSR. Second, the chromatographic peaks must be of acceptable quality for GC-EAD to work. Third, some insect species have antenna that are notoriously difficult to use with any one of these methods, often for unknown reasons.

One innovation that may help overcome some of these difficulties is chopper-modulated locked in amplified GC-EAD. This method reduces electrical noise by rapidly connecting and disconnecting (i.e., chopping) a signal. When combined with matched filtering of noise, the signal-to-noise ratios can be improved a 1000-fold (Myrick and Baker, 2012). Use of a Deans switch effluent chopper, rather than a mechanical chopper device, can further reduce noise from mechanoreceptors (Myrick and Baker, 2018).

3.5.3 Requirements for Structural Elucidation

Unless NMR spectroscopy is possible, several confirmatory analyses need to be conducted when relying on GC-based methods to identify compounds. It is critical to compare the retention time and mass spectrum between the natural product and a structurally confirmed reference standard, rather than rely on database matching. The GC method must allow possible isomers to separate, and the sample should be run on at least two columns with different stationary phases. Additional methods, such as GC-FTIR, would add another level of confirmation (e.g., Franke et al., 2009).

It is crucial to rule out all possible close structural isomers, if possible by finding literature data demonstrating differences with the natural product, or by synthesizing standards, before claiming that a semiochemical identity has been confirmed. Critically, it is essential not to rely on successful field bioassay results as sole confirmation that you have identified the correct isomer. For example, structural isomers of substituted pyrazines can elicit biological activity comparable with the natural orchid semiochemicals (Bohman and Peakall, 2014, Bohman et al., 2016b).

Once a candidate semiochemical with one or more chiral centers has been identified, the absolute configuration also needs to be elucidated. For any bioassay, it is important to test the correct, naturally occurring isomer, or to confirm that there is no difference in activity between isomers with behavioral bioassays. Chapter 1 provides additional advice on enantioselective GC methods and structure elucidation.

Within sexually deceptive orchids, several examples are already known where the identification of absolute stereochemistry has been crucial, and where stereoisomers elicit very different biological activity (Borg-Karlson et al., 2003, Ayasse et al., 2003, Xu et al., 2017, Bohman et al., 2018). Case studies testing stereoisomers in the field in other pollination systems appear to be few (but see Brandt et al., 2019).

3.6 CONCLUSION

Despite the thousands of floral volatiles already known, we predict that at least among highly specialized pollination systems, many new-to-science semiochemical systems await discovery. The lack of commercially available semiochemicals severely limits the opportunities for bioassays, and therefore the full experimental confirmation of their biological roles. Clearly, productive partnerships between biologists and chemists are crucial (Raguso et al., 2015, Bohman et al., 2016a), not only to achieve the challenging quest to elucidate and synthesize new natural products, but also to enable the development of cost-effective chemical methods to synthesize a broader range of biologically important compounds, in high purity, for extended experimental work.

One way to encourage more of these multi-disciplinary partnerships is for biologists to more fully embrace the bioassay-guided approaches presented in this chapter. Indeed, most methods outlined in Figure 3.1 can be performed in biological laboratories equipped with GC and HPLC instrumentation. Some laboratory-based bioassay techniques, such as GC-EAD, are already often implemented in biology labs. Alternatively, the chemists can be more involved, and work more closely with the biologists, in these areas of research. Regardless of how the work is shared, it is fundamental to include the whole multi-disciplinary team in both the process of planning as well as the execution of the entire project. Finally, it is crucial to fully recognize the vital contributions of all team members.

3.7 ACKNOWLEDGMENTS

We thank Eran Pichersky and Natalia Dudareva for the invitation to write this chapter. During the process we were aptly reminded that important ideas in science, such as the bioassay-guided framework, have a long history that should not be forgotten. Björn Bohman was supported by an

Australian Research Council (ARC) Discovery Early Career Researcher Award (DE 160101313). Rod Peakall gratefully acknowledges funding for his current research program from an ARC Discovery project (DP150102762).

REFERENCES

Ågren L., and A.-K. Borg-Karlson. 1984. Responses of *Argogorytes* (Hymenoptera: Sphecidae) males to odor signals from *Ophrys insectifera* (Orchidaceae). Preliminary EAG and chemical investigation. *Nova Acta Regiae Societatis Scientiarum* 3:111–117.

Andreadis S.S., K.R. Cloonan, A.J. Myrick, H.B. Chen and T.C. Baker. 2015. Isolation of a female-emitted sex pheromone component of the fungus gnat, *Lycoriella ingenua*, attractive to males. *Journal of Chemical Ecology* 41:1127–1136.

Armbruster W.S. 2017. The specialization continuum in pollination systems: Diversity of concepts and implications for ecology, evolution and conservation. *Functional Ecology* 31:88–100.

Arn H., E. Stadler and S. Rauscher. 1975. Electroantennagraphic detector—selective and sensitive tool in gaschromatographic analysis of insect pheromones. *Zeitschrift Fur Naturforschung Section C-A Journal of Biosciences* 30:722–725.

Ayasse M., F.P. Schiestl, H.F. Paulus, F. Ibarra and W. Francke. 2003. Pollinator attraction in a sexually deceptive orchid by means of unconventional chemicals. *Proceedings of the Royal Society of London Series B-Biological Sciences* 270:517–522.

Ayasse M., J. Stökl and W. Francke. 2011. Chemical ecology and pollinator-driven speciation in sexually deceptive orchids. *Phytochemistry* 72:1667–1677.

Bergström G., M. Appelgren, A.-K. Borg-Karlson, I. Groth, S. Strömberg and S. Strömberg. 1980. Studies on natural odoriferous compounds. 22. Techniques for the isolation/enrichment of plant volatiles in the analyses of *Ophrys* orchids (Orchidaceae). *Chemica Scripta* 16:173–180.

Bohman B., A. Karton, G.R. Flematti, A. Scaffidi and R. Peakall. 2018. Structure-activity studies of semiochemicals from the spider orchid *Caladenia plicata* for sexual deception. *Journal of Chemical Ecology* 44:436–443.

Bohman B., G.R. Flematti, R.A. Barrow, E. Pichersky and R. Peakall. 2016a. Pollination by sexual deception—it takes chemistry to work. *Current Opinion in Plant Biology* 32:37–46.

Bohman B., A. Karton, R.C.M. Dixon, R.A. Barrow and R. Peakall. 2016b. Parapheromones for thynnine wasps. *Journal of Chemical Ecology* 42:17–23.

Bohman B., A.M. Weinstein, R.D. Phillips, R. Peakall and G.R. Flematti. 2019. 2-(Tetrahydrofuran-2-yl) acetic acid and ester derivatives as long-range pollinator attractants in the sexually deceptive orchid *Cryptostylis ovata*. *Journal of Natural Products* 82:1107–1113.

Bohman B., and R. Peakall. 2014. Pyrazines attract *Catocheilus* thynnine wasps. *Insects* 5:474–487.

Bohman B., R.D. Phillips, G.R. Flematti, R.A. Barrow and R. Peakall. 2017. The spider orchid *Caladenia crebra* produces sulfurous pheromone mimics to attract its male wasp pollinator. *Angewandte Chemie International Edition* 56:8455–8458.

Bohman B., R.D. Phillips, M.H.M. Menz et al. 2014. Discovery of pyrazines as pollinator sex pheromones and orchid semiochemicals: Implications for the evolution of sexual deception. *New Phytologist* 203:939–952.

Borg-Karlson A.-K. 1979. Kemisk beteendestimulation hos *Eucera longicornis* hanar [Odour released behavior of *Eucera longicornis* males (Hymenoptera, Anthophoridae)] *Entomologisk tidskrift* 100:125–128.

Borg-Karlson A.-K. 1990. Chemical and ethological studies of pollination in the genus *Ophrys* (Orchidaceae). *Phytochemistry* 29:1359–1387.

Borg-Karlson A.-K., and J. Tengö. 1986. Odor mimetism—Key substances in *Ophrys lutea—Andrena* pollination relationship (Orchidaceae, Andrenidae). *Journal of Chemical Ecology* 12:1927–1941.

Borg-Karlson A.-K., G. Bergström and B. Kullenberg. 1987. Chemical basis for the relationship between *Ophrys* orchids and their pollinators. II. Volatile compounds of *O. insectifera* and *O. speculum* as insect mimetic attractants/excitants. *Chemica Scripta* 27:303–325.

Borg-Karlson A.-K., G. Bergström and I. Groth. 1985. Chemical basis for the relationship between *Ophrys* orchids and their pollinators. I. Volatile compounds of *Ophrys lutea* and *O. fusca* as insect mimetic attractants/excitants. *Chemica Scripta* 25:283–294.

Borg-Karlson A.-K., J. Tengö, I. Valterova et al. 2003. (S)-(+)-linalool, a mate attractant pheromone component in the bee *Colletes cunicularius*. *Journal of Chemical Ecology* 29:1–14.

Bower C.C. 1996. Demonstration of pollinator-mediated reproductive isolation in sexually deceptive species of *Chiloglottis* (Orchidaceae: Caladeniinae). *Aust. J. Bot.* 44:15–33.

Bower C.C. 2006. Specific pollinators reveal a cryptic taxon in the bird orchid, *Chiloglottis valida* sensu lato (Orchidaceae) in south-eastern Australia. *Australian Journal of Botany* 54:53–64.

Brandt K., S. Dötterl, R. Fuchs et al. 2019. Subtle chemical variations with strong ecological significance: Stereoselective responses of male orchid bees to stereoisomers of carvone epoxide. *Journal of Chemical Ecology* 45:464–473.

Brodmann J., R. Twele, W. Francke, G. Holzler, Q.H. Zhang and M. Ayasse. 2008. Orchids mimic green-leaf volatiles to attract prey-hunting wasps for pollination. *Current Biology* 18:740–744.

Coleman E. 1927. Pollination of *Cryptostylis leptochila* F.v. M. *Victorian Naturalist* 44:333–340.

Coleman E. 1938. Further observations on the pseudocopulation of the male *Lissopimpla semipunctata* Kirby (Hymenoptera parasitica) with the Australian orchid *Cryptostylis leptochila* F. v. M. *Proceedings of the Royal Entomological Society*, London, UK , 13:82–84.

Correvon H., and M. Pouyanne. 1916. Un curieux cos de mimetisme chez les ophrydees. *Journal of Applied Horticulture* 17:29–31, 41–42.

Cuervo M., D. Rakosy, C. Martel, S. Schulz and M. Ayasse. 2017. Sexual deception in the *Eucera*-pollinated *Ophrys leochroma*: A chemical intermediate between wasp- and *Andrena*-pollinated species. *Journal of Chemical Ecology* 43:469–479.

de Jager M.L., and R. Peakall. 2019. Experimental examination of pollinator-mediated selection in a sexually deceptive orchid. *Annals of Botany* 123:347–354.

Franke S., F. Ibarra, C.M. Schulz et al. 2009. The discovery of 2,5-dialkylcyclohexan-1,3-diones as a new class of natural products. *Proceedings of the National Academy of Sciences USA* 106:8877–8882.

Godfery M.J. 1925. The fertilisation of *Ophrys speculum*, *O. lutea*, and *O. fusca*. *Journal of Botany* 63:33–40.

Heiduk A., I. Brake, M. von Tschirnhaus et al. 2016. *Ceropegia sandersonii* mimics attacked honeybees to attract kleptoparasitic flies for pollination. *Current Biology* 26:2787–2793.

Jersáková J., S.D. Johnson and P. Kindlmann. 2006. Mechanisms and evolution of deceptive pollination in orchids. *Biological Reviews* 81:219–235.

Johnson S.D., and F.P. Schiestl. 2016. *Floral Mimicry*. Oxford, Oxford University Press.

Kullenberg B. 1956. On the scents and colors of *Ophrys* flowers and their specific pollinators among the Aculeate Hymenoptera. *Svensk Botanisk Tidskrift* 50:25–46.

Kullenberg B. 1961. Studies in *Ophrys* pollination. *Zoologiska Bidrag. Uppsala* 34:1–340.

Kullenberg B. 1973. Field experiments with chemical sexual attractants on Aculeate Hymenoptera males. II. *Zoon Supplement* 1:31–41.

Mant J., C. Brandli, N.J. Vereecken, C.M. Schulz, W. Francke and F.P. Schiestl. 2005. Cuticular hydrocarbons as sex pheromone of the bee *Colletes cunicularius* and the key to its mimicry by the sexually deceptive orchid, *Ophrys exaltata*. *Journal of Chemical Ecology* 31:1765–1787.

Milet-Pinheiro P., D. Navarro, S. Dötterl et al. 2015. Pollination biology in the dioecious orchid *Catasetum uncatum*: How does floral scent influence the behaviour of pollinators? *Phytochemistry* 116:149–161.

Myrick A.J., and T.C. Baker. 2012. Chopper-modulated locked in amplified gas chromatography—electroantennography part II: Signal processing and performance comparisons. *IEEE Sensors Journal* 12:2974–2983.

Myrick A.J., and T.C. Baker. 2018. Increasing signal-to-noise ratio in gas chromatography—electroantennography using a Deans switch effluent chopper. *Journal of Chemical Ecology* 44:111–126.

Ollerton J., R. Winfree and S. Tarrant. 2011. How many flowering plants are pollinated by animals? *Oikos* 120:321–326.

Peakall R. 1990. Responses of male *Zaspilothynnus trilobatus* Turner wasps to females and the sexually deceptive orchid it pollinates. *Functional Ecology* 4:159–167.

Peakall R., and M.R. Whitehead. 2014. Floral odour chemistry defines species boundaries and underpins strong reproductive isolation in sexually deceptive orchids. *Annals of Botany* 113:341–355.

Peakall R., and S.N. Handel. 1993. Pollinators discriminate among floral heights of a sexually deceptive orchid—Implications for selection. *Evolution* 47:1681–1687.

Peakall R., D. Ebert, J. Poldy et al. 2010. Pollinator specificity, floral odour chemistry and the phylogeny of Australian sexually deceptive *Chiloglottis* orchids: Implications for pollinator-driven speciation. *New Phytologist* 188:437–450.

Phillips R.D., D. Scaccabarozzi, B.A. Retter et al. 2014. Caught in the act: Pollination of sexually deceptive trap-flowers by fungus gnats in *Pterostylis* (Orchidaceae) *Annals of Botany* 113:629–641.

Pichersky E., and J. Gershenzon. 2002. The formation and function of plant volatiles: Perfumes for pollinator attraction and defense. *Current Opinion in Plant Biology* 5:237–243.

Poldy J., R. Peakall and R. Barrow. 2009. Synthesis of chiloglottones—semiochemicals from sexually deceptive orchids and their pollinators. *Organic and Biomolecular Chemistry* 7:4296–4300.

Policha T., A. Davis, M. Barnadas, B.T.M. Dentinger, R.A. Raguso and B.A. Roy. 2016. Disentangling visual and olfactory signals in mushroom-mimicking *Dracula* orchids using realistic three-dimensional printed flowers. *New Phytologist* 210:1058–1071.

Priesner E. 1973. Reaktionen von Riechrezeptoren männlicher Solitärbienen (Hymenoptera, Apoidea) auf Inhaltsstoffe von *Ophrys*-Blüten. *Zoon supplement* 1:43–54.

Raguso R.A., J.N. Thompson and D.R. Campbell. 2015. Improving our chemistry: Challenges and opportunities in the interdisciplinary study of floral volatiles. *Natural Product Reports* 32:893–903.

Reiter N., B. Bohman, G.R. Flematti and R.D. Phillips. 2018. Pollination by nectar-foraging thynnine wasps: Evidence of a new specialized pollination system for Australian orchids. *Botanical Journal of the Linnean Society* 188:327–337.

Reiter N., M. Freestone, G. Brown and R. Peakall. 2019. Pollination by sexual deception of fungus gnats (Keroplatidae and Mycetophilidae) in two clades of *Pterostylis* (Orchidaceae). *Botanical Journal of the Linnean Society* 190:101–116.

Schaffler I., K.E. Steiner, M. Haid et al. 2015. Diacetin, a reliable cue and private communication channel in a specialized pollination system. *Scientific Reports* 5:12779.

Schiestl F.P., and F. Marion-Poll. 2002. Detection of physiologically active flower volatiles using gas chromatography with electroantennography. In: Jackson JF, Linskens HF, Inman R eds. *Analysis of Taste and Aroma*. Berlin, Germany, Springer.

Schiestl F.P., and M. Ayasse. 2001. Post-pollination emission of a repellent compound in a sexually deceptive orchid: A new mechanism for maximising reproductive success? *Oecologia* 126:531–534.

Schiestl F.P., M. Ayasse, H.F. Paulus et al. 1999. Orchid pollination by sexual swindle. *Nature* 399:421–422.

Schiestl F.P., M. Ayasse, H.F. Paulus et al. 2000. Sex pheromone mimicry in the early spider orchid (*Ophrys sphegodes*): Patterns of hydrocarbons as the key mechanism for pollination by sexual deception. *Journal of Comparative Physiology A-Sensory Neural and Behavioral Physiology* 186:567–574.

Schiestl F.P., R. Peakall, J.G. Mant et al. 2003. The chemistry of sexual deception in an orchid-wasp pollination system. *Science* 302:437–438.

Schneider D. 1957. Electrophysiological investigation on the antennal receptors of the silk moth during chemical and mechanical stimulation. *Experientia* 13:89–91.

Ställberg-Stenhagen S., E. Stenhagen and G. Bergström. 1973. Analytical techniques in pheromone studies. *Zoon, Supplement* 1:77–82.

Stökl J., A. Strutz, A. Dafni et al. 2010. A deceptive pollination system targeting Drosophilids through olfactory mimicry of yeast. *Current Biology* 20:1846–1852.

Stökl J., J. Brodmann, A. Dafni, M. Ayasse and B.S. Hansson. 2011. Smells like aphids: Orchid flowers mimic aphid alarm pheromones to attract hoverflies for pollination. *Proceedings of the Royal Society B-Biological Sciences* 278:1216–1222.

Stoutamire W.P. 1974. Australian terrestrial orchids, thynnid wasps and pseudocopulation. *American Orchid Society Bulletin* 43:13–18.

Stoutamire W.P. 1983. Wasp-pollinated species of *Caladenia* in South-western Australia. *Australian Journal of Botany* 31:383–394.

Svensson G.P., T. Okamoto, A. Kawakita, R. Goto and M. Kato. 2010. Chemical ecology of obligate pollination mutualisms: Testing the "private channel" hypothesis in the *Breynia–Epicephala* association. *New Phytologist* 186:995–1004.

Tengö J. 1979. Odor-released behavior in *Andrena* male bees (Apoidea, Hymenoptera). *Zoon* 7:15–48.

Valterová I., J. Kunze, A. Gumbert et al. 2007. Male bumble bee pheromonal components in the scent of deceit pollinated orchids; unrecognized pollinator cues? *Arthropod-Plant Interactions* 1:137–145.

Whitehead M.R., and R. Peakall. 2013. Short-term but not long-term patch avoidance in an orchid-pollinating solitary wasp. *Behavioral Ecology* 24:162–168.

Xu H., B. Bohman, D.C.J. Wong et al. 2017. Complex sexual deception in an orchid is achieved by co-opting two independent biosynthetic pathways for pollinator attraction. *Current Biology* 27:1867–1877.

4 The Chemical Diversity of Floral Scent

Jette T. Knudsen and Jonathan Gershenzon

CONTENTS

4.1 INTRODUCTION

The practical approaches of how to collect, analyze, and identify floral scent compounds are described in Chapters 1–3. Because of commercial importance, the chemistry of floral scents has been studied for a long time. Such investigations have shown that fragrant flowers typically emit not just a single compound but a mixture of a few to over hundred different volatile compounds present in varying amounts. Furthermore, a large variety of compounds have been found in the floral scent of taxonomically related as well as diverse species. Ecological investigations (described in subsequent chapters) have indicated that these floral scent chemicals function either alone or in combination with visual and tactile cues to mediate interactions between plants and animals. The outcome of the interactions is in most cases beneficial to both parties, animals being rewarded with pollen, nectar, and oils and the emitting plants in being pollinated. However, deceptive pollination systems that involve floral scent as an attractant are also well known (Chapter 15).

The number of taxa with florally scented species is large, and while some have been investigated extensively, other taxa have not. In a series of papers, Roman Kaiser and his collaborators at Givaudan-Roure had taken us on journeys into the world of floral scent molecules found in the Orchid family (Kaiser, 1993a, b), and he and his collaborators have presented many astonishing new scent molecules from some other plant families and environments (Kaiser and Lamparsky,

1982; Kaiser and Nussbaumer, 1990; Kaiser and Tollsten, 1995; Kaiser, 1988, 1991, 1995, 1997a, 1997b, 2000, 2002; Schultz et al., 1999; Kaiser and Kraft, 2001). However, apart from many studies of medicinal plants (e.g., Bicchi and Joulian, 1990; Bozan et al., 1999; Brunke et al., 1993), most investigators have focused on species from one or a few plant families, either from a pollination biological or a taxonomical point of view. This chapter reviews the structures and abundances of the chemical compound groups in a wider range of plants (Knudsen et al., 2006), and discusses various aspects of variation in floral scent chemistry.

4.2 HOW DIVERSE ARE FLORAL SCENTS?

The volatile compounds emitted from flowers belong to several different classes, but are united by their low molecular weight (30–300 amu) and vapor pressure sufficient to be released and dispersed into the air under normal temperature regimes. The individual compound classes are widely distributed among the flowers of different species, probably reflecting the fact that the major biosynthetic pathways leading to them are present in all plants.

So far, more than 1700 compounds have been identified in the floral headspace of 990 taxa (most at the species level) belonging to 90 families and 38 orders (Knudsen et al., 2006). The majority of these taxa (78%) belong to the following 19 plant families, in each of which floral scent composition has been characterized from at least ten or more taxa listed in descending order of taxa (taxa number given in parentheses): Orchidaceae (417), Araceae (55), Arecaceae (40), Magnoliaceae (26), Rosaceae (24), Cactaceae (21), Rutaceae (21), Solanaceae (21), Caryophyllaceae (20), Nyctaginaceae (20), Fabaceae (18), Amaryllidaceae (17), Moraceae (15), Ranunculaceae (14), Asteraceae (13), Lecythidaceae (13), Oleaceae (13), Apiaceae (11), Rubiaceae (10). For details on species identity and references, see Knudsen et al. (2006).

4.2.1 CHEMICAL COMPOUND CLASSES IN FLORAL SCENTS

Most of the 1700 compounds reported in headspace samples of floral scent are lipophilic (Knudsen et al., 2006). The two largest groups (see Table 4.1) are terpenoids, synthesized by the mevalonate or methylerythritol phosphate pathway (Gershenzon and Kreis, 1999; Rodriguez-Concepcion and Boronat, 2002), and aliphatics, synthesized predominantly from fatty acids.

A total of 556 terpenoids have been identified and these include monoterpenes, sesquiterpenes, diterpenes, and irregular terpenes. The monoterpenes are divided about equally between compounds with acyclic or cyclic skeletons; the latter can be mono-, bi- or tricyclic skeletons (Table 4.1, Figures 4.1 and 4.2). Sesquiterpenes are also characterized by both acyclic and cyclic skeletons, but cyclic skeletons are much more common (Table 4.1, Figure 4.3). Because the enzymes producing terpene skeletons, the terpene synthases, often form multiple cyclic and acyclic products from a single substrate, either geranyl diphosphate or farnesyl diphosphate (Gershenzon and Kreis, 1999; Chen et al., 2003), dividing compounds into acyclic and cyclic categories does not necessarily follow strict biosynthetic criteria. A number of cyclic sesquiterpene skeletons are shown in Figure 4.3C–J. The irregular terpenes include compounds varying in number of carbon atoms from 8 to 18. Among these are apocarotenoids, which are biodegradation products of carotenoid compounds (C40) like β-carotene (Kaiser, 2002; Eugster et al., 1969; Eugster and Märki-Fischer, 1991) (Table 4.1, Figure 4.4A–C). Others like the C11 and C16 irregular terpenes 4,8-dimethyl-1,3,7-nonatriene and 4,8,12-trimethyl-1,3,7,11-tridecatetraene are acyclic homoterpenes derived from nerolidol and geranyllinalool, respectively (Donath and Boland, 1994, 1995) (Figure 4.4E and H).

The aliphatics include 528 compounds with chains having between one and 25 carbon atoms, the majority having between 2 and 17 carbon atoms (Table 4.1, Figure 4.5). The C6 aliphatic compounds include the well known green-leaf volatiles, like (Z)-3-hexenyl acetate (Figure 4.5D) found in vegetative as well as floral scents of numerous plants and probably playing a role in plant defense

TABLE 4.1

Distribution of Floral Scent Compounds According to Their Supposed Biosynthetic Origin

Compound Class	No. of Compounds
Aliphatics	
C1	5
C2	59
C3	33
C4	41
C5	59
C6	64
C7	33
C8	39
C9	19
C10	49
C11-C15	76
C16-C20	44
C21-C25	7
Sum	528
Benzenoids and Phenylpropanoids	
C6-C0	23
C6-C1	133
C6-C2	78
C6-C3	83
C6-C4, -C5, -C7	12
Sum	329
C5-branched Chain Compounds	
Saturated	40
Unsaturated	53
Sum	93
Terpenoids	
Monoterpenes	
Acyclic	
Regular	136
Irregular	11
Cyclic	
Menthanes	91
Bicyclo[2.2.1]	14
Bicyclo[3.1.0]	12
Bicyclo[3.1.1]	27
Bicyclo[4.1.0]	3
Tricyclic	1
Sum	295
Sesquiterpenes	
Acyclic	44
Cyclic	114
Sum	158

(*Continued*)

TABLE 4.1 (*Continued*)
Distribution of Floral Scent Compounds According to Their Supposed Biosynthetic Origin

Compound Class	No. of Compounds
Diterpenes	
Acyclic	4
Cyclic	2
Sum	6
Irregular Terpenes	
Apocarotenoid	52
C8	7
C9	2
C10	8
C11	10
C12	3
C13	5
C14, C16, C18	10
Sum	97
Nitrogen Compounds	
Ammonia	1
Acyclic	41
Cyclic	19
Sum	61
Sulfur Compounds	
Acyclic	37
Cyclic	4
Sum	41
Miscellaneous Cyclic Compounds	
Carbocyclic	60
Heterocyclic	51
Sum	111

Compounds in each class are divided according either to carbon chain length, saturatedness, or skeletal structure.

(Paré and Tumlinson, 1999). Another large group consists of benzenoids and phenylpropanoids, which are synthesized either from the phenyl propanoid pathway starting with deamination of phenylalanine or from an intermediate of the shikimate pathway prior to phenylalanine (Jarvis et al., 2000; Wildermuth et al., 2001). A total of 329 benzenoids and phenylpropanoid compounds have been identified in floral scents (Table 4.1, Figure 4.6). Another group is the C5-branched chain compounds (Table 4.1, Figure 4.7). These are probably derived directly from branched chain amino acids, but there is little direct evidence to support this assumption (Rowan et al., 1996). Nitrogen compounds, like indole, and sulfur-containing compounds are also most likely derived from amino acid metabolism (Frey et al., 2000). The nitrogen containing compounds (Table 4.1, Figure 4.8**A–F**) include cyclic and acyclic compounds, whereas the sulfur-containing compounds are mainly acyclic (Table 4.1, Figure 4.8**G–K**). The 111 miscellaneous compounds are either carbocyclic or heterocyclic, the latter including both carbon and other atoms in the cyclic structures (Table 4.1, Figure 4.9). The group contains compounds of uncertain biosynthetic origin, but many of these are derived from fatty acids or amino acids.

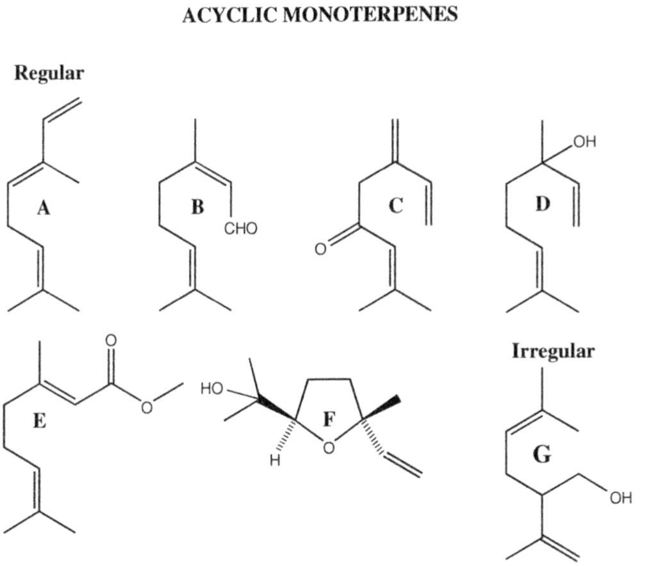

ACYCLIC MONOTERPENES

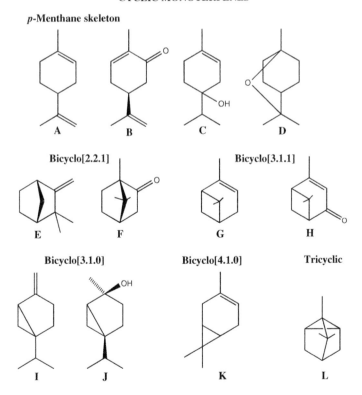

FIGURE 4.1 Acyclic monoterpenes in floral scents: **regular skeleton**, **A**, (*E*)-β-ocimene; **B**, neral; **C**, ipsdienone; **D**, linalool; **E**, methyl geranate; **F**, *trans*-linalool oxide (furanoid); **irregular skeleton**, **G**, lavandulol.

CYCLIC MONOTERPENES

FIGURE 4.2 Cyclic monoterpenes in floral scents: ***p*-menthane skeleton**, **A**, limonene; **B**, (+)-carvone; **C**, terpinen-4-ol; **D**, 1,8-cineole; **bicyclo[2.2.1] skeleton**, **E**, camphene; **F**, camphor; **bicyclo[3.1.1] skeleton**, **G**, α-pinene; **H**, verbenone; **bicyclo[3.1.0] skeleton**, **I**, sabinene; **J**, *cis*-sabinene hydrate; **bicyclo[4.1.0] skeleton**, **K**, 3-carene; **tricyclic skeleton**, **L**, tricyclene.

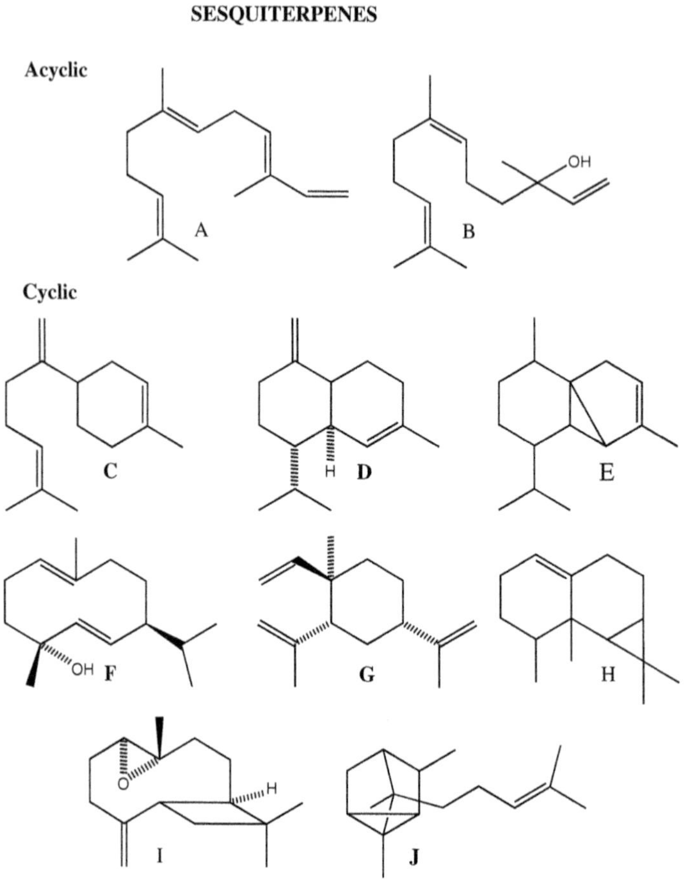

SESQUITERPENES

Acyclic

Cyclic

FIGURE 4.3 Sesquiterpenes in floral scents: **acyclic**, **A**, (*E, E*)-α-farnesene; **B** (*Z*)-nerolidol; **cyclic**, **C**, β-bisabolene; **D**, γ-cadinene; **E**, α-cubebene; **F**, germacrene D-4-ol; **G**, β-elemene; **H**, β-gurjunene; **I**, caryophyllene oxide; **J**, α-santalene.

4.2.2 FUNCTIONAL GROUPS OF FLORAL SCENT COMPOUNDS

Floral scent compounds range from nonpolar (e.g., alkanes, alkenes) to polar. The majority contains an oxygen function and are slightly to moderately polar. Esters, especially esters of aliphatic compounds and benzenoid and phenylpropanoids are the most common functional group found, followed by alcohols, ketones, ethers, aldehydes and acids (Table 4.2, Figures 4.1-4.7 and 2.9). Among nitrogen containing compounds amines and oximes (Figure 4.8**B** and **C**) are most common, while sulfides are most common among sulfur-containing compounds (Table 4.2, Figure 4.8**G**). In the miscellaneous cyclic compound group, furans and pyrans are common (Table 4.2, Figure 4.9**I–L**).

Within esters it is the alcohol part of the compound that is most variable (Table 4.2). Among the terpenes of floral scent, the percentage of oxygenated compounds is much higher for monoterpenes (81%) than for sesquiterpenes (38%). This difference may reflect the fact that many oxygenated sesquiterpenes are much less volatile than oxygenated monoterpenes and sesquiterpene hydrocarbons. Only mono-oxygenated sesquiterpenes are likely to have sufficient volatility to be found in floral headspace (Joulain and König, 1998). The lower frequency of oxygenated sesquiterpenes may also be a consequence of their inefficient collection by standard adsorbent methods and poor desorption

IRREGULAR TERPENES

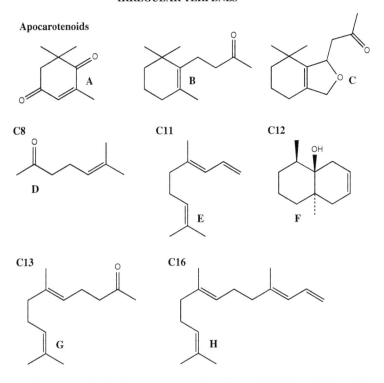

FIGURE 4.4 Irregular terpenes in floral scents: **apocarotenoids, A**, 2,6,6-trimethyl-2-cyclohexene-1,4-dione (4-oxoisophorone); **B**, dihydro-β-ionone; **C**, 7(11)-oxymegastigma-5(6)-en-9-one; **C8; D**, 6-methyl-5-hepten-2-one; **C11; E** (*E*)-4,8-dimethyl-1,3,7-nonatriene; **C12, F**, dehydrogeosmin; **C13, G** (*E*)-6,10-dimethyl-5,9-undecadien-2-one (geranylacetone); **C16, H** (*E, E*)-4,8,12-trimethyl-1,3,7,11-tridecatetraene.

ALIPHATICS

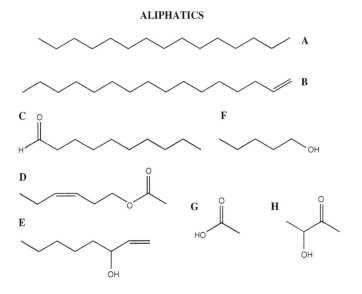

FIGURE 4.5 Aliphatic compounds in floral scents: **A**, pentadecane; **B**, 1-hexadecene; **C**, decanal; **D**, (*Z*)-3-hexenyl acetate; **E**, 1-octen-3-ol; **F**, pentanol; **G**, acetic acid; **H**, 3-hydroxy-2-butanone.

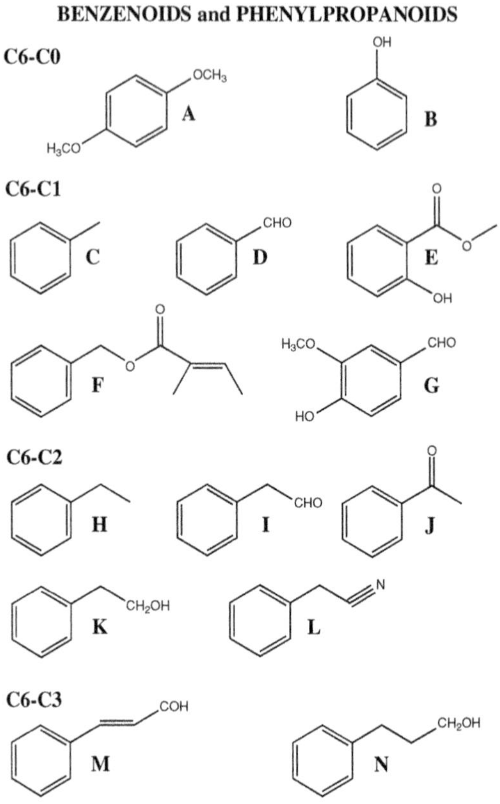

BENZENOIDS and PHENYLPROPANOIDS

FIGURE 4.6 Benzenoids and phenylpropanoids in floral scents: **C6-C0**, **A**, 1,4-dimethoxybenzene; **B**, phenol; **C6-C1**, **C**, methylbenzene; **D**, benzaldehyde; **E**, methyl 2-hydroxybenzoate (methyl salicylate); **F**, benzyl (*E*)-2-methyl-2-butenoate (benzyl tiglate); **G**, 4-hydroxy-3-methoxybenzaldehyde (vanillin); **C6-C2**, **H**, ethylbenzene; **I**, phenylacetaldehyde; **J**, acetophenone; **K**, 2-phenylethanol; **L**, 2-phenylacetonitrile; **C6-C3**, **M** (*E*)-cinnamic aldehyde; **N**, phenylpropanol.

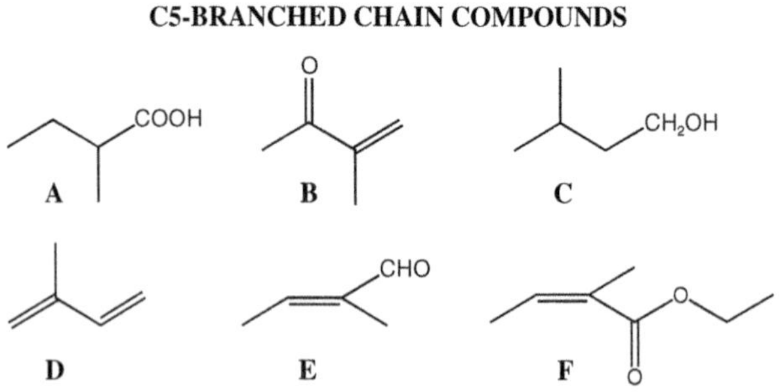

C5-BRANCHED CHAIN COMPOUNDS

FIGURE 4.7 C5-branched chain compounds in floral scents: **A**, 2-methylbutanoic acid; **B**, 3-methyl-3-buten-2-one; **C**, 3-methylbutanol; **D**, 2-methyl-1,3-butadiene; **E**, (*E*)-2-methyl-2-butenal; **F**, ethyl (*Z*)-2-methyl-2-butenoate (ethyl angelate).

with nonpolar solvents like hexane (Raguso and Pellmyr, 1998). At the other end of the spectrum, a number of nitrogenous and sulfurous compounds are probably lost during absorption or sample concentration because of their high volatility. Alternatively, they may not be detected by GC-MS, because of their proximity to the solvent peak. In many studies, the mass detector is first switched on after the solvent has eluted.

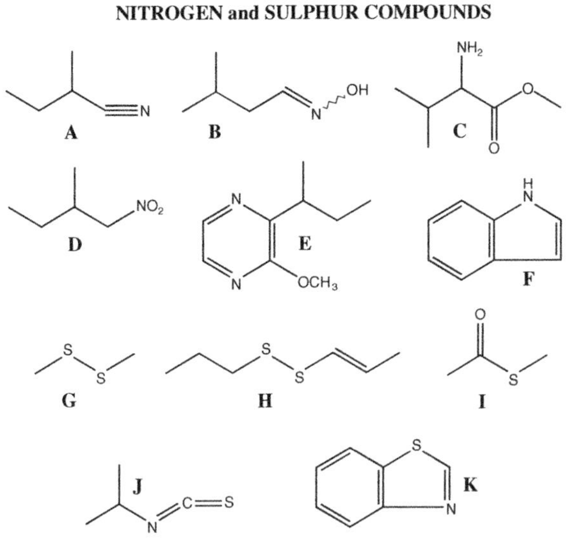

NITROGEN and SULPHUR COMPOUNDS

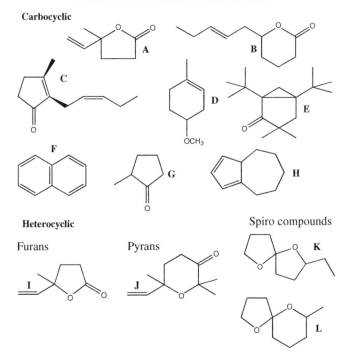

FIGURE 4.8 Nitrogen and sulfur compounds in floral scents: **A**, 2-methylbutylnitrile; **B**, 3-methylbutylaldoxime; **C**, methyl 2-amino-3-methylbutanoate (valine methyl ester); **D**, 1-nitro-2-methylbutane; **E**, 2-methoxy-3-*sec*-butylpyrazine; **F**, indole; **G**, dimethyldisulfide; **H**, propyl prop-2-enyl disulfide; **I**, methyl thioacetate; **J**, isopropyl isothiocyanate; **K**, benzothiazole.

MISCELLANEOUS CYCLIC COMPOUNDS

FIGURE 4.9 Miscellaneous cyclic compounds in floral scents: **carbocyclic**, **A**, 4-methyl-5-hexen-4-olide (γ-vinyl-γ-valerolactone); **B**, Δ-decalactone; **C**, *cis*-jasmone; **D**, 1-methyl-4-methoxy-1-cyclohexene; **E**, 1,5-*ditert*-butyl-3,3-dimethylbicyclo[3.1.0]hexan-2-one; **F**, naphthalene; **G**, 2-methylcyclopentanone; **H**, azulene; **heterocyclic**, **furans**, **I**, 5-methyl-5-ethenyldihydro-2(3H)-furanone; **pyrans**, **J**, 2,2,6-trimethyl-6-ethenyltetrahydro-3-pyranone; **spiro compounds**, **K**, chalcogran; **L**, 7-Methyl-1,6-dioxaspiro[4.5]decane.

TABLE 4.2
Distribution among the Major Chemical Compound Classes of Functional Groups Found in Floral Scent Compounds

Functional Groups	Aliphatics	Benzenoids and Phenylpropanoids	C5-branched Compounds	Monoterpenes	Sesquiterpenes	Diterpenes	Irregular Terpenes	Nitrogen Compounds	Sulfur Compounds	Miscellaneous	Total
No. of compounds	528	329	93	295	158	6	97	61	41	111	1719
Hydrocarbons		34	2	56	98	3	10				203
Alkanes	40									6	46
Alkenes	54									4	58
Acids	31	3	2								36
Aldehydes	52	21	6	16	7		3				105
Ketones	60	24	2	36	2		52			12	188
Alcohols	86	47	7	87	31	3	20			5	286
Esters	195	119	74	40	9		7	2		29	475
Ethers	9	48		60	11		4			8	140
Chloro	1	1									2
N-containing		28									28
Benzofurans		1									1
Benzopyrans		3									3
Ammonia								1			1
Amides								5			5
Amines								15			15
Nitriles								5			5
Nitro compounds								3			3
Oximes								11			11
Imidazole								2			2
Indole								3			3
Pyrazine								7			7
Pyrazole								1			1
Pyridine								5			5

(Continued)

TABLE 4.2 (Continued)
Distribution among the Major Chemical Compound Classes of Functional Groups Found in Floral Scent Compounds

Functional Groups	Aliphatics	Benzenoids and Phenyl propanoids	C5-branched Compounds	Monoterpenes	Sesquiterpenes	Diterpenes	Irregular Terpenes	Nitrogen Compounds	Sulfur Compounds	Miscellaneous	Total
Triazine								1			1
Isothiocyanates									3		3
Thiocyanates									1		1
Sulfides									25		25
Thioesters									3		3
Thiols									4		4
Sulfoxides									1		1
Thiazole									2		2
Thiafurans									2		2
Furans										27	27
Pyrans										8	8
Naphthalenes										3	3
Azulenes										1	1
Spiro compounds										8	8
Unknown							1				1

In compounds containing more than one functional group priority follows from highest to lowest: acids, aldehydes, ketones, alcohols, esters, ethers.

Most of the common constituents of floral scents are easy to identify by gas chromatographic retention times and mass spectra. However, sesquiterpene hydrocarbons are much more difficult than most others because of the large number of structures with similar mass spectra and retention times (see these excellent references: Adams (1989) and Joulain and König (1998)). Final confirmation frequently requires a reference standard for comparison, but for many sesquiterpenes, these are not available commercially, and so compounds are often reported as "unknown sesquiterpenes."

4.2.3   STEREOCHEMISTRY OF FLORAL SCENT COMPOUNDS

4.2.3.1   Geometrical Isomers

Geometrical or (Z)- and (E)- isomers are common in unsaturated aliphatics, phenylpropanoids, and terpenoid compounds, and often co-occur in the scent bouquet of a single species, e.g., geranial and neral (Figure 4.1**B**), which are the (E)- and (Z)- of isomer 3,7-dimethylocta-2,6-dienal (citral), respectively. Because (E)- and (Z)- isomers of a compound usually separate well under standard GC-conditions double bond geometry is reported often for floral scent compounds. The mass spectra of (E)- and (Z)- isomers usually differ slightly and the configuration of a compound can be determined with some confidence, especially, when both isomers are present. When geometrically pure reference compounds are available the configuration can be determined unambiguously.

The *cis*- and *trans*-prefixes are preferably not used for double bond geometry (Joulain and König, 1998), but rather to indicate through-space relationships of two substituents or two bulky parts of a molecule, e.g., *trans*-linalool oxide (furanoid) (Figure 4.1**F**).

4.2.3.2   Optical Isomers

Many floral scent compounds are chiral molecules with one or more asymmetric carbons. Enantiomers have identical values for all physical constants except that they interact with plane-polarized light in opposite directions. Special protocols using chiral GC-columns, multi-dimensional gas chromatography and enantioselective chemical synthesis are therefore necessary to separate and identify enantiomers (Bergström et al., 1991; Borg-Karlson et al., 1994a, 1994b, 1996; Mori, 1998). Diastereoisomers, unlike enantiomers, have different physical properties and usually separate on standard GC-columns, but like enantiomers they have identical mass spectra (e.g., Francke et al., 1979). Despite equal physical properties, enantiomers react differently with other chiral compounds and their biological characteristics may differ. For example to the human nose (R)-carvone smells like peppermint and (S)-carvone like caraway, while (4R)-*cis*-rose oxide has a threshold value ten times lower than its (4S)-enantiomer (Fráter et al., 1998).

While most components of floral scent are optically pure, some terpenes occur as mixtures of two enantiomers (Knudsen et al., 2006). As a general rule, enantiomers are likely to be produced by different enzymes. However, some terpene synthases have been described that produced both (–)- and (+)-enantiomers of single substances, such as an enzyme from the flowers of *Arabidopsis thaliana* that produces both enantiomers of sabinene and limonene in readily detectable amounts (Wise et al., 1998; Chen et al., 2003).

4.3   VARIATION IN FLORAL SCENT COMPOSITION

The qualitative and quantitative composition of a floral scent varies within and between species and is probably best defined as a product of phylogenetic constraints and balancing selection due to pollinator and florivore attraction as suggested by Raguso (2001).

At the intraspecific level floral scent composition may vary spatially and temporally within a flower, between flowers on the same plant (see below), between plants, and between populations (Olesen and Knudsen, 1994; Kite, 1995; Raguso and Pichersky, 1995; Azuma et al., 2001; Pettersson and Knudsen, 2001; Grison-Pigé et al., 2001; Knudsen, 2002; Dufaÿ et al., 2004; Dötterl et al., 2005). At the interspecific level, a few groups of plants have shown little variation between closely related taxa (Dahl et al., 1990; Ervik et al., 1999; Tollsten and Bergström, 1993; Tollsten and Knudsen, 1992), but in most groups each taxon produces its own specific floral scent blend (Azuma et al., 1997; Barkman et al., 1997; Gerlach and Schill, 1989; Gregg, 1983; Grison-Pigé et al., 2002a; Jürgens and Dötterl, 2004; Knudsen, 1999; Knudsen et al., 2001; Levin et al., 2001; Lindberg et al., 2000; Jürgens et al., 2003; Raguso et al., 2003; Thien et al., 1985; Toyoda et al., 1993; Whitten and Williams, 1992).

Floral scent production and composition is genetically determined (Raguso and Pichersky, 1995; Pichersky and Gang, 2000; Ishizaka et al., 2002), but environmental factors may also exert a great impact on the scent emitted. For example, water stress and high or low temperatures may decrease floral scent production or inhibit it completely (Loper and Berdel, 1978; Hansted et al., 1994; Jakobsen and Olsen, 1994; Nielsen et al., 1995; Knudsen et al., 1999).

4.3.1 COMPOSITION AND AMOUNT OF EMITTED FLORAL SCENTS

Floral scent bouquets may contain from one to more than a hundred compounds (Gerlach and Schill, 1991; Levin et al., 2001; Knudsen, 2002; Omata et al., 1999), but most species emit between 20 and 60 different compounds.

The amounts of floral scent produced vary from in the low picogram range to more than 30 µg/h (Kaiser, 1993b; Knudsen et al. 1993, 2004). The flowers or inflorescences of various beetle and moth pollinated species produce the largest amounts of fragrances, while more than half of the hummingbird pollinated plant analyzed did not produce any detectable scent (Knudsen et al., 2004).

The total amounts reported must be regarded as only approximate because rigorous quantitation of floral scent emission rates is seldom attempted and often not possible. Such an approach requires extended collections under physiological conditions, use of internal standard to control for losses during collection and sample processing and dose-response curves to calibrate the detector response of individual compounds (which requires adequate amounts of pure standards).

4.3.1.1 Temporal Variation

Floral scent emission may vary in time because emission follows a circadian rhythm or external stimuli such as light or temperature (e.g., Hills and Williams, 1990; Euler and Baldwin 1996; Loper and Lapiolo, 1971; Helsper et al., 1998; Kuanprasert et al., 1998; Altenburger and Matile, 1988; Dudareva et al., 2000; MacTavish et al., 2000; Pott et al., 2002). Variation in emission may also be caused by pollination or as a part of the developmental aging of the flower itself (e.g., Pham-Delegue et al., 1989; Matsumoto et al., 1993; Sazima et al., 1993; Tollsten, 1993; Patt et al., 1995; Schiestl et al., 1997; Awano et al., 1997; Lewis et al., 1998; Omata et al., 1999; Stránsky and Valterová, 1999). In many hermaphroditic as well as monoecious plant species there is, on individual plants, a temporal separation of the pistillate and staminate flowering phase with or without overlap in male and female function. In *Peltandra virginica* the temporal separation is accompanied by scent difference and the difference in scent was thought to signal oviposition and food sites, respectively (Patt et al., 1995).

4.3.1.2 Spatial Variation

Floral scent emission varies in space because scent may be emitted from specialized regions of floral tissues (Chapter 13, Vogel, 1990). However, different floral parts, like stamens and pistils, as well as pollen and nectar, may also participate in floral scent emission (Knudsen and Tollsten 1991; Sazima et al., 1993; Bergström et al., 1995; Ecroyd et al., 1995; Dobson et al., 1996; Dudareva et al.,

2000; Dobson and Bergström, 2000; Jürgens and Dötterl, 2004; Flamini et al., 2002; Raguso, 2004). Even the position within inflorescences may influence amount and composition of scent emitted by individual flowers (Ayasse et al., 2000; Moya and Ackerman, 1993), although such variation has so far only been reported in deceptive flowers.

4.3.2 Co-occurrence of Individual Compounds in Floral Scent

Some compounds in the floral scent of a species are normally found together with one or several other related compounds. This may be a consequence of a shared biosynthetic pathway or enzyme like a terpene synthase (Bohlmann et al., 2000; Chen et al., 2003; Tholl et al., 2005). However, co-occurrence may also be a result of phylogenetic influences and the co-occurring compounds may not necessarily be biosynthetically related.

4.3.2.1 Biosynthetic Relatedness

It is common to find series of compounds in a single floral scent mixture that are biosynthetically closely related. Among aliphatic compounds, series of hydrocarbons or methyl ketones with increasing numbers of carbon atoms are found, as are compounds of a single chain length present as aldehydes, ketones and alcohols (Croteau and Karp, 1991). Studies of *Clarkia breweri* flowers showed that methyleugenol and methylisoeugenol are produced by the action of a single enzyme (Dudareva et al., 2000). This could be true in other species as well. Methylisoeugenol has so far been reported in headspace samples of floral scent in 16 species from nine different plant families (four mono- and five eudicot families), but 12 of these (representing all nine families) also contained methyleugenol, a volatile reported from 45 species. In terpenoid compounds it has been demonstrated that several terpene synthases catalyze the production of more than one product (Gershenzon and Kreis, 1999; Chen et al., 2003; Tholl et al., 2005). Multiproduct terpene synthases may also explain the co-occurrence of some of the most commonly reported monoterpene hydrocarbons in floral scents, i.e., limonene (found in 71% of investigated plant families) (E)-β-ocimene (71%), myrcene (70%), α-pinene (67%) and β-pinene (59%). Other compounds that occur in more than 50% of the plant families are the terpenoids, linalool (70%), caryophyllene (52%) and 6-methyl-5-hepten-2-one (52%) and the aromatics benzaldehyde (64%), methyl salicylate (57%), benzyl alcohol (56%), and 2-phenylethanol (54%) (Knudsen et al., 2006. Geometrical- or optical isomers of a compound often co-occur with one of the isomers being much more common than the other.

Some compounds identified in floral scent blends may be produced as artifacts of sample collections. For example some of the sulfur-containing compounds identified in species of Cactaceae and other species probably are rearrangement products of dimethyl disulfide on the adsorbent trap (see Brunke et al., 1993). Another example are the β-ocimenols, which usually occur together with the corresponding β-ocimenes. Brunke et al. (1993) showed that these are artifacts produced on the active charcoal surface of one type of charcoal trap.

4.3.2.2 Phylogenetic Constraints

Because major biosynthetic pathways leading to the synthesis of the precursors of volatiles are present in all plant species, there may be few constraints on floral scent distribution at the level of compound classes. However, the musty-earthy odors of geosmin and dehydrogeosmin may be examples of compounds that are phylogenetically constrained to one group of land plants. Both compounds have been reported in many species of Cactaceae and geosmin also in *Dorstenia turneraefolia* (Moraceae) (Kaiser and Nussbaumer, 1990; Kaiser, 1991; Schlumpberger, 2002). However, the geosmin in *Dorstenia* may originate from epiphyllic algae (Schlumpberger, 2002) and not be produced by the plant itself.

Reconstructions of plant phylogenies based on floral scent compounds are still few (Azuma et al., 1997, 1999; Williams and Whitten, 1999; Lindberg et al., 2000; Barkman, 2001; Levin et al., 2003) and in most cases only the outermost branching pattern is consistent with phylogenetic trees based on either morphological or DNA sequence data. Mapping of the major compound classes found in headspace samples of floral scents onto the most parsimonious phylogenetic tree based on DNA sequence data revealed no groupings or discernible patterns (Knudsen et al., 2006). This indicates that at higher taxonomic levels, floral scent chemicals are too evolutionary labile to be useful for phylogenetic inference (Williams and Whitten, 1999; Barkman, 2001). However, this does not exclude the possibility that some floral scent chemicals are patterned phylogenetically at lower taxonomic levels (Barkman, 2001; Levin et al., 2003; Borchsenius et al., 2016).

4.4 ADDITIONAL FUNCTIONS OF FLORAL SCENTS

The primary function of floral scent in flowering plants is to attract and guide pollinators (Williams, 1983; Metcalf, 1987; Robacker et al., 1988; Dobson, 1994; Raguso, 2001). However, additional functions are ascribed to the presence of volatile chemicals in flowers (reviewed in Pichersky and Gershenzon (2002)), including defense and protection against abiotic stresses. These additional functions may help explain some of the abundance and variety of different constituents detected. Because flowers produce pollen and ovules for the next generation they have a high fitness value to the plant, and the finding of chemicals that defend flowers against herbivores and pathogens is not surprising. Apart from attracting pollinators floral scent chemicals may also increase the risk of herbivore attack on floral structures as shown in *Nicotiana attenuata*, and floral tissues may therefore require relatively more protection from enemies (Euler and Baldwin, 1996). In *N. attenuata* this was achieved through an herbivore induced increase in corolla pools of nicotine, a biocidal defense metabolite.

4.4.1 VOLATILES IN LEAVES AND FLOWERS

Most of the plant volatiles identified in defense against insect herbivores (Paré and Tumlinson, 1999) have also been identified in floral scents (Knudsen et al., 2006). Whether these compounds originate from floral or green parts of an inflorescence is difficult to determine (Andersson et al., 2002; Raguso et al., 2003; Levin et al., 2003), but if they originate from floral parts, their function may be different from that in vegetative parts. Some of the compounds shared by vegetative and floral samples are very common and may have a general function to signal the presence of a plant/flower, while the specific flower chemicals signal the presence of a certain flower species (see signal-noise discussion in Raguso, 2003). Thus chemicals common to the vegetative and floral tissues may have a dual function, defense against herbivores and attraction of pollinators. In the mutualism between *Derelomus chamaeropsis* (Curculionidae) and the dwarf palm *Chamaerops humilis* (Arecaceae) the leaves have overtaken the attractive function of the flowers by producing volatiles that attract the pollinating weevils (Dufaÿ et al., 2003). The attractive leaf volatiles so far identified in this mutualism are among those volatiles that have been found both in volatile samples of vegetative and floral tissues of other species. Another example of a dual function is ascribed to chemicals emitted by *Acacia* flowers. In the ant-*Acacia* mutualism there is a conflict during flowering between the use of guarding ants to deter herbivores and the need to attract pollinators. However, young first-day flowers contain some volatile chemicals that deter the guarding ants, thus allowing pollinating bees to visit the flowers (Willmer and Stone, 1997).

4.4.2 FLOWER VOLATILES AS MEDIATORS OF SEX AND BREEDING BEHAVIOR IN INSECTS

Representatives of all principal classes of floral scent chemicals have also been shown to be used by various insects, although not all these compounds are synthesized *de novo* by the insects (see references in El-Sayed, 2004). In insects, the compounds either mediate intraspecific interactions, i.e., function as pheromones, or interspecific interactions, i.e., function as allelochemicals. Some plants deceive insects to visit and pollinate their flowers by producing compounds that imitate pheromones or allelochemicals. The foul sulfur or nitrogen containing odors and thermogenesis in many species of Araceae imitate allelochemicals that elicit feeding or oviposition behavior in carrion and dung flies (Borg-Karlson et al., 1994a; Kite, 1995; Angioy et al., 2003; Seymour and Schultze-Motel, 1997; Seymour et al., 2003a, b; Stensmyr et al., 2002). Flowers of many deceptive orchid species produce compounds that mimic the sex pheromones of female insects (usually Hymenoptera) and elicit sexual behavior in males, which effect pollination through attempted copulation with the flowers (Chapter 15, Schiestl and Ayasse, 2002, Schiestl et al., 1999; Ayasse et al., 1997, 2003). The co-production of identical volatiles by most other insects and plants is probably just a coincidence reflecting a limited biosynthetic machinery, or limited sensory modalities in perceiving insects. Most of the chemicals probably constitute honest signals with different functions in plants and insects.

In male euglossine bees floral scent compounds are a primary reward, which the male bees collect during an extensive part of their lives. The fragrance is somehow exposed during brief premating encounters between the sexes and is suggested to serve as an indicator of male quality. Fragrance collecting behavior could thus have evolved through sexual selection (Eltz et al., 1999, 2003). In plants pollinated by scarab beetles (Scarabaeidae: *Dynastinae*), production of strong odors and heat coincide with the crepuscular/nocturnal behavior of the beetles (Ervik and Knudsen, 2003; Gottsberger, 1990; Gottsberger and Silberbauer-Gottsberger, 1991; Prance and Arias, 1975; Schatz, 1990).

The floral volatiles are suggested to function as a sex pheromone or an aggregation cue attracting both male and female scarab beetles, which produce a high degree of mutual dependence between the plants and their scarab pollinators. Floral scents have only been analyzed in a few scarab pollinated plant species, but all contained either methoxylated or hydroxylated compounds that otherwise are rare in floral scents, and some have been reported only in scarab pollinated plants (e.g., Azuma et al., 1997; Ervik et al., 1999; Knudsen et al., 2001, 2006; Ervik and Knudsen, 2003). In addition, floral scent is important in brood-site pollination mutualisms, systems in which plants breed their own pollinators (Sakai, 2002), perhaps in parity with the scarab-plant pollination mechanism. In some species of Annonaceae, Araceae, Arecaceae, Cyclanthaceae, Zamiaceae and other families both sexes of pollinators are attracted by the floral volatiles produced. Pollinators mate within the inflorescence/flower and females lay their eggs, usually in staminate flowers or floral parts other than the gynoecium, thus avoiding the destruction of developing ovules (Eriksson, 1994; Bernal and Ervik, 1996; Gottsberger, 1999; Anstett, 1999; Miyake and Yafuso, 2003; Terry et al., 2004). In the fig-fig wasp mutualism each *Ficus* species produces its own specific odor, which attracts its specific Agaonid pollinator (Grison-Pigé et al., 2002a, b). The occurrence of brood-site mutualisms in both gymnosperm and angiosperms points to an early origin of this interaction. However, further studies are needed to pin-point the exact role of floral volatiles in the evolution of these mutualisms.

4.5 CONCLUSIONS

The numbers of species in which floral scents have been analyzed and the numbers of compounds that have been identified, have more than doubled between 1992 and 2002 and over 1700 floral scent compounds are now known (Knudsen et al., 1993, 2006). However only about 2.5% of the angiosperms have been studied and many families are still totally unknown concerning floral scents. Thus we may expect to encounter more new compounds and perhaps even new compound classes as research continues. To date, the reported compounds can be classified into seven major groups, each of which has a wide distribution in the plant kingdom.

The composition of floral scent varies both quantitatively and qualitatively at many different levels of organization. Hence, in planning an investigation it is important to standardize environmental factors as well as the collection and analytical methods. Especially in interspecific comparisons, it is important that samples are collected using the same methods and that sufficient replicate samples are collected. All scent collection methods have advantages and disadvantages, and the choice of analytical procedures may introduce bias. In most studies of floral scent only a limited number of authentic reference compounds have been available to the researchers and so most compounds are tentatively identified by comparison of retention times and mass spectra with those in literature and in databases. However, such identifications can be quite erroneous (Webster et al., 1998), because the appearance of a mass spectrum is influenced by its concentration and the kind of instrument it is recorded on. In addition, mass spectra in spectral libraries are sometimes stored in abbreviated form including only a subset of all mass fragments. Agreement between Kovats retention index of a known compound (e.g., Adams, 1989) and that of a tentatively, identified compound can help support the identification (Heath and Dueben, 1998).

Knowledge of the chemistry of floral scent is critical in understanding its evolution and biological function. Research progress in this area should be fastest if collaborative studies are undertaken between biologists, biochemists, and analytical and synthetic chemists. The excellent series of studies on sexually deceptive orchids (Ayasse et al., 1997, 2003; Schiestl et al., 1999; Stránsky and Valterová, 1999; Schiestl et al., 2000; Schiestl and Ayasse, 2001, 2002) serve as an excellent example.

ACKNOWLEDGMENTS

We thank Natalia Dudareva and Eran Pichersky for the invitation to contribute to this book. J. T. K is grateful to her husband, Bertil Ståhl, who made it possible for her to write this chapter, while J. G. acknowledges the support of the Max Planck Society.

REFERENCES

Adams, R.P. 1989. *Identification of Essential Oils by Ion Trap Mass Spectroscopy*. Academic Press, San Diego, CA.

Altenburger, R., and P. Matile. 1988. Circadian rhythmicity of fragrance emission in flowers of *Hoyacarnosa* R. Br. *Planta* 174: 248–252.

Andersson, S., L.A. Nilsson, I. Groth, and G. Bergstrom. 2002. Floral scents in butterfly-pollinated plants: Possible convergence in chemical composition. *Bot. J. Linn. Soc.* 140: 129–153.

Angioy, A.-M., et al. 2003. Function of the heater: The dead horse arum revisited. *Proc. R. Soc. Lond. B.* (Suppl.) biology letters, 03b10216: 1.

Anstett, M.C. 1999. An experimental study of the interaction between the dwarf palm (*Chamaerops humilis*) and its floral visitor *Derelomus chamaeropsis* throughout the life cycle of the weevil. *Acta Oecol.* 20: 551–558.

Awano, K., T. Honda, T. Ogawa, S. Suzuki, and Y. Matsunaga. 1997. Volatile components of *Phalaenopsis schilleriana* Rehb. f, *Flav. Frag. J.* 12: 341–344.

Ayasse, M., F.P. Schiestl, H.F. Paulus, D. Erdmann, and W. Francke. 1997. Chemical communication in the reproductive biology of *Ophrys sphegodes*, Mitt. Dtsch. *Ges. Allg. Angew. Ent.* 11: 473–476.

Ayasse, M., F.P. Schiestl, H.F. Paulus, F. Ibarra, and W. Francke. 2003. Pollinator attraction in a sexually deceptive orchid by means of unconventional chemicals. *Proc. R. Soc. Lond. B.* 270: 517–522.

Ayasse, M., F.P. Schiestl, H.F. Paulus, C. Lofstedt, B. Hansson, F. Ibarra, and W. Francke. 2000. Evolution of reproductive strategies in the sexually deceptive orchid *Ophrys sphegodes*: How does flower-specific variation of odor signals influence reproductive success? *Evolution* 54: 1995–2006.

Azuma, H, L.B. Thien, and S. Kawano. 1999. Molecular phylogeny of Magnolia (Magnoliaceae) inferred from cpDNA sequences and evolutionary divergence of the floral scents. *J. Plant Res.* 112: 291–306.

Azuma, H., M. Toyota, and Y. Asakawa. 2001. Intraspecific variation of floral scent chemistry in *Magnolia kobus* DC (Magnoliaceae). *J. Plant Res.* 114: 411–422.

Azuma, H., M. Toyota, Y. Asakawa, et al. 1997. Chemical divergence in floral scents of Magnolia and allied genera (Magnoliaceae). *Pl. Spec. Biol.* 12: 69–83.

Barkman, T.J. 2001. Character coding of secondary chemical variation for use in phylogenetic analyses. *Biochem. Syst. Ecol.* 29: 1–20.

Barkman, T.J., J.H. Beaman, and D.H. Gage. 1997. Floral fragrance variation in Cypripedium: Implications for evolutionary and ecological studies. *Phytochemistry* 44: 875–882.

Bergström, G., H.E.M. Dobson, and I. Groth. 1995. Spatial fragrance patterns within the flowers of *Ranunculus acris* (Ranunculaceae). *Pl. Syst. Evol.* 195: 221–242.

Bergström, G., I. Groth, O. Pellmyr, P.K. Endress, L.B. Thien, A. Hubner, and W. Francke. 1991. Chemical basis for a highly specific mutualism: Chiral esters attract pollinating beetles in Eupomatiaceae. *Phytochemistry* 30: 3221–3225.

Bernal, R, and F. Ervik. 1996. Floral biology and pollination of the dioecious palm *Phytelephas seemannii* in Colombia: An adaptation to Staphylinid beetles. *Biotropica* 28: 682–696.

Bicchi, C., and D. Joulain. 1990. Headspace-gas chromatographic analysis of medicinal and aromatic plants and flowers. *Flav. Frag. J.* 5: 131–146.

Bohlmann, J., D. Martin, N.J. Oldham, and J. Gershenzo. 2000. Terpenoid secondary metabolism in *Arabidopsis thaliana*: CDNA cloning, characterization, and functional expression of α myrcene/(E)-β-ocimene synthase. *Arch. Biochem. Biophys.* 375: 261–269.

Borchsenius F, T. Lozada, and T. Knudsen. 2016. Reproductive isolation of sympatric forms of the neotropical understory palm *Geonoma macrostachys* var. *macrostachys* in Western Amazonia. *Bot. J. Linn. Soc.* 182:398–410.

Borg-Karlson, A.-K., C.R. Unelius, I. Valterová, and L.A. Nilsson. 1996. Floral fragrance chemistry in the early flowering shrub *Daphne mezereum*. *Phytochemistry* 41: 1477–1483.

Borg-Karlson, A.K., F.O. Englund, and C.R. Unelius. 1994a. Dimethyl oligosulphides, major volatiles released from *Sauromatum guttatum* and *Phallus impudicus*. *Phytochemistry* 35: 321–323.

Borg-Karlson, A.K., I. Valterová, and L.A. Nilsson. 1994b. Volatile compounds from flowers of six species in the family Apiaceae: Boquets for different pollinators. *Phytochemistry* 35: 111–119.

Bozan, B., T. Ozek, M. Kurkcuoglu, N. Kirimer, and K.H.C. Baser. 1999. The analysis of essential oil and headspace volatiles of the flowers of *Pelargonium endlicherianum* used as an anthelmintic in folk medicine. *Planta Med.* 65: 781–782.

Brunke, E.J., F.J. Hammerschmidt, and G. Schmaus. Teranishi, R. Buttery, R.G., and Sugisawa, H., Eds. 1993. Bioactive volatile compounds from plants. Flower scent of some traditional medical plants, Washington, DC, pp. 282.

Chen, F., D. Tholl, J.C. D'Auria, A. Farooq, E. Pichersky, and J. Gershenzon. 2003. Biosynthesis and emission of volatiles from *Arabidopsis* flowers. *Plant Cell* 15: 481–494.

Croteau, R., and F. Karp. Müller, P.M. and Lamparsky, D., Eds. 1991. *Perfumes: Art, Science and Technology. Origin of Natural Odorants*, Elsevier Applied Science, London, UK, pp. 101.

Dahl, A.E., A.B. Wassgren, and G. Bergström. 1990. Floral scents in *Hypecoum* Sect. *Hypecoum* (Papaveraceae): Chemical composition and relevance to taxonomy and mating system. *Biochem. Syst. Ecol.* 18: 157–168.

Dobson, H.E.M., and G. Bergström. 2000. The ecology and evolution of pollen odors. *Pl. Syst. Evol.* 222: 63–87.

Dobson, H.E.M., I. Groth, and G. Bergström. 1996. Pollen advertisement: Chemical contrasts between whole-flower and pollen odors. *Am. J. Bot.* 83: 877–885.

Dobson, H.E.M. 1994. Insect-plant interactions. In: *Floral Volatiles in Insect Biology*, Bernays, E.A., Ed., CRC Press, Boca Raton, FL, pp. 47–81.

Donath, J., and W. Boland. 1994. Biosynthesis of acyclic homoterpenes in higher plants parallels steroid hormone metabolism. *J. Plant Physiol.* 143: 473–478.

Donath, J., and W. Boland. 1995. Biosynthesis of acyclic homoterpenes: Enzyme selectivity and absolute configuration of the nerolidol precursor. *Phytochemistry* 39: 785–790.

Dötterl, A., L.M. Wolfe, and A. Jürgens. 2005. Qualitative and quantitative analyses of flower scent in *Silene latifolia*. *Phytochemistry* 66: 195–213.

Dudareva, D., B. Piechulla, and E. Pichersky. 2000. Biogenesis of floral scents. *Hort. Rev.* 24: 31–54.

Dufaÿ, M., M. Hossaert-McKey, and M.C. Anstett. 2003. When leaves act like flowers: How dwarf palms attract their pollinators. *Ecol. Lett.* 6: 28–34.

Dufaÿ, M., M. Hossaert-McKey, and M.C. Anstett. 2004. Temporal and sexual variation in leaf-produced pollinator-attracting odours in the dwarf palm. *Oecologia* 139: 392–398.

Ecroyd, C.E., R.A. Franich, H.W. Kroese, and D. Steward. 1995. Volatile constituents of *Dactylanthus taylorii* flower nectar in relation to flower pollination and browsing by animals. *Phytochemistry* 40: 1387–1389.

El-Sayed, A.M. 2004. The Pherobase: Database of insect pheromones and semiochemicals. http://www.pherobase.com

Eltz, T., D.W. Roubik, and M. Whitten. 2003. Fragrances, male display and mating behaviour of *Euglossa hemichlora*: A flight cage experiment. *Physiol. Entomol.* 28: 251–260.

Eltz, T., M. Whitten, D.W. Roubik, and K.E. Linsenmair. 1999. Fragrance collection, storage, and accumulation by individual male orchid bees. *J. Chem. Ecol.* 25: 157–176.

Eriksson, R. 1994. The remarkable weevil pollination of the neotropical *Carludovicoideae* (Cyclanthaceae). *Pl. Syst. Evol.* 189: 75–81.

Ervik, F., and J.T. Knudsen. 2003. Scarabs and water lilies: Faithful partners for the past 100 million years? *Linn. J. Biol.* 80: 539–543.

Ervik, F., L. Tollsten, and J.T. Knudsen. 1999. Floral scent chemistry and pollination ecology in phytelephan-toid palms (Arecaceae). *Pl. Syst. Evol.* 217: 279–297.

Eugster, C.H., and E. Märki-Fischer. 1991. The chemistry of rose pigments. *Angew. Chem.* 30: 654–672.

Eugster, C.H., H. Hürlimann, and H.J. Leuenberger. 1969. Crocetindialdehyd und crocetinhalbaldehyd als Blütenfarbstoffe von *Jacquinia angustifolia*, *Helv. Chem. Acta* 52: 89.

Euler, M., and I.T. Baldwin. 1996. The chemistry of defense and apparency in the corollas of *Nicotiana attenuata*. *Oecologia* 107: 102–112.

Flamini, G., P.L. Cioni, and I. Morelli. 2002. Differences in the fragrances of pollen and different floral parts of male and female flowers of *Laurus nobilis*. *J. Agric. Food Chem.* 50: 4647–4652.

Francke, W., G. Hindorf, and W. Reith. 1979. Mass-spectrometric fragmentation of alkyl-1,6-dioxaspiro[4.5] decanes. *Naturwissenschaften* 66: 619–620.

Fráter, G., J.A. Bajgrowicz, and P. Kraft. 1998. Fragrance chemistry. *Tetrahedron* 54: 7633–7703.

Frey, M., C. Stettner, P.W. Pare, E.A. Schmelz, J.H. Tumlinson, and A. Gierl. 2000. An herbivore elicitor acti-vates the gene for indole emission in maize. *Proc. Natl. Acad. Sci. USA* 97: 14801–14806.

Gerlach, G., and R. Schill. 1989. Fragrance analyses, an aid to taxonomic relationship of the genus *Coryanthes* (Orchidaceae). *Pl. Syst. Evol.* 168: 159–165.

Gerlach, G., and R. Schill. 1991. Composition of orchids scents attracting euglossine bees. *Bot. Acta* 104: 379–391.

Gershenzon, J., and W. Kreis. 1999. Biochemistry of terpenoids: Monoterpenes, sesquiterpenes, diterpenes, sterols, cardiac glycosides and steroid saponins. In: *Biochemistry of Plant Secondary Metabolism*, Wink, M., Ed., Sheffield Academic Press, Sheffield, UK, pp. 222–299.

Gottsberger, G. 1990. Flowers and beetles in the South American tropics. *Bot. Acta* 103: 360–365.

Gottsberger, G. 1999. Pollination and evolution in neotropical Annonaceae. *Pl. Spec. Biol.* 14: 143–152.

Gottsberger, G. and I. Silberbauer-Gottsberger. 1991. Olfactory and visual attraction of *Erioscelis emarginata* (Cyclocephalini, Dynastinae) to the inflorescences of *Philodendronselloum* (Araceae). *Biotropica* 23: 23–28.

Gregg, K.B. 1983. Variation in floral fragrances and morphology: Incipient speciation in Cycnoches? *Bot. Gaz.* 144: 566.

Grison-Pigé, L., J.M. Bessière and M. Hossaert-McKey. 2002a. Specific attraction of fig-pollinating wasps: Role of volatile compounds released by tropical figures. *J. Chem. Ecol.* 28:283–295.

Grison-Pigé, L., Hossaert-McKey, M., J.M. Greeff, and J.M. Bessiere. 2002b. Figure volatile compounds—A first comparative study. *Phytochemistry* 61: 61–71.

Grison-Pigé, L., J.M. Bessiere, T.C.J. Turlings, F. Kielberg, J. Roy, M.M. Hossaert-McKey. 2001. Limited intersex mimicry of floral odour in *Ficus carica*. *Func. Ecol.* 15: 551–558.

Hansted, L., H.B. Jakobsen, and C.E. Olsen. 1994. Influence of temperature on the rhythmic emission of vola-tiles from *Ribes nigrum* in situ. *Plant, Cell and Environ.* 17: 1069–1072.

Heath, R.R., B.D. Dueben., J.G. Millar, and K.F. Haynes. 1998. Analytical and preparative gas chromatogra-phy, In: *Methods in Chemical Ecology*, Millar, J. G. and Haynes, K.F., Eds, Kluwer Academic Publisher, Norwell, MA, pp. 85.

Helsper, J.P.F.G., H.J. Bouwmeester, A.F. Krol, and M.H. van Kampen. 1998. Circadian rhythmicity in emis-sion of volatile compounds by the flowers of *Rosa hybrida* L. cv. Honesty. *Planta* 207: 88–95.

Hills, H.G., and N.H. Williams. 1990. Fragrance cycle of *Clowesia rosea*. *Orquídea* (Méx.) 12: 19–22.

Ishizaka, H., H. Yamada, and K. Sasaki. 2002. Volatile compounds in the flowers of *Cyclamen persicum*, *C. purpurascens* and their hybrids. *Scient. Hort.* 94: 125–135.

Jakobsen, H.B. and C.E. Olsen. 1994. Influence of climatic factors on emission of volatiles in situ. *Planta* 192: 365–371.

Jarvis, A.P., O. Schaaf, and N.J. Oldham. 2000. 3-Hydroxy-3-phenylpropanoic acid is an intermediate in the biosynthesis of benzoic acid and salicyclic acid but benzaldehyde is not. *Planta* 212: 119–126.29.

Joulain, D. and W.A. König. 1998. *The Atlas of Spectral Data of Sesquiterpene Hydrocarbons*. E. B.-Verlag, Hamburg, Germany.

Jürgens, A., and S. Dötterl. 2004. Chemical composition of anther volatiles in Ranunculaceae: Genera-specific profiles in *Anemone, Aquilegia, Caltha, Pulsatilla, Ranunculus*, and *Trollius* species. *Am. J. Bot.* 91: 1669–1980.

Jürgens, A., T. Witt, and G. Gottsberger. 2003. Flower scent composition in *Dianthus* and *Saponaria* species (Caryophyllaceae) and its relevance for pollination biology and taxonomy. *Biochem. Syst. Ecol.* 31: 345–357.

Kaiser, R. 1988. Flavors and Fragrances: A world perspective. In: *New volatile constituents of Jasminum sambac (L.) Aiton, Proceeding 10th International Congress Essential. Oils* (Washington, DC. 1986), Lawrence, B.M. Mookherjee, B.D., and Willis, B.J., Eds, Elsevier Science Publisher, Amsterdam, The Netherlands, p. 669.

Kaiser, R. 1991. Perfumes: Art, science and technology. In: *Trapping, Investigation and Reconstitution of Flower Scents*. Müller, P.M. and Lamparsky, D., Eds, Elsevier Applied Science, London, UK, p. 213.

Kaiser, R. 1993a. Bioactive volatile compounds from plants. In: *On the Scents of Orchids*. Teranishi, R. Buttery, R.G., and Sugisawa, H., Eds, American Chemical Society, Washington DC, p. 240.

Kaiser, R. 1993b. *The Scents of Orchids*. Elsevier, Amsterdam, The Netherlands.

Kaiser, R. 1995. Flavours, fragrances and essential oils. New and uncommon volatile compounds in floral scents. *Proceeding of the 13th International Congress of Flavours, Fragrances and Essential Oils*, Baser, K.H.C., Ed., Istanbul, Istanbul, Turkey, 15–19 Oct 1995, p. 135.

Kaiser, R. 1997a. Environmental scents at the Ligurian coast. *Perfumer Flavorist* 22: 7.

Kaiser, R. 1997b. New or uncommon volatile compounds in the most diverse natural scents. *Revista Italiana Eppos* 18: 18.

Kaiser, R. 2000. Scents from rain forests. *Chimia* 54: 346–363.

Kaiser, R. 2002. ACS Symposium Series. In: *Carotenoid-Derived Aroma Compounds in Flower Scents*. Winterhalter, P. and Rouseff, R., Eds, American Chemical Society, Washington DC, p. 160.

Kaiser, R. and C. Nussbaumer. 1990. Dehydrogeosmin, a novel compound occurring in the flower scent of various species of Cactaceae. *Helvetica Chimica Acta* 73: 133–139.

Kaiser, R. and D. Lamparsky. 1982. *Proceedings of the 8th International Congress of Essential Oils* (Cannes 1980). Constituants azotés en trace de quelques absolues de fleurs et leurs head spaces correspondants. Fedarom, Grasse. p. 287.

Kaiser, R. and L. Tollsten. 1995. An introduction to the scent of cacti. *Flavour and Fragrance Journal* 10: 153–164.

Kaiser, R. and P. Kraft. 2001. Neue und ungewöhnliche Naturstoffe faszinierender Blütendüfte. *Chemie in unserer Zeit* 35: 8–22.

Kite, G.C. 1995. The floral odour of *Arum maculatum*. *Biochem. Syst. Ecol.* 23: 343.

Knudsen, J.T., and L. Tollsten. 1991. Floral scent and intrafloral scent differentiation in *Moneses* and *Pyrola* (Pyrolaceae). *Pl. Syst. Evol.* 177: 81–91.

Knudsen, J.T., and L. Tollsten. 1993. Trends in floral scent chemistry in pollination syndromes: Floral scent composition in moth-pollinated taxa. *Bot. J. Linnean Soc.* 113: 263–284.

Knudsen, J.T., L. Tollsten, and F. Ervik. 2001. Flower scent and pollination in selected neotropical palms. *Pl. Biol.* 3: 642–653.

Knudsen, J.T., L. Tollsten, and L.G. Bergström. 1993. A review: Floral scents - a check list of volatile compounds isolated by head-space techniques. *Phytochemistry* 33: 253–280.

Knudsen, J.T., R. Eriksson, J. Gershenzon, and B. Stahl. 2006. Diversity and distribution of floral scent. *Bot. Rev.* 72:1–120.

Knudsen, J.T., S. Andersson, and P. Bergman. 1999. Floral scent attraction in *Geonoma macrostachys*, an understorey palm of the Amazonian rain forest. *Oikos* 85: 409–418.

Knudsen, J.T. 1999. Floral scent chemistry in geonomoid palms (Palmae: Geonomeae) and its importance in maintaining reproductive isolation. *Mem. New York Bot. Gard.* 83: 141–168.

Knudsen, J.T. 2002. Variation in floral scent composition within and between populations of *Geonoma macrostachys* (Arecaceae) in the Western Amazon. *Am. J. Bot.* 89: 1772.

Knudsen, J.T., L. Tollsten, I. Groth, G. Bergstrom, and R.A., Raguso. 2004. Trends in floral scent chemistry in pollination syndromes: Floral scent composition in hummingbird-pollinated taxa. *Bot. J. Linnean Soc.* 146: 191–199.

Kuanprasert, N, A.R. Kuehnle, and C.S. Tang. 1998. Floral fragrance compounds of *Anthurium* (Araceae) species and hybrids. *Phytochemistry* 49: 521–528.

Levin, R.A., L.A. McDade, and R.A. Raguso. 2003. The systematic utility of floral and vegetative fragrance in two genera of Nyctaginaceae. *Syst. Biol.* 52: 334–351.

Levin, R.A., L.A. Raguso, and L.A. McDade. 2001. Fragrance chemistry and pollinator affinities in Nyctaginaceae. *Phytochemistry* 58: 429–440.

Lewis, J.A., C.J. Moore, M.T. Fletcher, R.A. Drew, and W. Kitching. 1988. Volatile compounds from flowers of *Spathiphyllum cannaefolium*. *Phytochemistry* 27: 2755–2757.

Lindberg, A.B., J.T. Knudsen, and J.M. Olesen. Totland, Ö., Ed. 2000. The Scandinavian Association for Pollination Ecology honours Knut Fægri. Independence of floral morphology and scent chemistry as trait groups in a set of Passiflora species, Det Norske Videnskaps-Akademi. I. Mat.-Naturv. Klasse, Ny Ser. p. 91.

Loper, G.M., and A.M. Lapiolo. 1971. Photoperiodic effects on the emanation of volatiles from alfalfa (*Medicago sativa* L.) florets. *Plant Physiol.* 49: 729–732.

Loper, G.M., and R.L. Berdel. 1978. Seasonal emanation of ocimene from alfalfa flowers with three irrigation treatments. *Crop Sci.* 18: 447–452.

MacTavish, H.S., N.W. Davies, and R.C. Menary. 2000. Emission of volatiles from brown Boronia flowers: Some comparative observations. *Ann. Bot.* 86: 347–354.

Matsumoto, F., H. Idetsuki, K. Harada, I. Nohara, and T. Toyoda. 1993. Volatile components of *Hedychium coronarium* Koenig flowers. *J. Ess. Oil Res.* 5: 123–133.

Metcalf, R.L. 1987. Plant volatiles as insects attractants. *CRC Crit. Rev. Plant Sci.* 45: 251–301.

Miyake, T., and M. Yafuso. 2003. Floral scents affect reproductive success in fly-pollinated *Alocasia odora* (Araceae). *Am. J. Bot.* 90: 370–376.

Mori, K. 1998. Methods in chemical ecology. In: *Separation of Enantiomers and Determination of Absolute Configuration, Chemical Methods.* Millar, J.G. and Haynes, K.F., Eds, Kluwer Academic Publisher, Norwell, MA, p. 295.

Moya, S., and J.D. Ackerman. 1993. Variation in the floral fragrance of *Epidendrum ciliare* (Orchidaceae). *Nord. J. Bot.* 13: 41–47.

Nielsen, J.K., H.B. Jakobsen, P. Friis, K. Hansen, J. Moller, Ce. Olsen. 1995. Asynchronous rhythms in the emission of volatiles from *Hesperis matronalis* flowers. *Phytochemistry* 38: 847–851.

Olesen, J.M., and J.T. Knudsen. 1994. Scent profiles of flower colour morphs of *Corydalis cava* (Fumariaceae) in relation to foraging behaviour of bumblebee queens (*Bombus terrestris*). *Biochem. Syst. Ecol.* 22: 231–237.

Omata, A., K. Yomogida, S. Nakamura, T. Ohta, Y. Izawa, S. Watanabe. 1991. The scent of Lotus flowers. *J. Ess. Oil Res.* 3: 221–227.

Paré, P.W., and J.H. Tumlinson. 1999. Plant volatiles as a defense against insect herbivores. *Plant Physiol.* 121: 325–331.

Patt, J.M., J.C. French, C. Schal, J. Lech, T.G. Hartman. 1995. The pollination biology of tuckahoe, *Peltandra virginica* (Araceae). *Am. J. Bot.* 82: 1230–1240.

Pettersson, S., and J.T. Knudsen. 2001. Floral scent and nectar production in *Parkia biglobosa* Jacq. (Leguminosae: Mimosoideae). *Bot. J. Linn. Soc.* 135: 97–106.

Pham-Delegue, M.H., P. Etievant, E. Guichard, C. Masson. 1989. Sunflower volatiles involved in honeybee discrimination among genotypes and flowering stages. *J. Chem. Ecol.* 15: 329–343.

Pichersky, E. and D.R. Gang. 2000. Genetics and biochemistry of secondary metabolites in plants: An evolutionary perspective. *Trends plant sci.* 5: 439–445.

Pichersky, E., and J. Gershenzon. 2002. The formation and function of plant volatiles: Perfumes for pollinator attraction and defense. *Curr Opin Plant Biol.* 5: 237–243.

Pott, M.B., E. Pichersky, and B. Piechulla. 2002. Evening specific oscillations of scent emission, SAMT enzyme activity, and SAMT mRNA in flowers of *Stephanotis floribunda*. *J. Plant Physiol.* 159: 925–934.

Prance, G.T., and J.R. Arias. 1975. A study of the floral biology of *Victoria amazonica* (Poepp.) Sowerby (Nymphaeaceae). *Acta Amazonica* 5: 109–139.

Raguso, R.A. 2001. Cognitive ecology of pollination. In: *Floral Scent, Olfaction, and Scent Driven Foraging Behavior.* Chittka, L. and Thomson, J.D., Eds, Cambridge University Press, Cambridge, MA, p. 83.

Raguso, R.A. 2003. Insect pheromone biochemistry and molecular biology. *Olfactory Landscapes and Deceptive Pollination: Signal, Noise and Convergent Evolution in Floral Scent.* Blomquist, G.J. and Vogt, R., Eds, Academic Press, New York, p. 631.

Raguso, R.A. 2004. Why are some floral nectars scented? *Ecology* 85: 1486–1494.

Raguso, R.A. and E. Pichersky. 1995. Floral volatiles from *Clarkia breweri* and *C. concinna* (Onagraceae): Recent evolution of floral scent and moth pollination. *Pl. Syst. Evol.* 194: 55–67.

Raguso, R.A., and O. Pellmyr. 1998. Dynamic headspace analysis of floral volatiles: A comparison of methods. *Oikos* 81: 238–254.

Raguso, R.A., R.A. Levin, S.E. Foose, M.W. Holmberg, and L.A. McDade. 2003. Fragrance chemistry, nocturnal rhythms and pollination "syndromes" in *Nicotiana*. *Phytochemistry* 63: 265–284.

Robacker, D.C., B.J.D. Meeuse, and E.H. Erickson. 1988. Floral aroma. How far will plants go to attract pollinators? *BioScience* 38: 390–398.

Rodriguez-Concepcion, M., and A. Boronat. 2002. Elucidation of the methylerythritol phosphate pathway for isoprenoid biosynthesis in bacteria and plastids. A metabolic milestone achieved through genomics. *Plant Physiol.* 130: 1079–1089.

Rowan, D.D., H.P. Lane, J.M. Allen, S. Fielder, and M.B. Hunt. 1996. Biosynthesis of 2-methylbutyl, 2-methyl-2-butenyl, and 2-methylbutanoate esters in Red Delicious and Granny Smith apples using deuterium-labeled substrates. *J. Agric. Food Chem.* 44: 3276–3285.

Sakai, S. 2002. A review of brood-site pollination mutualisms: Plants providing breeding sites for their pollinators. *J. Pl. Res.* 115: 161–168.

Sazima, M., S. Vogel, A. Cocucci, and G. Hausner. 1993. The perfume flowers of *Cyphomandra* (Solanaceae): Pollination by euglossine bees, bellow mechanism, osmophores, and volatiles. *Pl. Syst. Evol.* 187: 51–88.

Schatz, G.E. 1990. Reproductive ecology of tropical plants. In: *Some Aspects of Pollination biology in Central American forests*, Bawa, K.S. and Hadley, M., Eds, The Parthenon Publishing Group, Park Ridge, NJ, pp. 69–84.

Schiestl, F.P., and M. Ayasse. 2001. Post-pollination emission of a repellent compound in a sexually deceptive orchid: A new mechanism for maximising reproductive success. *Oecologia* 126: 531–534.

Schiestl, F.P., and M. Ayasse. 2002. Do changes in floral odor cause speciation in sexually deceptive orchids? *Pl. Syst. Evol.* 234: 111–119.

Schiestl, F.P., M. Ayasse., H.F. Paulus, et al. 1999. Orchid pollination by sexual swindle. *Nature* 399: 421–422.

Schiestl, F.P., M. Ayasse., H.F. Paulus, et al. 2000. Sex pheromone mimicry in the early spider orchid (*Ophryssphegodes*): Patterns of hydrocarbons as the mechanism for pollination by sexual deception. *J. Comp. Physiol. A*: 186, 567–574.

Schiestl, F.P., M. Ayasse, H.F. Paulus, D. Erdmann, and W. Francke. 1997. Variation of floral scent emission and postpollination changes in individual flowers of *Ophrys sphegodes* subsp. *sphegodes*. *J. Chem. Ecol.* 23: 2881–2895.

Schlumpberger, B.O. 2002. Dehydrogeosmin produzierende Kakteen: Untersuchungen zur Verbreitung, Duftstoff-Produktion und Bestäubung. Thesis. Verlag Grauer, Beuren and Stuttgart.

Schultz, K., R. Kaiser, and J.T. Knudsen. 1999. Cyclanthone and derivatives, new natural products in the flower scent of *Cyclanthus bipartitus* Poit. *Flav. Frag. J.* 14: 185–190.

Seymour, R.S., and P. Schultze-Motel. 1997. Heat-producing flowers. *Endeavour* 21: 125–129.

Seymour, R.S., M. Gibernau, and K. Ito. 2003a. Thermogenesis and respiration of inflorescences of the dead horse arum *Helicodiceros muscivorus*, a pseudo-thermoregulatory aroid associated with fly pollination. *Func. Ecol.* 17: 886–894.

Seymour, R.S., C.R. White, and M. Gibernau. 2003b. Heat reward for insect pollinators. *Nature* 426: 243–244.

Stensmyr, M.C., I. Urru, I. Collu, et al. 2002. Rotting smell of dead-horse arum florets. *Nature* 420: 625–626.

Stránsky, K., and I. Valterová. 1999. Release of volatiles during the flowering period of *Hydrosmeriviera*. *Phytochemistry* 52: 1387–1390.

Terry, I., C.J. Moore, G.H. Walter, et al. 2004. Association of cone thermogenesis and volatiles with pollinator specificity in *Macrozamia cycads*. *Pl. Syst. Evol.* 243:233–247.

Thien, L.B., P. Berhardt, G.W. Gibbs, O. Pellmyr, G. Bergstrom, I. Groth, G. McPherson. 1985. The pollination of *Zygogynum* (Winteraceae) by a moth, *Sabatinca* (Micropterigidae): An ancient association? *Science* 227: 540–543.

Tholl, D., F. Chen, J. Petri, J. Gershenzon, and E. Pichersky. 2005. Two sesquiterpene synthases are responsible for the complex mixture of sesquiterpenes emitted from *Arabidopsis* flowers. *Plant J.* 42: 757–771.

Tollsten, L. 1993. A multivariate approach to post-pollination changes in the floral scent of *Platanthera bifolia* (Orchidaceae). *Nord. J. Bot.* 13: 495–499.

Tollsten, L., and G. Bergström. 1993. Fragrance chemotypes in *Platanthera* (Orchidaceae)—The result of adaptation to pollinating moths? *Nord. J. Bot.* 13: 607–613.

Tollsten, L., and J.T. Knudsen. 1992. Floral scent in dioecious *Salix* (Salicaceae)—a cue determining the pollination system? *Pl. Syst. Evol.* 182: 229–237.

Toyoda, T., I. Nohara and T. Sato. Teranishi, R. Buttery, R.G., and Sugisawa, H., Eds. 1993. *Bioactive Volatile Compounds from Plants*. Headspace analysis of volatile compounds from various citrus blossoms, Washington, DC, p. 205.

Vogel, S. 1990. *The Role of Scent Glands in Pollination*. Amerind Publishing, New Delhi, India.

Webster, M., J.G. Millar, and K.F. Haynes. 1998. Mass spectrometry, In: *Methods in Chemical Ecology*, Millar, J.G. and Haynes, K., Eds, Kluwer Academic Publisher, Norwell, Massachusetts, p. 127.

Whitten, V.M., and N.H. Williams. 1992. Floral fragrances of *Stanhopea* (Orchidaceae). *Lindleyana* 7: 130–153.

Wildermuth, M.C., J. Dewdney, G. Wu, and F.M. Ausubel. 2001. Isochorismate synthase is required to synthesize salicylic acid for plant defence. *Nature* 414: 562–565.

Williams, N.H., and W.M. Whitten. 1999. Molecular phylogeny and floral fragrances of male euglossine bee-pollinated orchids: A study of *Stanhopea*. *Pl. Spec. Biol.* 14: 129–136.

Williams, N.H. 1983. Handbook of experimental pollination biology. In: *Floral Fragrances as Cues in Animal Behavior*, Jones, C.E. and Little, R.J., Eds, Van Nostrand Reinhold, New York, pp. 50–72.

Willmer, P.G., and G.N. Stone. 1997. How aggressive ant-guards assist seed-set in Acacia flowers. *Nature* 388: 165–167.

Wise, M.L., T.J. Savage, E. Katahira, and R. Croteau. 1998. Monoterpene synthases from Common Sage (*Salvia officinalis*). *Journal of Biological Chemistry* 273: 14891–14899.

5 Vegetative and Fruit Volatiles for Human Consumption

Bhagwat Nawade, Mossab Yahyaa, Efraim Lewinsohn, and Mwafaq Ibdah

CONTENTS

5.1 INTRODUCTION

Plant volatiles are specialized metabolites, synthesized in the cytoplasm and plastids of plant cells. They are generally odorous compounds (<C15) with low molecular mass (<300 Da), high vapor pressure, low boiling point, and a lipophilic moiety. Over the years >1700 volatile organic compounds (VOCs) compounds have been found to be emitted from more than 90 plant families (Chapter 4; Dudareva et al., 2013). A great diversity of volatiles are emitted from leaves, flowers, fruits, bark, and roots, as well as specialized storage structures (Loreto and Schnitzler, 2010). These plant volatiles are derived from several biochemical pathways, constituting about 1% of all plant specialized metabolites. Plant volatiles are often classified into major groups including the terpenoids, phenylpropanoids/benzenoids, fatty acid and amino acid derivatives (Dudareva et al., 2004). Volatiles are low molecular weight molecules that can be emitted either constitutively (e.g., floral volatiles attracting pollinators) or as a response to biotic (Dudareva et al., 2006, 2013; Unsicker et al., 2009) or abiotic stress (Loreto and Schnitzler, 2010). The vegetative volatiles directly deter or are toxic to herbivores and microorganisms. Indirectly, plant volatiles can protect the emitting plant by attracting natural enemies of herbivores *via* tritrophic interactions (Dicke et al., 2003; Heil, 2008). Plant volatiles also mediate intra and interplant signaling processes inducing the emission of other volatiles and the expression of defense genes in neighboring plants (Baldwin et al., 2006). The various bioactivities of plant volatiles including antimicrobial, antifungal, antiviral, anti-inflammatory, antimutagenic, anticarcinogenic, and antioxidant as well as other miscellaneous activities widen their applications in medicinal and pharmaceutical industries (Vaughn, 2001).

Essential oils are volatile, complex mixtures of compounds characterized by a strong flavor and aroma formed by plants as specialized metabolites. Flavor and aroma are the important attributes that determining acceptability of product by a consumer. It has recently been suggested that plant volatiles due their aroma, provides sensory clues as to the health and nutritional status of foodstuff (Goff and Klee, 2006). Aroma is the description of sensations induced by thousands of different volatiles providing various kinds of floral, fruity, minty, woody, mushroom, and other. aroma perceptions (Schwab et al., 2008). The perception of aroma is not only due to the additive effect of the different volatile compounds but also their interaction (Goff and Klee, 2006). Volatile compounds are important components of flavor and aroma in many crops. For example, apocarotenoids, a group of terpenoid flavor volatile compounds derived from carotenoid metabolism possess very strong flavor characteristics on the overall human perception. Among these, β-ionone, geranylacetone, pseudoionone, β-cyclocitral, geranial, and α-damascenone are reported from various fruits and vegetables. Therefore, concentrated efforts have been made for identification and characterization of volatile compounds from various crops and huge number of volatiles has been reported with their potential uses for various industries.

In this chapter, we discuss the composition and diversity of volatiles from some plants (Figure 5.1) that are used by humans for dietary and medicinal purposes, but this analysis is not limited to the parts of the plants consumed by people.

Bay leaves   Bay fruit   Carrot roots

Fennel fruits   Basil leaves

Nigella fruit   *Pistacia* leaves and galls

FIGURE 5.1   Photograph of bay leaf (*Laurus nobilis*) leaves and fruits, carrot (*Daucus carota*) roots, fennel (*Foeniculum vulgare*) fruits, sweet basil (*Ocimum basilicum*) leaves, *Nigella sativa* fruit (split into two halves) and *Pistacia palaestina* leaves with galls.

5.2 BAY (*LAURUS NOBILIS*)

Laurus nobilis L. (bay leaves) is a member of the Lauraceae family. Laurel is an evergreen tree or shrub growing wild or cultivated in temperate and warm parts of the word, particularly in the countries bordering on the Mediterranean region (Sharma et al., 2012). The chemically complex volatile natural compounds produced in different tissues of this aromatic plants have promising biological properties and have wider its potential applications in many areas, including agriculture, medical, food and pharmaceutical industries (Kilic et al., 2005).

5.2.1 LEAF

The chemical composition of bay leaves from different origins has been studied and reported to consist of 1,8-cineole as the major volatile in all the cases (Figure 5.2a). Apart from this, sabinene, α-terpinyl acetate, linalool, α-pinene, α-terpineol and methyl eugenol are among the major volatiles reported from bay leaf essential oils with varying concentrations from different locations. The 1,8-cineole performs important ecological functions, such as to repelling insects, deterring herbivores (Franks et al., 2012; Southwell et al., 2003). The phenylpropene derivatives eugenol, methyl eugenol, and elemicin are also reported in the bay leaf and these are responsible for the spicy aroma of the leaves and are extremely important factors determining its sensory quality. Eugenol and methyl eugenol have anesthetic, hypothermic, muscle-relaxant, anticonvulsant and anti-stress activities on humans, as well as antifungal, antibacterial, antinematodal, or toxicant roles against pathogens and insect herbivores (Tan and Nishida, 2012). The bicyclic monoterpenes α-pinene and

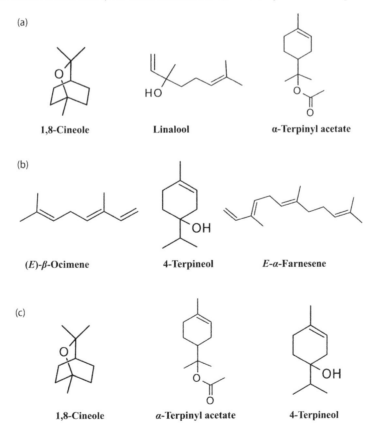

(a)

1,8-Cineole   Linalool   α-Terpinyl acetate

(b)

(*E*)-β-Ocimene   4-Terpineol   *E-α*-Farnesene

(c)

1,8-Cineole   *α*-Terpinyl acetate   4-Terpineol

FIGURE 5.2 Representative terpenoids biosynthesized by bay leaves (*Laurus nobilis*). (a) Bay leaves terpenoids, (b) Bay fruits terpenoids, and (c) Bay roots terpenoids.

β-pinene, among the frequently occurring volatiles in bay leaves, also have lipophilic, insecticidal, sedative, fungicidal and anticarcinogenic effects (Mercier et al., 2009).

Kilic et al. (2004) reported a higher concentration of odor contributing compounds in autumn (e.g., linalool, 2-4 fold; eugenol, 4-10-fold) as compared to July harvested leaf samples, leading to a better flavoring quality in the autumn samples despite their lower essential oil content. The (E)-isoeugenol content from leaves harvested in July had up to 6-8-fold more amounts as compared to leaves harvested in October (Kilic et al., 2004). Analysis of Iranian bay leaves from March, June, September and December months revealed that the June sample had a higher content of 1,8-cineole (40.25%) than others. During these months, δ-3-carene, camphor, camphene and sabinene were reported as the second major compounds respectively. No seasonal changes were found in the concentration of the volatiles eugenol, methyl eugenol and α-terpenyl acetate (Shokoohinia et al., 2014). Bendjersi et al. (2016) found some volatile compounds such as tricyclene, decanal, aceteugenol and germacrene-D-4-ol only when they used an extraction procedure utilizing a microwave oven, while p-cymene, trans-β-ocimene, pinocarvone, myrtenol, trans-carveol, carvone and β-selinene were detected only in hydrodistillation (HD). Similarly, Flamini et al. (2004) also detected β-elemene, spathulenol and epi-γ-eudesmol in HD fraction and δ-terpineol and borneol using microwave extraction. Caredda et al. (2002) observed remarkable differences in the contents of 1,8-cineole and methyl eugenol after the first and fourth hour of extraction (1,8-cineole, 30% vs. 2% and methyleugenol, 6.8% vs. 16.4%). A higher amount of sesquiterpenes were obtained when extraction was allowed to go beyond 90 min, while the monoterpene hydrocarbon extraction was almost completed at this point. Therefore, extraction time also has a major effect on the reported results of the volatile composition of L. nobilis oil (Marzouki et al., 2009).

The volatiles compositions from leaves of young and old shoots of the plant were significantly different. The leaves of young shoot contained higher amounts of β-pinene, α-pinene, linalool, α-terpineol, 2-hydroxy-1,8-cineole, and some sesquiterpenes, while 1,8-cineole, sabinene, sabinene hydrates, terpinene-4-ol, α-terpinyl acetate, eugenol, and eugenol methyl ether were found in higher concentration in leaves of old shoots (Kilic et al., 2004). In our own study, we also found that not only the leaf stage but also the gender of the plant significantly affected leaf volatiles composition. In general, all parts of the male plant had a consistently lower concentration of 1,8-cineole than female plants. In contrast, the leaves and flowers from male plants had considerably more δ-elemene than did the corresponding female organs (Yahyaa et al., 2015b). Apart from these factors, the environmental conditions, location and season during which the plants are collected, methods of drying, extraction and analytical conditions also contribute to the differences in major volatiles composition. We further characterized three terpene synthases from bay leaves and in vitro analysis revealed a monoterpene synthase belonging to the TPS-b clade catalyzing the formation of mostly 1,8-cineole; a sesquiterpene synthase belonging to the TPS-a clade catalyzing the formation of mainly cadinenes; and a diterpene synthase of the TPS-e/f clade catalyzing the formation of geranyllinalool (Yahyaa et al., 2015b).

5.2.2 FRUIT

The fruits of this dioecious plant are olive-like black berries. The composition of the volatiles from bay fruits is different from leaves somewhat but mostly quantitatively. The volatiles from different plant organs contained similar compounds, but the quantitative differences between all main compounds were quite large. Volatiles such as 1,8 cineole, sabinene α-terpinyl acetate, methyl eugenol, eugenol and linalool, which are the basic components of the essential oil of leaves, buds and flowers, were present in small quantity in the fruits (Marzouki et al., 2008). Kilic et al. (2004) reported (E)-β-ocimene as a major fruit volatile which is not present in Turkish bay leaves, while Hafizoğlu and Reunanen (1993) found 4-terpineol to be the main component in the fruit essential oil from Turkey (Figure 5.2b). Castilho et al. (2005) reported (E)-β-ocimene and germacrene-D, while Marzouki et al. (2009) also found (E)-β-ocimene as predominating volatiles in Portuguese and Tunisian bay fruit respectively. In our research, we analyzed green and black fruit separately and found (E)-β-ocimene, γ-murolene (E)-α-farnesene, γ-cardinene and δ-cardinene in green bay fruits. Conversely, 1,8 cineole (E)-β-ocimene (E)-α-farnesene, γ-cardinene and δ-cardinene were abundant in black fruits (Yahyaa et al., 2015b). The norisoprenoid volatiles such as 6-methyl-5-hepten-2-one (MHO), pseudoionone, and β-ionone were reported to present only in the fresh pericarp of the black bay fruits. These norisoprenoid volatiles are distributed in numerous fruits and considered an important and potent flavor, aroma and scent contributors in many fruits and fruit-based foods, due to extremely low odor thresholds (Yahyaa et al., 2015a).

5.2.3 ROOT

In a report on the volatile composition of bay roots, Yahyaa et al. (2015b) performed auto-headspace-SPME-GC-MS analysis of both male and female roots and found 1,8-cineole, α-terpinyl acetate, terpinene-4-ol, *p*-cymene and δ-cardinene were prominent volatiles (Figure 5.2C). However, most of the volatiles were obtained in higher concentration in female roots than male.

5.3 CARROT (*DAUCUS CAROTA*)

Carrot (*Daucus carota* L.) is major commercial root vegetable, consumed worldwide due to its characteristic flavor and beneficial health effects. Its flavor is based on a mixture of volatile and nonvolatile components. Among the volatiles, terpenoids are considered major determinants of carrot flavor and aroma. The complexity of carrot flavor and aroma arises from the large number of monoterpenes and sesquiterpenes, which make up approximately 98% of the total volatile mass (Alasalvar et al., 1999).

5.3.1 ROOT

Carrot root volatile analysis was performed in various studies and more than 90 volatile compounds have now been identified with mono- and sesquiterpenoids being the most abundant. Alasalvar et al. (1999) have utilized the headspace GC-MS method and obtained a total of 34 terpenoids from seven carrot cultivars and F$_1$ hybrids. Subsequently, various research groups reported the presence of monoterpenes such as α-pinene, β-pinene, β-myrcene, sabinene, limonene, γ-terpinene, *p*-cymene, terpinolene and bornylacetate, and sesquiterpenes such as β-caryophyllene, α-humulene (*E*)-γ-bisabolene, and (*Z*)-α-bisabolene in carrot roots (Alasalvar et al., 1999; Fukuda et al., 2013; Güler et al., 2015; Kjeldsen et al., 2001, 2003). Most of these studies indicated β-caryophyllene as the major sesquiterpene and terpinolene as among the main monoterpenes (Figure 5.3a).

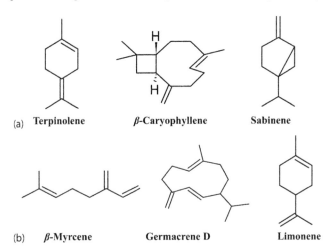

(a) Terpinolene   *β*-Caryophyllene   Sabinene

(b)   *β*-Myrcene   Germacrene D   Limonene

FIGURE 5.3 Representative terpenoids biosynthesized by carrot (*Daucus carota*) (a) Carrot roots terpenoids and (b) Carrot leaves terpenoids.

Gas chromatography-olfactometry (GC-O) analysis showed monoterpenes like sabinene, terpinolene, *p*-cymene, and β-myrcene contributing to the "carrot top" aroma, whereas sesquiterpenes like β-caryophyllene, and α-humulene contributed to the "spicy" and "woody" aromas (Fukuda et al., 2013; Kjeldsen et al., 2003). The norisoprenoid β-ionone was found responsible for the sweet note (Kjeldsen et al., 2003). Generally, the terpenes typically contribute to a harsh or bitter flavor, and such flavor attributes were indeed positively correlated with terpene content in different carrot genotypes (Kreutzmann et al., 2008).

In our research, we have focused on the identification and functional characterization of two carrot TPS genes namely *DcTPS1* and *DcTPS2* (Yahyaa et al., 2015c, 2016). In an *E. coli*-based expression system, the recombinant DcTPS1 protein catalyzed the formation of mainly the sesquiterpenes (*E*)-β-caryophyllene and α-humulene, while recombinant DcTPS2 was active as a monoterpene synthase, giving rise to geraniol as the main product generated from GPP (Yahyaa et al., 2015c). Furthermore, Iorizzo et al. (2016), by searching a carrot whole genome assembly, found a total of at least 30 TPS genes in the *D. carota* genome. However, a recent analysis by Keilwagen et al. (2017) expanded this number to 65 full-length putative TPS genes. In this study, the authors performed homology-based gene prediction using TPS gene models from five reference species. The large size of the carrot TPS gene family is not unexpected given the presence of a large number of terpenes in carrot tissues.

5.3.2 Leaf

Kainulainen et al. (1998) analyzed two carrot varieties and found myrcene, sabinene, *trans*-α-ocimene, limonene, germacrene-D, and *trans*-α-caryophyllene as major leaf volatiles. Ulrich et al. (2015) isolated VOCs from 10 carrot cultivars and reported β-myrcene, germacrene, β-caryophyllene, limonene, sabinene and α-pinene as major compounds (Figure 5.3b). Keilwagen et al. (2017) also reported β-myrcene, β-caryophyllene, germacrene-D and limonene as the most abundant compounds in leaves of a panel of 85 carrot cultivars and accessions. In total, 19 monoterpenes and 12 sesquiterpenes were identified in this study from leaves and a total of 30 quantitative trait loci (QTLs) were identified for 15 terpenoid volatiles like ocimene, sabinene, β-pinene, borneol and bornyl acetate. Recently, Koutouan et al. (2018) studied carrot leaf secondary metabolites potentially linked to the resistance to *A. dauci* and revealed that limonene, *cis*-β-farnesene, *cis*-α-bergamotene and *trans*-β-farnesene accumulated at a higher level in susceptible genotype H1. This result may indicate that these compounds do not confer resistance, while the accumulation of camphene, α-pinene, α-humulene may favor resistance in carrot. There are some reports indicating that carrot pests use cues including the volatile terpenoids to identify their host plant (Nehlin et al., 1994; Nissinen et al., 2005). Further understanding of the physiological roles of volatile compounds from different cultivars and their interactions with pests or diseases may lead to better integrated pest management systems through herbivore induced resistance.

5.4 FENNEL (*FOENICULUM VULGARE*)

5.4.1 Fruit

Fennel (*Foeniculum vulgare* Mill.) is often classified as both an herb and a vegetable and has been used for thousands of years due to its medicinal and aromatic properties. This perennial umbelliferous (*Apiaceae*) herb is native from the Circum-Mediterranean region and now is cultivated throughout the world. Mature fennel fruit and its essential oil are used in the food, pharmaceutical, perfumery and cosmetics industries. Especially in the food industry, fennel has been used since ancient times as a flavoring agent in food products such as fish and meat dishes, savory formulations,

Fennel fruit terpenoids

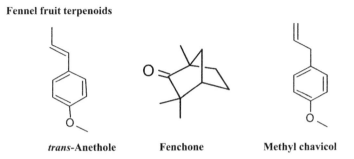

trans-Anethole   Fenchone   Methyl chavicol

FIGURE 5.4   Representative terpenoids biosynthesized by Fennel (*Foeniculum vulgare*) fruits.

liqueurs, confectionery, pickles, sauces, cheese etc. The myriad properties of fennel seeds and oil are attributed to its volatile composition (Senatore et al., 2013).

Fennel essential oil is characterized by the presence of at least 87 volatile compounds that include *trans*-anethole, fenchone, methyl chavicol, eugenol, limonene as major components while 3-methylbutanol, linalool, α-pinene, mycerene, *cis*-anethole, and thymol are among the minor ones (Figure 5.4). The relative concentrations of these compounds vary considerably depending on factors such as geographical origin (Díaz-Maroto et al., 2006).

Damjanović et al. (2005) identified a total of 28 compounds from fennel seed extracts, with major compounds being *trans*-anethole, fenchone, and methylchavicol. Similarly, *trans*-anethole was detected in higher concentration among 76 volatiles identified by Fang et al. (2006), and also in 18 lines from Italy analyzed by Tognolini et al. (2007). Moreover, many studies also confirmed that the *trans*-anethole is a major volatile fraction in fennel seeds (Diao et al., 2014; Pavela et al., 2016; Rahimmalek et al., 2014; Salami et al., 2016; Telci et al., 2009; Zhao et al., 2012). Fennel populations with estragole as the dominant profiles have also been identified in Spain (García-Jiménez et al., 2000), Israel (Barazani et al., 2002) and Italy (Napoli et al., 2010). In addition, it was found that pinenes, fenchone, estragole, myrcene and camphene were present in high concentrations in Egyptian cultivated fennels, while wild plants growing in Egypt showed a much higher level of limonene (Shahat et al., 2012) and α-phellandrene was the dominant compound detected in wild fennel plants from Bari, Italy (Piccaglia and Marotti, 2001).

Fennel plants with high levels of *trans*-anethole (75%–83%) and fenchone (up to 5%) are classified as sweet fennels, while the fennel plants with less *trans*-anethole (<55%) higher estragole concentrations, and at least 10% fenchone are considered bitter. *O*-methyltransferase that act on chavicol and *t*-anol *in vitro* to produce estragole and *t*-anethole respectively were identified in bitter fennel (Gross et al., 2002). Furthermore, the presence of a biallelic genetic system with partial dominance for high estragole content was inferred and a reverse correlation between estragole and *t*-anethole content was observed, suggesting a common biosynthetic precursor for both (Gross et al., 2006, 2009).

Gas chromatography-olfactometry (GC-O) analysis, sensory evaluation the aroma extract dilution analysis (AEDA) procedure, and odor activity values (OAV) determination identified the key aroma compounds to include *trans*-anethol (spicy, grassy), estragole (spicy), *p*-anisaldehyde (sweet), δ-cadinene (woody), α-pinene (pine, grassy), α-phellandrene (spicy), α-cubebene (grassy), linalool (floral) and limonene (green) (Xiao et al., 2017). *trans*-Anethole (anise, licorice), estragole (anise, licorice, sweet), fenchone (mint, camphor, warm), and 1-octen-3-ol (mushroom) were the most intense odor compounds, and *trans*-anethole and estragole represented the characteristic anise, sweet, and licorice notes of this spice (Díaz-Maroto et al., 2005).

5.5 BASIL (*OCIMUM BASILICUM*)

5.5.1 Leaf

Basil (*Ocimum basilicum* L.; Lamiaceae), initially native to India and known also as sweet basil or simply basil, is commonly used as a herb and a typical ingredient of the healthy diet. It is cultivated in several regions of Asia, Africa, South America, and the Mediterranean. The essential oil of this plant is an excellent source for many volatiles and aromatic chemicals in the preparation of food, perfume, and cosmetics, as well as in traditional medicine. The essential oil typically extracted from its leaves, has been demonstrated to have antimicrobial, antifungal, insect-repelling, anticonvulsant, and antioxidant, activities (Mohamed et al., 2016).

Researchers from different parts of the world have reported chemical composition of the leaf essential oil of *O. basilicum* and several chemotypes have been reported. Large diversity has been recorded in sweet basil volatile profiles from different regions; for instance linalool 1,8-cineole and eugenol were main compounds in Egyptian sweet basil (Edris and Farrag, 2003; El-Soud et al., 2015), while in southern India the main components were (*E*)-methyl cinnamate, linalool and camphor (Jirovetz et al., 2003) (Figure 5.5). Sweet basil cultivated in the Mongolian desert was reported to possess linalool (*Z*)-α-bergamotene and methyl chavicol as major compounds (Shatar et al., 2007).

The major compounds of Algerian basil leaves were linalool, linalyl acetate and elemol (Hadj Khelifa et al., 2012). In Greece, sweet basil was enriched in 1,8-cineole/eucalyptol, α-pinene and camphor (Alexopoulos et al., 2011), while α-terpineol and β-caryophyllene were dominant volatiles in basil plants from Burkina Faso (Bayala et al., 2014). Geraniol and 1,8-cineole were the major compounds reported from Oman basil (Hanif et al., 2011), as well as Poland basil (Nurzynska-Wierdak et al., 2013). Among 12 varieties of *Ocimum* from Colombia, ten were characterized by the presence of a high percentage of methyl cinnamate (Viña and Murillo, 2003). Similarly (*E*)-methyl cinnamate and (*Z*)-methyl cinnamate were recorded as major compounds from French Polynesia basil leaf (Adam et al., 2009). In another study on a germplasm collection of 18 *O. basilicum* accessions from Spain, methyl chavicol and linalool were predominated (Pascual-Villalobos and Ballesta-Acosta, 2003). Methyl chavicol was also a main principal compound of basils from Turkey (Chalchat and Özcan, 2008), Iran (Shirazi et al., 2014), Thailand (Bunrathep et al., 2007). Interestingly, Purkayastha and Nath (2006) reported that camphor, followed by limonene and β-selinene, were the major compounds in Northeast Indian sweet basil. In conclusion, sweet basil leaf volatiles consist mostly of terpenes (monoterpene and sesquiterpene) and phenylpropenes. Linalool and 1,8 cineole are among major monoterpenes, and germacrene-D and R-bergamotene are the predominate sesquiterpenes. The major phenylpropenes are eugenol and methylchavicol.

Various environmental factors are known to influence the essential oil content and composition in sweet basil. These include tissue type (Chalchat and Özcan, 2008), phenophase (Fischer et al., 2011), season (Vani et al., 2009), abiotic factors such as salinity (Hassanpouraghdam et al., 2011),

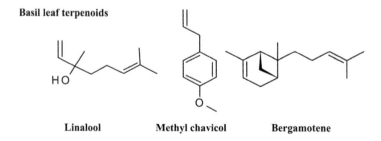

Basil leaf terpenoids

Linalool   Methyl chavicol   Bergamotene

FIGURE 5.5   Representative terpenoids biosynthesized by basil (*Ocimum basilicum* L.) leaves.

water stress (Khalid, 2006), solar irradiance (Chang et al., 2008), light quality (Carvalho et al., 2016) temperature (Chang et al., 2005), and drying and storage methods (Calín-Sánchez et al., 2012; Di Cesare et al., 2003) as well as extraction methods (Boggia et al., 2015; Carro et al., 2013). Fischer et al. (2011) analyzed leaves from different positions on the main stem and found higher levels of eugenol in younger leaves while in older leaves methyl-eugenol levels predominated, while linalool was lower in mature leaves than in younger leaves. They reported that the sweet basil aroma was significantly affected by the position of the leaf rather than by the leaf age or maturation process.

The basil plants synthesize most of the volatile phenylpropanoids and terpenes in the four-celled peltate glands found on both sides, and the volatiles are stored in the sac surrounding the gland (Gang et al., 2001; Werker et al., 1993). Genes involved in the biosynthesis of these volatiles have been reported and characterized, including eugenol synthase (Louie et al., 2007), linalool, geraniol, terpinolene, fenchol, β-Myrcene, cadinene, selinene, zingiberene, and germacrene-D synthases (Iijima et al., 2004), chavicol and eugenol methyltransferases (Anand et al., 2016; Gang et al., 2002; Koeduka et al., 2006; Lewinsohn et al., 2000). The developmental regulation of phenylpropene pathway was reported by (Gang et al., 2002; Lewinsohn et al., 2000), where a higher level of chavicol *O*-methyltransferase activity was found in younger leaves than in mature ones, and basal regions of both young and mature tissues contained greater transcript levels of these biosynthetic genes than in apical regions.

5.6 NIGELLA (*NIGELLA SATIVA*)

5.6.1 FRUIT/SEED

Nigella sativa L. (Ranunculaceae), commonly known as black cumin, is native to south and southwest Asia. It is widely cultivated in Mediterranean countries; middle Europe, and western Asia. Black cumin seeds and oils are used as a condiment and herbal medicine all over the world for the treatment and prevention of a number of diseases including asthma, diarrhea and dyslipidaemia (Burits and Bucar, 2000). Due to their pungent aroma, the seeds are added to food, especially to bread and bakery products, in Turkey and Arabic countries.

Various studies on seed volatiles have confirmed the presence of diversity in chemotypes with olefinic and oxygenated monoterpenes, mainly thymohydroquinone, *p*-cymene, thymoquinone, anethole, γ-terpinene and α-thujene (Kalidasu et al., 2017) (Figure 5.6).

Thirty components were identified from Iranian *N. sativa* seeds by microwave-assisted extraction and hydrodistillation (HD) methods. Monoterpene hydrocarbons were noticeably dominated by *p*-cymene (Abedi et al., 2017). In another study of Iranian fennel using hydrodistillation, a *trans*-anethole chemotype was reported (Nickavar et al., 2003). This study found thirty-two compounds, with high levels of *trans*-anethole (38.3%) and *p*-cymene (14.8%). Burits and Bucar (2000) reported that thymoquinone (27%–57%) and *p*-cymene (7%–15.7%) were the major compounds in essential oils of black cumin seeds from Austria. Black cumin essential oils from Algeria (Benkaci–Ali et al.,

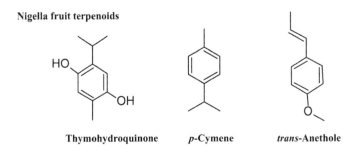

FIGURE 5.6 Representative terpenoids biosynthesized by *Nigella sativa* L. fruits.

2007), Bangladesh (Liu et al., 2013), and India (Kalidasu et al., 2017; Singh et al., 2005) were found to be a *p*-cymene/thymoquinone chemotype. A chemotype with 33.0% *p*-cymene and 26.8% thymol was reported from Morocco (Moretti et al., 2004). Polish black cumin had 60.2% *p*-cymene and 12.9% α-terpinene (Wajs et al., 2008). The main constituents of Tunisian black cumin were *p*-cymene (49.48%), α-thujene (18.93%), α-pinene (5.44%), and β-pinene (4.31%) (Harzallah et al., 2011). Recently Farag et al. (2017) used HS-SPME to extract oils from 12 Nigella species from different origins and found 34 volatile constituents that included thymoquinone, *p*-cymene and (*E*)-anethole. Thymoquinone content varied from 99.7% in *N. sativa* samples from Syria to only 1.8% in samples from India. This study also revealed that roasted seeds and fixed oil preparations were less enriched in thymoquinone and more dominated by its immediate precursor *p*-cymene as compared to fresh seeds. They suggested this may be attributed to differences in solubility, as *p*-cymene is less polar than thymoquinone which could account for its higher abundance in the nonpolar fixed oil matrix. Kıralan (2014) investigated the oxidation stability of volatile compounds in Nigella seed oil at 60°C and 100°C and found that some volatiles (e.g., 1,8-cineole, camphor, and thymol) were lost rapidly during thermal oxidation while the levels of thymoquinone, α-longipinene, 4-terpineol, carvacrol, and isolongifolene decreased more slowly and remained more stable during storage. Similarly, a progressive reduction in α-thujene, limonene, β-pinene, γ-terpinene and isolongifolene was observed with the enhancement of the heating time in the microwave extraction method (Kiralan and Kiralan, 2015). This suggests that the reported volatile composition of *N. sativa* seeds is influenced by the extraction method as well as the specific cultivars examined.

For better understanding the mechanisms of volatiles accumulation and localization in black cumin seed we studied the changes in volatile oil composition during seed development (Botnick et al., 2012). γ-Terpinene and α-thujene are the major monoterpenes accumulated in immature seeds but upon seed development the former is gradually replaced by *p*-cymene, carvacrol, thymohydroquinone and thymoquinone. The monoterpene γ-terpinene was detected in developing seeds at 30 days after anthesis (DAA) and its concentration dramatically increasing by 40 DAA to 170 µg/g FW. After 55 DAA, it began a marked decrease, reaching 30 µg/g FW at 60 DAA and continuing to decrease till at seed maturation its concentration was 15 µg/g FW. Interestingly, *p*-cymene accumulation began at 55 DAA, markedly behind γ-terpinene. The concentrations of thymohydroquinone and carvacrol steadily increased starting at 50 DAA and peaking upon seed maturation. Thymoquinone was detectable at 50 DAA, reaching maximal levels at 65 DAA and decreasing afterwards. Based on these observation, Lastly, Botnick et al. (2012) proposed that *p*-cymene serves as a precursor of thymoquinone in *N. sativa* seed.

5.7 PISTACIA (*PISTACIA PALAESTINA*)

5.7.1 LEAF

Pistacia palaestina Boiss. (Anacardiaceae), is a common deciduous wild tree common in Israel and the Eastern Mediterranean basin. It is a sibling species of *P. terebinthus*, the terebinth tree, which is widely distributed in Western Mediterranean basin (Zohary, 1972). The plant is deeply rooted in human history in the region and is traditionally used in folk medicine with modern implications (Golan-Goldhirsh, 2009). Its resin is used in the local folk medicine as a diuretic, stimulant, laxative, and aphrodisiac while the leaves decoction is employed as a diuretic, antihypertensive and antidiabetic remedy (Flamini et al., 2004). *Pistacia* has been used traditionally in medicine, for food preservation and as aromas, cosmetics, and varnishes due to their high oleoresin content (Ulukanli et al., 2014). The oleoresins of leaves and galls of terebinth may provide protection from herbivores and microorganisms (Rand et al., 2014).

Analysis of ten *P. palaestina* trees has revealed the presence of 20 terpenoids in leaves including 14 monoterpenes and 6 sesquiterpenes. The monoterpenes were the main leaf volatile compounds and their concentrations varied between 3 and 20 µg/g FW, while sesquiterpene levels ranged from

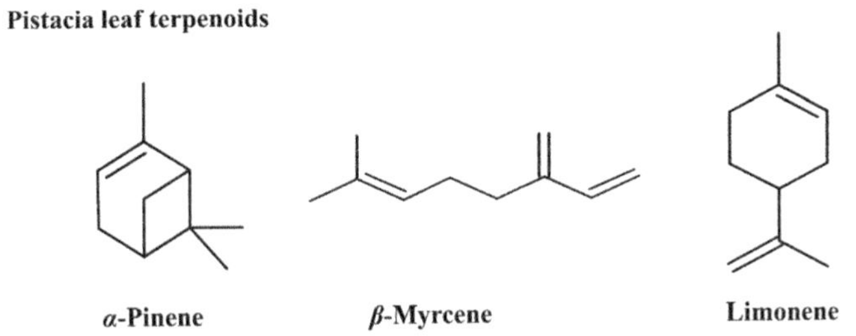

Pistacia leaf terpenoids

α-Pinene   β-Myrcene   Limonene

FIGURE 5.7   Representative terpenoids biosynthesized by *Pistacia palaestina* Boiss leaves.

1.6 to 28 μg/g FW (Rand et al., 2014). The results showed a substantial polymorphism in terpene compositions among different *P. palaestina* trees growing in natural maquis ecosystems in Israel. In an earlier study, a bulked sample originating from ten *P. palaestina* trees identified monoterpenes-rich oils with α-pinene, myrcene, and limonene as main monoterpenes while the principal sesquiterpene was α-caryophyllene (Flamini et al., 2004) (Figure 5.7). Rand et al. (2017) reported that young intact leaves accumulate higher levels of monoterpenes than later developmental stages and leaves showed a gradual decrease in monoterpenes during development, albeit they did not completely disappear.

5.7.2   INSECT INDUCED GALLS

Plants of the genus *Pistacia* are the obligate hosts for specialized gall-forming aphids (Homoptera: Fordini) (Inbar et al., 2004). The most notable aphid is *Baizongia pistaciae* L. that induces large (25 cm long or more), banana or horn like galls on the terminal buds of *P. palaestina* branches reviewed by (Wool, 2012). The incipient galls are developed on the leaflets, but eventually take over the entire apex of the branch, modifying the tree architecture. Each gall may support thousands of phloem-feeding aphids for nearly eight months during the dry season (spring-fall).

The galls induced by *B. pistaciae* and other related aphid species (Fordini) that develop on wild *Pistacia* trees contain higher and more varied levels of tannins, mono-, sesqui- and triterpenes (Caputo et al., 1979; Flamini et al., 2004; Rand et al., 2014; Rostás et al., 2013). We have recently shown that *B. pistaciae* galls accumulate significantly higher levels of monoterpenes than noncolonized leaves from the same tree, while leaves generally accumulate higher levels of sesquiterpenes as compared to galls (Flamini et al., 2004; Rand et al., 2014). Besides the substantial differences in terpene compositions and levels noted among trees, the terpene composition in galls generally differed from that of the supporting tree (Rand et al., 2014). These galls possess higher rates of monoterpene synthase activity and the products formed *in vitro* were not always identical in enzymatic extracts from galls as compared to leaves (Rand et al., 2017). Our studies indicate that galls possess independent and augmented potential biosynthetic capacities to produce monoterpenes and that the terpenes are not merely being mobilized to the galls from distant tissues (Rand et al., 2017).

5.8   CONCLUSION

Overall this chapter highlights the variation in the volatile profile of some aromatic and fruit crops. This variability is present due to wide distribution of these crops throughout the world and as a result of natural evolution as well as concentrated and selective breeding efforts put forth to identify new chemotypes. Identification of key volatiles from each crop that carries the specific function has increased the potential applications of these compounds in various areas. Over recent decades with the availability of advanced analytical techniques has enabled to the characterization of essential oils of various crops but the understanding about the physiological and biochemical basis of their production, till is in process. Henceforth our research is directed for characterization of genes and enzymes responsible for volatile production which will help to understand the biochemical pathways and control mechanisms for their synthesis. The natural diversity present in volatile compositions and understanding of the molecular basis these traits can be exploited for breeding special chemotypes with specific aromas and flavor for niche markets.

REFERENCES

Abedi, A.-S., M. Rismanchi, M. Shahdoostkhany, A. Mohammadi and A.M. Mortazavian. 2017. Microwave-assisted extraction of *Nigella sativa* L. essential oil and evaluation of its antioxidant activity. *Journal of Food Science and Technology* 54: 3779–3790.

Adam, F., I. Vahirua-Lechat, E. Deslandes, J.M. Bessiere and C. Menut. 2009. Aromatic plants of french polynesia. III. Constituents of the essential oil of leaves of *Ocimum basilicum* L. *Journal of Essential Oil Research* 21: 237–240.

Alasalvar, C., J. Grigor and P. Quantick. 1999. Method for the static headspace analysis of carrot volatiles. *Food Chemistry* 65: 391–397.

Alexopoulos, A., A. Kimbaris, S. Plessas, et al. 2011. Antibacterial activities of essential oils from eight Greek aromatic plants against clinical isolates of *Staphylococcus aureus. Anaerobe* 17: 399–402.

Anand, A., R.H. Jayaramaiah, S.D. Beedkar, et al. 2016. Comparative functional characterization of eugenol synthase from four different Ocimum species: Implications on eugenol accumulation. *Biochimica et Biophysica Acta (BBA)-Proteins and Proteomics* 1864: 1539–1547.

Baldwin, I.T., R. Halitschke, A. Paschold, C.C. Von Dahl and C.A. Preston. 2006. Volatile signaling in plant-plant interactions: "Talking trees" in the genomics era. *Science* 311: 812–815.

Barazani, O., Y. Cohen, A. Fait, et al. 2002. Chemotypic differentiation in indigenous populations of *Foeniculum vulgare* var. vulgare in Israel. *Biochemical Systematics and Ecology* 30: 721–731.

Bayala, B., I.H.N. Bassole, C. Gnoula, et al. 2014. Chemical composition, antioxidant, anti-inflammatory and anti-proliferative activities of essential oils of plants from Burkina Faso. *PLoS One* 9: e92122.

Bendjersi, F.Z., F. Tazerouti, R. Belkhelfa-Slimani, B. Djerdjouri and B.Y. Meklati. 2016. Phytochemical composition of the Algerian *Laurus nobilis* L. Leaves extracts obtained by solvent-free microwave extraction and investigation of their antioxidant activity. *Journal of Essential Oil Research* 28: 202–210.

Benkaci–Ali, F., A. Baaliouamer, B.Y. Meklati and F. Chemat. 2007. Chemical composition of seed essential oils from Algerian *Nigella sativa* extracted by microwave and hydrodistillation. *Flavour and Fragrance Journal* 22: 148–153.

Boggia, R., P. Zunin, V. Hysenaj, A. Bottino and A. Comite. 2015. Dehydration of basil leaves and impact of processing composition. In: *Processing and Impact on Active Components in Food*, Elsevier, London, pp. 645–653.

Botnick, I., W. Xue, E. Bar, et al. 2012. Distribution of primary and specialized metabolites in *Nigella sativa* seeds, a spice with vast traditional and historical uses. *Molecules* 17: 10159–10177.

Bunrathep, S., C. Palanuvej and N. Ruangrungsi. 2007. Chemical compositions and antioxidative activities of essential oils from four *Ocimum* species endemic to Thailand. *Journal of Health Research* 21: 201–206.

Burits, M. and F. Bucar. 2000. Antioxidant activity of *Nigella sativa* essential oil. *Phytotherapy Research* 14: 323–328.

Calín-Sánchez, Á., K. Lech, A. Szumny, A. Figiel and Á.A. Carbonell-Barrachina. 2012. Volatile composition of sweet basil essential oil (*Ocimum basilicum* L.) as affected by drying method. *Food Research International* 48: 217–225.

Caputo, R., L. Mangoni, P. Monaco and G. Palumbo. 1979. Triterpenes from the galls of *Pistacia palestina. Phytochemistry* 18: 896–898.

Caredda, A., B. Marongiu, S. Porcedda and C. Soro. 2002. Supercritical carbon dioxide extraction and characterization of *Laurus nobilis* essential oil. *Journal of Agricultural and Food Chemistry* 50: 1492–1496.

Carro, M.D., C. Ianni and E. Magi. 2013. Determination of terpenoids in plant leaves by GC-MS: Development of the method and application to *Ocimum basilicum* and *Nicotiana langsdorffii. Analytical Letters* 46: 630–639.

Carvalho, S.D., M.L. Schwieterman, C.E. Abrahan, T.A. Colquhoun and K.M. Folta. 2016. Light quality dependent changes in morphology, antioxidant capacity, and volatile production in sweet basil (*Ocimum basilicum*). *Frontiers In Plant Science* 7: 1328.

Castilho, P.C., M. do Céu Costa, A. Rodrigues and A. Partidário. 2005. Characterization of laurel fruit oil from Madeira Island, Portugal. *Journal of the American Oil Chemists' Society* 82: 863–868.

Chalchat, J.-C. and M.M. Özcan. 2008. Comparative essential oil composition of flowers, leaves and stems of basil (*Ocimum basilicum* L.) used as herb. *Food Chemistry* 110: 501–503.

Chang, X., P. Alderson and C. Wright. 2005. Effect of temperature integration on the growth and volatile oil content of basil (*Ocimum basilicum* L.). *The Journal of Horticultural Science and Biotechnology* 80: 593–598.

Chang, X., P.G. Alderson and C.J. Wright. 2008. Solar irradiance level alters the growth of basil (*Ocimum basilicum* L.) and its content of volatile oils. *Environmental and Experimental Botany* 63: 216–223.

Damjanović, B., Ž. Lepojević, V. Živković and A. Tolić. 2005. Extraction of fennel *(Foeniculum vulgare* Mill.) seeds with supercritical CO₂: Comparison with hydrodistillation. *Food Chemistry* 92: 143–149.

Di Cesare, L.F., E. Forni, D. Viscardi and R.C. Nani. 2003. Changes in the chemical composition of basil caused by different drying procedures. *Journal of Agricultural and Food Chemistry* 51: 3575–3581.

Diao, W.-R., Q.-P. Hu, H. Zhang and J.-G. Xu. 2014. Chemical composition, antibacterial activity and mechanism of action of essential oil from seeds of fennel (*Foeniculum vulgare* Mill.). *Food Control* 35: 109–116.

Díaz-Maroto, M.C., I.J. Díaz-Maroto Hidalgo, E. Sánchez-Palomo and M.S. Pérez-Coello. 2005. Volatile components and key odorants of fennel (*Foeniculum vulgare* mill.) and thyme (*Thymus vulgaris* L.) oil extracts obtained by simultaneous distillation–extraction and supercritical fluid extraction. *Journal of Agricultural and Food Chemistry* 53: 5385–5389.

Díaz-Maroto, M.C., M.S. Pérez-Coello, J. Esteban and J. Sanz. 2006. Comparison of the volatile composition of wild fennel samples (*Foeniculum vulgare* Mill.) from Central Spain. *Journal of Agricultural and Food Chemistry* 54: 6814–6818.

Dicke, M., R.M. van Poecke and J.G. de Boer. 2003. Inducible indirect defence of plants: From mechanisms to ecological functions. *Basic and Applied Ecology* 4: 27–42.

Dudareva, N., A. Klempien, J.K. Muhlemann and I. Kaplan. 2013. Biosynthesis, function and metabolic engineering of plant volatile organic compounds. *New Phytologist* 198: 16–32.

Dudareva, N., E. Pichersky and J. Gershenzon. 2004. Biochemistry of plant volatiles. *Plant Physiology* 135: 1893–1902.

Dudareva, N., F. Negre, D.A. Nagegowda and I. Orlova. 2006. Plant volatiles: Recent advances and future perspectives. *Critical Reviews in Plant Sciences* 25: 417–440.

Edris, A.E. and E.S. Farrag. 2003. Antifungal activity of peppermint and sweet basil essential oils and their major aroma constituents on some plant pathogenic fungi from the vapor phase. *Food/Nahrung* 47: 117–121.

El-Soud, N.H.A., M. Deabes, L.A. El-Kassem and M. Khalil. 2015. Chemical composition and antifungal activity of *Ocimum basilicum* L. essential oil. *Open Access Macedonian Journal of Medical Sciences* 3: 374.

Fang, L., M. Qi, T. Li, Q. Shao and R. Fu. 2006. Headspace solvent microextraction-gas chromatography–mass spectrometry for the analysis of volatile compounds from *Foeniculum vulgare* Mill. *Journal of Pharmaceutical and Biomedical Analysis* 41: 791–797.

Farag, M.A., D.M. El-Kersh, D.M. Rasheed and A.G. Heiss. 2017. Volatiles distribution in *Nigella* species (black cumin seeds) and in response to roasting as analyzed via solid-phase microextraction (SPME) coupled to chemometrics. *Industrial Crops and Products* 108: 564–571.

Fischer, R., N. Nitzan, D. Chaimovitsh, B. Rubin and N. Dudai. 2011. Variation in essential oil composition within individual leaves of sweet basil (*Ocimum basilicum* L.) is more affected by leaf position than by leaf age. *Journal of Agricultural and Food Chemistry* 59: 4913–4922.

Flamini, G., A. Bader, P.L. Cioni, A. Katbeh-Bader and I. Morelli. 2004. Composition of the essential oil of leaves, galls, and ripe and unripe fruits of Jordanian *Pistacia palaestina* Boiss. *Journal of Agricultural and Food Chemistry* 52: 572–576.

Franks, S.J., G.S. Wheeler and C. Goodnight. 2012. Genetic variation and evolution of secondary compounds in native and introduced populations of the invasive plant *Melaleuca quinquenervia. Evolution: International Journal of Organic Evolution* 66: 1398–1412.

Fukuda, T., K. Okazaki and T. Shinano. 2013. Aroma characteristic and volatile profiling of carrot varieties and quantitative role of terpenoid compounds for carrot sensory attributes. *Journal of Food Science* 78: S1800–S1806.

Gang, D.R., J. Wang, N. Dudareva, et al. 2001. An investigation of the storage and biosynthesis of phenylpropenes in sweet basil. *Plant Physiology* 125: 539–555.

Gang, D.R., N. Lavid, C. Zubieta, et al. 2002. Characterization of phenylpropene *O*-methyltransferases from sweet basil: Facile change of substrate specificity and convergent evolution within a plant *O*-methyltransferase family. *The Plant Cell* 14: 505–519.

García-Jiménez, N., M.J. Péerez-Alonso and A. Velasco-Negueruela. 2000. Chemical composition of fennel oil, *Foeniculum vulgare* Miller, from Spain. *Journal of Essential Oil Research* 12: 159–162.

Goff, S.A. and H.J. Klee. 2006. Plant volatile compounds: Sensory cues for health and nutritional value? *Science* 311: 815–819.

Golan-Goldhirsh, A. 2009. Bridging the gap between ethnobotany and biotechnology of Pistacia. *Israel Journal of Plant Sciences* 57: 65–78.

Gross, M., D.M. Joel, Y. Cohen, et al. 2006. Ontogenesis of mericarps of bitter fennel (*Foeniculum vulgare* Mill. var. *vulgare*) as related to *t*-anethole accumulation. *Israel Journal of Plant Sciences* 54: 309–316.

Gross, M., E. Lewinsohn, Y. Tadmor, et al. 2009. The inheritance of volatile phenylpropenes in bitter fennel (*Foeniculum vulgare* Mill. var. *vulgare*, Apiaceae) chemotypes and their distribution within the plant. *Biochemical Systematics and Ecology* 37: 308–316.

Gross, M., J. Friedman, N. Dudai, et al. 2002. Biosynthesis of estragole and *t*-anethole in bitter fennel (*Foeniculum vulgare* Mill. var. *vulgare*) chemotypes. Changes in SAM: Phenylpropene *O*-methyltransferase activities during development. *Plant Science* 163: 1047–1053.

Güler, Z., F. Karaca and H. Yetisir. 2015. Identification of volatile organic compounds (VOCs) in different colour carrot (*Daucus carota* L.) cultivars using static headspace-gas chromatography-mass spectrometry. *Cogent Food and Agriculture* 1: 1117275.

Hadj Khelifa, L., M. Brada, F. Brahmi, et al. 2012. Chemical composition and antioxidant activity of essential oil of *Ocimum basilicum* leaves from the northern region of Algeria. *Topclass Journal of Herbal Medicine* 1: 53–58.

Hafizoğlu, H. and M. Reunanen. 1993. Studies on the components of *Lauras nobilis* from Turkey with special reference to Laurel berry fat. *Lipid/Fett* 95: 304–308.

Hanif, M.A., M.Y. Al-Maskari, A. Al-Maskari, et al. 2011. Essential oil composition, antimicrobial and antioxidant activities of unexplored Omani basil. *Journal of Medicinal Plants Research* 5: 751–757.

Harzallah, H.J., B. Kouidhi, G. Flamini, A. Bakhrouf and T. Mahjoub. 2011. Chemical composition, antimicrobial potential against cariogenic bacteria and cytotoxic activity of Tunisian *Nigella sativa* essential oil and thymoquinone. *Food Chemistry* 129: 1469–1474.

Hassanpouraghdam, M., G. Gohari, S. Tabatabaei, M. Dadpour and M. Shirdel. 2011. NaCl salinity and Zn foliar application influence essential oil composition of basil (*Ocimum basilicum* L.). *Acta Agriculturae Slovenica* 97: 93.

Heil, M. 2008. Indirect defence via tritrophic interactions. *New Phytologist* 178: 41–61.

Iijima, Y., R. Davidovich-Rikanati, E. Fridman, et al. 2004. The biochemical and molecular basis for the divergent patterns in the biosynthesis of terpenes and phenylpropenes in the peltate glands of three cultivars of basil. *Plant Physiology* 136: 3724–3736.

Inbar, M., M. Wink and D. Wool. 2004. The evolution of host plant manipulation by insects: Molecular and ecological evidence from gall-forming aphids on Pistacia. *Molecular Phylogenetics and Evolution* 32: 504–511.

Iorizzo, M., S. Ellison, D. Senalik, et al. 2016. A high-quality carrot genome assembly provides new insights into carotenoid accumulation and asterid genome evolution. *Nature Genetics* 48: 657.

Jirovetz, L., G. Buchbauer, M.P. Shafi and M.M. Kaniampady. 2003. Chemotaxonomical analysis of the essential oil aroma compounds of four different Ocimum species from southern India. *European Food Research and Technology* 217: 120–124.

Kainulainen, P., J. Tarhanen, K. Tiilikkala and J. Holopainen. 1998. Foliar and emission composition of essential oil in two carrot varieties. *Journal of Agricultural and Food Chemistry* 46: 3780–3784.

Kalidasu, G., G.S. Reddy, S.S. Kumari, A.L. Kumari and A. Sivasankar. 2017. Secondary volatiles and metabolites from *Nigella sativa* L. seed. *Indian Journal of Natural Products and Resources* 8: 151–158.

Keilwagen, J., H. Lehnert, T. Berner, et al. 2017. The terpene synthase gene family of carrot (*Daucus carota* L.): Identification of QTLs and candidate genes associated with terpenoid volatile compounds. *Frontiers in Plant Science* 8: 1930.

Khalid, K.A. 2006. Influence of water stress on growth, essential oil, and chemical composition of herbs (*Ocimum* sp.). *International Agrophysics* 20: 289–296.

Kilic, A., H. Hafizoglu, H. Kollmannsberger and S. Nitz. 2004. Volatile constituents and key odorants in leaves, buds, flowers, and fruits of *Laurus nobilis* L. *Journal of Agricultural and Food Chemistry* 52: 1601–1606.

Kilic, A., H. Kollmannsberger and S. Nitz. 2005. Glycosidically bound volatiles and flavor precursors in *Laurus nobilis* L. *Journal of Agricultural and Food Chemistry* 53: 2231–2235.

Kıralan, M. 2014. Changes in volatile compounds of black cumin (*Nigella sativa* L.) seed oil during thermal oxidation. *International Journal of Food Properties* 17: 1482–1489.

Kiralan, M. and S.S. Kiralan. 2015. Changes in volatile compounds of black cumin oil and hazelnut oil by microwave heating process. *Journal of the American Oil Chemists Society* 92: 1445–1450.

Kjeldsen, F., L.P. Christensen and M. Edelenbos. 2001. Quantitative analysis of aroma compounds in carrot (*Daucus carota* L.) cultivars by capillary gas chromatography using large-volume injection technique. *Journal of Agricultural and Food Chemistry* 49: 4342–4348.

Kjeldsen, F., L.P. Christensen and M. Edelenbos. 2003. Changes in volatile compounds of carrots (*Daucus carota* L.) during refrigerated and frozen storage. *Journal of Agricultural and Food Chemistry* 51: 5400–5407.

Koeduka, T., E. Fridman, D.R. Gang, et al. 2006. Eugenol and isoeugenol, characteristic aromatic constituents of spices, are biosynthesized via reduction of a coniferyl alcohol ester. *Proceedings of the National Academy of Sciences* 103: 10128–10133.

Koutouan, C., V. Le Clerc, R. Baltenweck, et al. 2018. Link between carrot leaf secondary metabolites and resistance to *Alternaria dauci*. *Scientific Reports* 8: 13746.

Kreutzmann, S., A.K. Thybo, M. Edelenbos and L.P. Christensen. 2008. The role of volatile compounds on aroma and flavour perception in coloured raw carrot genotypes. *International Journal of Food Science and Technology* 43: 1619–1627.

Lewinsohn, E., I. Ziv-Raz, N. Dudai, et al. 2000. Biosynthesis of estragole and methyl-eugenol in sweet basil (*Ocimum basilicum* L.). Developmental and chemotypic association of allylphenol *O*-methyltransferase activities. *Plant Science* 160: 27–35.

Liu, X., J.H. Park, A. Abd El-Aty, et al. 2013. Isolation of volatiles from *Nigella sativa* seeds using microwave-assisted extraction: Effect of whole extracts on canine and murine CYP1A. *Biomedical Chromatography* 27: 938–945.

Loreto, F. and J.-P. Schnitzler. 2010. Abiotic stresses and induced BVOCs. *Trends in Plant Science* 15: 154–166.

Louie, G.V., T.J. Baiga, M.E. Bowman, et al. 2007. Structure and reaction mechanism of basil eugenol synthase. *PLoS One* 2: e993.

Marzouki, H., A. Piras, B. Marongiu, A. Rosa and M.A. Dessi. 2008. Extraction and separation of volatile and fixed oils from berries of *Laurus nobilis* L. by supercritical CO_2. *Molecules* 13: 1702–1711.

Marzouki, H., A. Piras, K.B.H. Salah, et al. 2009. Essential oil composition and variability of *Laurus nobilis* L. growing in Tunisia, comparison and chemometric investigation of different plant organs. *Natural Product Research* 23: 343–354.

Mercier, B., J. Prost and M. Prost. 2009. The essential oil of turpentine and its major volatile fraction (α-and β-pinenes): A review. *International Journal of Occupational Medicine and Environmental Health* 22: 331–342.

Mohamed, M., M. Ibrahim, H. Wahba and K. Khalid. 2016. Research article yield and essential oil of sweet basil affected by chemical and biological fertilizers. *Research Journal of Medicinal Plant* 10: 246–253.

Moretti, A., L.F. D'Antuono and S. Elementi. 2004. Essential oils of *Nigella sativa* L. and *Nigella damascena* L. seed. *Journal of Essential Oil Research* 16: 182–183.

Napoli, E.M., G. Curcuruto and G. Ruberto. 2010. Screening the essential oil composition of wild Sicilian fennel. *Biochemical Systematics and Ecology* 38: 213–223.

Nehlin, G., I. Valterová and A.-K. Borg-Karlson. 1994. Use of conifer volatiles to reduce injury caused by carrot psyllid, *Trioza apicalis*, Förster (Homoptera, Psylloidea). *Journal of Chemical Ecology* 20: 771–783.

Nickavar, B., F. Mojab, K. Javidnia and M.A.R. Amoli. 2003. Chemical composition of the fixed and volatile oils of *Nigella sativa* L. from Iran. *Zeitschrift für Naturforschung C* 58: 629–631.

Nissinen, A., M. Ibrahim, P. Kainulainen, K. Tiilikkala and J.K. Holopainen. 2005. Influence of carrot psyllid (*Trioza apicalis*) feeding or exogenous limonene or methyl jasmonate treatment on composition of carrot (*Daucus carota*) leaf essential oil and headspace volatiles. *Journal of Agricultural and Food Chemistry* 53: 8631–8638.

Nurzynska-Wierdak, R., B. Borowski, K. Dzida, G. Zawislak and R. Kowalski. 2013. Essential oil composition of sweet basil cultivars as affected by nitrogen and potassium fertilization. *Turkish Journal of Agriculture and Forestry* 37: 427–436.

Pascual-Villalobos, M. and M. Ballesta-Acosta. 2003. Chemical variation in an *Ocimum basilicum* germplasm collection and activity of the essential oils on *Callosobruchus maculatus*. *Biochemical Systematics and Ecology* 31: 673–679.

Pavela, R., M. Žabka, J. Bednář, J. Tříska and N. Vrchotová. 2016. New knowledge for yield, composition and insecticidal activity of essential oils obtained from the aerial parts or seeds of fennel (*Foeniculum vulgare* Mill.). *Industrial Crops and Products* 83: 275–282.

Piccaglia, R. and M. Marotti. 2001. Characterization of some Italian types of wild fennel (*Foeniculum vulgare* Mill.). *Journal of Agricultural and Food Chemistry* 49: 239–244.

Purkayastha, J. and S.C. Nath. 2006. Composition of the camphor-rich essential oil of *Ocimum basilicum* L. native to Northeast India. *Journal of Essential Oil Research* 18: 332–334.

Rahimmalek, M., H. Maghsoudi, M. Sabzalian and A. Ghasemi Pirbalouti. 2014. Variability of essential oil content and composition of different Iranian fennel (*Foeniculum vulgare* Mill.) accessions in relation to some morphological and climatic factors. *Journal of Agricultural Science and Technology* 16: 1365–3174.

Rand, K., E. Bar, M. Ben-Ari, E. Lewinsohn and M. Inbar. 2014. The mono-and sesquiterpene content of aphid-induced galls on *Pistacia palaestina* is not a simple reflection of their composition in intact leaves. *Journal of Chemical Ecology* 40: 632–642.

Rand, K., E. Bar, M.B. Ari, et al. 2017. Differences in monoterpene biosynthesis and accumulation in *Pistacia palaestina* Leaves and aphid-induced galls. *Journal of Chemical Ecology* 43: 143–152.

Rostás, M., D. Maag, M. Ikegami and M. Inbar. 2013. Gall volatiles defend aphids against a browsing mammal. *BMC Evolutionary Biology* 13: 193.

Salami, M., M. Rahimmalek, M.H. Ehtemam, et al. 2016. Essential oil composition, antimicrobial activity and anatomical characteristics of *Foeniculum vulgare* Mill. fruits from different regions of Iran. *Journal of Essential Oil Bearing Plants* 19: 1614–1626.

Schwab, W., R. Davidovich-Rikanati and E. Lewinsohn. 2008. Biosynthesis of plant-derived flavor compounds. *The Plant Journal* 54: 712–732.

Senatore, F., F. Oliviero, E. Scandolera, et al. 2013. Chemical composition, antimicrobial and antioxidant activities of anethole-rich oil from leaves of selected varieties of fennel [*Foeniculum vulgare* Mill. ssp. *vulgare* var. *azoricum* (Mill.) Thell]. *Fitoterapia* 90: 214–219.

Shahat, A.A., F.M. Hammouda, K.A. Shams and M.A. Saleh. 2012. Comparative chemical analysis of the essential oil of wild and cultivated fennel (*Foeniculum vulgare* Mill). *Journal of Essential Oil Bearing Plants* 15: 314–319.

Sharma, A., J. Singh and S. Kumar. 2012. Bay leaves. In: Handbook of Herbs and Spices (2nd edition), Peter K.V., ed., Volume 1, Woodhead Publishing, Cambridge, pp. 73–85.

Shatar, S., S. Altantsetseg, I. Sarnai, et al. 2007. Chemical composition of the essential oil of *Ocimum basilicum* cultivated in Mongolian Desert-Gobi. *Chemistry of Natural Compounds* 43: 726–727.

Shirazi, M.T., H. Gholami, G. Kavoosi, V. Rowshan and A. Tafsiry. 2014. Chemical composition, antioxidant, antimicrobial and cytotoxic activities of *Tagetes minuta* and *Ocimum basilicum* essential oils. *Food Science and Nutrition* 2: 146–155.

Shokoohinia, Y., A. Yegdaneh, G. Amin and A. Ghannadi. 2014. Seasonal variations of *Laurus nobilis* L. leaves volatile oil components in Isfahan, Iran. *Research Journal of Pharmacognosy* 1: 1–6.

Singh, G., P. Marimuthu, H. Murali and A. Bawa. 2005. Antioxidative and antibacterial potentials of essential oils and extracts isolated from various spice materials. *Journal of Food Safety* 25: 130–145.

Southwell, I.A., M.F. Russell, C.D. Maddox and G.S. Wheeler. 2003. Differential metabolism of 1, 8–cineole in insects. *Journal of Chemical Ecology* 29: 83–94.

Tan, K.H. and R. Nishida. 2012. Methyl eugenol: Its occurrence, distribution, and role in nature, especially in relation to insect behavior and pollination. *Journal of Insect Science* 12: 56.

Telci, I., I. Demirtas and A. Sahin. 2009. Variation in plant properties and essential oil composition of sweet fennel (*Foeniculum vulgare* Mill.) fruits during stages of maturity. *Industrial Crops and Products* 30: 126–130.

Tognolini, M., V. Ballabeni, S. Bertoni, et al. 2007. Protective effect of *Foeniculum vulgare* essential oil and anethole in an experimental model of thrombosis. *Pharmacological Research* 56: 254–260.

Ulrich, D., T. Nothnagel and H. Schulz. 2015. Influence of cultivar and harvest year on the volatile profiles of leaves and roots of carrots (*Daucus carota* spp. *sativus* Hoffm.). *Journal of Agricultural and Food Chemistry* 63: 3348–3356.

Ulukanli, Z., S. Karabörklü, B. Öztürk, M. Çenet and M. Balcilar. 2014. Chemical composition, antibacterial and insecticidal activities of the essential oil from the *Pistacia terebinthus* L. Spp. Palaestina (B oiss.) (A nacardiaceae). *Journal of Food Processing and Preservation* 38: 815–822.

Unsicker, S.B., G. Kunert and J. Gershenzon. 2009. Protective perfumes: The role of vegetative volatiles in plant defense against herbivores. *Current Opinion in Plant Biology* 12: 479–485.

Vani, S.R., S. Cheng and C. Chuah. 2009. Comparative study of volatile compounds from genus *Ocimum*. *American Journal of Applied Sciences* 6: 523.

Vaughn, S.F. 2001. Plant volatiles. In: *Encyclopedia of life sciences, Wiley Interscience*. John Wiley & Sons, New York, Volume 1, pp. 1–6.

Viña, A. and E. Murillo. 2003. Essential oil composition from twelve varieties of basil (*Ocimum* spp.) grown in Colombia. *Journal of the Brazilian Chemical Society* 14: 744–749.

Wajs, A., R. Bonikowski and D. Kalemba. 2008. Composition of essential oil from seeds of *Nigella sativa* L. cultivated in Poland. *Flavour and Fragrance Journal* 23: 126–132.

Werker, E., E. Putievsky, U. Ravid, N. Dudai and I. Katzir. 1993. Glandular hairs and essential oil in developing leaves of *Ocimum basilicum* L. (Lamiaceae). *Annals of Botany* 71: 43–50.

Wool, D. 2012. Autecology of *Baizongia pistaciae* (L.): A monographical study of a galling aphid. *Israel Journal of Entomology* 41: 67–93.

Xiao, Z., J. Chen, Y. Niu and F. Chen. 2017. Characterization of the key odorants of fennel essential oils of different regions using GC–MS and GC–O combined with partial least squares regression. *Journal of Chromatography B* 1063: 226–234.

Yahyaa, M., A. Berim, T. Isaacson, et al. 2015a. Isolation and functional characterization of carotenoid cleavage dioxygenase-1 from *Laurus nobilis* L. (bay laurel) fruits. *Journal of Agricultural and Food Chemistry* 63: 8275–8282.

Yahyaa, M., Y. Matsuba, W. Brandt, et al. 2015b. Identification, functional characterization, and evolution of terpene synthases from a basal dicot. *Plant Physiology* 169: 1683–1697.

Yahyaa, M., D. Tholl, G. Cormier, et al. 2015c. Identification and characterization of terpene synthases potentially involved in the formation of volatile terpenes in carrot (*Daucus carota* L.) roots. *Journal of Agricultural and Food Chemistry* 63: 4870–4878.

Yahyaa, M., M. Ibdah, S. Marzouk and M. Ibdah. 2016. Profiling of the terpene metabolome in carrot fruits of wild (*Daucus carota* L. ssp. *carota*) accessions and characterization of a geraniol synthase. *Journal of Agricultural and Food Chemistry* 66: 2378–2386.

Zhao, N.N., L. Zhou, Z.L. Liu, S.S. Du and Z.W. Deng. 2012. Evaluation of the toxicity of the essential oils of some common Chinese spices against *Liposcelis bostrychophila*. *Food Control* 26: 486–490.

Zohary, M. 1972. Flora Palaestina: Part 2. Platanaceae to Umbelliferae. *Jerusalem, Israel Academy of Sciences and Humanities* 656.

Section II

Biochemistry, Molecular Biology, and Evolution of Plant Volatiles

6 The Role of Transcriptome Analysis in Shaping the Discovery of Plant Volatile Genes
Past, Present, and Future

Darren C. J. Wong, Rod Peakall, and Eran Pichersky

CONTENTS

6.1   INTRODUCTION

6.1.1   THE DIVERSITY OF PLANT VOLATILES

Since antiquity, humans have been sourcing specific plants and organs (e.g., flowers, fruits, leaves, and bark) for use in numerous applications such as flavorings, fragrances, preservatives, and herbal remedies, among others (Pichersky et al. 2006). Among the specialized compounds responsible for the desired properties of such plant tissues are low-molecular-weight lipophilic compounds with high vapor pressures, called plant volatiles (hereafter PVs). Beyond their utility to humans, PVs play vital roles at the plant, species and ecosystem levels. For example, PVs serve as signals for the attraction of beneficial animal pollinators, mutualistic microbes, seed dispersers, and predators/parasitoids. They also confer protection to the plant against abiotic and biotic stress (see elsewhere in this book).

The number of PV compounds already identified across the plant kingdom is estimated to be in the thousands (Chapter 4). With methods for qualitative and quantitative analysis of PVs constantly improving, and with an ever increasing number of plant species being analyzed in detail (Pichersky et al. 2006), the number of identified PVs will continue to grow rapidly. Yet, despite the sheer diversity of PVs, the biosynthetic origin of most compounds can be categorized into one of four major biosynthetic classes (Figure 6.1): (1) terpenoids, (2) phenylpropanoids/benzenoids, (3) fatty acid derivatives, and (4) amino-acid derivatives (Dudareva et al. 2004, 2013).

6.1.2   THE SPATIAL AND TEMPORAL MODULATION OF VOLATILE BIOSYNTHESIS AND EMISSIONS

It is now widely established that synthesis, storage, and emission of PVs are both spatially and temporally regulated (Dudareva et al. 2004; Pichersky et al. 2006). For example, compared to other organs such as leaves or roots, flowers tend to produce the most diverse and most abundant volatiles. Furthermore, within the flower, tissue-specific emission is a common feature, with petals often the primary source of volatiles, although other tissues such as sepals, stamens, pistils, or nectaries may also be involved (Muhlemann et al. 2014). In some plant species such as those belonging to the Cannabaceae, Lamiaceae, and Solanaceae families, specialized glandular trichomes on leaves and flowers are well-known cell factories that synthesize, release, and store large quantities of diverse volatiles (Chapter 13). Internal structures such as the resin ducts of coniferous plants are also the site of PV synthesis (Chapter 13).

The biosynthesis and/or emission of PVs is also strongly influenced by the developmental stage of the plant, flower or fruit. For example, diverse blends of PVs are released from the ripe fruits of many plants to attract seed dispersers (Chapter 10). A diurnal rhythm of volatile biosynthesis and emissions have also been observed in the flowers and leaves of some plant species (Dudareva et al. 2013).

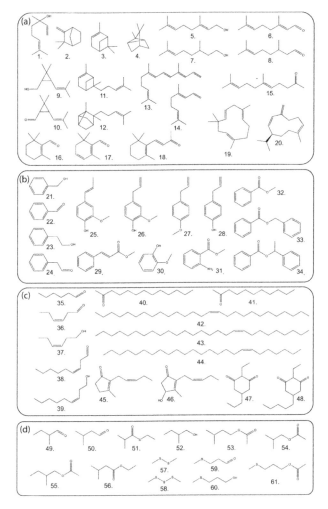

FIGURE 6.1 The diversity and examples of volatiles released from plants. Volatile (a) terpenoids, (b) benze-noids and phenylpropanoids, (c) fatty acid derivatives, and (d) branched-chain and sulfurous amino-acid deriva-tives. 1, Linalool; 2, Camphene; 3, α-pinene; 4, Tricyclene; 5, Geraniol; 6, Geranial; 7, Citronellol; 8, Citronellal; 9, trans-Chrysanthemol; 10, trans-Chrysanthemal; 11, endo-α-Bergamotene; 12, α-Santalene; 13, (E, E)-4,8,12-trimethyl-1,3,7,11-tridecatetraene; 14, (E)-4,8-dimethyl-1,3,7-nonatriene; 15, Geranylacetone; 16, β-ionone; 17, β-cyclocitral; 18, Safranal; 19, α-humulene; 20, Germacrene-D; 21, Benzylalcohol; 22, Benzaldehyde; 23, Phenylethanol; 24, Phenylacetaldehyde; 25, Isoeugenol; 26, Eugenol; 27, Methylchavicol; 28, Chavicol; 29, Methylcinnamate; 30, Guaiacol; 31, Methylanthranilate; 32, Methylbenzoate; 33, Benzylbenzoate; 34, Phenylethylbenzoate; 35, Hexanal; 36, (Z)-3-hexenal; 37, (Z)-3-hexenol; 38, (Z)-3-nonenal; 39, (Z)-3-nonenol; 40, 2-undecanone; 41, 2-tridecanone; 42, (Z)-12-heptacosene; 43, (Z)-9-heptacosene; 44, (Z)-7-pentacosene; 45, cis-Jasmone; 46, Jasmolone; 47, 2-ethyl-5-propylcyclohexan-1,3-dione; 48, 2-ethyl-5-pentylcyclohexan-1,3-dione; 49, 2-methylbutanal; 50, 3-methylbutanal; 51, ethyl 2-methylpropanoate; 52, 2-methylbutanol; 53, 3-methylbutyl acetate; 54, 2-methylpropyl acetate; 55, 2-methylbutyl acetate; 56, ethyl 3-methylbutanoate; 57, Dimethyldisulfide; 58, Dimethyltrisulfide; 59, 3-(methylthio)propanal; 60, 3-(methylthio)propanol; 61, 3-(methylthio)propyl acetate.

This rhythm is likely regulated by the circadian clock and light signals. The biosynthesis and emis-sion of volatiles can also be influenced by a complex interplay between environmental/abiotic (e.g., light irradiance, temperature, drought, and nutrient status) and biotic (e.g., pathogen infection and herbivory) factors (Loreto et al. 2014). Plant hormones (e.g., ethylene, jasmonic acid, abscisic acid) are also key components of PV signaling and regulation in plants (Broekgaarden et al. 2015).

6.1.3 The Landscape of Plant Volatile Gene Discovery: From Functional Genomics to Systems Biology

Given the critical importance of PVs, an understanding of the molecular mechanisms underpinning their biosynthesis and regulation has been a prime research interest over the last several decades. Early work focused on the identification of enzymatic activities involved in volatile formation (Schilmiller et al. 2012; Fridman and Pichersky 2005). Later, with the application and continued improvements of DNA sequencing technologies from the late 1990s, work shifted to the identification of the genes underpinning the enzymes involved (Pichersky et al. 2006). Now, nearly two decades later, vast sequence data (e.g., whole-genome and transcriptomes) are available for a wide range of plants, including non-model species known for their rich diversity of unique PVs. Efforts such as the 1,000 Plant Genomes Project (1 KP) alone have achieved the sequencing and analysis of thousands of plant transcriptomes (Matasci et al. 2014). And now, more ambitious plans are afoot (Cheng et al. 2018).

In parallel, remarkable advances in metabolomics have also been achieved, particularly in various separation techniques linked to mass spectrometry (MS). Consequently, identifying and quantifying not only target PVs and their precursors/derivatives, but also entire metabolomes, is now possible (Schauer and Fernie 2006). The ongoing revolution in genomics, transcriptomics, proteomics, and metabolomics (Figure 6.2) now allows large-scale gene-function prediction

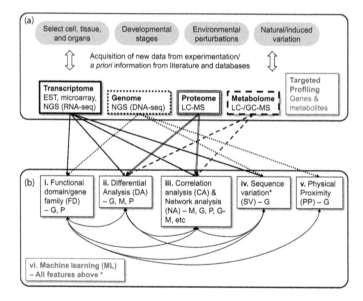

FIGURE 6.2 A typical workflow of plant volatile gene discovery using functional genomics and systems biology approaches. (a) Gene discovery often begins with the resource building phase which encompasses the acquisition of new data from experimentation (e.g., select cell, tissue, and organs) using targeted profiling and multi-omics approaches such as transcriptome (bold), genome (narrow dash), proteome (double lines), and metabolome (broad lines). When available, *a priori* information from the literature and databases can also be used. (b) The data/information from (a) especially those from multi-omics platforms enable a variety of analysis for hypothesis generation and candidate prioritization for downstream functional analysis (see edges connecting each omics dataset and type of analysis). The possible analysis often includes the annotation of functional domains/gene family of transcripts (FD) and performing a variety of statistical analyses such as differential analysis (DA), correlation analysis (CA), advanced network analysis (NA) from multi-omics datasets, sequence variation analysis (SV), and physical proximity analysis (PP), among others. Recent state-of-the-art methods namely machine learning (ML) is capable of seamlessly integrating multiple analytical features (see above) for robust gene predictions, however, their efficacy have yet to be thoroughly evaluated for plant volatile-related genes. Often multiple analysis is performed in parallel to maximize the success of hypothesis generation and candidate prioritization for downstream functional analysis (see edges connecting each analysis type). G, gene; P, protein; M, metabolite.

through systems biology or systems-based approaches (Rhee and Mutwil 2014). Below, we outline the main approaches used in hypothesis generation and candidate gene prioritization, and highlight some recent successes in the identification of new plant enzymes involved in PV biosynthesis.

6.2 SPEARHEADING PLANT VOLATILE GENE DISCOVERIES WITH RNA-SEQ TRANSCRIPTOMES

Two early approaches, expressed sequence tag (EST) and microarray analyses, enabled the identification of many genes encoding the enzymes and transcriptional regulators of secondary metabolites (SM) (Ohlrogge and Benning 2000; Fridman and Pickersky 2005). This was in part due to their ability to provide quantitative gene expression estimates in many metabolically active tissues (Schilmiller et al. 2012; Fridman and Pickersky 2005). RNA sequencing (RNA-seq) has now become the method of choice over EST and microarrays for several reasons. It is applicable to any organism, is very cost-effective (Egan et al. 2012), and offers unprecedented accuracy and sensitivity for detection of lowly-expressed genes and unknown alternative spliced isoforms. When used for differential expression analysis, RNA-seq brings higher sensitivity, a broader dynamic range, and is more robust to technical variation (Oshlack et al. 2010). Furthermore, a reference genome is not a pre-requisite, enabling transcriptome investigations to be performed in nonmodel plants via *de novo* transcriptome assembly. It also provides an alternative means to reduced-representation approaches for single nucleotide polymorphism (SNP) genotyping (Scheben et al. 2017).

6.2.1 ANNOTATION OF FUNCTIONAL DOMAINS/GENE FAMILIES

With RNA-seq transcriptomes, assignment of putative gene functions to the thousands of transcripts is a crucial first step (Figure 6.2b(i)). Tools like Basic Local Alignment Search Tool (BLAST) provide rapid and reliable annotation of whole transcriptomes against any reference species or database of interest. This homology-based functional annotation approach transfers existing knowledge about a gene to another gene on the basis of sequence similarity (Bolger et al. 2018). Alternatively, screening for functional domains of interests (deduced from *a priori* multiple sequence alignments of known functional enzymes) provides a sensitive approach to sequence annotation. This latter approach may even allow sequence diversity at individual residues to be identified. Collectively, by taking advantage of putative homology, or on the basis of biochemical principles, candidate PV genes can be shortlisted for further functional work.

6.2.2 TRANSCRIPTOME DIFFERENTIAL EXPRESSION ANALYSES

Differential expression (DE) analysis is perhaps the most important statistical approach in many transcriptome analysis studies (Figure 6.2b(ii)). In the simplest case, regardless of the type of platform used (microarray or RNA-seq), such analysis reveals a list of genes with associated fold changes and statistical significance (Oshlack et al. 2010). In the context of PV gene discovery, it is of particular interest to determine the genes that show significant expression differences (e.g., higher/lower, presence/absence) between active and nonactive tissues and treatments. Such findings alone can often suffice for candidate prioritization, albeit still requiring additional downstream analysis and functional characterization (Wong et al. 2017).

Some of the possible contrasts of active and nonactive tissues include comparisons of different tissue types (Xu et al. 2017; Wong et al. 2018, 2019), development stages (Widhalm et al. 2015; Adebesin et al. 2017), and populations resulting from different crosses (Galpaz et al. 2018). For example, the complete biosynthesis of β-citronellol from geraniol was recently elucidated in an Australian native orchid, *Caladenia plicata* (Xu et al. 2017), a species lacking any prior molecular and biochemical study. In this case, DE analysis between the transcriptomes of the active sepal tips and nonactive floral tissue led to the prioritization and later confirmation of a geraniol synthase gene and a double-bond reductase in the biosynthesis of geraniol and citronellal, respectively.

6.2.3 Gene Co-expression Analyses and Omics Data Integration

Gene co-expression analysis (GCA) (Figure 6.2b(iii)) is based on the "guilt-by-association" principle whereby genes sharing common functions (e.g., belonging to the same pathway) are often co-ordinately regulated across a wide range or conditions. Many key discoveries in plant metabolism, development, and gene regulation, have been made possible with GCA in both *Arabidopsis* and nonmodel plant species (Rhee and Mutwil 2014; Usadel et al. 2009). Using the databases that curate and analyze publicly available transcriptomic datasets in the hundreds to thousands across multiple species, GCA can now be performed for a wide range of plant species. Gene-function inference via GCA can be made with strong statistical confidence regardless of the type of similarity metrics used. However, when using these simple correlation approaches to match gene and metabolite profiles, sufficient variation in gene and the target metabolite profiles is needed. Furthermore, to allow its full potential to be realized, gene-to-metabolite correlations with large-scale transcriptome and metabolome data are desirable.

6.2.4 New Opportunities for High-Resolution Sequence Variation Analysis with RNA-Seq

RNA-seq SNP genotyping, which targets protein-coding regions with functional variants, provides a promising alternative for SNP genotyping and QTL map construction compared to reduced-representation approaches, such as genotyping-by-sequencing, which are unable to avoid highly repetitive noncoding regions (Figure 6.2b(iv)). When coupled with gene expression, this approach can shed new insights into the functional context of SNP variation (Scheben et al. 2017). However, presently it seems the full potential of RNA-seq SNP genotyping for PV gene identification is yet to be realized (but see discussion of melon PV flavor genes in Section 6.3.6).

6.3 LESSONS FROM SUCCESS STORIES IN THE DISCOVERY OF PLANT VOLATILE-RELATED GENES: FROM TARGETED STUDIES TO LARGE-SCALE MULTI-OMICS EFFORTS

6.3.1 Leveraging Genome Sequence Data with Transcriptome, Targeted Genes, and Metabolite Profiling: Lessons from Studies of Terpenes

Well annotated genomes are particularly suited for the discovery of PV genes that belong to mid- to large-sized gene families such as the terpene synthase (TPS; see Chapter 8). TPSs synthesize the backbone of terpene molecules, including those of volatile mono-, sesqui-, and diterpenes (Pichersky and Raguso 2018). One key contributing factor to this diversity is the common presence of scores of encoded TPS genes per genome and the ability of many TPS enzymes to produce multiple products from a single substrate. However, functionally validating the biochemical activities of each predicted TPS nonetheless remains a huge challenge due to the potentially large number of genes involved.

One case illustrates successful TPS validation, despite the challenges. In the cultivated apple (*Malus domestica*), only 10 of the 55 putative *TPS* genes were predicted to be functional based on known structural features (Nieuwenhuizen et al. 2013). Furthermore, only eight potentially functional TPS enzymes were supported by expressed sequence tags from the apple cv. "Royal Gala." Further biochemical characterization of these eight TPS genes revealed that they account for the majority of terpene volatiles produced in Royal Gala apples (e.g., linalool, α-pinene, germacrene-D, and (E)-β-caryophyllene). For example, elevated expression of *MdLIS-RG1*, a linalool synthase, was correlated with high floral and fruit linalool production.

6.3.2 INTEGRATING TRANSCRIPTOME ANALYSIS WITH TARGETED GENE AND METABOLITE PROFILING OF SELECT TISSUES/ORGANS AND DEVELOPMENTAL STAGES: LESSONS FROM THE GARDEN PETUNIA

The garden petunia plant (*Petunia hybrida*) has a long history as a genetic model for understanding the molecular basis of as floral PV formation (Dudareva et al. 2013). Although genomes of several petunia species (i.e., *P. axillaris* and *P. inflata*) and their hybrids are now available (Bombarely et al. 2016), to date the majority of PV gene discoveries were made possible using functional genomics approaches, before these genomes became available.

The thousands of ESTs available for petunia from active PV-producing tissues (e.g., corolla limbs of mature two-day-old flowers) have been indispensable for mining DNA sequences containing functional domain(s) deduced to be relevant based on biochemical principles or *a priori* background information (i.e., homology to known enzymes). For example, to isolate the gene responsible for the synthesis of benzylbenzoate and phenylethylbenzoate, ESTs were screened for sequences containing key structural features of the BAHD family of acyltransferases. One candidate gene was isolated and later confirmed to encode a benzoyl-CoA:benzyl alcohol/phenylethanol benzoyltransferase (BPBT) protein responsible for benzylbenzoate and phenylethylbenzoate formation (Boatright et al. 2004). Similarly, the identification of a phenylacetaldehyde synthase (PAAS) responsible for the synthesis of phenylacetaldehyde from phenylalanine was possible by mining flower ESTs for genes putatively encoding aminotransferases, decarboxylases, and amine oxidases (Kaminaga et al. 2006).

Transcriptome mining using one or a few specific sequences from other systems (commonly *Arabidopsis*), under the assumption that the homologs (in petunia) should share similar enzymatic activities, has also been effective for prioritizing candidate genes. The core β-oxidative pathway of benzoic acid biosynthesis is one example where its components have been fully elucidated using this approach. For example, querying the petal-specific EST database using two *Arabidopsis* MULTIFUNCTIONAL PROTEIN (*AtAIM1* and *AtMFP2*) involved in peroxisomal fatty acid β-oxidation led to the discovery of the *Petunia hybrida* bifunctional cinnamoyl-CoA hydratase-dehydrogenase (PhCHD) (Qualley et al. 2012).

Recent discoveries in petunia are built upon RNA-seq analysis. In two landmark discoveries, *de novo* assembled RNA-seq transcriptomes of petal corolla tissues of young buds and mature flowers (2 days post-anthesis, d.p.a.), the developmental stages that correspond to the least and highest volatile emission, respectively, were employed to unravel the molecular mechanisms involved in the transport of a key PV precursor within the cell, and the resultant volatile products from the cell.

The amino acid, phenylalanine, is the gateway precursor for the synthesis of all benzenoids and phenylpropanoids. In the first discovery, a series of sequential hypotheses were tested to identify the gene(s) responsible for its transport from the plastid to the cytosol: Are there bacterial phenylalanine transporter homologs in the expressed transcriptome? If so, are they significantly upregulated (i.e., differentially expressed) in 2 d.p.a. flowers compared to buds? Do they show co-regulation with known structural genes? Do they possess putative plastid-targeting signal. Three full-length transcripts sharing moderate identity (22%–24%) with a bacterial phenylalanine exporter were identified and one satisfied the latter three criteria. The candidate was then confirmed by various biochemical experiments as a plastidial cationic amino-acid transporter (PhpCAT) (Widhalm et al. 2015).

In the second discovery, a similar hypothesis-driven functional genomics approach was employed to identify candidate genes responsible for benzenoid floral volatile emission in petunia (Adebesin et al. 2017). The authors hypothesized that rather than merely emitted by diffusion, volatile emission required a transporter (Hwang et al. 2016). One candidate belonging to the adenosine triphosphate-binding cassette G subfamily (ABCG), PhABCG1, was shown to be localized to the plasma membrane and targeted by the R2R3-type MYB family transcription factor, ODORANT1 which regulates volatile benzenoids in petunia cv Mitchell flowers (Van Moerkercke et al. 2012). Targeted metabolite profiling of transgenic lines of PhABCG1-RNAi constructs and of similar constructs of other candidate genes were performed to test their involvement in PV emission. Only PhABCG1-RNAi lines significantly reduced the total volatile emission and concomitantly increased the internal pools of various volatile phenylpropanoid/benzenoid (e.g., benzaldehyde, benzylbenzoate, phenylethanol, phenylacetaldehyde, methylbenzoate, and isoeugenol). Furthermore, radiolabel transport assays using labeled methylbenzoate, benzyl alcohol, menthol (terpene), and a sclareol analog (diterpene) revealed that PhABCG1 specifically transports the tested phenylpropanoid/benzenoid compounds only.

Finally, a comprehensive RNA-seq transcriptome atlas of petunia cv. Mitchell tissues revealed 25 *P. hybrida* APETALA2/Ethylene Response Factor (PhERFs) that were highly expressed in the petal. These candidates were screened for the ability to regulate the expression of genes directly involved in volatile phenylpropanoid/benzenoid production as well as upstream precursor supply pathways using virus-induced gene silencing (Liu et al. 2017). Of the 25 ERFs screened, five are potentially implicated in the regulation of volatile phenylpropanoid/benzenoid biosynthesis. Silencing of *PhERF6* in particular resulted in the upregulation of many genes involved in the latter two pathways and *vice versa* when ERF6 was overexpressed in transient assays. Suppression of *PhERF6* also increased emissions and internal pools of volatile phenylpropanoid/benzenoid constituents (e.g., benzaldehyde, benzylbenzoate, methylbenzoate, isoeugenol) in the flower. Gene expression analysis also showed that *PhERF6* was induced by ethylene and was co-ordinately regulated with volatile phenylpropanoid/benzenoid production in the flowers.

6.3.3 INTEGRATING TRANSCRIPTOME ANALYSIS WITH TARGETED GENE AND METABOLITE PROFILING OF SPECIALIZED TISSUE/CELL TYPES: LESSONS FROM SWEET BASIL

In the sweet basil (*Ocimum basilicum*) variety "EMX," leaf trichomes accumulate very high levels of the phenylpropene methylchavicol in the peltate glandular trichomes. To identify the genes involved in methylchavicol biosynthesis, cell-specific EST libraries from these trichomes were constructed and mined for sequences bearing homology to O-methyltransferases (Gang et al. 2002). One sequence bearing homology to Type I methyltransferases (MT) was identified and confirmed to catalyze the O-methylation of chavicol to form methylchavicol. The gene *ObCVOMT* was also shown to be highly expressed in the trichome of methylchavicol-accumulating varieties and devoid in a nonaccumulating variety. In the variety "SW," leaf trichomes accumulate very high levels of the phenylpropene eugenol with no detectable methylchavicol. Mining the glandular trichome EST libraries of this variety for putative oxidoreductases enabled the discovery of isoeugenol synthase (*ObEGS1*), a PIP family of NADPH-dependent reductases, responsible for the formation of eugenol from coniferyl acetate (Koeduka et al. 2006).

6.3.4 COMPARATIVE TRANSCRIPTOMICS ANALYSIS AND TARGETED GENE AND METABOLITE PROFILING: HARNESSING WITHIN- AND BETWEEN-SPECIES GENETIC VARIATION IN FLORICULTURAL AND AGRICULTURAL CROP PLANTS

The *Phalaenopsis* orchids are iconic house plants. However, modern-day varieties have been selectively breed to provide stunning, colorful, and long-lasting floral displays at the expense of the complete loss of floral scent production. *Phalaenopsis bellina* and *P. aphrodite* are two wild species with contrasting floral scent profiles. Flowers of *P. bellina* are highly scented and are dominated by the monoterpenes linalool, geraniol, and their derivatives, while *P. aphrodite* flowers are scentless to humans (Chuang et al. 2018).

To understand the molecular mechanisms underpinning this contrasting scent phenotype, RNA-seq transcriptomes of four flower developmental stages representing the onset, increase, peak, and decline of monoterpene emission were compared with that of *P. aphrodite* flowers at the floral bud and full-blossom stages (Chuang et al. 2018). One gene encoding geranyl diphosphate synthase (PbGDPS) and three predicted monoterpene synthases (PbTPS5, 9, and 10) were DE (i.e., highly upregulated in *P. bellina* and lowly expressed/absent in *P. aphrodite*). Mining of candidate TFs that share similar DE pattern revealed eight candidate TFs, of which five (i.e., PbbHLH4, PbbHLH6, PbbZIP4, PbERF1, and PbNAC1) were later confirmed to transactivate the promoters of *GDPS*, *TPS5*, and *TPS10* in dual-luciferase assays. Furthermore, all five TFs were able to induce the biosynthesis of floral terpenes (between 10- and 950-fold) in the scentless *P. aphrodite* when overexpressed in transient expression experiments.

Another example of the use of the comparative transcriptomic analysis in elucidating terpene biosynthesis comes from the rose (*Rosa*). The rose is perhaps best known as an iconic ornamental garden plant and as a cut flower, but is also an important floriculture crop that provides essential oils for the fragrance, food, and cosmetic industries. A typical rose scent is dominated by various monoterpene alcohols and 2-phenylethanol. In a breakthrough discovery, a novel terpene synthase-independent route to monoterpenes was recently reported (Magnard et al. 2015). The discovery was made by comparing the volatile, transcriptome, and gene expression of several rose varieties with distinct scent profiles. An unexpected enzyme that belongs to the Nudix hydrolase family (RhNUDX1) was prioritized and later confirmed to be involved in geraniol biosynthesis.

Expression of *RhNUDX1* was also shown to be tissue (petal)-specific, to increase concomitantly with scent emissions as the flower develops to maturity, and to be positively correlated with the occurrence of monoterpene alcohols (e.g., geraniol, nerol, citronellol) in ten varieties with contrasting scent profiles. Stable and transient transformation in two other heavily scented rose varieties (i.e., *R. chinensis* cv. Old Blush and *R. hybrida* The McCartney Rose) with RNA interference constructs unequivocally demonstrated that downregulation of *RhNUDX1* expression significantly reduced the total monoterpene content in petal tissues (due to decrease in geraniol derivatives) compared to the controls. All other volatile classes assayed were not affected. Further biochemical assays show that RhNUDX1 uses GPP as the substrate, and hydrolyses one phosphate group to form geranyl monophosphate. Thus, formation of geraniol in rose petals must require a yet unidentified second phosphatase.

6.3.5 Integrating Proteomics Techniques with Transcriptome and Targeted Gene and Metabolite Profiling

Large-scale investigations of proteins using techniques such as shotgun proteomics provide an additional set of tools for enzyme and pathway discovery. When transcriptome datasets (e.g., large-scale EST or RNA-seq) are coupled with matching proteome datasets of the same tissues/treatments, the combined transcriptome-proteome dataset enables the identification of expressed proteins in the metabolically active tissue of interest. This approach has the potential to detect critical enzymes and pathways by finding correlations between candidate gene expression, observed enzyme activities, and target metabolite profiles.

Work on the biosynthesis of methylcinnamate in sweet basil (Kapteyn et al. 2007) offers a good example of this approach. In this case, four basil varieties were used for comparisons—the MC variety that accumulates high levels of methylcinnamate and the three varieties (i.e., EMX, SW, and SD) that produced little to none of this compound. Methylcinnamate biosynthesis had been hypothesized to involve either MT and/or an acyltransferase activities. Of the two possible routes, cell-free protein extracts from the leaves showed cinnamic acid carboxyl methyltransferase (CCMT) activities, thus supporting the route involving MT enzymes. Formation of methylcinnamate was readily detectable when extracts were incubated with cinnamoyl-CoA (or labeled cinnamate) and S-adenosyl methionine while labeled methylcinnamate remained undetected when incubated with cinnamoyl-CoA and labeled methanol. In this context, comparative transcriptomics of glandular trichomes indicated salicylic acid carboxyl methyltransferase (SAMT)-like proteins that were abundant in MC, but absent in the other three varieties, as candidates.

Further transcriptome analysis narrowed the focus to three highly expressed genes belonging to the SABATH family of carboxyl MTs. The corresponding proteins were confirmed to be expressed in the glandular trichomes using shotgun proteomics and their abundance correlated well with observed gene expression and CCMT activity. Each candidate gene was expressed in *E. coli* and tested with 20 potential substrates including cinnamate and other structurally related molecules. The best substrates for methylation by all three tested recombinant proteins were cinnamate and *p*-coumarate and thus designated as cinnamate/*p*-coumarate carboxyl methyltransferases (ObCCMT1—3) (Kapteyn et al. 2007).

6.3.6 THE UTILITY OF GENE CO-EXPRESSION ANALYSIS IN MODEL AND NONMODEL PLANT SYSTEMS

There are several approaches to gene co-expression analysis (GCA) for gene discovery, which can be categorized based on the type of dataset (i.e., condition-independent or condition-dependent/specific) used, and the type of analysis to be performed (i.e., guide/bait gene or non-targeted approach) (Usadel et al. 2009). The goal of condition-independent GCA analysis is to provide an overview of hundreds or even thousands of inferred gene-to-gene relationships across multiple experimental sets. The often contrasting sets may include samples from different organs/tissues, developmental stages, stress treatments (abiotic and biotic), hormone treatments, etc. Conversely, the goal of condition-specific GCA is to highlight dynamic gene relationships (i.e., relationships enhanced only under specific conditions) that might otherwise be lost when performing condition-independent GCA (Obayashi et al. 2011). Application of the guide-genes approach requires *a priori* knowledge of the guide-gene function(s) as opposed to the nontargeted approach where all genes are considered and functional relationships are detected on the basis of identifying clusters matching the expression profiles of interests (Usadel et al. 2009).

The use of condition-independent guide-gene GCA to identify PV genes is illustrated by work on the biosynthesis of the C_{16}-homoterpene (*E, E*)-4,8,12-trimethyltrideca-1,3,7,11-tetraene homoterpene (TMTT) (Lee et al. 2010). In *Arabidopsis*, the homoterpenes TMTT is amongst the most common volatiles emitted by the night-scented flowers or from above ground plant tissues (e.g., leaves) damaged by herbivory. Key *a priori* knowledge included: stable-isotope precursor feeding experiments had established (E, E)-geranyllinalool as the precursor, the identification of geranyllinalool synthase (GES) in the biosynthesis of (E, E)-geranyllinalool, and the prior characterization of P450 enzymes known to catalyze C-C bond cleavage during tertiary alcohol oxidation (Mizutani and Ohta 2010).

With this knowledge, two widely used plant gene co-expression databases were queried with the guide gene, GES, revealing many highly co-expressed candidate genes. These included genes encoding P450 enzymes, flavin-dependent monooxygenases, dioxygenases, and peroxidases. In subsequent experiments, two P450 genes were selected for further characterization on the basis that they closely matched the expected co-expression profile with GES. Of these two enzymes, the gene encoding CYP82G1 enzyme was confirmed as a TMTT synthase. Later the recombinant CYP82G1 enzyme showed narrow substrate specificity for (E, E)-geranyllinalool (Lee et al. 2010).

In nonmodel plants, including most crop species, large-scale datasets in the hundreds to thousands of samples/experiments and spanning a variety of conditions, are rarely available (Obayashi et al. 2018). Thus, studies exploiting gene co-expression analysis are often performed on a smaller "condition-specific" scale that often comprise a strategic set of organs/tissues or stress treatments. Surprisingly, widespread use of this condition-specific co-expression analysis to aid the discovery of volatile pathway genes is still limited (Xu et al. 2018; Li et al. 2018). However, the elucidation of pyrethrin biosynthesis offers an exemplar case.

The flowers of *Tanacetum cinerariifolium*, a member of the daisy family, synthesize natural pesticides called pyrethrins—esters bearing a monoterpenoid acid (chrysanthemic acid or pyrethric acid) and a jasmonic acid-derived alcohol moiety (pyrethrolone, cinerolone, or jasmolone). In two recent studies, condition-specific co-expression analysis was used to identify candidate genes involved in several predicted intermediate steps of pyrethrin biosynthesis. They include the identification of an alcohol dehydrogenase (TcADH2) and an aldehyde dehydrogenase (TcALDH1) involved in the oxidation of *trans*-chrysanthemol to *trans*-chrysanthemic acid (Xu et al. 2018), and the identification of jasmolone hydroxylase (TcJMH) involved in the hydroxylation of jasmone to jasmolone (Li, Zhou, and Pichersky 2018). In both studies, RNA-seq transcriptomes assembled from leaf tissue, and flowers at different developmental stages (Stages 1–5) and tissue types (ray and disk florets), were used in the co-expression analysis using the guide genes, TcCDS and TcGLIP. These two genes were already known to be involved in pyrethrin biosynthesis, and were found to be highly expressed in the flowers while expression was barely detectable in the leaf. Within the flowers, a strong differential expression between the tissue types (i.e., expression higher in disk vs. ray florets) and to a lesser extent during flower development (i.e., expression higher in later stages of flower development) was also observed. When GCA identified multiple candidates for specific steps—as was the case for the ADH and JMH genes—*in vitro* biochemical reactions and heterologous gene expression studies with multiple candidate genes were carried out to identify the correct genes.

6.3.7 INTEGRATING POPULATION GENOMICS AND MULTI-OMICS DATASETS FOR LARGE-SCALE PLANT VOLATILE GENE DISCOVERY

The use of quantitative trait loci (QTL) analysis to identify genomic regions associated with a given phenotypic trait has a long history. A wide range of different types of genetic markers have been employed in the process, including: RFLPs, AFLPs, and SNPs (Scheben et al. 2017). High-resolution QTL mapping via the development of ultra-high-density single-nucleotide polymorphism (SNP) markers, in the tens to hundreds of thousands, in now achievable with RNA-seq SNP genotyping approaches, offering the potential to rapidly find candidate genes with high precision.

For example, in a melon (*Cucumis melo*) recombinant inbred line (RIL) ("PI 414723" x "Dulce"), RNA-seq-based QTL and expression-QTL (eQTL) mapping combined with large-scale genotypic, metabolomic, and transcriptomic data has helped to identify the genetic basis of key fruit-quality traits (Galpaz et al. 2018). In total, 241 QTLs were found to be associated with 129 fruit-quality traits. Of these, 91 were related to aroma. In one specific example, variation in levels of S-methyl thioacetate, a key volatile imparting sulfurous-tropical fruit notes to some melon varieties, was traced to a single QTL. A gene (CmThAT1) annotated as a thiolase was located within close proximity to the QTL. Biochemical *in vitro* assays established that CmThAT1 encodes a ThAT involved in the production of S-methyl thioacetate and S-methyl propanethioate from methanethiol and the respective acyl-CoA precursors. While the coding sequences of CmThAT1 from "PI 414723" and "Dulce" are identical, polymorphisms found in their 5′ and 3′ UTR were associated positively with the levels of CmThAT1 expression and S-methyl thioacetate across the RIL population.

6.4 FUTURE DIRECTIONS

RNA-seq transcriptomes are poised to provide a key genetic resource for gene discovery, even in nonmodel system lacking substantial genomic resources. By considering a few key steps when designing experiments (see guidelines in Section 6.2 and examples in Section 6.3), workflows for PV gene discovery in any plant system can be tailored to the available resources, capabilities, and *a priori* background knowledge (Figure 6.2). In the most generalized scenario, such as the case of a nonmodel plant producing unique volatile(s) of interest but with little or no background information available, volatiles and transcriptomes of select tissues or organs, and at various developmental

stages, can first be profiled. Genes that encode enzymes or TFs that match the qualitative or quantitative profile of the target metabolite(s), and maybe homologous to proteins of known function, can
provide excellent starting candidates for further functional validation. Below, several other promising approaches for enhancing gene discovery are also emerging.

6.4.1 Harnessing TF Overexpression or Knockdown Coupled with Transcriptome (or Multi-omics) Analysis

In plants, structural gene and metabolic networks of the general phenylpropanoid pathways that
generate an enormous array of specialized metabolites (flavonoid, anthocyanin, phenylpropanoid/
benzenoid volatiles inclusive) are often regulated by select TFs that belong to the R2R3-MYB and
bHLH families (Allan and Espley 2018; Davies et al. 2012). Some studies have used overexpression
of these TFs (with known roles) as a tool to discover novel genes. TF overexpression may accelerate
the discovery of volatile pathway genes by activating multiple candidate genes simultaneously and
bypassing the need for more complex analysis that takes into account the spatial and temporal variation in volatile gene expression. The opposite effect may also occur when target TFs are knocked
down. For example, silencing of PH4, a petunia R2R3-MYB TF known for its role in activating
vacuolar acidification, revealed a new role of the gene in floral volatile emission in petunia (Cna'ani
et al. 2015).

6.4.2 Harnessing Single-Cell or Single-Cell Type Isolation Methods with Next-Generation Transcriptome (or Multi-omics) Analysis

To date, most of our understanding of the biosynthesis and regulation of PVs originates from investigations conducted at the tissue/organ levels. However, the "dilution" and "masking" due to the
uncertain homogeneity of the underlying tissues/cells may mask the unique metabolic capacities of
individual cell types. Recent advances in isolation methods (e.g., laser microdissection; fluorescent-
activated cell sorting, FACS of marked cell populations; and Isolation of Nuclei in Tagged Cell
Types, INTACT) now present exciting new opportunities to study the diverse metabolic facets of
plant single-cell (sc) or sc types rich in PVs and secondary metabolites (e.g., resin ducts) that were
previously inaccessible to multi-omics methods (Libault et al. 2017).

In one study, fine-scale spatiotemporal dynamics of constitutive and induced conifer defense was
successfully unmasked with the use of laser microdissection coupled with RNA-seq and targeted
metabolite analysis (Celedon et al. 2017). Transcriptome analysis of white spruce (*Picea glauca*)
cortical resin duct cells, phenolic cells, and phloem cells against bulk bark tissues under normal
and methyl jasmonate-induced conditions revealed that only a small proportion of expressed transcriptome (~6%) appeared to be cell-/tissue-specific. Spatially restricted SM gene expression (e.g.,
TPSs and P450s) that are otherwise confounded by "dilution" and "masking" effects in bulk bark
tissues were also uncovered. In PV-rich tissues that are accessible to sc-type isolation methods and
transcriptome profiling (e.g., fruits and trichomes), spatial and temporal transcriptome maps are
now becoming available (Voo et al. 2012; Pattison et al. 2015; Shinozaki et al. 2018; Livingston
et al. 2019). Beyond the primary goal of teasing apart cell/tissue-specific metabolism and developmental processes, these studies provide a key resource for future structural and regulatory PV gene
discoveries.

The exciting potential application of single-cell datasets was demonstrated by one recent study
in tomato which provided a high-resolution spatiotemporal transcriptome map encompassing six
major fruit tissues and five pericarp cell/tissue types in the equatorial regions of the fruit spanning ten developmental stages (Shinozaki et al. 2018). Pericarp cell/tissue-dependent expression of
several known PV genes were revealed for the first time. For example, the gene encoding catechol-
O-methyltransferase (CTOMT1) involved in the biosynthesis of guaiacol via methylation of catechol (Mageroy et al. 2012) was expressed exclusively in the vascular tissues. Conversely, the gene

encoding a 13-lipoxygenease, TOMLOXC involved in the formation of various C_5 and C_6 volatiles (Shen et al. 2014), was shown to be predominantly expressed in collenchyma cells and to a lesser extent in parenchyma cells.

6.4.3 HARNESSING THE PHYSICAL PROXIMITY OF GENES WITH GENE CO-EXPRESSION ANALYSIS

In plants, most of the known SM pathways are randomly distributed across the genome; however, some newly discovered pathways show clustering and co-expression phenomenon (Nützmann et al. 2016). This discovery of gene clusters has been aided by the growing number of sequenced plant genomes with high-quality sequence assemblies and comprehensive transcriptome catalogues (Figure 6.2b(v)). Therefore, the utility of gene proximity and co-expression mining may prove informative for revealing novel PV genes that are organized in gene clusters. Indeed, at least for the terpene biosynthetic pathway genes, this seem highly feasible.

For example, TPS and CYP genes are often found co-located together far more frequently than expected by chance and volatile production may require both two types of genes. Indeed, formation of the volatile DMNT in the belowground tissues of *Arabidopsis* was recently demonstrated to involve the degradation of triterpenes by CYP genes (Sohrabi et al. 2015). The responsible gene encodes CYP705A1, a cytochrome P450 monooxygenase involved in the cleavage of arabidiol triterpene precursor to produce DMNT and a nonvolatile C_{19}-ketone product. This gene is clustered next to *PENTACYCLIC TRITERPENE SYNTHASE 1* gene encoding an arabidiol synthase (ABDS) and shared coordinated regulation in the roots.

6.4.4 HARNESSING EXPRESSION QUANTITATIVE TRAIT LOCI (eQTL) FOR CANDIDATE PRIORITIZATION

Compared to classical QTL analysis, expression-QTL (eQTL) analysis aims to discover genetic variants that contribute to the variation in gene expression (the quantitative trait) at a genome-wide scale across precise genetic backgrounds (e.g., RILs). Such approaches promise to greatly assist the linking of genes to various phenotypic traits (Kliebenstein 2009). Now with RNA-seq, genome-wide sequence and expression information can be simultaneously obtained (Scheben et al. 2017), allowing eQTL analysis to be performed for a wider range of species than ever before.

In a recent study, eQTL was used as supporting evidence for prioritizing candidate genes associated with the production of multiple fatty acid-derived flavor volatiles in tomato fruits (Garbowicz et al. 2018). Through large-scale lipid profiling of *S. pennellii* introgression lines (IL) fruit pericarp and leaf tissues, and by utilizing additional genetic mapping resources (e.g., backcross inbred lines and sub-IL populations), tissue-specific lipid metabolite QTLs hotspots were first identified. A hypothesis-driven search for lipid metabolism-related genes within these hotspots was then conducted, resulting in the identification of some lipid-related genes that were supported by eQTLs exhibiting higher expression in *S. pennellii*-derived lines compared to wild-type *S. lycopersicum* var. "M82." In particular, one gene encoding class III lipase family (*LIP1*), with >600-fold greater expression in *S. pennellii*, was also positively correlated with high levels of fatty acid-derived volatiles during fruit development and ripening. The role of LIP1 was subsequently confirmed using *SpLIP1*-silenced and overexpression lines.

6.4.5 HARNESSING STATE-OF-THE-ART MACHINE LEARNING ALGORITHMS: PUTTING ALL THE PIECES TOGETHER?

Efforts to integrate the diverse information into a seamless metric for candidate prioritization or gene-function prediction are still challenging (Figure 6.2b(vi)). Nonetheless, several recent studies employing the use of machine learning techniques demonstrate the promise of the future (Moore et al. 2019).

When compared to general metabolism (GM) pathway genes, PV genes, like many SM pathway genes, share many common genomic and transcriptomic features. These features can be categorized into five main categories: expression/co-expression, evolution/conservation, epigenetic modification, gene duplication, gene function, and gene networks. One recent study, which has taken these factors into account, integrated multiple features into a prediction model which allowed SM pathway genes to be distinguished with high accuracy from GM genes, outperforming even the best-performing single feature (i.e., gene family size) (Moore et al. 2019). A confidence measure, called an SM score, was also introduced, providing a global estimate of SM gene content with potential application to any plant genome. When applied to the *Arabidopsis* genome, >1,000 genes that were previously annotated as unknown functions are now functionally ascribed as likely being SM. Considering that some of the most informative features (e.g., gene duplication, evolutionary rate, and gene expression) are derived from genome and transcriptome datasets that are becoming widely available, we expect that predicting gene function with machine learning will soon become routine and accessible even to nonmodel systems.

Although the binary classification of enzyme/genes as SM or GM is likely an oversimplification, it may offer a meaningful first step for PV candidate gene prioritization and function prediction. This is exemplified by the ability to correctly classify *A. thaliana* TPS genes (Chen et al. 2011) as being involved in SM. Twenty-six *A. thaliana* TPS genes not universally distributed among angiosperms were subjected to SM gene prediction. The machine learning model correctly classified all but one TPS as being SM-related. This included correcting the classification for four potentially mis-annotated TPSs previously annotated as being involved in GM in other databases. However, the same algorithm mis-classified an *Arabidopsis* aromatic aldehyde synthase (AtAAS) gene, involved in the conversion of phenylalanine to phenylacetaldehyde, a volatile specialized metabolite (Gutensohn et al. 2011), as GM genes. This misclassification was due to multiple confounding features possessed by this gene (i.e., broadly expression pattern, has high connectivity in gene networks, and belongs to a small gene family).

6.4.6 MAXIMIZING THE SUCCESS OF PLANT VOLATILE GENE DISCOVERY

The bulk of the PV gene discoveries that we have highlighted in this chapter were made by utilizing various combinations of datasets and analysis types (Figure 6.2a and b). Each one of these combinations has its own underlying experimental and analytical complexities, and each brings different potential for the discovery of PV-related genes (Figure 6.3). For example, PV gene discoveries made with the most common combinations 2–5 (full circle) employ highly specific methods with the target of one to a few gene discoveries per study. Their widespread applicability is in part due to the ongoing advancements in the acquisition of new "omics" data and maturity of most statistical and bioinformatics analysis associated with the underlying analyses, thus making them preferred options. Conversely, PV gene discoveries made with combinations 6–8 (dash circle) have yet to be fully exploited. These approaches are often state-of-the-art and highly complex, but hold great promise for enabling multiple gene discoveries to be made in a single study (i.e., multiple genes both within and/or between pathways).

It is clear that many success stories in PV gene discovery, both in the past and more recently, have been made possible by the integration of findings from multiple types of analyses. In all these cases, background knowledge from earlier studies combined with innovative additional experimentation and data analysis (Figure 6.2a and b) was indispensable to the hypothesis generation and PV gene prioritization. This in turn was followed up by comprehensive functional characterization of the candidate genes. With the rapidly expanding capability to catalogue whole genomes, transcriptomes, metabolomes, and proteomes at unprecedented experimental scales and decreasing costs, for both model and nonmodel organisms alike, we predict exponential growth in PV gene discovery (Figure 6.3).

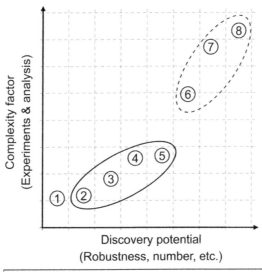

1. FD
2. FD + DA (e.g. Adebesin et al. 2017, Xu et al. 2017)
3. FD + DA + CA (e.g. Magnard et al. 2015, Xu et al. 2018)
4. FD + PP (e.g. Sohrabi et al. 2015)
5. FD + DA + NA (e.g. Lee et al. 2010)
6. FD + DA (*HST) + NA (e.g. Shinozaki et al. 2018)
7. FD + DA + NA + SV (e.g. Galpaz et al. 2018)
8. ML (e.g. Moore et al. 2019)

FIGURE 6.3   Relationship between the complexity of integrating various analyses afforded by multi-omics approaches and their combined potential for maximizing the success of hypothesis generation and candidate prioritization. Although there are many different approaches in which plant volatile genes were investigated, four common (full circle) and three uncommon (dash circle) combination of dataset and analysis types (see components of each combination in rectangle box) constitute the bulk of plant volatile gene discoveries. See captions of Figure 6.2 for abbreviations. Example(s) of studies from each combination are indicated. *HST, Hi-res spatiotemporal transcriptome.

6.5   CONCLUSION

Despite the extraordinary progress, there are still tens of thousands of PVs, both known and unknown, whose biosynthesis and regulation remains to be elucidated. Nonetheless, large-scale PV gene discovery now seems possible with strategic experimental designs that take full advantage of multi-omics systems-based comparative approaches. In the future, it is also anticipated that machine learning will revolutionize how PV candidate gene prioritization and functional predictions are made. Furthermore, this revolution will not just be limited to model systems already rich in genetic resources, but also to nonmodel species which are often rich in PV diversity.

ACKNOWLEDGMENTS

Darren Wong was supported by an Australian Research Council (ARC) Discovery project (DP150102762) to Rod Peakall and Eran Pichersky.

REFERENCES

Adebesin, Funmilayo, Joshua R. Widhalm, Benoît Boachon, François Lefèvre, Baptiste Pierman, Joseph H. Lynch, Iftekhar Alam, et al. 2017. "Emission of Volatile Organic Compounds from Petunia Flowers Is Facilitated by an ABC Transporter." *Science* 356 (6345): 1386–88. doi:10.1126/science. aan0826.

Allan, Andrew C., and Richard V. Espley. 2018. "MYBs Drive Novel Consumer Traits in Fruits and Vegetables." *Trends in Plant Science* 23 (8): 693–705. doi:10.1016/j.tplants.2018.06.001.

Boatright, Jennifer, Florence Negre, Xinlu Chen, Christine M. Kish, Barbara Wood, Greg Peel, Irina Orlova, David Gang, David Rhodes, and Natalia Dudareva. 2004. "Understanding In Vivo Benzenoid Metabolism in Petunia Petal Tissue." *Plant Physiology* 135 (4): 1993–2011. doi:10.1104/pp.104.045468.

Bolger, Marie E., Borjana Arsova, and Björn Usadel. 2018. "Plant Genome and Transcriptome Annotations: From Misconceptions to Simple Solutions." *Briefings in Bioinformatics* 19 (3): 437–49. doi:10.1093/bib/bbw135.

Bombarely, Aureliano, Michel Moser, Avichai Amrad, Manzhu Bao, Laure Bapaume, Cornelius S. Barry, Mattijs Bliek, et al. 2016. "Insight into the Evolution of the Solanaceae from the Parental Genomes of Petunia Hybrida." *Nature Plants* 2 (June): 1–9. doi:10.1038/nplants.2016.74.

Broekgaarden, Colette, Irene A. Vos, Corné M.J. Pieterse, Saskia C.M. Van Wees, and Lotte Caarls. 2015. "Ethylene: Traffic Controller on Hormonal Crossroads to Defense." *Plant Physiology* 169 (December): pp. 2371–2379. doi:10.1104/pp.15.01020.

Celedon, Jose M., Macaire M.S. Yuen, Angela Chiang, Hannah Henderson, Karen E. Reid, and Jörg Bohlmann. 2017. "Cell-Type- and Tissue-Specific Transcriptomes of the White Spruce (*Picea glauca*) Bark Unmask Fine-Scale Spatial Patterns of Constitutive and Induced Conifer Defense." *Plant Journal* 92 (4): 710–26. doi:10.1111/tpj.13673.

Chen, Feng, Dorothea Tholl, Jörg Bohlmann, and Eran Pichersky. 2011. "The Family of Terpene Synthases in Plants: A Mid-Size Family of Genes for Specialized Metabolism That Is Highly Diversified throughout the Kingdom." *Plant Journal* 66 (1): 212–229. doi:10.1111/j.1365-313X.2011.04520.x.

Cheng, Shifeng, Michael Melkonian, Stephen A. Smith, Samuel Brockington, John M. Archibald, Pierre Marc Delaux, Fay Wei Li, et al. 2018. "10KP: A Phylodiverse Genome Sequencing Plan." *GigaScience* 7 (3): 1–9. doi:10.1093/gigascience/giy013.

Chuang, Yu Chen, Yi Chu Hung, Wen Chieh Tsai, Wen Huei Chen, and Hong Hwa Chen. 2018. "PbbHLH4 Regulates Floral Monoterpene Biosynthesis in *Phalaenopsis* Orchids." *Journal of Experimental Botany* 69 (18): 4363–77. doi:10.1093/jxb/ery246.

Cna'ani, Alon, Ben Spitzer-Rimon, Jasmin Ravid, Moran Farhi, Tania Masci, Javiera Aravena-Calvo, Marianna Ovadis, and Alexander Vainstein. 2015. "Two Showy Traits, Scent Emission and Pigmentation, Are Finely Coregulated by the MYB Transcription Factor PH4 in Petunia Flowers." *New Phytologist* 208 (3): 708–14. doi:10.1111/nph.13534.

Davies, Kevin M., Nick W. Albert, and Kathy E. Schwinn. 2012. "From Landing Lights to Mimicry: The Molecular Regulation of Flower Colouration and Mechanisms for Pigmentation Patterning." *Functional Plant Biology* 39 (8): 619–38. doi:10.1071/FP12195.

Dudareva, Natalia, Antje Klempien, Joëlle K. Muhlemann, and Ian Kaplan. 2013. "Biosynthesis, Function and Metabolic Engineering of Plant Volatile Organic Compounds." *New Phytologist* 198 (1): 16–32. doi:10.1111/nph.12145.

Dudareva, Natalia, Eran Pichersky, and Jonathan Gershenzon. 2004. "Biochemistry of Plant Volatiles." *Plant Physiology* 135 (August): 1893–1902. doi:10.1104/pp.104.049981.1.

Egan, Ashley N., Jessica Schlueter, and David M. Spooner. 2012. "Applications of Next-Generation Sequencing in Plant Biology." *American Journal of Botany* 99 (2): 175–85. doi:10.3732/ajb.1200020.

Fridman, Eyal, and Eran Pichersky. 2005. "Metabolomics, Genomics, Proteomics, and the Identification of Enzymes and Their Substrates and Products." *Current Opinion in Plant Biology* 8 (3): 242–48. doi:10.1016/j.pbi.2005.03.004.

Galpaz, Navot, Itay Gonda, Doron Shem-Tov, Omer Barad, Galil Tzuri, Shery Lev, Zhangjun Fei, et al. 2018. "Deciphering Genetic Factors That Determine Melon Fruit-Quality Traits Using RNA-Seq-Based High-Resolution QTL and EQTL Mapping." *The Plant Journal* 94 (1): 169–91. doi:10.1111/tpj.13838.

Gang, David R, Noa Lavid, Chloe Zubieta, Feng Chen, Till Beuerle, Efraim Lewinsohn, Joseph P Noel, and Eran Pichersky. 2002. "Characterization of Phenylpropene O-Methyltransferases from Sweet Basil: Facile Change of Substrate Specificity and Convergent Evolution within a Plant O-Methyltransferase Family." *The Plant Cell* 14 (2): 505–19. doi:10.1105/tpc.010327.1.

Garbowicz, Karolina, Zhongyuan Liu, Saleh Alseekh, Denise Tieman, Mark Taylor, Anastasiya Kuhalskaya, Itai Ofner, et al. 2018. "Quantitative Trait Loci Analysis Identifies a Prominent Gene Involved in the Production of Fatty Acid-Derived Flavor Volatiles in Tomato." *Molecular Plant* 11 (9): 1147–65. doi:10.1016/j.molp.2018.06.003.

Gutensohn, Michael, Antje Klempien, Yasuhisa Kaminaga, Dinesh A. Nagegowda, Florence Negre-Zakharov, Jung Hyun Huh, Hongli Luo, et al. 2011. "Role of Aromatic Aldehyde Synthase in Wounding/Herbivory Response and Flower Scent Production in Different Arabidopsis Ecotypes." *Plant Journal* 66 (4): 591–602. doi:10.1111/j.1365-313X.2011.04515.x.

Hwang, Jae Ung, Won Yong Song, Daewoong Hong, Donghwi Ko, Yasuyo Yamaoka, Sunghoon Jang, Sojeong Yim, et al. 2016. "Plant ABC Transporters Enable Many Unique Aspects of a Terrestrial Plant's Lifestyle." *Molecular Plant* 9 (3): 338–55. doi:10.1016/j.molp.2016.02.003.

Kaminaga, Yasuhisa, Jennifer Schnepp, Greg Peel, Christine M. Kish, Gili Ben-Nissan, David Weiss, Irina Orlova, et al. 2006. "Plant Phenylacetaldehyde Synthase Is a Bifunctional Homotetrameric Enzyme That Catalyzes Phenylalanine Decarboxylation and Oxidation." *Journal of Biological Chemistry* 281 (33): 23357–66. doi:10.1074/jbc. M602708200.

Kapteyn, Jeremy, Anthony V. Qualley, Zhengzhi Xie, Eyal Fridman, Natalia Dudareva, and David R. Gang. 2007. "Evolution of Cinnamate/p-Coumarate Carboxyl Methyltransferases and Their Role in the Biosynthesis of Methylcinnamate." *The Plant Cell* 19 (10): 3212–29. doi:10.1105/tpc.107.054155.

Kliebenstein, Dan. 2009. "Quantitative Genomics: Analyzing Intraspecific Variation Using Global Gene Expression Polymorphisms or EQTLs." *Annual Review of Plant Biology* 60 (1): 93–114. doi:10.1146/annurev.arplant.043008.092114.

Koeduka, T., E. Fridman, D. R. Gang, D. G. Vassao, B. L. Jackson, C. M. Kish, I. Orlova, et al. 2006. "Eugenol and Isoeugenol, Characteristic Aromatic Constituents of Spices, Are Biosynthesized via Reduction of a Coniferyl Alcohol Ester." *Proceedings of the National Academy of Sciences* 103 (26): 10128–33. doi:10.1073/pnas.0603732103.

Lee, Sungbeom, Somayesadat Badieyan, David R. Bevan, Marco Herde, Christiane Gatz, and Dorothea Tholl. 2010. "Herbivore-Induced and Floral Homoterpene Volatiles Are Biosynthesized by a Single P450 Enzyme (CYP82G1) in Arabidopsis." *Proceedings of the National Academy of Sciences of the United States of America* 107 (49): 21205–10. doi:10.1073/pnas.1009975107.

Li, Wei, Fei Zhou, and Eran Pichersky. 2018. "Jasmone Hydroxylase, a Key Enzyme in the Synthesis of the Alcohol Moiety of Pyrethrin Insecticides." *Plant Physiology* 177 (4): 1498–1509 doi:10.1104/pp.18.00748.

Libault, Marc, Lise Pingault, Prince Zogli, and John Schiefelbein. 2017. "Plant Systems Biology at the Single-Cell Level." *Trends in Plant Science* 22 (11): 949–60. doi:10.1016/j.tplants.2017.08.006.

Liu, Fei, Zhina Xiao, Li Yang, Qian Chen, Lu Shao, Juanxu Liu, and Yixun Yu. 2017. "PhERF6, Interacting with EOBI, Negatively Regulates Fragrance Biosynthesis in Petunia Flowers." *New Phytologist* 215 (4): 1490–1502. doi:10.1111/nph.14675.

Livingston, Samuel J, Teagen D. Quilichini, Judith K. Booth, Darren C.J. Wong, Kim H. Rensing, Jessica Laflamme-Yonkman, Simone D. Castellarin, Joerg Bohlmann, Jonathan E. Page, and A. Lacey Samuels. 2020. "Cannabis Glandular Trichomes Alter Morphology and Metabolite Content during Flower Maturation." *The Plant Journal* 101 (1), 37–56. doi:10.1111/tpj.14516.

Loreto, Francesco, Marcel Dicke, Jörg Peter Schnitzler, and Ted C.J. Turlings. 2014. "Plant Volatiles and the Environment." *Plant, Cell and Environment* 37 (8): 1905–8. doi:10.1111/pce.12369.

Mageroy, Melissa H., Denise M. Tieman, Abbye Floystad, Mark G. Taylor, and Harry J. Klee. 2012. "A Solanum Lycopersicum Catechol-O-Methyltransferase Involved in Synthesis of the Flavor Molecule Guaiacol." *Plant Journal* 69 (6): 1043–51. doi:10.1111/j.1365-313X.2011.04854.x.

Magnard, Jean-Louis, Aymeric Roccia, Jean-Claude Caissard, Philippe Vergne, Pulu Sun, Romain Hecquet, Annick Dubois, et al. 2015. "Biosynthesis of Monoterpene Scent Compounds in Roses." *Science (New York, N.Y.)* 349 (6243): 81–83. doi:10.1126/science.aab0696.

Matasci, Naim, Ling-Hong Hung, Zhixiang Yan, Eric J. Carpenter, Norman J. Wickett, Siavash Mirarab, Nam Nguyen, et al. 2014. "Data Access for the 1,000 Plants (1KP) Project." *GigaScience* 3 (1): 17. doi:10.1186/2047-217X-3-17.

Mizutani, Masaharu, and Daisaku Ohta. 2010. "Diversification of P450 Genes During Land Plant Evolution." *Annual Review of Plant Biology* 61 (1): 291–315. doi:10.1146/annurev-arplant-042809-112305.

Moerkercke, Alex Van, Carlos S. Galván-Ampudia, Julian C. Verdonk, Michel A. Haring, and Robert C. Schuurink. 2012. "Regulators of Floral Fragrance Production and Their Target Genes in Petunia Are Not Exclusively Active in the Epidermal Cells of Petals." *Journal of Experimental Botany* 63 (8): 3157–71. doi:10.1093/jxb/ers034.

Moore, Bethany M, Peipei Wang, Pengxiang Fan, Bryan Leong, Craig A Schenck, John P Lloyd, Melissa D Lehti-Shiu, Robert L Last, Eran Pichersky, and Shin-Han Shiu. 2019. "Robust Predictions of Specialized Metabolism Genes through Machine Learning." *Proceedings of the National Academy of Sciences of the United States of America* 116 (6): 2344–53. doi:10.1073/pnas.1817074116.

Muhlemann, Joëlle K., Antje Klempien, and Natalia Dudareva. 2014. "Floral Volatiles: From Biosynthesis to Function." *Plant, Cell and Environment* 37 (8): 1936–49. doi:10.1111/pce.12314.

Nieuwenhuizen, Niels J, Sol A Green, Xiuyin Chen, Estelle J D Bailleul, Adam J Matich, Mindy Y Wang, and Ross G Atkinson. 2013. "Functional Genomics Reveals That a Compact Terpene Synthase Gene Family Can Account for Terpene Volatile Production in Apple." *Plant Physiology* 161 (2): 787–804. doi:10.1104/pp.112.208249.

Nützmann, Hans Wilhelm, Ancheng Huang, and Anne Osbourn. 2016. "Plant Metabolic Clusters—From Genetics to Genomics." *New Phytologist* 211 (3): 771–89. doi:10.1111/nph.13981.

Obayashi, Takeshi, Yuichi Aoki, Shu Tadaka, Yuki Kagaya, and Kengo Kinoshita. 2018. "ATTED-II in 2018: A Plant Coexpression Database Based on Investigation of the Statistical Property of the Mutual Rank Index." *Plant & Cell Physiology* 59 (1): e3. doi:10.1093/pcp/pcx191.

Obayashi, Takeshi, Kozo Nishida, Kota Kasahara, and Kengo Kinoshita. 2011. "ATTED-II Updates: Condition-Specific Gene Coexpression to Extend Coexpression Analyses and Applications to a Broad Range of Flowering Plants." *Plant & Cell Physiology* 52 (2): 213–19. doi:10.1093/pcp/pcq203.

Ohlrogge, John, and Christoph Benning. 2000. "Unraveling Plant Metabolism by EST Analysis." *Current Opinion in Plant Biology* 3 (3): 224–28. doi:10.1016/S1369-5266(00)00068-6.

Oshlack, Alicia, Mark D Robinson, and Matthew D Young. 2010. "From RNA-Seq Reads to Differential Expression Results." *Genome Biology* 11 (12): 220. doi:10.1186/gb-2010-11-12-220.

Pattison, Richard J., Fabiana Csukasi, Yi Zheng, Zhangjun Fei, Esther van der Knaap, and Carmen Catalá. 2015. "Comprehensive Tissue-Specific Transcriptome Analysis Reveals Distinct Regulatory Programs during Early Tomato Fruit Development." *Plant Physiology* 168 (4): 1684–1701. doi:10.1104/pp.15.00287.

Pichersky, Eran, Joseph P Noel, and Natalia Dudareva. 2006. "Biosynthesis of Plant Volatiles: Nature's Diversity and Ingenuity." *Science (New York, N.Y.)* 311 (5762): 808–11. doi:10.1126/science.1118510.

Pichersky, Eran, and Robert A. Raguso. 2018. "Why Do Plants Produce so Many Terpenoid Compounds?" *The New Phytologist* 220 (3): 692–702. doi:10.1111/nph.14178.

Qualley, Anthony V., Joshua R. Widhalm, Funmilayo Adebesin, Christine M. Kish, and Natalia Dudareva. 2012. "Completion of the Core β-Oxidative Pathway of Benzoic Acid Biosynthesis in Plants." *Proceedings of the National Academy of Sciences of the United States of America* 109 (40): 16383–88. doi:10.1073/pnas.1211001109.

Rhee, Seung Yon, and Marek Mutwil. 2014. "Towards Revealing the Functions of All Genes in Plants." *Trends in Plant Science* 19 (4): 212–21. doi:10.1016/j.tplants.2013.10.006.

Schauer, Nicolas, and Alisdair R. Fernie. 2006. "Plant Metabolomics: Towards Biological Function and Mechanism." *Trends in Plant Science* 11 (10): 508–16. doi:10.1016/j.tplants.2006.08.007.

Scheben, Armin, Jacqueline Batley, and David Edwards. 2017. "Genotyping-by-Sequencing Approaches to Characterize Crop Genomes: Choosing the Right Tool for the Right Application." *Plant Biotechnology Journal* 15 (2): 149–61. doi:10.1111/pbi.12645.

Schilmiller, Anthony L., Eran Pichersky, and Robert L. Last. 2012. "Taming the Hydra of Specialized Metabolism: How Systems Biology and Comparative Approaches Are Revolutionizing Plant Biochemistry." *Current Opinion in Plant Biology* 15 (3): 338–44. doi:10.1016/j.pbi.2011.12.005.

Shen, Jiyuan, Denise Tieman, Jeffrey B. Jones, Mark G. Taylor, Eric Schmelz, Alisa Huffaker, Dawn Bies, Kunsong Chen, and Harry J. Klee. 2014. "A 13-Lipoxygenase, TomloxC, Is Essential for Synthesis of C5 Flavour Volatiles in Tomato." *Journal of Experimental Botany* 65 (2): 419–28. doi:10.1093/jxb/ert382.

Shinozaki, Yoshihito, Philippe Nicolas, Noe Fernandez-Pozo, Qiyue Ma, Daniel J. Evanich, Yanna Shi, Yimin Xu, et al. 2018. "High-Resolution Spatiotemporal Transcriptome Mapping of Tomato Fruit Development and Ripening." *Nature Communications* 9 (1): 364. doi:10.1038/s41467-017-02782-9.

Sohrabi, Reza, Jung-Hyun Huh, Somayesadat Badieyan, Liva Harinantenaina Rakotondraibe, Daniel J. Kliebenstein, Pablo Sobrado, and Dorothea Tholl. 2015. "In Planta Variation of Volatile Biosynthesis: An Alternative Biosynthetic Route to the Formation of the Pathogen-Induced Volatile Homoterpene DMNT via Triterpene Degradation in Arabidopsis Roots." *The Plant Cell* 27 (3): 874–90. doi:10.1105/tpc.114.132209.

Usadel, Björn, Takeshi Obayashi, Marek Mutwil, Federico M Giorgi, George W Bassel, Mimi Tanimoto, Amanda Chow, Dirk Steinhauser, Staffan Persson, and Nicholas J Provart. 2009. "Co-Expression Tools for Plant Biology: Opportunities for Hypothesis Generation and Caveats." *Plant, Cell & Environment* 32 (12): 1633–51. doi:10.1111/j.1365-3040.2009.02040.x.

Voo, S. S., H. D. Grimes, and B. M. Lange. 2012. "Assessing the Biosynthetic Capabilities of Secretory Glands in Citrus Peel." *Plant Physiology* 159 (1): 81–94. doi:10.1104/pp.112.194233.

Widhalm, Joshua R., Michael Gutensohn, Heejin Yoo, Funmilayo Adebesin, Yichun Qian, Longyun Guo, Rohit Jaini, et al. 2015. "Identification of a Plastidial Phenylalanine Exporter That Influences Flux Distribution through the Phenylalanine Biosynthetic Network." *Nature Communications* 6 (September): 8142. doi:10.1038/ncomms9142.

Wong, Darren C.J., Ranamalie Amarasinghe, Vasiliki Falara, Eran Pichersky, and Rod Peakall. 2019. "Duplication and Selection in β-Ketoacyl-ACP Synthase Gene Lineages in the Sexually Deceptive *Chiloglottis* (Orchidaceace)." *Annals of Botany* 123 (6): 1053–66. doi:10.1093/aob/mcz013.

Wong, Darren C J, Ranamalie Amarasinghe, Eran Pichersky, and Rod Peakall. 2018. "Evidence for the Involvement of Fatty Acid Biosynthesis and Degradation in the Formation of Insect Sex Chiloglottones in Sexually Deceptive *Chiloglottis* Orchids." *Frontiers in Plant Science* 9 (June): 389. doi:10.3389/fpls.2018.00839.

Wong, Darren C J, Eran Pichersky, and Rod Peakall. 2017. "The Biosynthesis of Unusual Floral Volatiles and Blends Involved in Orchid Pollination by Deception: Current Progress and Future Prospects." *Frontiers in Plant Science* 8: 1955. doi:10.3389/fpls.2017.01955.

Xu, Haiyang, Björn Bohman, Darren C.J. Wong, Claudia Rodriguez-Delgado, Adrian Scaffidi, Gavin R. Flematti, Ryan D. Phillips, Eran Pichersky, and Rod Peakall. 2017. "Complex Sexual Deception in an Orchid Is Achieved by Co-Opting Two Independent Biosynthetic Pathways for Pollinator Attraction." *Current Biology* 27 (13): 1867–77. doi:10.1016/j.cub.2017.05.065.

Xu, Haiyang, Gaurav D. Moghe, Krystle Wiegert-Rininger, Anthony L. Schilmiller, Cornelius S. Barry, Robert L. Last, and Eran Pichersky. 2018. "Coexpression Analysis Identifies Two Oxidoreductases Involved in the Biosynthesis of the Monoterpene Acid Moiety of Natural Pyrethrin Insecticides in *Tanacetum cinerariifolium*." *Plant Physiology* 176 (1): 524–37. doi:10.1104/pp.17.01330.

7 Flux Distribution Dynamics at the Interface of Central Carbon Metabolism and Terpenoid Volatile Formation

Bernd Markus Lange

CONTENTS

7.1 INTRODUCTION

The metabolic network in any given plant cell consists of thousands, if not tens of thousands, of biochemical reactions. Even when attempting to capture merely the essence of these reactions as schemes or maps, one is still left with a puzzling complexity. As an example, a node map, which only shows signature intermediates and end products, is given in Figure 7.1 for a leaf mesophyll cell of the model plant *Arabidopsis thaliana* L. Such complexity cannot be understood intuitively, and researchers are therefore developing computational approaches to simulate how individual reactions and pathways impact the overall metabolic network. In Flux Balance Analysis (FBA), net flux distribution is constrained by experimentally determined inputs and outputs (external fluxes), mass balances (conservation of mass across all reactions), and thermodynamic feasibility (Gibbs free energy) (Orth et al., 2010). FBA probes the probable solution space to identify metabolic flux distributions that are optimized toward achieving an objective, such as the accumulation of biomass. In metabolic flux analysis (MFA), isotopically labeled precursors are supplied and the labeling patterns of metabolic end products determined. Fluxes are then calculated by optimizing the fitting of internal fluxes to the measurements of labeling and external fluxes (Allen, 2016). This approach generally leads to more accurate flux estimations compared to FBA but the scope is often limited due to limitations in quantifying labeling of a larger number of metabolites. When detailed biochemical information is available, it is possible to combine kinetic rate equations with the known stoichiometry of metabolic reactions to predict fluxes accurately down to the levels of individual metabolic intermediates (kinetic models), but such models can usually only be generated for a fairly limited number of reactions (Rios-Estepa and Lange, 2007). While mathematical modeling of metabolic fluxes in plants is becoming more common overall, I am only aware of two classes of volatiles for which these computational approaches have been employed: (1) metabolites of the benzenoid/phenylpropanoid pathway in petunia flowers, and (2) terpenoids, which are the focus of the present chapter.

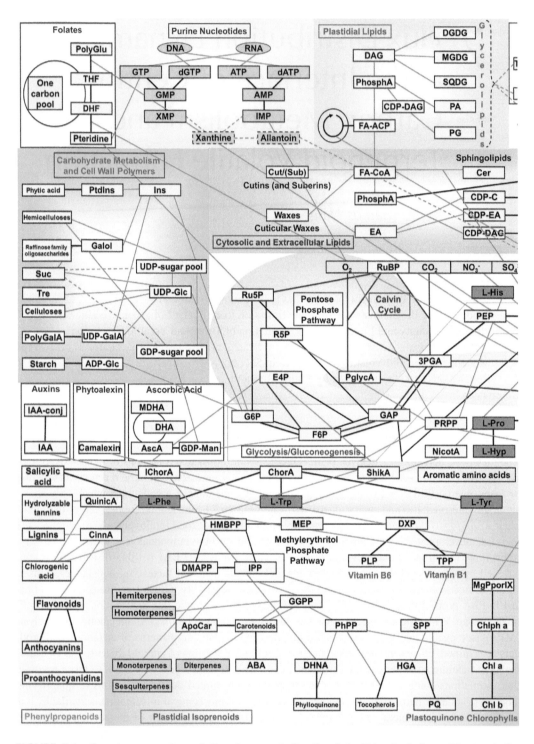

FIGURE 7.1 Overview map of metabolism in mesophyll cells of *Arabidopsis thaliana*. The metabolite classes in purple boxes (various volatile and semi-volatile terpenoids) are the focus of the present chapter.

(*Continued*)

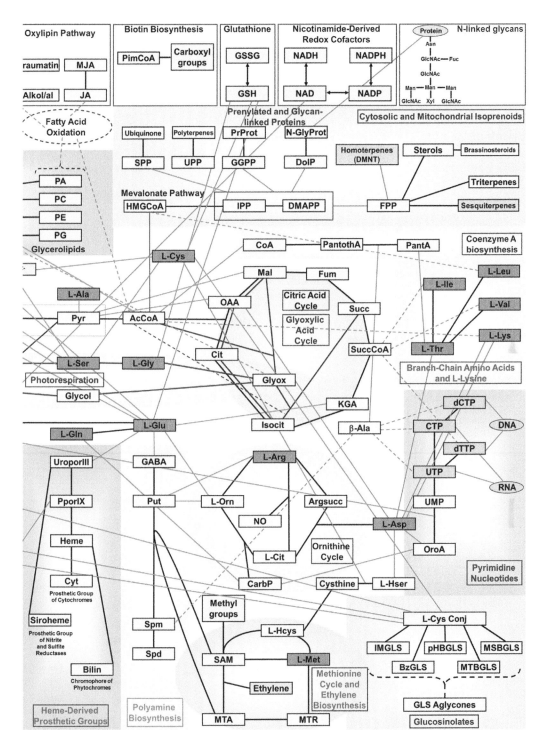

FIGURE 7.1 (Continued)   Overview map of metabolism in mesophyll cells of *Arabidopsis thaliana*. The metabolite classes in purple boxes (various volatile and semi-volatile terpenoids) are the focus of the present chapter.

Terpenoids (also referred to as isoprenoids) constitute the largest and structurally most diverse class of plant metabolites. If volatiles and semi-volatiles are defined as evaporating between 5% and 95% by weight during six months at ambient temperature (Võ and Morris, 2014), terpenoids up to about C15, with two or fewer polar functional groups, fit this interpretation of the term. I will discuss the biosynthesis of several classes of volatile terpenoids in general but also mention specific metabolites the structures of which are provided in Figure 7.2.

The biosynthesis of terpenoids is based on the modular assembly of two C5 intermediates, isopentenyl diphosphate (IPP) and dimethylallyl diphosphate (DMAPP). In plants, two spatially separated pathways generate IPP and DMAPP: the enzymes of the methylerythritol 4-phosphate (MEP) pathway are localized in plastids, while those of the mevalonate (MVA) pathway are found in the

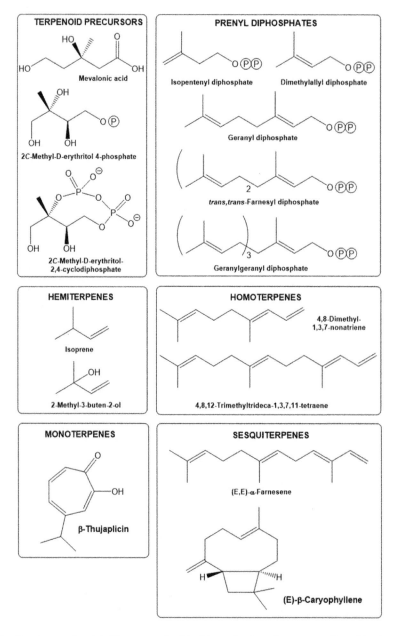

FIGURE 7.2   Structures of metabolites, organized by biosynthetic pathways, mentioned in the narrative.

cytosol, ER and peroxisomes (Hemmerlin et al., 2012). The implications resulting from the operation of and crosstalk between these precursor pathways for terpenoid volatile biosynthesis are discussed in Section 7.2.

Isoprene is a hemiterpene consisting of a single C5 unit. This metabolite is emitted in very large quantities by various trees and shrubs and accounts for about one-third of all hydrocarbons released into the atmosphere (about 600 Tg per year; Guenther et al., 2006). It is thus not surprising that isoprene has significant impacts on the chemistry of the troposphere, the lowest layer of earth's atmosphere (Sharkey et al., 2008). A model accounting for the dynamics underlying isoprene emission from poplar foliage is introduced in Section 7.3.

Terpenoid volatile biosynthesis is a highly dynamic process, with remarkable organ-, tissue- and cell type-specificity, which is regulated by various developmental and environmental signals (Nagegowda, 2010). Examples of the differences in flux distribution through terpenoid biosynthetic pathways under various experimental conditions in *Arabidopsis* are reviewed in Section 7.4.

Larger amounts of terpenoid volatiles, with important roles in constitutive and/or inducible defenses, can be accumulated in secretory structures, among which resin ducts and glandular trichomes are the most common (Lange, 2015). Various types of mathematical models have been developed to better understand the metabolic specialization of secretory cell types, and an overview of recent progress is given in Section 7.5.

7.2 CONTRIBUTION OF TWO PRECURSOR PATHWAYS TO TERPENOID VOLATILE PRODUCTION

Research in the 1950s with yeasts and animals established the enzymatic reactions constituting the MVA pathway, which was long assumed to be the universal source of precursors for terpenoid biosynthesis; a reevaluation beginning in the late 1980s ultimately led to the discovery of the MEP pathway, which was found to operate in certain microbes and plants (Hemmerlin et al., 2012). An elegant approach to investigate the relative contribution of the MVA and MEP pathways in the biosynthesis of terpenoids relies on feeding isotopically labeled precursors and analyzing labeling patterns in the relevant end products.

The biosynthesis of isoprene was investigated in four evolutionarily distant species (*Chelidonium majus*, *Platanus* × *acerifolia*, *Populus nigra*, and *Salix viminalis*) and its precursor DMAPP was demonstrated to be generated predominantly via the MEP pathway in all cases (Schwender et al., 1997; Zeidler et al., 1997) (Table 7.1). This is also consistent with the MEP origin of the terpenoid moiety of a terpenyl phenyl ether of the liverwort *Tricholea tomentella*. Studies on monoterpene biosynthesis are in agreement with the stipulation that the MEP pathway is the primary source of precursors (Table 7.1 and citations therein), with the only currently known exception being monoterpenes accumulated in fruit and, to a lesser extent, leaves of *Fragaria* × *ananassa* (strawberry) (Hampel et al., 2006).

The source of precursors for sesquiterpene biosynthesis can vary substantially. Sesquiterpenes of some liverworts (*Conocephalum conicum*, *Fossombronia pusilla*, and *Ricciocarpus natans*) are derived from precursors generated via the MVA pathway (Adam and Croteau, 1998; Spiteller et al., 2002; Thiel and Adam, 2002), whereas IPP and DMAPP for others (*Lepidolaena pusilla* and *Tricicolea tomentella*) are produced by the MEP pathway (Barlow et al., 2001, 2003). Only two gymnosperms have been investigated thus far, and these studies indicate that precursors for β-thujaplicin (monoterpene of *Cupressus lusitanica*) and 2-methyl-3-buten-2-ol (hemiterpenoid of and *Pinus ponderosa*) were both formed via the MEP pathway (Fujita et al., 2000; Zeidler and Lichtenthaler, 2001) (Table 7.1). In the angiosperm lineage, the MVA pathway was determined to be the primary source of precursors for sesquiterpenes in *Atractylodes lancea* (Asteraceae) (Wang et al., 2015), *Tanacetum vulgare* (Asteraceae) (Umlauf et al., 2004), and *Fragaria* × *ananassa* (Rosaceae) (Hampel et al., 2006). The MEP pathway was the primary (essentially exclusive) route

TABLE 7.1

Contribution of MEP and MVA Pathways to Terpenoid Volatile Biosynthesis Based on Incorporation of Labeled Precursors

Plant Species	Metabolite Class(es) Investigated	Relevant Precursor Pathways	Comments	References
Liverworts				
Conocephaluym conicum (snake liverwort)	Monoterpenes (borneol, bornyl acetate)	MEP (exclusively)	Feeding of [1-^{13}C]glucose; inference of precursor pathway from labeling patterns in end products	Adam et al. (1998)
	Sesquiterpene (cubebanol)	MVA (exclusively)		
	Sesquiterpene (cubebanol)	MVA (exclusively)	Feeding of [1-^{13}C]1-deoxy-ᴅ-xylulose; breakdown to acetate and incorporation via MVA pathway	Thiel and Adam (2002)
Fossombronia pusilla (common frillwort)	Sesquiterpene (Geosmin)	MVA (exclusively)	Feeding of [5,5-^{2}H$_2$]1-desoxy-ᴅ-xylulose or [4,4,6,6,6-^{2}H$_5$]mevalonolactone	Spiteller et al. (2002)
Lepidolaena hodgsoniae (leafy liverwort)	Sesquiterpenoid (hodgsonox)	MEP (exclusively)	Feeding of [1-^{13}C]glucose; inference of precursor pathway from labeling patterns in end products	Barlow et al. (2003)
Ricciocarpos natans (fringed heartwort)	Sesquiterpene (ricciocarpin A)	MVA	Feeding of [1-^{13}C]glucose; inference of precursor pathway from labeling patterns in end products	Adam et al. (1998)
Tricocolea tomentella (woolywort)	Hemiterpenes (tricholein with C5 prenyl unit, deoxytomentellin with C10 prenyl unit)	MEP (exclusively)	Feeding of [1-^{13}C]glucose; inference of precursor pathway from labeling patterns in end products	Barlow et al. (2001)
Gymnosperms				
Cupressus lusitanica (Mexican white cedar)	Monoterpene (thujaplicine)	MEP (primarily)	Feeding of [1-^{13}C]glucose, [2-^{13}C]glucose or [U-^{13}C]glucose/unlabeled glucose; inference of precursor pathway from labeling in end products	Fujita et al. (2000)
Pinus ponderosa (ponderosa pine)	Hemiterpene (2-methyl-3-buten-2-ol)	MEP (primarily)	Feeding of [1-^{2}H$_1$]1-deoxy-ᴅ-xylulose or [2-^{13}C] deoxy-ᴅ-xylulose	Zeidler and Lichtenthaler (2001)

(Continued)

TABLE 7.1 (Continued)
Contribution of MEP and MVA Pathways to Terpenoid Volatile Biosynthesis Based on Incorporation of Labeled Precursors

Plant Species	Metabolite Class(es) Investigated	Relevant Precursor Pathways	Comments	References
Apiaceae				
Daucus carota (wild carrot)	Monoterpenes (terpinolene, myrcene)	MEP	Feeding of [5,5-^{2}H$_2$]1-deoxy-D-xylulose or [5,5-^{2}H$_2$]mevalolactone to roots	Hampel et al. (2005a)
	Sesquiterpenes (β-caryophyllene)	MEP and MVA		
Asteraceae				
Atractylodes lancea ("Cangzhu" in Chinese)	Sesquiterpenoids (hinesol, β-eudesmol, atracylone)	MVA (primarily)	Induction of volatile production by bacterial endophyte; treatment with fosmidomycin (MEP pathway inhibitor) or mevinolin (MVA pathway inhibitor)	Wang et al. (2015)
Matricaria recutita (German chamomile)	Sesquiterpenes (bisabololoxide A, chamazulene)	MEP (primarily) and MVA (contribution to terminal C5 unit)	Feeding of [1-^{13}C]glucose; inference of precursor pathway from labeling patterns in end products	Adam and Zapp (1998)
Matricaria recutita (German chamomile)	Sesquiterpenes (bisabololoxide A, bisabololoxide B, chamazulene)	MEP (primarily) and MVA (contribution to terminal C5 unit)	Feeding of [1-^{13}C]1-deoxy-D-xylulose	Adam et al. (1999)
Solidago canadensis (Canadian goldenrod)	Sesquiterpene (germacrene D)	MEP (predominantly)	Feeding with [5,5-^{1}H$_2$]1-deoxy-D-xylulose-5-phosphate, [5-^{13}C]mevalonolactone or [1-^{13}C]glucose	Steliopoulos et al. (2002)
Tanacetum vilgare (tansy)	Monoterpenes (artemisia ketone, camphor, β-thujone)	MEP	Feeding of [1-^{13}C]glucose; inference of precursor pathway from labeling patterns in end products	Umlauf et al. (2004)
	Sesquiterpenes (germacrene D	MVA		
Fabaceae				
Phaseolus lunatus (lima bean)	Monoterpenes (ocimene, linalool)	MEP (primarily)	Feeding of [5,5-^{2}H$_2$]1-deoxy-D-xylulose or [4,4,6,6,6-^{2}H$_5$]mevalolactone	Bartram et al. (2006)
	Sesquiterpene (caryophyllene)	MEP and MVA (mixed)		
	Homoterpene (4,8-dimethyl-nona-1,3,7-triene; C15-derived)	MEP and MVA (mixed)		

(Continued)

TABLE 7.1 (Continued)
Contribution of MEP and MVA Pathways to Terpenoid Volatile Biosynthesis Based on Incorporation of Labeled Precursors

Plant Species	Metabolite Class(es) Investigated	Relevant Precursor Pathways	Comments	References
Lamiaceae				
Mentha citrata (bergamot mint)	Monoterpenes (linalyl acetate)	MEP	Feeding of [6,6-^{2}H$_2$]glucose; inference of precursor pathway from labeling patterns in end products	Fowler et al. (1999)
Mentha × piperita (peppermint)	Monoterpenes (limonene, pulegone, menthone)	MEP (exclusively)	Feeding of [^{14}C]pyruvic acid, [^{14}C]acetyl-CoA or [^{14}C]mevalonolactone	McCaskill and Croteau (1995)
	Sesquiterpenes (caryophyllene, humulene, germacrene D)	MEP (exclusively)		
Mentha × piperita (peppermint)	Monoterpenes (geraniol, thymol, pulegone, menthone)	MEP	Feeding of [1-^{13}C]glucose or [U-^{13}C]glucose/unlabeled glucose; inference of precursor pathway from labeling patterns in end products	Eisenreich et al. (1997)
Myrtaceae				
Eucalyptus globulus (Tasmanian blue gum)	Monoterpenes (cineol)	MEP	Feeding of [4-^{2}H$_1$]1-deoxy-D-xylulose or [2-^{13}C],4-^{2}H$_1$]1-deoxy-D-xylulose	Rieder et al. (2000)
Oleaceae				
Syringa vulgaris (common lilac)	Monoterpenes (lilac alcohol, lilac aldehyde)	MEP	Feeding with [5,5-^{1}H$_2$]1-deoxy-D-xylulose-5-phosphate	Burkhardt and Mosandl (2003)
Papaveraceae				
Chelidonium majus (greater calandine)	Isoprene	MEP	Feeding of [1-^{2}H$_1$]1-deoxy-D-xylulose	Zeidler et al. (1997)
Plantaginaceae				
Antirrhinum majus (snapdragon)	Monoterpenes (myrcene, ocimene, linalool)	MEP (exclusively)	Feeding with [5,5-^{1}H$_2$]1-deoxy-D-xylulose-5-phosphate, [5-^{13}C]mevalonolactone	Dudareva et al. (2005)
	Sesquiterpenes (nerolidol)	MEP (exclusively)		

(Continued)

TABLE 7.1 (*Continued*)

Contribution of MEP and MVA Pathways to Terpenoid Volatile Biosynthesis Based on Incorporation of Labeled Precursors

Plant Species	Metabolite Class(es) Investigated	Relevant Precursor Pathways	Comments	References
Rosaceae				
Fragaria × ananassa fruit (garden strawberry)	Monoterpenes ((*S*)-linalool, (-)-α-pinene) Sesquiterpenes (nerolidol)	MVA	Feeding of [5,5-^{2}H$_2$]1-deoxy-D-xylulose or [5,5-^{2}H$_2$]mevalolactone to roots	Hampel et al. (2006)
Fragaria × ananassa leaf (garden strawberry)	Monoterpenes ((*R*)-linalool, (*S*)-linalool)	MEP and MVA (no quantitative data)		
Salicaceae				
Populus nigra (black poplar)	Isoprene	MEP and MVA (no quantitative data)	Feeding of [1-^{2}H$_1$]1-deoxy-D-xylulose or [2-^{13}C] mevalolactone	Schwender et al. (1997)
Salix nigra (black poplar)	Isoprene	MEP	Feeding of [1-^{2}H$_1$]1-deoxy-D-xylulose	Zeidler et al. (1997)
Vitaceae				
Vitis vinifera (common grapevine)	Monoterpenes ((*S*)-linalool, (*E*)-β-ocimene) Sesquiterpene ((*E, E*)-α-farnesene) C15-derived homoterpene (4,8-dimethyl-nona-1,3,7-triene)	MEP MEP and MVA (mixed) MEP and MVA (mixed)	Feeding of [5,5-^{2}H$_2$]1-deoxy-D-xylulose or [5,5-^{2}H$_2$]mevalolactone to roots	Hampel et al. (2005b)

for the formation of sesquiterpene precursors in *Mentha* × *piperita* (Lamiaceae) (McCaskill and Croteau, 1995) and *Antirrhinum majus* (Plantaginaceae) (Dudareva et al., 2005).

In other studies, the sesquiterpene skeletons were of mixed MEP and MVA pathway origin, with the contribution of the MEP pathway predominating, as demonstrated for *Daucus carota* (Apiaceae) (Hampel et al., 2005a), *Matricaria recutita* (Asteraceae) (Adam and Zapp, 1998; Adam et al., 1998), *Solidago canadensis* (Asteraceae) (Steliopoulos et al., 2002), *Phaseolus lunatus* (Fabaceae) (Bartram et al., 2006), and *Vitis vinifera* (Vitaceae) (Hampel et al., 2005b) (Table 7.1). Additional complexity was recognized when researchers were able to dissect the source of differential labeling of each individual carbon atom of sesquiterpenes. In flowers of *Matricaria recutita* (German chamomile), sesquiterpenes contained two C5 units that were derived almost exclusively from the MEP pathway, whereas a mixed contribution from both the MEP and MVA pathways was determined for the terminal C5 unit (Adam and Zapp, 1998; Adam et al., 1999). A dynamic, treatment-dependent, allocation of precursors was reported for plantlets of *Phaseolus lunatus* (lima bean): the C30-derived homoterpene 4,8-dimethyl-nona-1,3,7-triene (DMNT) was formed primarily from the MVA pathway following insert feeding, whereas the MEP pathway was the predominant source for DMNT precursors after elicitation with a fungal cell wall preparation (Bartram et al., 2006). Currently available evidence thus indicates that it will likely be difficult to formulate simple rules about the roles of the MVA and MEP pathways for specific terpenoids, as their contribution appears to be a dynamic parameter that is dependent on several internal and external factors. Such knowledge is therefore essential when attempting to build mathematical models for the metabolic networks underlying terpenoid volatile biosynthesis.

Pokhilko et al. (2015) developed a kinetic model of the MEP pathway to compare how various assumptions affect the predicted accumulation of terpenoid end products in *Arabidopsis*. I will use this study to delineate what kinds of inferences are employed in modeling and emphasize the need for experimental testing. The authors measured transcript abundance for the MEP pathway genes 1-deoxyxylulose 5-phosphate synthase (DXS) and 4-hydroxy-3-methylbut-2-enyl diphosphate reductase (HDR), as well as the carotenoid biosynthetic gene phytoene synthase (PSY) (Figure 7.3a), and for all of these genes a diurnal pattern of expression was noted. The protein abundance of DXS and 1-deoxyxylulose 5-phosphate reductoisomerase (DXR) were assessed as well. Transgenic lines in which DXS expression was increased by three-fold at mid-day (compared to the appropriate wild-type controls) served as an experimental resource for this project (Figure 7.3a). Model parameters that were inferred from the literature included the (1) ordinary differential equations that capture enzyme catalysis, including the catalytic constant (K_{cat}) and Michaelis constant (K_m) of each enzyme/substrate pair, (2) allosteric regulation and post-transcriptional regulation of DXS activity by IPP and DMAPP (through modification of the ordinary differential equation for this enzymatic reaction), (3) steady-state concentrations of each MEP pathway intermediate, and (4) overall flux through the MEP pathway (Figure 7.3a). Substrate availability in source and sink tissues was incorporated by modeling carbon partitioning in these cell types separately, and diurnal dynamics were modeled by allowing components of the circadian oscillator to exert control over the expression of the DXS, HDR and PSY genes (Figure 7.3b).

Simulations with the above-mentioned parameters (and variations of parameters to test for the sensitivity of flux) were then used to model the diurnal fluctuations of metabolites and flux through the MEP pathway (Figure 7.3c). Other predictions pertained to the effects of increased DXS transcript abundance in transgenics on MEP pathway flux (Figure 7.3d). The authors presented evidence that the feedback regulation of DXS adjusts flux through the MEP pathway depending on precursor supply and their modeling hinted at the possibility that product demand, through consumption of the MEP pathway products IPP and DMAPP, could increase flux by decreasing feedback down-regulation of DXS (Pokhilko et al., 2015).

The reader might ask how one can make predictions when so many modeling parameters are educated guesses rather than measured values under the experimental conditions evaluated in this study. That is certainly a valid concern but, while absolute values might not be accurate, certain

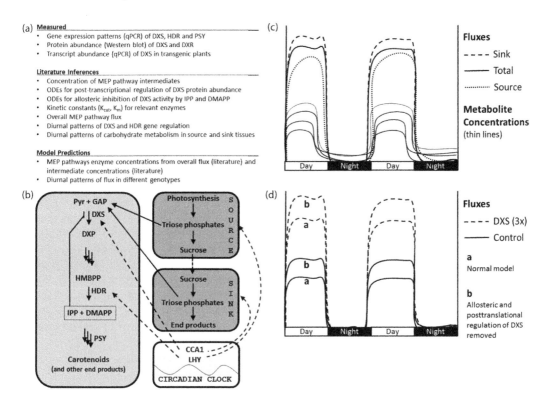

FIGURE 7.3 Kinetic model of the MEP pathway in *Arabidopsis* (Pokhilko et al., 2015). (a) Listing of modeling parameters. (b) Outline of model scope, with gray-shaded submodels (dark gray, source-sink tissues that provide precursors through photosynthetic activity; light gray, the genes of the circadian clock that regulate genes of the MEP pathway; medium gray, MEP pathway with primary end products). (c) Diurnal patterns of accumulation (no absolute scale shown) of intermediates of the MEP pathway (from top to bottom: glyceraldehyde 3-phosphate, 1-deoxy-D-xylulose 5-phosphate, IPP/DMAPP, and 2C-methyl-D-erythritol-2,4-cyclodiphosphate). Fluxes were calculated for source tissues (black dotted line), sink tissues (black broken line) and both cell types combined (solid black line). (d) MEP pathway flux calculated for transgenic lines overexpressing the DXS gene (broken line) and wild-type controls (black line). Modeling outcomes were significantly different when feedback regulation of DXS was considered (a) or omitted (b) in simulations. For details see text

general trends can be inferred that still provide interesting insights. However, it is still essential to integrate modeling into an iterative cycle that uses information gained from modeling to design new experiments, which then test the first-generation model predictions and suggest adjustments for the development of a second-generation model (and so on). For example, the predictions by Pokhilko et al. (2015) about the effects of DXS feedback regulation would be testable by using transgenics that express an engineered, feedback-insensitive DXS (Banerjee et al., 2016). Another important consideration for modeling is the scope of the model. Pokhilko et al. (2015) considered DXS as a master regulator, with the assumption that other enzymes of the MEP pathway play nonlimiting roles. This view might be too simplistic, as the control of flux through the MEP pathway is likely to be shared by several enzymes (Lange et al., 2015). However, since information is lacking about the mechanistic underpinnings of how each enzyme activity of the MEP pathway is controlled under specific experimental conditions, this study helps with defining the upper and lower bounds for flux through individual reactions. Further research will now be needed to test the accuracy of modeling predictions. It is also important to note that, at present, we should generally not regard metabolic models as directly reflecting reality but rather as a tool to explore possibilities that are currently too difficult to fully address experimentally.

7.3   MODELING THE BIOCHEMISTRY OF ISOPRENE
EMISSION FROM POPLAR FOLIAGE

A large body of literature on the physiological and biochemical dynamics underlying isoprene emission has emerged since the early 1990s, and the interested reader is referred to an excellent review article for further details (Sharkey and Monson, 2017). I will only highlight a few landmark discoveries here (with apologies for not mentioning all relevant studies). The first evidence that the foliage of some plants emits copious quantities of isoprene was presented in the 1950s (Sanadze, 1957). It was later demonstrated that carbon dioxide assimilated through photosynthesis was quickly incorporated into isoprene (Sanadze et al., 1972). However, high CO_2 concentrations were found to suppress isoprene emission (Sanadze, 1990). During the 1960s, it was recognized that the emission of isoprene was highly dependent on temperature (Sanadze and Kalanadze, 1966). Isoprene was reported to be emitted by undisturbed leaves of eucalyptus (*Eucalyptus obliqua*), but monoterpene (not isoprene) emission became prevalent when leaves where subjected to mechanical stress (such as movement) (Rasmussen, 1972). It was then realized that isoprene emission was light-saturated during most of the daylight hours in live oak (*Quercus virginiana*), indicating that temperature (and not light) was the dominating factor (Tingey et al., 1979). Water limitation was shown to decrease photosynthetic activity, while isoprene emission was less sensitive to that treatment in kudzu (*Pueraria lobata*) (Sharkey and Loreto, 1993). The MEP pathway was determined as the primary source of precursors for isoprene biosynthesis in several species (Schwender et al., 1997; Zeidler et al., 1997) (as mentioned above in Section 7.2). During the early 1990s, the enigmatic enzymatic activity for the conversion of DMAPP to isoprene, termed isoprene synthase, was discovered and characterized from aspen (*Populus tremuloides*) (Silver and Fall, 1991, 1995).

Ghirardo et al. (2014) developed a terpenoid pathway-specific MFA model for isoprene-emitting (IE) gray poplar (*Populus* × *canescens*) and transgenics of the same species in which isoprene synthase expression is downregulated (NE), thereby decreasing isoprene emissions to below 5% of those in wild-type plants. For both IE and NE plants, the authors determined, among other parameters, (1) the concentration and *in vitro* specific activity of DXS, (2) the incorporation of label from $^{13}CO_2$ into 2C-methyl-D-erythritol-2,4-cyclodiphosphate (MEcPP; intermediate of the MEP pathway), DMAPP, isoprene, chlorophyll a and b, and carotenoids (lutein and β-carotene) and the concentrations of these metabolites under standard growth conditions, and (3) the concentrations of MEcPP and DMAPP under conditions of varying temperature, light intensity and carbon dioxide availability (Figure 7.4a). The experiments confirmed the notion that the MEP pathway is responsible for providing precursors for isoprene production from $^{13}CO_2$, but also established that a tight correlation of MEcPP and plastidial DMAPP concentrations with isoprene emission (Ghirardo et al., 2014). The authors then compared *in vivo* carbon fluxes through the MEP pathway and into terpenoid end products in IE and NE plants (Figure 7.4b, c). The previously reported allosteric inhibition of DXS by DMAPP (Banerjee et al., 2013) was suggested to be an important factor in the reduced flux through the MEP pathway in NE plants (compared to IE controls). However, modeling indicated that additional regulation has to be invoked to explain the extremely low MEP pathway flux in NE. An interesting observation was that the decreased flux of carbon into isoprene in NE versus IE plants (0.5 and 26.1 nmol·C·m^{-2}·s^{-1}, respectively) was accompanied by a slightly increased flux into chlorophylls and carotenoids (0.4 and 0.01 nmol C m^{-2} s^{-1}, respectively). The increased availability of DMAPP in NE leaves might explain why flux is increased through the carotenoid pathway, but it does not help with rationalizing the observation that the overall flux into terpenoids is dramatically reduced in NE leaves. An intriguing possibility is that the observed lower isoprene emission might lead to a redirection of carbon flux into other metabolic pathways, which can ultimately lead to greater biomass (Behnke et al., 2012; Ryan et al., 2014), but this hypothesis requires further experimental support.

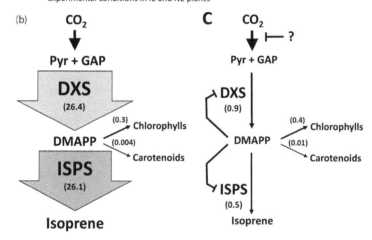

(a) **Measured**
- Labeling of MEcPP, DMAPP, isoprene from $^{13}CO_2$ in IE and NE plants
- Concentrations of MEcPP and DMAPP under various conditions (temperature, light intensity, CO_2 concentration)
- DXS protein concentration in IE and NE crude extracts
- DXS specific activity in IE and NE crude extracts
- Concentrations of terpenoid end products (chlorophylls, carotenoids)
- Counts of palisade and spongy mesophyll cells
- Transcript abundance of DXS, DXR, CMK, HDR, HMGR and MK in IE and NE plants

Literature Inferences
- Number of chloroplasts and their volume to calculate plastidial DMAPP concentration

Model Predictions
- Carbon fluxes into isoprene and other terpenoid end products under various experimental conditions in IE and NE plants

FIGURE 7.4 Metabolic flux analysis of isoprene biosynthesis in poplar leaves (Ghirardo et al., 2014). (a) Listing of modeling parameters. Visual representation of flux (in nmol C m^{-2} s^{-1}) toward isoprene in wild-type plants (b) and transgenic plants expressing a construct of the suppression of isoprene synthase (c). For details see text.

7.4 ORGAN-SPECIFICITY AND ENVIRONMENTAL REGULATION OF TERPENOID VOLATILE FORMATION IN *ARABIDOPSIS*

There are currently no published models available that compare flux toward volatiles in different tissues or organs of the same species, and an assessment of dynamic changes in flux when plants experience various stresses is also lacking in the literature. I am therefore using previously published data obtained as part of experiments with the model plant *Arabidopsis* to visualize the distribution of carbon flux into different terpenoids (purely based on the concentrations of end products). Please note that the goal of this exercise is not to provide a comprehensive description of the entire body of work that has been published to date but to emphasize, using a few exemplary studies, the enormous plasticity of metabolism under different experimental situations.

In leaves of plants maintained under growth conditions with no apparent stress, the plastidial MEP pathway provides precursors for chlorophylls (677 µg·g^{-1}) and carotenoids (498 µg·g^{-1}) as major constituents, with α-tocopherol (170 µg·g^{-1}) and plastoquinone (140 µg·g^{-1}) as minor products (Fraser et al., 2000); sterols are the dominant end products derived from the MVA pathway (1,900 µg·g^{-1}) (Feldman et al., 2015), with ubiquinone as a less abundant product (200 µg·g^{-1}) (Fraser et al., 2000) (Figure 7.5a). Under these conditions, *Arabidopsis* leaves emit only trace levels of volatiles. However, feeding by insects (e.g., larvae of the crucifer specialist *Plutella xylostella* (diamonback moth)) leads to the emission of a blend of volatiles. The C16 homoterpene 4,8,12-trimethyltrideca-1,3,7,11-tetraene (TMTT), which is derived from a C20 precursor by cytosolically localized enzymes, represents the primary herbivore-induced volatile (4.8 ng·g^{-1}·h^{-1}), while (E, E)-α-farnesene (sesquiterpene) is released at lower rates (0.47 ng·g^{-1}·h^{-1}) (Herde et al., 2008) (Figure 7.5a). It would now be informative to investigate if volatile emissions from leaves under herbivore attack affect flux distribution through other branches of terpenoid biosynthesis or carbon metabolism in general.

Sterols are abundant metabolites in *Arabidopsis* flowers with concentrations of ~3,900 μg·g^{-1} (DeBolt et al., 2009). Because of the white color of the petals, it can be assumed that chlorophyll (green) and carotenoid (orange/red) contents are comparatively low in this organ but I could not find reliable information on quantitative measurements in the literature. Interestingly, flowers constitutively release a blend of volatiles consisting of primarily sesquiterpenes (21.7 ng·h^{-1} from 70 inflorescences) and lower quantities of monoterpenes (4.6 ng h^{-1} from 70 inflorescences) (Tholl et al., 2005)

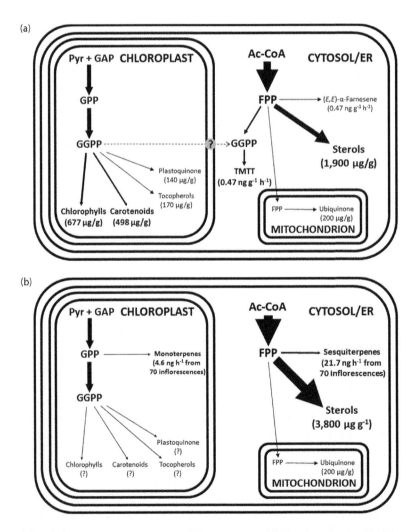

FIGURE 7.5 Flux distribution (indicated by thickness of reaction arrow) toward volatile and nonvolatile terpenoids in different *Arabidopsis* organs. (a) Leaves; (b) flowers; (c) seeds; and (d) roots. Abbreviations: Ac-CoA, acetyl-coenzyme A; ER, endoplasmic reticulum; FPP, farnesyl diphosphate; GAP, glyceraldehyde 3-phosphate; GGPP, geranylgeranyl diphosphate; GPP, geranyl diphosphate; Pyr, pyruvic acid. For details see text.   (*Continued*)

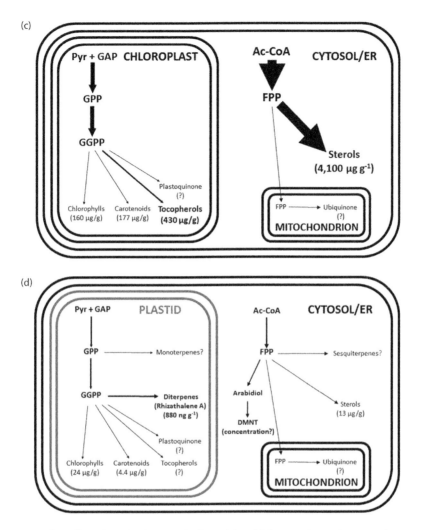

FIGURE 7.5 (Continued)   Flux distribution (indicated by thickness of reaction arrow) toward volatile and nonvolatile terpenoids in different *Arabidopsis* organs. (a) Leaves; (b) flowers; (c) seeds; and (d) roots. Abbreviations: Ac-CoA, acetyl-coenzyme A; ER, endoplasmic reticulum; FPP, farnesyl diphosphate; GAP, glyceraldehyde 3-phosphate; GGPP, geranylgeranyl diphosphate; GPP, geranyl diphosphate; Pyr, pyruvic acid. For details see text.

(Figure 7.5b). Constitutively emitted (E)-β-caryophyllene, which is formed in the *Arabidopsis* floral stigma, has antimicrobial activity and appears to serve as a direct defense against microbial pathogens such as *Pseudomonas synringae* (Huang et al., 2012). An important and as yet unresolved question pertains to how much of the overall flux is dedicated to the constitutive formation of floral terpenoid volatiles.

Mature seeds of *Arabidopsis* contain relatively low quantities of chlorophylls (160 $\mu g \cdot g^{-1}$) and carotenoids (177 $\mu g \cdot g^{-1}$) but comparatively higher concentrations of tocopherols (430 $\mu g \cdot g^{-1}$) (Gonzalez-Jorge et al., 2013; Zhang et al., 2014). However, the signature class of terpenoids of this organ are sterols, which accumulate to significant levels (>4,100 $\mu g \cdot g^{-1}$) (Chen et al., 2007) (Figure 7.5c). I am not aware of any studies reporting on the release of terpenoid volatiles from *Arabidopsis* seeds and their impact on flux distribution is therefore unknown. Experiments to assess the responses of seeds to pathogen attack would seem to be particularly informative for gaining insight into the capacity of this organ for terpenoid volatile production.

The inoculation of *Arabidopsis* roots with the oomycete root rot pathogen, *Pythium irregulare*, caused the upregulation of a pathway that leads to the initial formation of arabidiol (C30), with subsequent cleavage and further metabolization into nonvolatile 14-apo-arabidiol-related metabolites and the volatile homoterpene, (*E*)-4,8-dimethyl-1,3,7-nonatriene (DMNT) (Sohrabi et al., 2015). Quantitative data on these products are unfortunately not available. The emission of larger quantities of semi-volatile diterpenoids (880 ng·g⁻¹), was reported to occur as a response to root herbivory by the fungus gnat (*Bradysia* spp.), and one of these, the diterpenoid rhizathalene A, was shown to be a determinant of resistance (Vaughan et al., 2013) (Figure 7.5d). Interestingly, enzymes for the biosynthesis of root-specific monoterpenes and sesquiterpenes have been characterized (Chen et al., 2004; Ro et al., 2006), but the products of these enzymatic activities have not been measurable *in vivo*. The concentrations of chlorophylls (24 µg·g⁻¹), carotenoids (4.4 µg·g⁻¹) and sterols (~13 µg·g⁻¹) are comparatively low in *Arabidopsis* roots (Kobayashi et al., 2013; Frescatada-Rosa et al., 2014) and it would thus appear that the herbivore- or pathogen-induced production of volatiles and semi-volatiles is a significant contributor to the distribution of carbon across terpenoid metabolic pathways in this organ.

In summary, the examples discussed above provide evidence for significant variation of the concentrations of different terpenoids across different *Arabidopsis* organs and tissues. Volatiles are emitted constitutively by some organs (e.g., *Arabidopsis* flowers) but their formation can also be induced by pathogen attack or herbivore feeding. The concentrations of volatiles can be substantial at an infection site, which likely has as yet unknown impacts on other pathways. The ability to produce volatiles is important both as a direct defense but also for acquired immunity and plant–plant communication (Howe and Jander, 2008), and more experimental efforts in this area would be desirable, thereby informing models to assess dynamic changes in overall flux distribution.

7.5 MODELING TERPENOID VOLATILE METABOLISM IN SECRETORY STRUCTURES

Larger quantities of plant secondary (or specialized) metabolites often accumulate in specialized anatomical structures such as glandular trichomes, resin ducts, and laticifers (Chapter 13; Lange, 2015). The emphasis of the present chapter is placed on the discussion of emerging mathematical models developed in recent years to better understand metabolic flux in these secretory tissues. A more detailed account of other aspects of metabolism in glandular trichomes, including functional genomics for gene discovery and metabolic pathway regulation, is provided in Chapter 11. The first simulations of metabolic pathway flux in glandular trichomes were based on kinetic models. Rios-Estepa et al. (2008) employed mathematical modeling to help explain monoterpene profiles of peppermint glandular trichomes based on feedback regulation of a particular enzyme involved in the corresponding biosynthetic pathway. This model was then expanded in scope and depth to be brought to bear in investigations into the determinants of oil yield and composition in peppermint across various genotypes and experimental conditions (Rios-Estepa et al., 2010; Lange et al., 2011). These efforts have been reviewed recently (Lange and Rios-Estepa, 2014) and will therefore not be discussed in the present article, where the focus is on comparing genome-scale models of metabolism in secretory structures. Such models start with a reconstruction of the hundreds (in some cases thousands) of reactions deemed relevant for a certain biological context; measured parameters (e.g., gene expression patterns or proteomics data) are employed to constrain the solution space; and FBA is then used to compute, using linear programming, the flux through each reaction by finding the optimal solution to satisfy an objective function (e.g., producing experimentally determined quantities of metabolites) (Oberhardt et al., 2009).

The first genome-scale models for secretory structures were recently developed based on glandular trichome-specific data sets obtained with several species (Zager and Lange, 2018). Outcomes of this meta-analysis, using four species as examples, are visualized as flux maps in Figure 7.6. The abundance of transcripts related to oxygenic photosynthesis was sufficiently high in glandular trichomes of the painted tongue (*Salpiglossis sinuata*) (Moghe et al., 2017) and rough cocklebur (*Xanthium strumarium*) (Li et al., 2016) to warrant the assumption that photosynthetic processes contribute to the assembly of metabolic end products in both species. However, it has been shown for type VI glandular trichomes of tomato that, despite active photosynthesis therein, the most abundant end products, flavonoids and terpenoids, are generated primarily from an imported oligosaccharide, and it is therefore important to consider various levels of photosynthetic contribution to metabolism; furthermore, there is evidence that metabolically produced carbon dioxide might be recycled in photosynthetic glandular trichomes through the reactions of an incomplete Calvin–Benson cycle (Balcke et al., 2017). For modeling purposes (Zager and Lange, 2018), it was thus assumed that a combination of photosynthetic carbon fixation, carbon dioxide recycling and import of a transport oligosaccharide (sucrose) served as carbon sources for the reactions that lead to the accumulation of metabolic end products in glandular trichomes of painted tongue and rough cocklebur. In contrast, metabolism in glandular trichomes of peppermint (*Mentha × piperita*) and hop (*Humulus lupulus*) appears to be dependent entirely on an imported carbon source (Kavalier et al., 2011; Clark et al., 2013; Johnson et al., 2017), which has significant implications for flux distribution, as discussed in the following paragraph.

Genome-scale models suggest that carbon flux in photosynthetic glandular trichomes is distributed between the Calvin–Benson cycle and glycolysis, and the recycling of both redox cofactors (NADP/NADPH) and energy equivalents (ADP/ATP) is directly coupled to oxygenic photosynthesis (Zager and Lange, 2018) (Figure 7.6a, b). Flux distribution across central carbon metabolism is very similar in glandular trichomes of painted tongue and rough cocklebur. However, the distribution of flux into secondary (specialized) metabolism is quite different. In painted tongue, flavonoids (derived from phenylalanine), acyl sugars (assembled from oligosaccharides and branched-chain amino acids or fatty acids as precursors) and terpenoids (derived from acetyl-CoA or pyruvate/glyceraldehyde 3-phosphate) are formed as metabolic end products (Wollenweber et al., 2005; Moghe et al., 2017); in contrast, sesquiterpene lactones from acetyl-CoA are the primary end products accumulated in glandular trichomes of rough cocklebur (Chen et al., 2013), and these differences are reflected in the relevant fluxes (Figure 7.6a, b).

A prominent feature of metabolism in nonphotosynthetic glandular trichomes, based on genome-scale modeling outcomes (Zager and Lange, 2018), is that the oxidative pentose phosphate pathway provides reducing equivalents (NADPH), while the citric acid cycle and oxidative phosphorylation are highly active to regenerate ATP (Figure 7.6c, d); interestingly, ethanolic fermentation was demonstrated to occur as an alternate pathway to regenerate ATP in peppermint glandular trichomes (Johnson et al., 2017). Monoterpenoids are the primary end products of metabolism in peppermint glandular trichomes and flux through the MEP pathway is predicted to be very high (Johnson et al., 2017). In hop glandular trichomes, prenylated flavonoids are assembled from phenylalanine and terpenoid intermediates of MEP pathway origin; bitter acids, the predominant metabolic end products, are derived partly from branched-chain amino acids and DMAPP generated via the MEP pathway (Lange and Turner, 2013). Based on a meta-analysis of transcriptome data, it was concluded that the abundance of transcripts related to primary metabolic pathways were dramatically reduced in nonphotosynthetic glandular trichomes (<40% of metabolism-related transcripts) when compared to photosynthetic cell types (70–90% of metabolism-related transcripts) (Zager and Lange, 2018). In peppermint glandular trichomes, 55% of the metabolism-related transcripts are involved in terpenoid biosynthesis, while in hop roughly 90% of metabolism-related transcripts support the biosynthesis of bitter acids and prenyl flavonoids, indicating an extraordinarily high degree of metabolic specialization in these nonphotosynthetic cell types.

Metabolite profiling and cell type-specific transcriptome data have also been integrated into models of metabolism in the epithelial cells surrounding resin ducts in pine needles (Turner et al., 2019). Expectedly, the predicted flux distribution depended on the assumptions regarding the contribution of photosynthesis. Transcriptome data sets provided evidence for the expression of

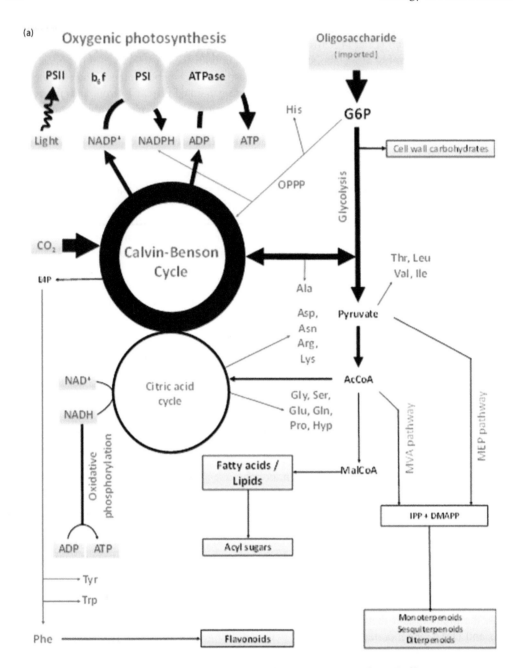

FIGURE 7.6 Flux distribution in secretory cell types. Examples are taken from the literature on secretory cells of glandular trichomes (a, *Salpiglossis sinuata*; b, *Xanthium stramonium*; c, *Mentha × piperita*; d, *Humulus lupulus*) and epithelial cells of resin ducts (e, *Pinus taeda*). Abbreviations: AcCoA, acetyl-coenzyme A; ADP, adenosine diphosphate; Ala, alanine; Arg, arginine; Asn, asparagine; Asp, aspartate; ATP, adenosine triphosphate; CO_2, carbon dioxide; DMAPP, dimethylallyl diphosphate; E4P, erythrose 4-phosphate; G6P, glucose 6-phosphate; Gln, glutamine; Glu, glutamate; Gly, glycine; His, histidine; Hyp, hydroxyproline; Ile, isoleucine; IPP, isopentenyl diphosphate; Leu, leucine; Lys, lysine; MalCoA, malonyl-coenzyme A; NAD^+, nicotinamide adenine dinucleotide (oxidized form); $NADP^+$, nicotinamide adenine dinucleotide phosphate (oxidized form); NADH, nicotinamide adenine dinucleotide (reduced form); NADPH, nicotinamide adenine dinucleotide phosphate (reduced form); OPPP, oxidative pentose phosphate pathway; Phe, phenylalanine; Pro, proline; Ser, serine; Thr, threonine; Trp, tryptophan; Tyr, tyrosine; Val, valine. For details see text.   *(Continued)*

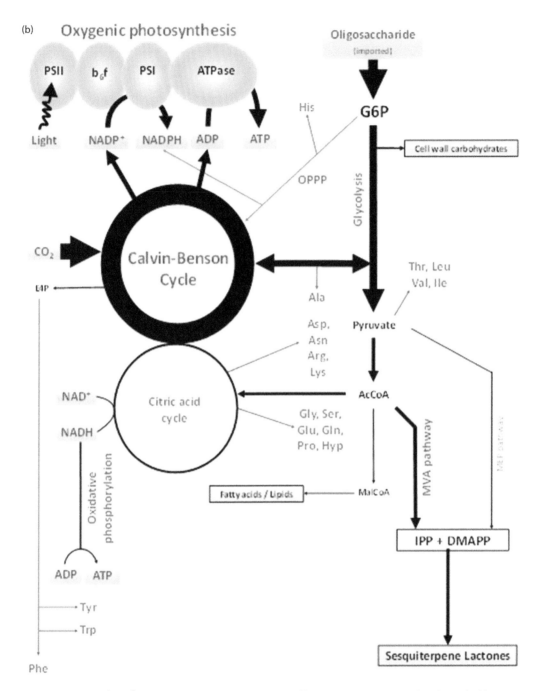

FIGURE 7.6 (Continued)   Flux distribution in secretory cell types. Examples are taken from the literature on secretory cells of glandular trichomes (a, *Salpiglossis sinuata*; b, *Xanthium stramonium*; c, *Mentha × piperita*; d, *Humulus lupulus*) and epithelial cells of resin ducts (e, *Pinus taeda*).

(c)

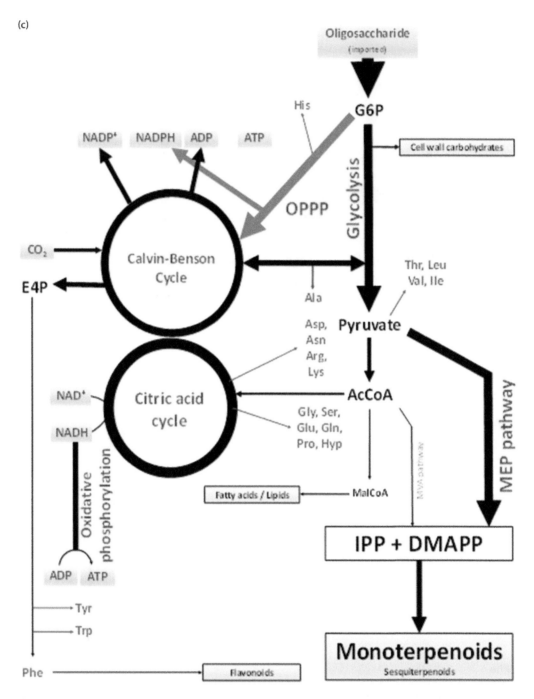

FIGURE 7.6 (Continued)   Flux distribution in secretory cell types. Examples are taken from the literature on secretory cells of glandular trichomes (a, *Salpiglossis sinuata*; b, *Xanthium stramonium*; c, *Mentha* × *piperita*; d, *Humulus lupulus*) and epithelial cells of resin ducts (e, *Pinus taeda*).

(d)

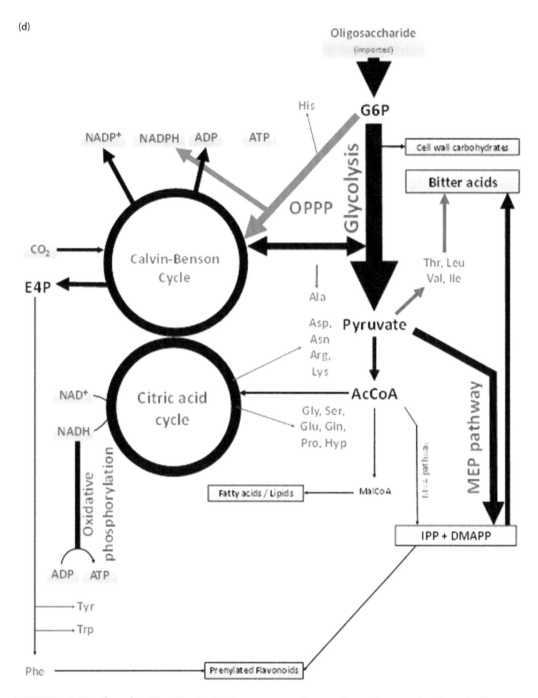

FIGURE 7.6 (Continued) Flux distribution in secretory cell types. Examples are taken from the literature on secretory cells of glandular trichomes (a, *Salpiglossis sinuata*; b, *Xanthium stramonium*; c, *Mentha × piperita*; d, *Humulus lupulus*) and epithelial cells of resin ducts (e, *Pinus taeda*).

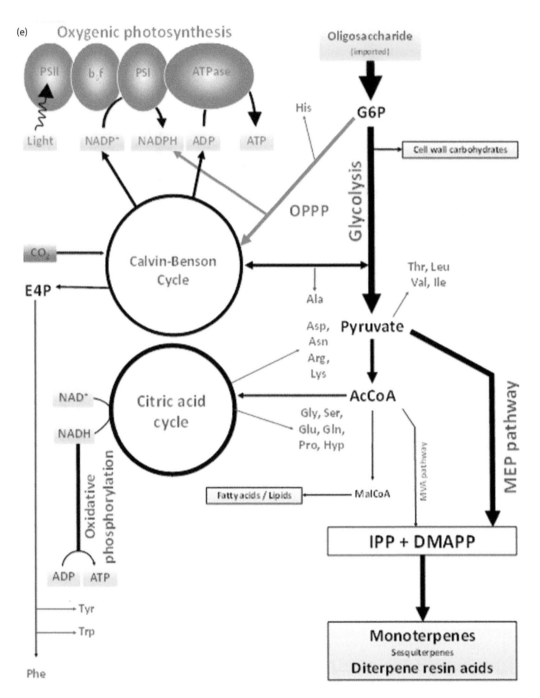

FIGURE 7.6 (Continued) Flux distribution in secretory cell types. Examples are taken from the literature on secretory cells of glandular trichomes (a, *Salpiglossis sinuata*; b, *Xanthium stramonium*; c, *Mentha × piperita*; d, *Humulus lupulus*) and epithelial cells of resin ducts (e, *Pinus taeda*).

genes related to photosynthesis in epithelial cells, while older microscopic data on the ultrastructure indicated a lack of functional photosystems therein (Cheniclet and Cardé, 1985; Charon et al., 1987). This discrepancy was not entirely resolved (Turner et al., 2019), but it has to be assumed that a certain bias was introduced during cell collection and/or mRNA amplification, thereby leading to an overrepresentation of photosynthetic transcripts. These minor shortcomings notwithstanding, models that simulated varying levels of photosynthetic activity in epithelial cells predicted a flux distribution mostly resembling that of peppermint glandular trichomes, which accumulate comparable end products (terpenoids) (Figure 7.6e). The outcomes of these simulations suggest that there might be similarities across specialized cell types with regard to flux distribution, and it will now be important to further investigate this intriguing hypothesis experimentally.

7.6 CONCLUSIONS AND OPPORTUNITIES FOR FUTURE RESEARCH

The first generation of kinetic and genome-scale models of plant secretory cell types indicated a unique flux distribution that reflects a high level of metabolic specialization. Models have aided in identifying parameters that determine their unique biochemistry. Examples include the discovery of a glandular trichome-specific pair of ferredoxin and ferredoxin NADP+ reductase isoforms, hypothesized to support high flux through the reductive steps of terpenoid biosynthesis (Johnson et al., 2017); the possibility that oxidative phosphorylation and ethanolic fermentation operate to generate energy equivalents (Johnson et al., 2017); the demonstration that feedback control determined the composition of secreted products (Rios-Estepa et al., 2008); and the identification of the genetic and developmental determinants of secreted product yields (Rios-Estepa et al., 2010; Lange et al., 2011). These studies have addressed some previously unanswered questions but have also raised awareness of unresolved issues:

- There is, for example, a lack of experimental studies that address to which extent photosynthetic processes contribute to the generation of a carbon source for metabolism in some species. While a sensitivity analysis can evaluate the general impacts of different levels of photosynthetic activity on overall metabolism, a quantitative understanding would greatly improve the accuracy of modeling predictions.
- Along the same lines, it would also be desirable to perform feeding experiments with appropriate isotope-labeled precursors and measuring label accumulation in metabolic intermediates and end products. This would enable performing MFA with much tighter constraints than those currently used in FBA.
- While models of metabolism in secretory cell types take compartmentation of reactions into account, transport fluxes are generally constrained by estimated values. To further improve the accuracy of simulations, the measurement of fluxes would be desirable for both intracellular and secretory transport into a storage cavity.
- Finally, it is now possible to develop models that link metabolism in different cell types and tissues, which will be highly valuable to more precisely model the dependency of secretory cell types on a carbon source provided by the underlying or surrounding tissue.

REFERENCES

Adam, K.P., and Croteau, R. 1998. Monoterpene biosynthesis in the liverwort *Conocephalum conicum*: Demonstration of sabinene synthase and bornyl diphosphate synthase. *Phytochemistry* 49, 475–480.

Adam, K.P., and Zapp, J. 1998. Biosynthesis of the isoprene units of chamomile sesquiterpenes. *Phytochemistry* 48, 953–959.

Adam, K.P., Thiel, R., and Zapp, J. 1999. Incorporation of 1-[1-13C]deoxy-D-xylulose in chamomile sesquiterpenes. *Arch. Biochem. Biophys.* 369, 127–132.

Adam, K.P., Thiel, R., Zapp, J., and Becker, H. 1998. Involvement of the mevalonic acid pathway and the glyceraldehyde–pyruvate pathway in terpenoid biosynthesis of the liverworts *Ricciocarpos natans* and *Conocephalum conicum*. *Arch. Biochem. Biophys.* 354, 181–187.

Allen, D.K. 2016. Quantifying plant phenotypes with isotopic labeling & metabolic flux analysis. *Curr. Opin. Biotechnol.* 37, 45–52.

Balcke, G.U., Bennewitz, S., Bergau, N., Athmer, B., Henning, A., Majovsky, P., Jiménez-Gómez, J.M., Hoehenwarter, W., and Tissier, A. 2017. Multi-omics of tomato glandular trichomes reveals distinct features of central carbon metabolism supporting high productivity of specialized metabolites. *Plant Cell* 29, 960–983.

Banerjee, A., Preiser, A.L., and Sharkey, T.D. 2016. Engineering of recombinant poplar deoxy-D-xylulose-5-phosphate synthase (PtDXS) by site-directed mutagenesis improves its activity. *PLoS One* 11, e0161534.

Banerjee, A., Wu, Y., Banerjee, R., Li, Y., Yan, H., and Sharkey, T.D. 2013. Feedback inhibition of deoxy-D-xylulose-5-phosphate synthase regulates the methylerythritol 4-phosphate pathway. *J. Biol. Chem.* 288, 16926–16936.

Barlow, A.J., Becker, H., and Adam, K.P. 2001. Biosynthesis of the hemi- and monoterpene moieties of isoprenyl phenyl ethers from the liverwort *Trichocolea tomentella*. *Phytochemistry* 57, 7–14.

Barlow, A.J., Lorimer, S.D., Morgan, E.R., and Weavers, R.T. 2003. Biosynthesis of the sesquiterpene hodgsonox from the New Zealand liverwort *Lepidolaena hodgsoniae*. *Phytochemistry* 63, 25–29.

Bartram, S., Jux, A., Gleixner, G., and Boland, W. 2006. Dynamic pathway allocation in early terpenoid biosynthesis of stress-induced lima bean leaves. *Phytochemistry* 67, 1661–1672.

Behnke, K., Grote, R., Brüggemann, N., Zimmer, I., Zhou, G., Elobeid, M., Janz, D., Polle, A., and Schnitzler, J.P. 2012. Isoprene emission-free poplars – A chance to reduce the impact from poplar plantations on the atmosphere. *New Phytol.* 194, 70–82.

Burkhardt, D., and Mosandl, A. 2003. Biogenetic studies in *Syringa vulgaris* L.: Bioconversion of 18O(2H)-labeled precursors into lilac aldehydes and lilac alcohols. *J. Agric. Food Chem.* 51, 7391–7395.

Charon, J., Launay, J., and Cardé, J.P. 1987. Spatial organization and volume density of leucoplasts in pine secretory cells. *Protoplasma* 138, 45–53.

Chen, F., Hao, F., Li, C., Gou, J., Lu, D., Gong, F., Tang, H., and Zhang, Y. 2013. Identifying three ecological chemotypes of *Xanthium strumarium* glandular trichomes using a combined NMR and LC-MS method. *PLoS One.* 8, e76621.

Chen, F., Ro, D.K., Petri, J., Gershenzon, J., Bohlmann, J., Pichersky, E., and Tholl, D. 2004. Characterization of a root-specific Arabidopsis terpene synthase responsible for the formation of the volatile monoterpene 1,8-cineole. *Plant Physiol.* 135, 1956–1966.

Chen, Q., Steinhauer, L., Hammerlindl, J., Keller, W., and Zou, J. 2007. Biosynthesis of phytosterol esters: Identification of a sterol o-acyltransferase in Arabidopsis. *Plant Physiol.* 145, 974–984.

Cheniclet, C., and Cardé, J.P. 1985. Presence of leucoplasts in secretory cells and of monoterpenes in the essential oil: A correlative study. *Isr. J. Bot.* 34, 219–238.

Clark, S.M., Vaitheeswaran, V., Ambrose, S.J., Purves, R.W., and Page, J.E. 2013. Transcriptome analysis of bitter acid biosynthesis and precursor pathways in hop (*Humulus lupulus*). *BMC Plant Biol.* 13, article 12.

DeBolt, S., Scheible, W.R., Schrick, K., Auer, M., Beisson, F., Bischoff, V.,et al. 2009. Mutations in UDP-glucose:sterol glucosyltransferase in Arabidopsis cause transparent testa phenotype and suberization defect in seeds. *Plant Physiol.* 151, 78–87.

Dudareva, N., Andersson, S., Orlova, I., Gatto, N., Reichelt, M., Rhodes, D., Boland, W., and Gershenzon, J. 2005. The nonmevalonate pathway supports both monoterpene and sesquiterpene formation in snapdragon flowers. *Proc. Natl. Acad. Sci. USA* 102, 933–938.

Eisenreich, W., Sagner, S., Zenk, M.H., and Bacher, A. 1997. Monoterpenoid essential oils are not of mevalonoid origin. *Tetrahedron Lett.* 38, 3889–3892.

Feldman, M.J., Poirier, B.C., and Lange, B.M. 2015. Misexpression of the Niemann-Pick disease type C1 (NPC1)-like protein in Arabidopsis causes sphingolipid accumulation and reproductive defects. *Planta* 242, 921–933.

Fowler, D.J., Hamilton, J.T.G., Humphrey, A.J., and O'Hagan, D. 1999. Plant terpene biosynthesis. The biosynthesis of linalyl acetate in *Mentha* citrate. *Tetrahedron Lett.* 19, 3803–3806.

Fraser, P.D., Pinto, M.E., Holloway, D.E., and Bramley, P.M. 2000. Technical advance: Application of high-performance liquid chromatography with photodiode array detection to the metabolic profiling of plant isoprenoids. *Plant J.* 24, 551–558.

Frescatada-Rosa, M., Stanislas, T., Backues, S.K., Reichardt, I., Men, S., Boutté, Y., Jürgens, G., Moritz, T., Bednarek, S.Y., and Grebe, M. 2014. High lipid order of Arabidopsis cell-plate membranes mediated by sterol and DYNAMIN-RELATED PROTEIN 1A function. *Plant J.* 80, 745–757.

Fujita, K., Yamagushi, T., Itose, R., and Sakai, K. 2000. Biosynthetic pathway of β-thujaplicin in the *Cupressus lusitanica* cell culture. *J. Plant Physiol.* 156, 462–467.

Ghirardo, A., Wright, L.P., Bi, Z., Rosenkranz, M., Pulido, P., Rodríguez-Concepción, M., Niinemets, Ü., Brüggemann, N., Gershenzon, J., and Schnitzler, J.P. 2014. Metabolic flux analysis of plastidic isoprenoid biosynthesis in poplar leaves emitting and nonemitting isoprene. *Plant Physiol.* 165, 37–51.

Gonzalez-Jorge, S., Ha, S.H., Magallanes-Lundback, M., Gilliland, L.U., Zhou, A., Lipka, A.E., et al. 2013. Carotenoid cleavage dioxygenase 4 is a negative regulator of β-carotene content in Arabidopsis seeds. *Plant Cell.* 25, 4812–4826.

Guenther, A., Karl, T., Harley, P., Wiedinmyer, C., Palmer, P.I., and Geron, C. 2006. Estimates of global terrestrial isoprene emissions using MEGAN (Model of Emissions of Gases and Aerosols from Nature). *Atmosph. Chem. Phys.* 6, 3181–3210.

Hampel, D., Mosandl, A., and Wüst, M. 2005a. Biosynthesis of mono- and sesquiterpenes in carrot roots and leaves (*Daucus carota* L.): Metabolic cross talk of cytosolic mevalonate and plastidial methylerythritol phosphate pathways. *Phytochemistry* 66, 305–311.

Hampel, D., Mosandl, A., and Wüst, M. 2005b. Induction of de novo volatile terpene biosynthesis via cytosolic and plastidial pathways by methyl jasmonate in foliage of *Vitis vinifera* L. *J. Agric. Food Chem.* 53, 2652–2657.

Hampel, D., Mosandl, A., and Wüst, M. 2006. Biosynthesis of mono- and sesquiterpenes in strawberry fruits and foliage: 2H labeling studies. *J. Agric. Food Chem.* 54, 1473–1478.

Hemmerlin, A., Harwood, J.L., and Bach, T.J. 2012. A raison d'être for two distinct pathways in the early steps of plant isoprenoid biosynthesis? *Prog. Lipid Res.* 51, 95–148.

Herde, M., Gärtner, K., Köllner, T.G., Fode, B., Boland, W., Gershenzon, J., Gatz, C., and Tholl, D. 2008. Identification and regulation of TPS04/GES, an Arabidopsis geranyllinalool synthase catalyzing the first step in the formation of the insect-induced volatile C16-homoterpene TMTT. *Plant Cell* 20,1152–1168.

Howe, G.A., and Jander, G. 2008. Plant immunity to insect herbivores. *Annu. Rev. Plant Biol.* 59, 41–66.

Huang, M., Sanchez-Moreiras, A.M., Abel, C., Sohrabi, R., Lee, S., Gershenzon, J., and Tholl, D. 2012. The major volatile organic compound emitted from Arabidopsis thaliana flowers, the sesquiterpene (E)-β-caryophyllene, is a defense against a bacterial pathogen. *New Phytol.* 193, 997–1008.

Johnson, S.R., Lange, I., Srividya, N., and Lange, B.M. 2017. Bioenergetics of monoterpenoid essential oil biosynthesis in nonphotosynthetic glandular trichomes. *Plant Physiol.* 175, 681–695.

Kavalier, A.R., Litt, A., Ma, C., Pitra, N.J., Coles, M.C., Kennelly, E.J., and Matthews, P.D. 2011. Phytochemical and morphological characterization of hop (*Humulus lupulus* L.) cones over five developmental stages using high performance liquid chromatography coupled to time-of-flight mass spectrometry, ultrahigh performance liquid chromatography photodiode array detection, and light microscopy techniques. *J. Agric. Food Chem.* 59, 4783–4793.

Kobayashi, K., Sasaki, D., Noguchi, K., Fujinuma, D., Komatsu, H., Kobayashi, M., et al. 2013. Photosynthesis of root chloroplasts developed in Arabidopsis lines overexpressing GOLDEN2-LIKE transcription factors. *Plant Cell Physiol.* 54, 1365–1377.

Lange, B.M. 2015. The evolution of plant secretory structures and emergence of terpenoid chemical diversity. *Annu. Rev. Plant Biol.* 66, 139–159.

Lange, B.M., and Rios-Estepa, R. 2014. Kinetic modeling of plant metabolism and its predictive power – Peppermint essential oil biosynthesis as an example. *Methods Mol. Biol.* 1083, 287–311.

Lange, B.M., and Turner, G.W. 2013. Terpenoid biosynthesis in trichomes – Current status and future opportunities. *Plant Biotechnol. J.* 11, 2–22.

Lange, B.M., Mahmoud, S.S., Wildung, M.R., Turner, G.W., Davis, E.M., Lange, I., Baker, R.C., Boydston, R.A., and Croteau, R.B. 2011. Improving peppermint essential oil yield and composition by metabolic engineering. *Proc. Natl. Acad. Sci.USA* 108, 16944–16949.

Lange, I., Poirier, B.C., Herron, B.K., and Lange, B.M. 2015. Comprehensive assessment of transcriptional regulation facilitates metabolic engineering of isoprenoid accumulation in Arabidopsis. *Plant Physiol.* 169, 1595–1606.

Li, Y., Gou, J., Chen, F., Li, C., and Zhang, Y. 2016. Comparative transcriptome analysis identifies putative genes involved in the biosynthesis of xanthanolides in *Xanthium strumarium* L. *Front. Plant Sci.* 7, 1317.

McCaskill, D., and Croteau, R. 1995. Isoprenoid synthesis in peppermint (*Mentha × piperita*): development of a model system for measuring flux of intermediates through the mevalonic acid pathway in plants. *Biochem. Soc. Trans.* 23, 290S.

Moghe, G.D., Leong, B.J., Hurney, S.M., Jones, D.A., and Last R.L. 2017. Evolutionary routes to biochemical innovation revealed by integrative analysis of a plant-defense related specialized metabolic pathway. *Elife* 6, e28468.

Nagegowda, D.A. 2010. Plant volatile terpenoid metabolism: Biosynthetic genes, transcriptional regulation and subcellular compartmentation. *FEBS Lett.* 584, 2965–2973.

Oberhardt, M.A., Chavali, A.K., and Papin, J.A. 2009. Flux balance analysis: Interrogating genome-scale metabolic networks. *Methods Mol. Biol.* 500, 61–80.

Orth, J.D., Thiele, I., and Palsson, B.Ø. 2010. What is flux balance analysis? *Nature Biotechnol.* 28, 245–248.

Pokhilko, A., Bou-Torrent, J., Pulido, P., Rodríguez-Concepción, M., and Ebenhöh, O. 2015. Mathematical modelling of the diurnal regulation of the MEP pathway in Arabidopsis. *New Phytol.* 206, 1075–1085.

Rasmussen, R.A. 1972. What do the hydrocarbons from trees contribute to air pollution? *J. Air Poll. Control Assoc.* 22, 537–543.

Rieder, C., Jaun, B., and Arigoni, D. 2000. On the early steps of cineol biosynthesis in *Eucalyptus globulus*. *Helv. Chim. Acta.* 83, 2504–2513.

Rios-Estepa, R., and Lange, B.M. 2007. Experimental and mathematical approaches to modeling plant metabolic networks. *Phytochemistry* 68, 2351–2374.

Rios-Estepa, R., Lange, I., Lee, J.M., and Lange, B.M. 2010. Mathematical modeling-guided evaluation of biochemical, developmental, environmental and genotypic determinants of essential oil composition and yield in peppermint leaves. *Plant Physiol.* 152, 2105–2119.

Rios-Estepa, R., Turner, G.W., Lee, J.M., Croteau, R.B., and Lange, B.M. 2008. A systems biology approach identifies biochemical mechanisms regulating monoterpenoid essential oil composition in peppermint. *Proc. Natl. Acad. Sci. USA* 105, 2818–2823.

Ro, D.K., Ehlting, J., Keeling, C.I., Lin, R., Mattheus, N., and Bohlmann, J. 2006. Microarray expression profiling and functional characterization of AtTPS genes: Duplicated Arabidopsis thaliana sesquiterpene synthase genes At4g13280 and At4g13300 encode root-specific and wound-inducible (Z)-gamma-bisabolene synthases. *Arch. Biochem. Biophys.* 448, 104–116.

Ryan, A.C., Hewitt, C.N., Possell, M., Vickers, C.E., Purnell, A., Mullineaux, P.M., Davies, W.J., and Dodd, I.C. 2014. Isoprene emission protects photosynthesis but reduces plant productivity during drought in transgenic tobacco (*Nicotiana tabacum*) plants. *New Phytol.* 201, 205–216.

Sanadze, G.A. 1957. The nature of gaseous substances emitted by leaves of *Robinia pseudoacacia*. Soobshch. *Akad. Nauk. Gruz SSR.* 19, 83–86.

Sanadze, G.A. 1990. The principle scheme of photosynthetic carbon conversion in cells of isoprene releasing plants. *Curr. Res. Photosyn.* IV, 231–237.

Sanadze, G.A. Dshaiari, G.I., and Tevadze, I.M. 1972. Incorporation into the isoprene molecule of carbon from $^{13}CO_2$ assimilated during photosynthesis. *Soviet Plant Physiol.* 19, 17–20.

Sanadze, G.A., and Kalanadze, A.N. 1966. Light and temperature curves of the evolution of C_5H_8. *Fiziol. Rast.* 13, 17–20.

Schwender, J., Zeidler, J., Gröner, R., Müller, C., Focke, M., Braun, S., Lichtenthaler, F.W., and Lichtenthaler, H.K. 1997. Incorporation of 1-deoxy-D-xylulose into isoprene and phytol by higher plants and algae. *FEBS Lett.* 414, 129–134.

Sharkey, T.D., and Loreto, F. 1993. Water stress, temperature, and light effects on the capacity for isoprene emission and photosynthesis of kudzu leaves. *Oecologia.* 95, 328–333.

Sharkey, T.D., and Monson, RK. 2017. Isoprene research – 60 years later, the biology is still enigmatic. *Plant Cell Environ.* 40, 1671–1678.

Sharkey, T.D., Wiberley, A.E., and Donohue AR. 2008. Isoprene emission from plants: Why and how. *Ann. Bot.* 101, 5–18.

Silver, G.M., and Fall, R. 1991. Enzymatic synthesis of isoprene from dimethylallyl diphosphate in aspen leaf extracts. *Plant Physiol.* 97, 1588–1591.

Silver, G.M., and Fall, R. 1995. Characterization of aspen isoprene synthase, an enzyme responsible for leaf isoprene emission to the atmosphere. *J. Biol. Chem.* 270, 13010–13016.

Sohrabi, R., Huh, J.H., Badieyan, S., Rakotondraibe, L.H., Kliebenstein, D.J., Sobrado, P., and Tholl, D. 2015. In planta variation of volatile biosynthesis: An alternative biosynthetic route to the formation of the pathogen-induced volatile homoterpene DMNT via triterpene degradation in Arabidopsis roots. *Plant Cell.* 27, 874–890.

Spiteller, D., Jux, A., Piel, J., and Boland, W. 2002. Feeding of [5,5-2H(2)]-1-desoxy-D-xylulose and [4,4,6,6,6-2H(5)]-mevalolactone to a geosmin-producing *Streptomyces* sp. and *Fossombronia pusilla*. *Phytochemistry.* 61, 827–834.

Steliopoulos, P., Wüst, M., Adam, K.P., and Mosandl, A. 2002. Biosynthesis of the sesquiterpene germacrene D in *Solidago canadensis*: 13C and (2)H labeling studies. *Phytochemistry.* 60, 13–20.

Thiel, R., and Adam, K.P. 2002. Incorporation of [1-(13)C]1-deoxy-D-xylulose into isoprenoids of the liverwort *Conocephalum conicum*. *Phytochemistry.* 59, 269–274.

Tholl, D., Chen, F., Petri, J., Gershenzon, J., and Pichersky, E. 2005. Two sesquiterpene synthases are responsible for the complex mixture of sesquiterpenes emitted from Arabidopsis flowers. *Plant J.* 42, 757–771.

Tingey, D.T., Manning, M., Grothaus, L.C., and Burns, W.F. 1979. The influence of light and temperature on isoprene emission rates from live oak. *Phys. Plant.* 47, 112–118.

Turner, G.W., Parrish, A.N., Zager, J.J., Fischedick, J.T., and Lange, B.M. 2019. Assessment of flux through oleoresin biosynthesis in epithelial cells of loblolly pine resin ducts. *J. Exp. Bot.* 70, 217–230.

Umlauf, D., Zapp, J., Becker, H., and Adam, K.P. 2004. Biosynthesis of the irregular monoterpene artemisia ketone, the sesquiterpene germacrene D and other isoprenoids in *Tanacetum vulgare* L. (Asteraceae). *Phytochemistry.* 65, 2463–2470.

Vaughan, M.M., Wang, Q., Webster, F.X., Kiemle, D., Hong, Y.J., Tantillo, D.J., et al. 2013. Formation of the unusual semivolatile diterpene rhizathalene by the Arabidopsis class I terpene synthase TPS08 in the root stele is involved in defense against belowground herbivory. *Plant Cell.* 25, 1108–1125.

Võ, U.U.T., and Morris, M.P. 2014. Nonvolatile, semivolatile, or volatile: Redefining volatile for volatile organic compounds. *J. Air Waste Manag. Assoc.* 64, 661–669.

Wang, X.M., Yang, B., Ren, C.G., Wang, H.W., Wang, J.Y., and Dai, C.C. 2015. Involvement of abscisic acid and salicylic acid in signal cascade regulating bacterial endophyte-induced volatile oil biosynthesis in plantlets of *Atractylodes lancea. Physiol. Plant.* 153, 30–42.

Wollenweber, E., Dörsam, M., Dörr, M., Roitman, J.N., and Valant-Vetschera, K.M. 2005. Chemodiversity of surface flavonoids in Solanaceae. *Z. Naturforsch. C.* 60, 661–670.

Zager, J.J., and Lange, B.M. 2018. Assessing flux distribution associated with metabolic specialization of glandular trichomes. *Trends Plant Sci.* 23, 638–647.

Zeidler, J., and Lichtenthaler, H.K. 2001. Biosynthesis of 2-methyl-3-buten-2-ol emitted from needles of *Pinus ponderosa* via the non-mevalonate DOXP/MEP pathway of isoprenoid formation. *Planta.* 213, 323–326.

Zeidler, J.G., Lichtenthaler, H.K., May, H.U., and Lichtenthaler, F.W. 1997. Is isoprene emitted by plants synthesized via the novel isopentenyl pyrophosphate pathway? *Z. Naturforsch. C.* 52, 15–23.

Zhang, W., Liu, T., Ren, G., Hörtensteiner, S., Zhou, Y., Cahoon, E.B., and Zhang. C. 2014. Chlorophyll degradation: The tocopherol biosynthesis-related phytol hydrolase in Arabidopsis seeds is still missing. *Plant Physiol.* 166, 70–79.

8 Floral Scent Metabolic Pathways and Their Regulation

Joseph H. Lynch, Eran Pichersky, and Natalia Dudareva

CONTENTS

8.1 INTRODUCTION

Floral scent is a diverse blend of low molecular weight, mostly lypophilic compounds (Chapter 4). Volatiles emitted from flowers play crucial roles in plant reproductive and evolutionary success by contributing to attraction of pollinators and serving as defense compounds against pathogens and florivores (Muhlemann, Klempien, and Dudareva 2014). Although floral volatiles are extraordinarily diverse, numbering in the thousands of distinct compounds, they are produced by only a few major biochemical pathways (terpenoid, lipoxygenase, amino acid and phenylpropanoid/benzenoid pathways). Their diversity is mainly derived from specific enzymatic derivatizations, which increase the volatility of compounds at the final step of their formation (Dudareva, Pichersky, and Gershenzon 2004). In the past couple of decades significant progress has been made in understanding the biosynthesis of floral volatiles and biochemical and molecular mechanisms that regulate their formation and, more recently, their release. Several model plant species with powerful floral scent such as *Clarkia breweri*, snapdragon (*Antirrhinum majus*), *Petunia hybrida*, rose (Rose spp.), *Stephanotis floribunda*, and *Nicotiana suaveolens*, as well as *Arabidopsis thaliana* with its slight floral scent, have been used to identify, isolate, and characterize enzymes and genes involved in the biosynthesis and emission of floral volatiles. Moreover, as detailed elsewhere in this book, investigations of the production in vegetative and fruit volatiles that also occur in flowers have significantly contributed to our understanding of the biosynthesis of such chemicals.

8.2 BIOCHEMICAL PATHWAYS

8.2.1 BIOSYNTHESIS OF VOLATILE TERPENES

Terpenoids, the largest class of floral volatiles (Chapter 4) and plant specialized metabolites in general, include such well-known common constituents of floral scents as the monoterpenes linalool, limonene, myrcene, ocimene, geraniol, and the sesquiterpenes farnesene, neralidol, caryophyllene, and germacrene. As discussed elsewhere, but summarized here with special emphasis on regulation and subcellular organization, all terpenoids (except isoprene) originate through the condensation of the universal five-carbon building blocks, isopentenyl diphosphate (IPP) and dimethylallyl diphosphate (DMAPP), which are derived from two alternative compartmentally separated pathways, the mevalonic acid (MVA) and methylerythritol phosphate (MEP) pathways.

The MEP pathway operates in the plastids, as a full set of the corresponding enzymes exists only in plastids (Hsieh et al. 2008). In contrast, subcellular localization of the MVA pathway is more complex. Recent data suggest that the MVA pathway is distributed between the cytosol, endoplasmic reticulum and peroxisomes (Pulido, Perello, and Rodriguez-Concepcion 2012). The MVA pathway consists of six enzymatic reactions and is initiated by a stepwise condensation of three molecules of acetyl-CoA, whereas the MEP pathway involves seven enzymatic steps and begins with the condensation of pyruvate and glyceraldehyde 3-phosphate (Vranová, Coman, and Gruissem 2013; Tholl 2015). The two pathways are connected via a metabolic "cross-talk," which is species- and/or organ-specific (Laule et al. 2003; Schuhr et al. 2003; Dudareva et al. 2005; Henry et al. 2015, 2018). Such metabolic connectivity of the pathways allows the MEP pathway, which often has a higher carbon flux than the MVA route (Dudareva et al. 2005; Wu et al. 2006), to support biosynthesis of cytosolically formed terpenoids (Laule et al. 2003; Dudareva et al. 2005). For example, the plastid-localized MEP pathway in snapdragon flowers provide a major IPP and DMAPP source for cytosolic sesquiterpene biosynthesis, while the MVA pathway appears to be inactive (Dudareva et al. 2005).

While the MVA pathway produces only IPP, the MEP pathway results in the synthesis of both IPP and DMAPP at a 6:1 ratio (Nakamura et al. 2001). Thus, both pathways rely on isopentenyl diphosphate isomerase (IDI), which reversibly converts IPP to DMAPP (Nakamura et al. 2001) and controls the equilibrium between them. In *Catharanthus roseus* and *Arabidopsis thaliana* genomes a single and two distinct IDI genes were identified, respectively, which are transcribed as splice variants. In both plants, the "long" proteins are transported into mitochondria and/or chloroplasts whereas the "short" proteins, lacking the targeting signal, are located in the peroxisomes (Phillips et al. 2008; Guirimand, et al. 2012a, b). It should also be noted that recently, using tobacco and *Arabidopsis* leaves as model systems, it was shown that there is a cytosolic pool of isopentenyl phosphate (IP), which can be phosphorylated to IPP and serves as a metabolically available carbon source for production of both mono- and sesquiterpenes (Henry et al. 2015, 2018).

The condensation of DMAPP with one or more IPP molecules gives rise to geranyl diphosphate (GPP), farnesyl diphosphate (FPP), and geranylgeranyl diphosphate (GGPP), the *trans*-prenyl diphosphates that are, respectively, the common substrates of mono-, sesqui-, and diterene synthases (Chen et al. 2011). Geranylfarnesyl diphosphate (GFPP) is also made in some plants and is the substrate of sesterterpene synthases (Liu et al. 2016). In the *Solanum* genus of Solanaceae, the *cis*-prenyl diphosphates neryl diphosphate (NPP, C_{10}), z,z-FPP, and nerylneryl diphosphate (NNPP, C_{20}) are also synthesized and serve as the substrates for some terpene synthases (TPSs) (Sallaud et al. 2009; Akhtar et al. 2013; Matsuba et al. 2015).

The *trans*- and *cis*-prenyltransferases – GPPS, FPPS, GGPPS, GFPPS, NDPS1 (CPT1), zz-FPPS, and NNPPS (CPT2) – responsible for the synthesis of the prenyl diphosphates not only determine the chain-length of their products, but also control precursor flux toward various classes of terpenoids by catalyzing the branch-point reactions (Liang, Ko, and Wang 2002; Akhtar et al. 2013). In general, while FPPS is localized in both cytosol and mitochondria and GGPPS may be found in these compartments as well as in plastids, the remaining prenyl-diphosphate synthases appear to be plastidial (Sommer et al. 1995; Akhtar et al. 2013; Liu et al. 2016). All are homodimeric enzymes with the exception of GPPS, which may form both homodimers and heterodimers, depending on plant species. The heterodimer usually consists of a large GGPPS subunit interacting with small subunit, which is itself catalytically inactive but required to confer GPPS activity of the large subunit (Burke, Wildung, and Croteau 1999; Tholl et al. 2004).

The genome of each plant species contains multiple terpene synthase genes with related sequences, encoding enzymes with distinct functions. The most exceptional property of these enzymes is the ability of many of them to synthesize multiple products from a single prenyl disphosphate precursor (Degenhardt, Köllner, and Gershenzon 2009). For example, only two sesquiterpene synthases are responsible for nearly all of the 20 sesquiterpenes emitted by *Arabidopsis* flowers (Tholl et al. 2005). Some of TPSs can also use more than one substrate (Tholl 2006; Bleeker et al. 2011). In these cases, their subcellular localization and the availability of a particular substrate determine the type of product formed (Nagegowda et al. 2008). A noteworthy TPS-independent pathway for monoterpene biosynthesis was recently described in rose, where geraniol formation relies on the action of GPP diphosphohydrolase of the Nudix family (Magnard et al. 2015).

In addition to a wide range of volatile terpenoids formed directly by TPSs, terpenoid diversity is further increased by other enzymes that are capable of modifying the TPS products. The first is hydroxylation, a generally irreversible modification almost always catalyzed by enzymes that belong to the cytochrome P450 oxidoreductase family. Examples include cytochrome P450-catalyzed 3-hydroxylation of limonene to produce *trans*-isopiperitenol, a volatile flavor compound found in mint (Lupien et al. 1999). A 6-hydroxylation of limonene by another P450 enzyme is the first step in the biosynthesis of another volatile spice, carvone, in the caraway (*Carum carvi*) fruit (Bouwmeester et al. 1999). In addition to hydroxylation, dehydrogenation, acylation, methylation, or other reactions can modify the terpene backbone thus increasing their volatility and altering their olfactory properties, as detailed in Section 8.2.4. Besides mono-, sesqui- and small number of diterpenes, the scent bouquet of some flowers often contains irregular volatile terpenoids (C_8 to C_{18}). They constitute a minor class of floral terpenoids (~7% of all floral terpenoids), which are formed via a three-step modification including a dioxygenase cleavage, enzymatic transformation and acid-catalyzed conversion into volatile compounds (Winterhalter and Rouseff 2001). The dioxygenase cleavage step itself can already result in volatile products, such as α- and β-ionone, geranylacetone and pseudoionone, as was found to be the case in petunia flowers (Simkin el al. 2004a, b). Although these compounds are present at very low concentrations in the scent, they contribute significantly to the overall scent bouquet due to their low odor threshold (Baldwin et al. 2000).

8.2.2 BIOSYNTHESIS OF VOLATILE PHENYLPROPANOIDS AND RELATED COMPOUNDS

Phenylalanine (Phe) derivatives constitute the second largest class of volatile compounds in plants and are synthesized via a complex series of branched pathways. These compounds all maintain the six-carbon aromatic ring of Phe, and can be generally divided into three classes according to the length of their side chains. Of the phenylpropanoids (which have a three-carbon side chain and thus designated C_6–C_3 compounds), only those that are reduced at the C9 position and sometimes esterified are volatile. Phe-derived benzenoid compounds (C_6–C_1), as well as the C_6–C_2 phenylpropanoid-related compounds, are also volatile and common constituents of floral scent (Knudsen, Tollsten, and Bergström 1993).

The first committed step in the biosynthesis of most phenylpropanoid and benzenoid compounds, as well as lignin, is catalyzed by the well-known and widely distributed enzyme Phe ammonia lyase (PAL; EC 4.7.1.5), which deaminates Phe to produce *trans*-cinnamic acid. The complete pathways for formation of the volatile phenylpropenes (C_6–C_3), such as eugenol, isoeugenol, methyleugenol, isomethyleugenol, chavicol, and methylchavicol, have now been elucidated in *Clarkia breweri* (Koeduka et al. 2008), *Petunia hybrida* (Koeduka et al. 2006, 2008) and *Ocimum basilicum* (Koeduka et al. 2006; Vassão et al. 2006). To eliminate the oxygen functionality at C-9 position, coniferyl alcohol, a precursor of lignin, is first converted to coniferyl acetate by coniferyl alcohol acetyltransferase (Dexter et al. 2007) prior to its reduction to eugenol and isoeugenol by eugenol synthase or isoeugenol synthase, respectively (Koeduka et al. 2006, 2008). Similarly, coumaryl acetate serves as the biosynthetic precursor of chavicol in basil (Vassão et al. 2006).

The formation of benzenoids (C_6-C_1) from cinnamic acid requires the shortening of the side chain by two carbon atoms, for which several routes have been shown to be active. A CoA-dependent β-oxidative pathway was recently fully elucidated in *Petunia hybrida* and *Arabidopsis thaliana* and appears to be analogous to that operating in catabolism of fatty acids and certain branched-chain amino acids. This pathway begins with import of cinnamic acid into the peroxisomes via the COMATOSE transporter (Bussell et al. 2014) and its activation to a CoA ester (Klempien et al. 2012), which is then subjected to the action of a bifunctional hydratase/dehydrogenase (Qualley et al. 2012) and thiolase (Van Moerkercke et al. 2009) to produce benzoyl-CoA. Benzyl-CoA and free benzoic acid released by a thioesterase are then exported to the cytosol by a yet unknown mechanism (Adebesin et al. 2018).

An alternative non-β-oxidative pathways, leading to the formation of benzaldehyde followed by its oxidation to benzoic acid, may also operate in some plants, although the biochemical steps leading to benzaldehyde formation are still in question. However, genetic evidence suggests that benzaldehyde formation in some members of the *Capsella* genus and in *Petunia axillaris* relies on the cinnamoyl-CoA intermediate produced by cinnamate:CoA ligase (CNL) (Sas et al. 2016; Amrad et al. 2016). In addition, *Arabidopsis* cytosolic nonspecific aldehyde oxidase and snapdragon mitochondrial NADP$^+$-dependent aldehyde dehydrogenase converting benzaldehyde to benzoic acid were isolated and characterized (Ibdah et al. 2009; Long et al. 2009).

C_6-C_2 phenylpropanoid-related compounds constitute approximately 24% of all described phenylalanine-derived volatile compounds (Chapter 4). In petunia and rose petals, formation of phenylacetaldehyde occurs directly from Phe via a PAL-independent route and depends on phenylacetaldehyde synthase, a bifunctional enzyme that catalyzes an unusual combined decarboxylation amine-oxidation reaction (Kaminaga et al. 2006). However, in roses an alternative route has also been proposed, where Phe is first subjected to deamination by an aromatic amino acid aminotransferase followed by decarboxylation of the formed phenylpyruvate intermediate. While an aromatic amino acid aminotransferase has been characterized from rose petals (Hirata et al. 2012), no phenylpyruvate decarboxylase has been identified so far. Phenylacetaldehyde can be reduced to phenylethanol in a reaction catalyzed by NADH-dependent phenylacetaldehyde reductase as was shown in roses (Sakai et al. 2007). In addition, a second alternative route for phenylethanol formation, independent of phenylacetaldehyde, was proposed via phenylpyruvate and phenyllactic acid intermediates (Watanabe et al. 2002).

As is the case for terpenoids, phenylpropanoids and related compounds are subject to a number of modifications, such as hydroxylation, acetylation, and methylation, and significant progress has been made in the discovery of the catalysts involved in production of these downstream products (see Section 8.2.4).

8.2.3 BIOSYNTHESIS OF VOLATILE FATTY ACID DERIVATIVES

Volatile fatty acid derivatives including saturated and unsaturated short-chain alcohols, aldehydes, and esters represent the third class of floral volatiles, which derives from the degradation of the unsaturated C_{18} fatty acids linolenic and linoleic acid. After entering the lipoxygenase (LOX) pathway (Chapter 9), the fatty acids undergo the stereospecific oxygenation to form 9-hydroperoxy and 13-hydroperoxy intermediates (Feussner and Wasternack 2002), which are further metabolized via either of two branches of the LOX pathway yielding volatile compounds. The allene oxide synthase branch uses only the 13-hydroperoxy intermediate and leads to the formation of jasmonic acid (JA), which in turn is converted to the volatile ester methyl jasmonate by JA carboxyl methyltransferase (Seo et al. 2001; see below). In contrast, the hydroperoxide lyase branch acts on both types of hydroperoxide fatty acid derivatives, converting 13-hydroperoxides and 9-hydroperoxides into short-chain C_6- and C_9-volatile aldehydes, respectively, which are often reduced to alcohols by alcohol dehydrogenases followed by further conversion to the corresponding esters (D'Auria et al. 2007). Although LOX pathway genes were originally characterized in non-floral tissues, their expression in flowers has recently been more thoroughly analyzed (Shi et al. 2018).

8.2.4 MODIFICATION REACTIONS

Modifications such as hydroxylation, reduction/oxidation, methylation, and acetylation enhance volatility of compounds, and subsequently increase the diversity of volatiles. A large portion of floral volatiles contain one or more methylated hydroxyl groups (i.e., a methoxyl group), which are the products of the O-methyltransferase (OMT) family of enzymes that use S-adenosyl-L-methionine (SAM) as the methyl donor. Several plant OMTs have been functionally characterized due to their involvement in floral scent biosynthesis. For example, two distinct but very similar enzymes, eugenol and chavicol OMTs, were shown to be responsible for 4-hydroxyl methylation of eugenol and chavicol, resulting in methyleugenol and methychavicol formation (Wang et al. 1997; Gang et al. 2002; Figure 8.1c). Two successive methylations of orcinol (3,5-dioxytoluene) also catalyzed by two very similar OMTs, orcinol OMTs (OOMT1 and OOMT2) lead to the formation of dimethoxy-orcinol, a major scent compound in many hybrid roses (Lavid et al. 2002). Chinese rose (*Rosa chinensis*) flowers make a similar compound with three methoxyl groups, 1,3,5-trimethoxybenzene, which is synthesized from 1,3,5-trihydroxybenzene. OOMT1 and OOMT2 can catalyze the methylation of the second and third intermediates (1-methoxy,3,5-dihydroxybenzene and 1,3-dimethoxy,5-hydroxybenzene) but not the methylation of 1,3,5-trihydroxybenzene, also known as phloroglucinol (Lavid et al. 2002). The enzyme that methylates the latter compound, phloroglucinol OMT, and the gene encoding this protein have been isolated and characterized from rose petals (Wu et al. 2004). An O-methyltransferase is also responsible for the final step in the biosynthesis of veratrole (1, 2-dimethoxybenzene) from guaiacol in white campion (*Silene latifolia*) flowers (Akhtar and Pichersky 2013).

Another wide-spread group of methylated fragrant compounds includes methyl esters such as methyl benzoate, methyl cinnamate, methyl jasmonate, and methyl salicylate, which are formed by carboxyl methyltransferases, many of which belong to the SABATH family (D'Auria, Chen, and Pichersky 2003), which transfer the methyl group of SAM to a free carboxyl group of corresponding acids (Figure 8.1c). The enzymes capable of methylating benzoic, jasmonic, and salicylic acids have been identified from several plant species (Effmert et al. 2005). While benzoic acid carboxyl methyltransferase (BAMT), the enzyme responsible for the snapdragon floral volatile methyl benzoate, methylates only benzoic acid (Murfitt et al. 2000), the SAM:benzoic acid/salicylic acid carboxyl methyltransferases (BSMT) and SAM:salicylic acid carboxyl methyltransferases (SAMT) are able to methylate both salicylic and benzoic acids (Effmert et al. 2005).

Acylation of alcohols, most often with an acetyl moiety but also with larger acyls such butanoyl or benzoyl acyls, leads to the formation of volatiles esters, and is also common in floral scents. Enzymes from the BAHD superfamily of acyltransferases (D'Auria 2006) were shown to be responsible for the biosynthesis of acetylated scent compounds. Examples include acetyl-coenzyme A:benzyl alcohol acetyltransferase (BEAT) from *C. breweri* flowers, which produces benzyl acetate (Dudareva et al. 1998), and acetyl-coenzyme A:geraniol acetyltransferase from *Rose hybrida* that generates geranyl acetate from geraniol (Shalit et al. 2003), to name a few. The BAHD acyltransferases often show wide substrate specificity for both the acyl moiety and the alcohol moiety. For example, the petunia BPBT enzyme can transfer a benzoyl moiety to two alcohol substrates – benzyl alcohol and phenylethanol, producing benzylbenzoate and phenylethyl benzoate, respectively (Boatright et al. 2004). Similarly, an acyltransferase from ripening strawberry (*Fragaria* spp.) fruit can use a series of acyl moieties such as acetyl, butanyl, and hexanyl, and transfer them to various alcohols such as heptanol, octanol, and geraniol (Beekwilder et al. 2004).

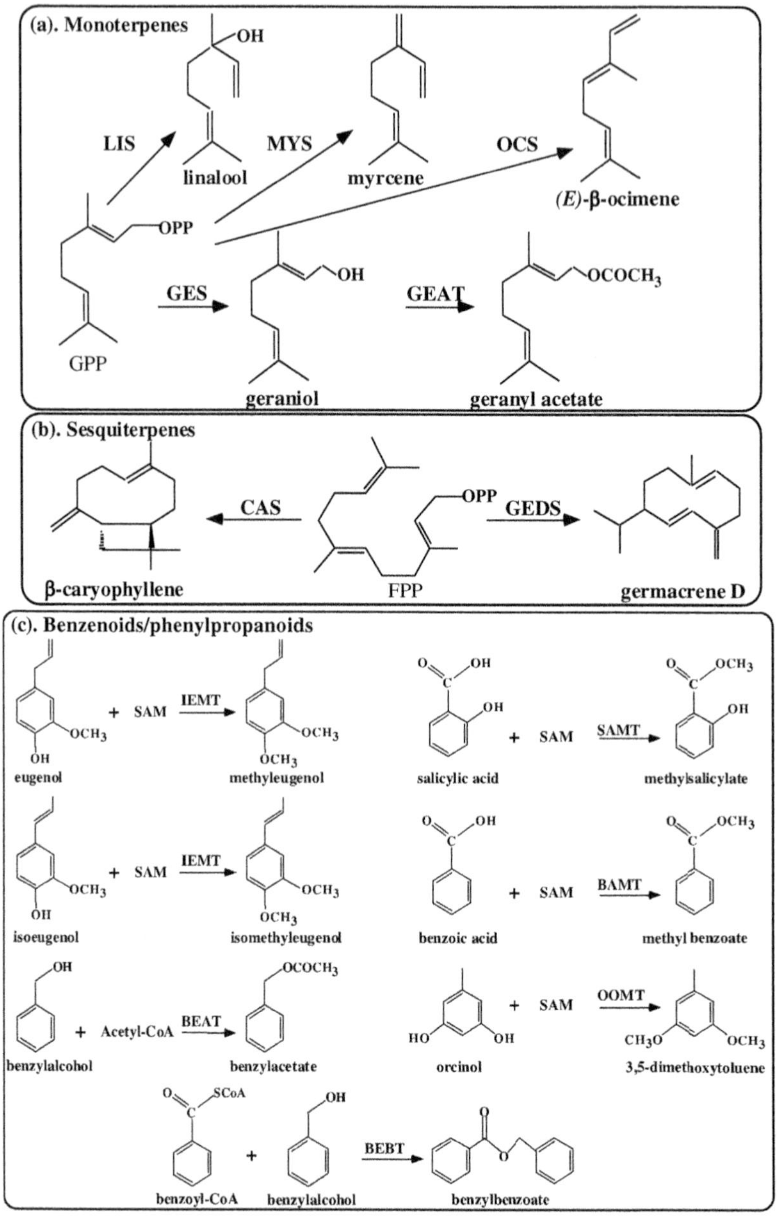

FIGURE 8.1 Examples of biochemical reactions that produce floral volatiles, and the enzymes that catalyze these reactions. (a) Biosynthesis of monoterpenes and monoterpene esters. GPP, geranyldiphosphate; LIS, linalool synthase; GES, geraniol synthase; OCS, ocimene synthase; GEAT, geraniol:acetyl-CoA acetyltransferase. (b) Biosynthesis of sesquiterpenes. CAS, β-caryophyllene synthase; GEDS, germacrene D synthase. (c) Biosynthesis of terpenoid and benzenoid volatiles. IEMT – isoeugenol/eugenol O-methyltransferase; SAMT, salicylic acid methyltransferase; BAMT, benzoic acid methyltransferase; BEAT, acetyl-CoA:benzyl alcohol acetyltransferase; OOMT, orcinol O-methyltransferase, BEBT, benzoyl-CoA:benzyl alcohol benzoyl transferase.

Volatile alcohols and aldehydes are common constituents of floral scents. Their interconversion in plant tissue is catalyzed by NADP/NAD-dependent oxidoreductases, which often have broad substrate specificity. It has been shown that benzyl alcohol, a major floral scent component in many Nicotianeae species and elsewhere, is derived from benzaldehyde in a reversible reaction catalyzed by a member of the NADP/NAD-dependent oxidoreductases family (Boatright et al. 2004). Similarly, 2-phenylethanol is derived from phenylacetaldehyde via an NADH-dependent reductase (Tieman et al. 2007). Some terpene alcohols, including geraniol and carveol, can also be converted to aldehydes by nonspecific dehydrogenases (Bouwmeester et al. 1998). In addition, a double-bond reductase has recently been identified that converts geranial to β-citronellal (Xu et al. 2017).

8.3 GENES RESPONSIBLE FOR SCENT PRODUCTION

In the last three decades, investigations into floral scent in many laboratories have resulted in the characterization of a large number of genes encoding enzymes responsible for the synthesis of scent compounds. The initial breakthrough began in 1996, when the (S)–linalool synthase (LIS) gene was isolated from *Clarkia breweri* flowers using a classical biochemical approach through enzyme purification (Dudareva et al. 1996). Since then, the development of functional genomic technologies has led to identification and characterization of hundreds of "scent" genes from scores of plant species. Many of these genes have turned out to belong to gene families, groups of genes that encode evolutionarily and structurally related enzymes (Table 8.1). For example, LIS is a member of the terpene synthase family, which includes monoterpene, sesquiterpene, and diterpene synthases (Dudareva, Pichersky, and Gershenzon 2004). Other scent genes belong to families that make some nonvolatile compounds. For example, BEAT belongs to the BAHD acyltranferases which includes members involved in the biosynthesis of anthocyanin pigments and some phytoalexins (Suzuki et al. 2004). SAMT and BAMT belong to the SABATH family of methyltransferases, which includes several members involved in the biosynthesis of the defense compound caffeine and the pigment compound bixin (Bouvier, Dogbo, and Camara 2003). Yet even with all the progress made recently, it is clear that most genes and enzymes involved in the biosynthesis of scent compounds remain unknown. Moreover, the recent finding that emission of volatiles to the atmosphere is a biologically mediated process (Adebesin et al. 2017; see also Chapter 14) increases the number of genes to be discovered to fully understand biology of floral scent.

TABLE 8.1

Examples of Gene Families with Members Encoding Floral Scent Biosynthetic Enzymes

Family	Example of Enzyme	Product	Plant
Terpene synthases	Linalool synthase (LIS)	Linalool	*Arabidopsis thaliana*
Methyltransferases (Type I)	Orcinol *O*-methyltransferase (OOMT)	Dimethoxytoluene	Rose (*Rosa hybrida*)
Methyltransferases (Type II)	Benzoic acid Methyltransferase (BAMT)	Methyl benzoate	Snapdragon (*Antirrhinum majus*)
Acyltransferases (BAHD)	Acetyl-CoA:benzyl alcohol acetyltransferase (BEAT)	Benzylacetate	*Clarkia breweri*
Carotenoid cleavage dioxygenases (CCD)	Carotenoid cleavage dioxygenases1	β-Ionone	*Petunia hybrida*

8.4   REGULATION OF SCENT BIOSYNTHETIC PATHWAYS

8.4.1   SPATIAL REGULATION

It has been found that floral volatiles are generally synthesized *de novo* in the organs from which they are emitted and tend to be immediately released rather than accumulate inside the floral tissue. Therefore, peak scent biosynthesis tends to be concurrent with peak emission, typically when the flowers are ready for pollination. Within the flowers, the petals are the principal, but not exclusive, emitters of volatiles. While the same floral scent components are often emitted from all parts of the flower (although not necessarily at the same amount or rate), sometimes specific compounds may be emitted from only a subset of the floral organs (Dobson, Bergström, and Groth 1990; Dobson, Groth, and Bergström 1996; Pichersky et al. 1994; Verdonk et al. 2003). In addition, as discussed in Chapters 6 and 13, in some species, orchids for example, floral volatiles are emitted from highly specialized areas, called "scent glands," or osmophores, within the flower (Francisco and Ascensão 2013; Pansarin, Pansarin, and Sazima 2014).

Tissue- and organ-specificity of scent emission is regulated at the level of scent biosynthetic gene expression and enzyme activity. Indeed, many of the scent biosynthetic genes isolated so far show a very specific expression profile, with the highest level found in the scent producing parts of the flower (see for example, Dudareva et al. 1996, 2003; Colquhoun et al. 2010). Within scent-emitting organs, formation of volatiles is often restricted to specific cell types or layers. In snapdragon flowers for example, biosynthesis of methyl benzoate by BAMT is restricted to the inner epidermal layer of the upper and lower petal lobes, the parts of the petals that face the approaching bees during landing (Kolosova et al. 2001b). Similar cell-specific expression of scent biosynthetic genes was also reported for roses and *Clarkia breweri* (Dudareva et al. 1996; Dudareva and Pichersky 2000; Bergougnoux et al. 2007).

8.4.2   TEMPORAL REGULATION

8.4.2.1   Regulation of Developmental Emission of Scent Compounds

In flowers of most species that produce foral scent, volatile emission increases during the early stages of floral development, peaks when the flowers are ready for pollination, and sharply decreases thereafter (Wang et al. 1997; Dudareva et al. 1998, 2000, 2003; Boatright et al. 2004; Lavid et al. 2002). In contrast, the activities of volatile-synthesizing enzymes show two different developmental patterns. One group of enzymes, exemplified by *C. breweri* LIS and SAMT (Dudareva and Pichersky 2000), snapdragon BAMT (Dudareva et al. 2000) and rose geraniol acetyltransferase (Shalit et al. 2003), display high activity in young flowers that declines in old flowers, but remains relatively high (30% to 50% from the maximum level) even though emissions of corresponding volatile compounds has practically ceased. The second group of enzymes, represented by *C. breweri* IEMT, BEAT (Dudareva and Pichersky 2000) and BEBT (D'Auria, Chen, and Pichersky 2002), and petunia BPBT (Boatright et al. 2004), shows little or no decline in activity at the end of the lifespan of the flower, although again, emissions of corresponding volatile compounds did decline substantially. The only enzyme identified so far that does not follow these two developmental patterns is rose OOMT. Its activity peaks sharply during flower maturation and decreases to almost undetectable level in old flowers (Lavid et al. 2002). The reason for decreasing volatile biosynthesis and emission in spite of high levels of activity of scent biosynthetic enzymes in old flowers is generally unknown, but in the case of snapdragon, it has been found that the target for the regulation of developmental production of volatiles includes the level of supplied substrate in the cell. For example, the low emission of methyl benzoate in old flowers is due to low levels of benzoic acid in petal tissue (Dudareva et al. 2000).

Analysis of expression of genes encoding scent biosynthetic enzymes over flower development revealed that they typically peak 1 to 2 days ahead of enzyme activity and emission of the corresponding compound. The concurrent temporal changes in activities of enzymes responsible for the final steps of volatile formation, enzyme protein levels, and the expression of corresponding structural genes suggest that the temporal biosynthesis of volatiles is regulated largely at the level of gene expression (Shalit et al. 2003; Lavid et al. 2002; Dudareva et al. 1996, 1998, 2000, 2003; D'Auria, Chen, and Pichersky 2002; Boatright et al. 2004). More recent data revealed that transcriptional regulation of volatile biosynthetic pathways is not limited to their last biochemical step, but rather shared by one or multiple intermediate steps (Colquhoun and Clark 2011; Muhlemann et al. 2012).

A common property of floral scent bouquets is the presence of compounds produced by more than one biochemical pathway, formation of which requires coordinated regulation. Comparative analysis of benzenoid and monoterpene emission in snapdragon (*Antirrhinum majus*) flowers revealed that the orchestrated emission of phenylpropanoid and isoprenoid compounds is regulated upstream of individual metabolic pathways (Dudareva et al. 2003), implying that transcription factors (TFs) control scent emission. Significant breakthroughs have been made in petunia flowers, where several TFs controlling the flux toward volatile production and emission have recently been isolated. The first TF identified was ODORANT1 (ODO1), an R2R3-type MYB TF. *ODO1* is exclusively expressed in petunia petals and controls a major portion of the shikimate pathway as well as the entry points into both the Phe (i.e., chorismate mutase) and phenylpropanoid (i.e., PAL) pathways, thus regulating the precursor availability for the volatile phenylpropanoid/benzenoid compounds (Verdonk et al. 2005). In addition to biosynthetic genes, ODO1 controls volatile emission by activating the promoter of the PhABCG1 transporter, which is localized at the plasma membrane and shown to transport phenylpropanoid and benzenoid compounds (Van Moerkercke et al. 2012; Adebesin et al. 2017). In petunia flowers, ODO1 is positively regulated by EMISSION OF BENZENOIDS II (EOBII), another R2R3-type MYB TF, which binds to the *ODO1* promoter via a putative MYB binding site. EOBII has partially overlapping targets with ODO1 and also activates the promoter of the biosynthetic gene isoeugenol synthase (Spitzer-Rimon et al. 2010; Van Moerkercke et al. 2012; Van Moerkercke, Haring, and Schuurink 2011; Colquhoun et al. 2010). Another recently identified TF is petunia EOBI, a flower-specific R2R3-type TF, which acts downstream of EOBII and upstream of ODO1 (Spitzer-Rimon et al. 2012; Van Moerkercke, Haring, and Schuurink 2011; Fenske and Imaizumi 2016). These three identified TFs form a complex transcriptional regulatory network, in which EOBII transcriptionally activates *EOBI*, and both EOBI and EOBII activate the expression of the distantly related *ODO1*, which in turn negatively affects *EOBI* transcription (Figure 8.2). In contrast, a MYB4 TF was found to be a repressor for only a single enzyme in the phenylpropanoid pathway, cinnamate-4-hydroxylase, thus controlling the flux toward phenylpropanoid volatile compounds, like isoeugenol and eugenol, in petunia flowers (Colquhoun and Clark 2011).

Transcriptional regulation of floral terpenoid pathways remains elusive. MYC2, a basic helix-loop-helix TF, was recently identified in *Arabidopsis* inflorescences and shown to activate the expression of two sesquiterpene synthase genes *TPS11* and *TPS21* via the gibberellic and jasmonic acid signaling pathways (Hong et al. 2012). Additionally, overexpression of the *Production of Anthocyanin Pigment1* TF in roses resulted in upregulation of both terpenoid and phenylpropanoid pathways (Zvi et al. 2012). However, it remains unknown whether promoters of genes involved in terpenoid formation are natural targets for this TF.

8.4.2.2 Regulation of Rhythmic Emission of Scent Compounds

Release of floral volatiles over a daily light/dark cycle is often governed by a circadian clock and/or light, apparently as an adaption to the temporal activity of their respective pollinators (Helsper et al. 1998; Kolosova et al. 2001a; Effmert et al. 2005). Although rhythmic emission allows plants to conserve valuable carbon and energy during times of the day when their primary pollinators are inactive, some flowers continuously emit volatiles at a constant level (Pichersky et al. 1994). In addition, some plants emit one set of compounds during the day and another at

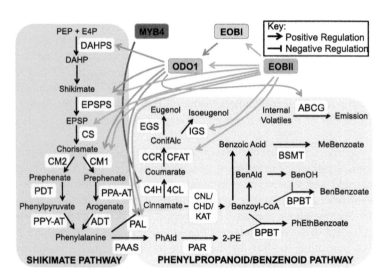

FIGURE 8.2 Overview of transcription factors involved in regulation of phenylpropanoid/benzenoid biosynthesis in petunia flowers. An abbreviated biosynthetic network for benzenoid, phenylpropanoid, and related volatiles is shown, along with transcription factors and their known targets. Abbreviations: BenAld, benzaldehyde; BenBenzoate, benzyl benzoate; BenOH, benzyl alcohol; ConifAlc, coniferyl alcohol; E4P, erythrose-4-phosphate; EPSP, 5-enolpyruvylshikimate-3-phosphate; MeBenzoate, methyl benzoate; DAHP, 3-deoxy-D-arabinoheptulosonate 7-phosphate; 2-PE, 2-phenylethanol; PEP, phosphoenolpyruvate; PhAld, phenylacetaldehyde; PhEthBenzoate, phenylethyl benzoate; ADT, arogenate dehydratase; BPBT, benzoyl-CoA:benzylalcohol/2-phenylethanol benzoyltransferase; BSMT, benzoic acid/salicylic acid carboxyl methyltransferase; C4H, cinnamate 4-hydroxylase; CCR, cinnamoyl-CoA reductase; CFAT, coniferyl alcohol acetyltransferase; CHD, cinnamoyl-CoA hydratase-dehydrogenase; 4CL, 4-coumarate:CoA ligase; CM, chorismate synthase; CNL, cinnamate:CoA ligase; CS, chorismate synthase; DAHPS, DAHP synthase; EGS, eugenol synthase; EPSPS, EPSP synthase; IGS, isoeugenol synthase; KAT, 3-ketoacyl-CoA thiolase; PAAS, PhAld synthase; PAL, phenylalanine ammonia lyase; PAR; PhAld reductase; PDT, prephenate dehydratase; PPA-AT, prephenate aminotransferase; PPY-AT, phenylpyruvate aminotransferase.

night (Matile and Altenburger 1988; Loughrin et al. 1992). It has also been found that in some flowers, certain compounds are emitted in a rhythmic manner during a 24-hr period, while others are not (Jakobsen and Olsen 1994; Nielsen et al. 1995). In some cases, as in *Dianthus inoxianus*, the amounts of total emitted volatiles do not change over the day/night cycle, yet the relative abundances of compounds contributing to pollinator attraction vary according to visitor activity (Balao et al. 2011).

Plants pollinated at night often exhibit a circadian, endogenously controlled rhythmicity in their nocturnal emission patterns. This nocturnal rhythmicity is maintained upon exposure to continuous light or dark (Matile and Altenburger 1988; Altenburger and Matile 1990; Fenske et al. 2015). In contrast, diurnal rhythmicity in emission of volatile compounds by plants was reported to be noncircadian and controlled by irradiation levels (Jakobsen and Olsen 1994; Altenburger and Matile 1990; Jakobsen et al. 1994). However, investigations in *Rosa hybrida* cv. Honesty and snapdragon *Antirrhinum majus* revealed that their scent emissions display "free-running" cycles in the absence of environmental cues (in continuous dark or continuous light), indicating that diurnal rhythmicity in emission of floral volatiles in these species is controlled by a circadian clock (Helsper et al. 1998; Kolosova et al. 2001a). In addition, it has been shown that within the same scent bouquet oscillating emission of some compounds is regulated by light, while emission of others is under control of endogenous circadian clock (Hendel-Rahmanim et al. 2007; Colquhoun et al. 2010).

The role of circadian clock in rhythmic emission of floral volatiles was recently investigated in petunia flowers (Fenske et al. 2015). The isolation of the first *P. hybrida* clock TF Late Elongated Hypocotyl (PhLHY) revealed that it regulates the timing of volatile emission by temporally controlling the expression profiles of genes primarily responsible for precursor availability, including TFs, *EOBI*, *EOBII* and *ODO1*, as well as enzyme-encoding genes *EPSPS*, *CM1*, *ADT* and *PAL*. It was shown that PhLHY directly binds to evening elements in the promoters of *ODO1*, *EPSPS* and *isoeugenol synthase 1* (Fenske et al. 2015). Whether PhLHY interacts with promoters of other scent biosynthetic genes remains to be determined.

8.4.3 REGULATION OF FLORAL SCENT AFTER POLLINATION

The scent of many flowers is markedly reduced soon after pollination. Such quantitative and/or qualitative post-pollination changes in floral bouquets, first shown in orchids (Tollsten 1993), lower the attractiveness of these flowers, as well as increase the overall reproductive success of the plant by directing pollinators to the unpollinated flowers (Neiland and Wilcock 1998). Research into post-pollination changes in floral scent emission in snapdragon and petunia flowers revealed that the decrease in emission begins only after pollen tubes reach the ovary, suggesting that successful fertilization is a prerequisite for the reduction of floral scent after pollination (Negre et al. 2003). In petunia flowers, reduced methyl benzoate emission after pollination is the result of transcriptional downregulation of the cognate gene in an ethylene-dependent manner (Negre et al. 2003; Underwood et al. 2005). In snapdragon flowers, however, the post-pollination decrease in methyl benzoate emission largely depends on reduced BAMT activity and substrate availability (Negre et al. 2003).

8.4.4 ROLE OF SUBSTRATES IN THE REGULATION OF SCENT FORMATION

In some cases, the nature of the product and the efficiency of its formation is determined by the availability of substrates for the final reaction, especially when that final reaction is catalyzed by an enzyme with broad substrate specificity (Negre et al. 2003; Boatright et al. 2004). For example, multiple carboxyl methyltransferases with dual specificity – benzoic acid and salicylic acid – have been characterized from at least ten plant species (Effmert et al. 2005), and analysis of the scent profiles in these plants revealed that scent composition depends mostly on the substrate availability. Indeed, in petunia flowers, which emit high levels of methyl benzoate and only trace amounts of methyl salicylate, the apparent catalytic efficiencies (k_{cat}/K_m ratio) of the isolated BSMTs are 40- to 75-fold higher with salicylic acid than with benzoic acid (Negre et al. 2003). However, there was a very small internal pool of free salicylic acid, indicating that the enzymes could not produce methyl salicylate *in planta* due to the lack of substrate. On the other hand, the level of benzoic acid (~7 mM) was in the range of K_m values for benzoic acid, suggesting that these enzymes are involved in methyl benzoate emission (Negre et al. 2003), which was also confirmed genetically (Underwood et al. 2005).

Heterologous expression in petunia flowers of rose alcohol acetyltransferase (RhAAT), which catalyzes the formation of geranyl acetate from geraniol and acetyl-CoA in rose flowers (Shalit et al. 2003), presents another example demonstrating that the spectrum of emitted volatiles depends on substrate availability and biochemical properties of the enzyme responsible for compound formation (Guterman et al. 2006). As petunia flowers do not produce geraniol, the introduced RhAAT used the endogenous phenylethyl alcohol and benzylalcohol and significantly increased the levels of emitted benzyl acetate and phenylethyl acetate in transgenic flowers (Guterman et al. 2006). However, feeding of transgenic flowers with geraniol, the preferred substrate in the *in vitro* assays, or 1-octanol, another potential RhAAT substrate (Shalit et al. 2003), resulted in the production of their respective acetate esters. Due to the compartmentalized nature of the plant cell, substrate availability in the compartment where the corresponding scent biosynthetic enzymes are present, rather than the measured total concentration of the substrate in the cell, may constitute the major determinant of the level of emission of volatiles. Consequently, the transporters themselves surely influence the levels of volatile emission as well (Widhalm et al. 2015; Lynch et al. 2017).

8.5 CONCLUSIONS

The emission of floral volatiles is a broadly adoptive strategy evolved in flowering plants to increase reproductive success. The information conveyed by floral volatiles depends on amount, composition and context, and it elicits distinct behavioral responses in various pollinators. Small differences in scent composition often distinguish closely related species, and result in the attraction of different pollinators, thus ensuring genetic isolation of the species. The diversity of the volatiles produced, as well as the corresponding diversity in biosynthetic routes and regulatory mechanisms, often reflects variations in ecological niches and selective pressures that had led to the separate evolutionary history of each species. However, despite such variability, many common themes are evident. As described in this chapter, generally the volatiles are derived from a limited number of metabolic pathways, with the exceptional diversity of the final products resulting from modifications of the respective core scaffolds.

Although much progress has been made toward understanding the biosynthesis of volatiles and its regulation, many questions remain unanswered. In addition to the numerous individual genes and their encoded enzymes that have yet to be identified and characterized, a key pressing question is how the orchestrated emission of various volatiles is coordinated. In addition, many floral volatiles exhibit antimicrobial and antifungal activities, as well as serving as deterrents to undesirable floral visitors. However, how flowers balance attractive and defensive functions of floral volatiles and what are the molecular mechanisms responsible for the protection of reproductive organs and subsequent overall plant survival in natural ecosystems remain unknown. Furthermore, the recent discovery that emission is an active process has opened a new field of research into the individual mechanisms and the responsible genes that are involved in this ultimate step of volatile production.

ACKNOWLEDGMENTS

We apologize to all of the authors whose contribution to this field was not cited due to space restrictions. This work was supported by grant IOS-1655438 from the National Science Foundation and by the USDA National Institute of Food and Agriculture Hatch project 177845 to ND.

REFERENCES

Adebesin, Funmi, Joshua R. Widhalm, Joseph H. Lynch, Rachel M. McCoy, and Natalia Dudareva. 2018. "A Peroxisomal Thioesterase Plays Auxiliary Roles in Plant β-Oxidative Benzoic Acid Metabolism." *Plant Journal* 93: 905–16.

Adebesin, Funmilayo, Joshua R. Widhalm, Benoît Boachon, François Lefèvre, Baptiste Pierman, Joseph H. Lynch, Iftekhar Alam, et al. 2017. "Emission of Volatile Organic Compounds from *Petunia* Flowers Is Facilitated by an ABC Transporter." *Science* 356: 1386–88.

Akhtar, Tariq A., Yuki Matsuba, Ines Schauvinhold, Geng Yu, Hazel A. Lees, Samuel E. Klein and Eran Pichersky. 2013. "The Tomato *cis*–prenyltransferase Gene Family." *The Plant Journal* 72: 640–52.

Akhtar, Tariq A., and Eran Pichersky. 2013. "Veratrole Biosynthesis in White Campion." *Plant Physiology* 162: 52–62.

Altenburger, Rolf, and Philippe Matile. 1990. "Further Observations on Rhythmic Emission of Fragrance in Flowers." *Planta* 180: 194–97.

Amrad, Avichai, Michel Moser, Therese Mandel, Michel de Vries, Robert C. Schuurink, Loreta Freitas, and Cris Kuhlemeier. 2016. "Gain and Loss of Floral Scent Production through Changes in Structural Genes during Pollinator-Mediated Speciation." *Current Biology* 26: 3303–12.

Balao, Francisco, Javier Herrera, Salvador Talavera, and Stefan Dötterl. 2011. "Spatial and Temporal Patterns of Floral Scent Emission in *Dianthus inoxianus* and Electroantennographic Responses of Its Hawkmoth Pollinator." *Phytochemistry* 72: 601–09.

Baldwin, Elizabeth A., John W. Scott, Christine K. Shewmaker, and Wolfgang Schuch. 2000. "Flavor Trivia and Tomato Aroma: Biochemistry and Possible Mechanisms for Control of Important Aroma Components." *HortScience* 35: 1013–22.

Beekwilder, Jules, Mayte Alvarez-Huerta, Evert Neef, Francel W.A. Verstappen, Harro J. Bouwmeester, and Asaph Aharoni. 2004. "Functional Characterization of Enzymes Forming Volatile Esters from Strawberry and Banana." *Plant Physiology* 135: 1865–78.

Bergougnoux, Véronique, Jean-Claude Caissard Frédéric, Jean-Louis Magnard Gabriel, Scalliet J. Mark, Philippe Hugueney, and Sylvie Baudino. 2007. "Both the Adaxial and Abaxial Epidermal Layers of the Rose Petal Emit Volatile Scent Compounds." *Planta* 226: 853–66.

Bleeker, Petra M., Eleni A. Spyropoulou, Paul J. Diergaarde, Hanne Volpin, Michiel T.J. De Both, Philipp Zerbe, Joerg Bohlmann, et al. 2011. "RNA-Seq Discovery, Functional Characterization, and Comparison of Sesquiterpene Synthases from *Solanum lycopersicum* and *Solanum habrochaites* Trichomes." *Plant Molecular Biology* 77: 323–36.

Boatright, Jennifer, Florence Negre, Xinlu Chen, Christine M. Kish, Barbara Wood, Greg Peel, Irina Orlova, David Gang, David Rhodes, and Natalia Dudareva. 2004. "Understanding In Vivo Benzenoid Metabolism in *Petunia* Petal Tissue." *Plant Physiology* 135: 1993–2011.

Bouvier, Florence, Odette Dogbo, and Bilal Camara. 2003. "Biosynthesis of the Food and Cosmetic Plant Pigment Bixin (Annatto)." *Science* 300: 2089–91.

Bouwmeester, Harro J, Jonathan Gershenzon, Maurice C.J.M. Konings, and Rodney Croteau. 1998. "Biosynthesis of the Monoterpenes Limonene and Carvone in the Fruit of Caraway." *Plant Physiology* 117: 901–12.

Bouwmeester, Harro J, Maurice C.J.M. Konings, Jonathan Gershenzon, Frank Karp, and Rodney Croteau. 1999. "Cytochrome P-450 Dependent (+)-Limonene-6-Hydroxylation in Fruits of Caraway (*Carum carvi*)." *Phytochemistry* 50: 243–48.

Burke, Charles C., Mark R. Wildung, and Rodney Croteau. 1999. "Geranyl Diphosphate Synthase: Cloning, Expression, and Characterization of This Prenyltransferase as a Heterodimer." *Proceedings of the National Academy of Sciences of the United States of America* 96: 13062–67.

Bussell, John D., Michael Reichelt, Andrew A.G. Wiszniewski, Jonathan Gershenzon, and Steven M. Smith. 2014. "Peroxisomal ATP-Binding Cassette Transporter COMATOSE and the Multifunctional Protein Abnormal INFLORESCENCE MERISTEM are Required for the Production of Benzoylated Metabolites in Arabidopsis Seeds." *Plant Physiology* 164: 48–54.

Chen, Feng, Dorothea Tholl, Jörg Bohlmann, and Eran Pichersky. 2011. "The Family of Terpene Synthases in Plants: A Mid-Size Family of Genes for Specialized Metabolism That Is Highly Diversified throughout the Kingdom." *The Plant Journal* 66: 212–29.

Colquhoun, Thomas A., Julian C. Verdonk, Bernardus C.J. Schimmel, Denise M. Tieman, Beverly A. Underwood, and David G. Clark. 2010. "Petunia Floral Volatile Benzenoid/ Phenylpropanoid Genes Are Regulated in a Similar Manner." *Phytochemistry* 71: 158–67.

Colquhoun, Thomas A., and David G. Clark. 2011. "Unraveling the Regulation of Floral Fragrance Biosynthesis." *Plant Signaling & Behavior* 6: 378–81.

D'Auria, John C. 2006. "Acyltransferases in Plants: A Good Time to Be BAHD." *Current Opinion in Plant Biology* 9: 331–40.

D'Auria, John C., Eran Pichersky, Andrea Schaub, Armin Hansel, and Jonathan Gershenzon. 2007. "Characterization of a BAHD Acyltransferase Responsible for Producing the Green Leaf Volatile (Z)-3-Hexen-1-yl Acetate in *Arabidopsis thaliana*." *The Plant Journal* 49: 194–207.

D'Auria, John C., Feng Chen, and Eran Pichersky. 2002. "Characterization of an Acyltransferase Capable of Synthesizing Benzylbenzoate and Other Volatile Esters in Flowers and Damaged Leaves of *Clarkia breweri*." *Plant Physiology* 130: 466–76.

D'Auria, John C., Feng Chen, and Eran Pichersky. 2003. "The SABATH Family of MTS in *Arabidopsis thaliana* and Other Plant Species." In *Integrative Phytochemistry: From Ethnobotany to Molecular Ecology*, edited by John Romeo, 37: 253–83. Elsevier.

Degenhardt, Jörg, Tobias G. Köllner, and Jonathan Gershenzon. 2009. "Monoterpene and Sesquiterpene Synthases and the Origin of Terpene Skeletal Diversity in Plants." *Phytochemistry* 70: 1621–37.

Dexter, Richard, Anthony Qualley, Christine M. Kish, Choong Je Ma, Takao Koeduka, Dinesh A. Nagegowda, Natalia Dudareva, Eran Pichersky, and David Clark. 2007. "Characterization of a Petunia Acetyltransferase Involved in the Biosynthesis of the Floral Volatile Isoeugenol." *Plant Journal* 49: 265–75.

Dobson, Heidi E.M., Gunnar Bergström, and Inga Groth. 1990. "Differences in Fragrance Chemistry Between Flower Parts of *Rosa rugosa* Thunb. (Rosaceae)." *Israel Journal of Plant Sciences* 39: 143–56.

Dobson Heidi E.M., Inga Groth, and Gunnar Bergström. 1996. "Pollen Advertisement: Chemical Contrasts between Whole-Flower and Pollen Odors." *American Journal of Botany* 83: 877–85.

Dudareva, Natalia, Susanna Andersson, Irina Orlova, Nathalie Gatto, Michael Reichelt, David Rhodes, Wilhelm Boland, and Jonathan Gershenzon. 2005. "The Nonmevalonate Pathway Supports Both Monoterpene and Sesquiterpene Formation in Snapdragon Flowers." *Proceedings of the National Academy of Sciences of the United States of America* 102: 933–38.

Dudareva, Natalia, Leland Cseke, Victoria M. Blanc, and Eran Pichersky. 1996. "Evolution of Floral Scent in Clarkia: Novel Patterns of S-Linalool Synthase Gene Expression in the *C. breweri* Flower." *The Plant Cell* 8: 1137–48.

Dudareva, Natalia, John C. D'Auria, Kyoung Hee Nam, Robert A. Raguso, and Eran Pichersky. 1998. "Acetyl-CoA:Benzylalcohol Acetyltransferase – an Enzyme Involved in Floral Scent Production in *Clarkia breweri*." *The Plant Journal* 14: 297–304.

Dudareva, Natalia, Diane Martin, Christine M. Kish, Natalia Kolosova, Nina Gorenstein, Jenny Fäldt, Barbara Miller, and Jörg Bohlmann. 2003. "*E*-β-Ocimene and Myrcene Synthase Genes of Floral Scent Biosynthesis in Snapdragon: Function and Expression of Three Terpene Synthase Genes of a New Terpene Synthase Subfamily." *The Plant Cell* 15: 1227–41.

Dudareva, Natalia, Lisa M. Murfitt, Craig J. Mann, Nina Gorenstein, Natalia Kolosova, Christine M. Kish, Connie Bonham, and Karl Wood. 2000. "Developmental Regulation of Methyl Benzoate Biosynthesis and Emission in Snapdragon Flowers." *The Plant Cell* 12: 949–61.

Dudareva, Natalia, and Eran Pichersky. 2000. "Biochemical and Molecular Genetic Aspects of Floral Scents." *Plant Physiology* 122: 627–33.

Dudareva, Natalia, Eran Pichersky, and Jonathan Gershenzon. 2004. "Biochemistry of Plant Volatiles." *Plant Physiology* 135: 1893–902.

Effmert, Uta, Sandra Saschenbrecker, Jeannine Ross, Florence Negre, Chris M. Fraser, Joseph P. Noel, Natalia Dudareva, and Birgit Piechulla. 2005. "Floral Benzenoid Carboxyl Methyltransferases: From *In Vitro* to *In Planta* Function." *Phytochemistry* 66: 1211–30.

Fenske, Myles P., Kristen D. Hewett Hazelton, Andrew K. Hempton, Jae Sung Shim, Breanne M. Yamamoto, Jeffrey A. Riffell, and Takato Imaizumi. 2015. "Circadian Clock Gene LATE ELONGATED HYPOCOTYL Directly Regulates the Timing of Floral Scent Emission in *Petunia*." *Proceedings of the National Academy of Sciences of the United States of America* 112: 9775–80.

Fenske, Myles P., and Takato Imaizumi. 2016. "Circadian Rhythms in Floral Scent Emission." *Frontiers in Plant Science* 7: 462.

Feussner, Ivo, and Claus Wasternack. 2002. "The Lipoxygenase Pathway." *Annual Review of Plant Biology* 53: 275–97.

Francisco, Ana, and Lia Ascensão. 2013. "Structure of the Osmophore and Labellum Micromorphology in the Sexually Deceptive Orchids *Ophrys bombyliflora* and *Ophrys tenthredinifera* (Orchidaceae)." *International Journal of Plant Sciences* 174: 619–36.

Gang, David R., Noa Lavid, Chloe Zubieta, Feng Chen, Till Beuerle, Efraim Lewinsohn, Joseph P. Noel, and Eran Pichersky. 2002. "Characterization of Phenylpropene *O*-Methyltransferases from Sweet Basil Facile Change of Substrate Specificity and Convergent Evolution within a Plant *O*-Methyltransferase Family." *The Plant Cell* 14: 505–19.

Guirimand, Grégory, Anthony Guihur, Michael A. Phillips, Audrey Oudin, Gaëlle Glévarec, Samira Mahroug, Céline Melin, et al. 2012a. "Triple Subcellular Targeting of Isopentenyl Diphosphate Isomerases Encoded by a Single Gene." *Plant Signaling & Behavior* 7: 1495–97.

Guirimand, Grégory, Anthony Guihur, Michael A. Phillips, Audrey Oudin, Gaëlle Glévarec, Céline Melin, Nicolas Papon, et al. 2012b. "A Single Gene Encodes Isopentenyl Diphosphate Isomerase Isoforms Targeted to Plastids, Mitochondria and Peroxisomes in *Catharanthus roseus*." *Plant Molecular Biology* 79: 443–59.

Guterman, Inna, Tania Masci, Xinlu Chen, Florence Negre, Eran Pichersky, Natalia Dudareva, David Weiss, and Alexander Vainstein. 2006. "Generation of Phenylpropanoid Pathway-Derived Volatiles in Transgenic Plants: Rose Alcohol Acetyltransferase Produces Phenylethyl Acetate and Benzyl Acetate in Petunia Flowers." *Plant Molecular Biology* 60: 555–63.

Helsper, Johannes P.F.G., Jacques A. Davies, Harro J. Bouwmeester, Aafke F. Krol, and Maria H. van Kampen. 1998. "Circadian Rhythmicity in Emission of Volatile Compounds by Flowers of *Rosa hybrida* L. Cv. Honesty." *Planta* 207: 88–95.

Hendel-Rahmanim, Keren, Tania Masci, Alexander Vainstein, and David Weiss. 2007. "Diurnal Regulation of Scent Emission in Rose Flowers." *Planta* 226: 1491–99.

Henry, Laura K., Suzanne T. Thomas, Joshua R. Widhalm, Joseph H. Lynch, Thomas C. Davis, Sharon A. Kessler, Jörg Bohlmann, Joseph P. Noel, and Natalia Dudareva. 2018. "Contribution of Isopentenyl Phosphate to Plant Terpenoid Metabolism." *Nature Plants* 4: 721–29.

Henry, Laura K., Michael Gutensohn, Suzanne T. Thomas, Joseph P. Noel, and Natalia Dudareva. 2015. "Orthologs of the Archaeal Isopentenyl Phosphate Kinase Regulate Terpenoid Production in Plants." *Proceedings of the National Academy of Sciences of the United States of America* 112: 10050–55.

Hirata, Hiroshi, Toshiyuki Ohnishi, Haruka Ishida, Kensuke Tomida, Miwa Sakai, Masakazu Hara, and Naoharu Watanabe. 2012. "Functional Characterization of Aromatic Amino Acid Aminotransferase Involved in 2-Phenylethanol Biosynthesis in Isolated Rose Petal Protoplasts." *Journal of Plant Physiology* 169: 444–51.

Hong, Gao-Jie, Xue-Yi Xue, Ying-Bo Mao, Ling-Jian Wang, and Xiao-Ya Chen. 2012. "Arabidopsis MYC2 Interacts with DELLA Proteins in Regulating Sesquiterpene Synthase Gene Expression." *The Plant Cell* 24: 2635–48.

Hsieh, Ming-Hsiun, Chiung-Yun Chang, Shih-Jui Hsu, and Ju-Jiun Chen. 2008. "Chloroplast Localization of Methylerythritol 4-Phosphate Pathway Enzymes and Regulation of Mitochondrial Genes in IspD and IspE Albino Mutants in Arabidopsis." *Plant Molecular Biology* 66: 663–73.

Ibdah, Mwafaq, Ying-Tung Chen, Curtis G. Wilkerson, and Eran Pichersky. 2009. "An Aldehyde Oxidase in Developing Seeds of Arabidopsis Converts Benzaldehyde to Benzoic Acid." *Plant Physiology* 150: 416–23.

Jakobsen, Henrik B., Palle Friis, Jens K. Nielsen, and Carl E. Olsen. 1994. "Emission of Volatiles from Flowers and Leaves of *Brassica napus In Situ*." *Phytochemistry* 37: 695–99.

Jakobsen, Henrik Byrial, and Carl Erik Olsen. 1994. "Influence of Climatic Factors on Emission of Flower Volatiles *in Situ*." *Planta* 192: 365–71.

Kaminaga, Yasuhisa, Jennifer Schnepp, Greg Peel, Christine M. Kish, Gili Ben-Nissan, David Weiss, Irina Orlova, et al. 2006. "Plant Phenylacetaldehyde Synthase is a Bifunctional Homotetrameric Enzyme that Catalyzes Phenylalanine Decarboxylation and Oxidation." *Journal of Biological Chemistry* 281: 23357–66.

Klempien, Antje, Yasuhisa Kaminaga, Anthony Qualley, Dinesh A. Nagegowda, Joshua R. Widhalm, Irina Orlova, Ajit Kumar Shasany, et al. 2012. "Contribution of CoA Ligases to Benzenoid Biosynthesis in Petunia Flowers." *The Plant Cell* 24: 2015–30.

Knudsen, Jette T., Lars Tollsten, and L. Gunnar Bergström. 1993. "Floral Scents – A Checklist of Volatile Compounds Isolated by Head-Space Techniques." *Phytochemistry* 33: 253–80.

Koeduka, Takao, Eyal Fridman, David R. Gang, D. G. Vassao, Brenda L. Jackson, Christine M. Kish, Irina Orlova, et al. 2006. "Eugenol and Isoeugenol, Characteristic Aromatic Constituents of Spices, Are Biosynthesized via Reduction of a Coniferyl Alcohol Ester." *Proceedings of the National Academy of Sciences of the United States of America* 103: 10128–33.

Koeduka, Takao, Gordon V. Louie, Irina Orlova, Christine M. Kish, Mwafaq Ibdah, Curtis G. Wilkerson, Marianne E. Bowman, et al. 2008. "The Multiple Phenylpropene Synthases in Both *Clarkia breweri* and *Petunia hybrida* Represent Two Distinct Protein Lineages." *Plant Journal* 54: 362–74.

Kolosova, Natalia, Nina Gorenstein, Christine M. Kish, and Natalia Dudareva. 2001a. "Regulation of Circadian Methyl Benzoate Emission in Diurnally and Nocturnally Emitting Plants." *The Plant Cell* 13: 2333–47.

Kolosova, Natalia, Debra Sherman, Dale Karlson, and Natalia Dudareva. 2001b. "Cellular and Subcellular Localization of *S*-Adenosyl-L-Methionine:Benzoic Acid Carboxyl Methyltransferase, the Enzyme Responsible for Biosynthesis of the Volatile Ester Methylbenzoate in Snapdragon Flowers." *Plant Physiology* 126: 956–64.

Laule, Oliver, Andreas Furholz, Hur-Song Chang, Tong Zhu, Xun Wang, Peter B. Heifetz, Wilhelm Gruissem, and Markus Lange. 2003. "Crosstalk between Cytosolic and Plastidial Pathways of Isoprenoid Biosynthesis in *Arabidopsis thaliana*." *Proceedings of the National Academy of Sciences of the United States of America* 100: 6866–71.

Lavid, Noa, Jihong Wang, Moshe Shalit, Inna Guterman, Einat Bar, Till Beuerle, Naama Menda, et al. 2002. "*O*-Methyltransferases Involved in the Biosynthesis of Volatile Phenolic Derivatives in Rose Petals." *Plant Physiology* 129 (4): 1899–907.

Liang, Po-Huang, Tzu-Ping Ko, and Andrew H.–J. Wang. 2002. "Structure, Mechanism and Function of Prenyltransferases." *European Journal of Biochemistry* 269: 3339–54.

Liu, Yan, Shi-Hong Luo, Axel Schmidt, Guo-Dong Wang, Gui-Ling Sun, Marcus Grant, Ce Kuang, et al. 2016. "A Geranylfarnesyl Diphosphate Synthase Provides the Precursor for Sesterterpenoid (C_{25}) Formation in the Glandular Trichomes of the Mint Species *Leucosceptrum canum*." *The Plant Cell* 28: 804–22.

Long, Michael C., Dinesh A. Nagegowda, Yasuhisa Kaminaga, Kwok Ki Ho, Christine M. Kish, Jennifer Schnepp, Debra Sherman, Henry Weiner, David Rhodes, and Natalia Dudareva. 2009. "Involvement of Snapdragon Benzaldehyde Dehydrogenase in Benzoic Acid Biosynthesis." *The Plant Journal* 59: 256–65.

Loughrin, John H., Thomas R. Hamilton-Kemp, Harold R. Burton, Roger A. Andersen, and David F. Hildebrand. 1992. "Glycosidically Bound Volatile Components of *Nicotiana sylvestris* and *N. suaveolens* Flowers." *Phytochemistry* 31: 1537–40.

Lupien, Shari, Frank Karp, Mark Wildung, and Rodney Croteau. 1999. "Regiospecific Cytochrome P450 Limonene Hydroxylases from Mint (*Mentha*) Species: cDNA Isolation, Characterization, and Functional Expression of (–)-4S-Limonene-3-Hydroxylase and (–)-4S-Limonene-6-Hydroxylase." *Archives of Biochemistry and Biophysics* 368: 181–92.

Lynch, Joseph H., Irina Orlova, Chengsong Zhao, Longyun Guo, Rohit Jaini, Hiroshi Maeda, Tariq Akhtar, et al. 2017. "Multifaceted Plant Responses to Circumvent Phe Hyperaccumulation by Downregulation of Flux through the Shikimate Pathway and by Vacuolar Phe Sequestration." *The Plant Journal* 92: 939–50.

Magnard, Jean-Louis, Aymeric Roccia, Jean-Claude Caissard, Philippe Vergne, Pulu Sun, Romain Hecquet, Annick Dubois, et al. 2015. "Biosynthesis of Monoterpene Scent Compounds in Roses." *Science* 349: 81–83.

Matile, Philippe, and Rolf Altenburger. 1988. "Rhythms of Fragrance Emission in Flowers." *Planta* 174: 242–47.

Matsuba, Yuki, Jiachen Zi, A. Daniel Jones, Reuben J. Peters, and Eran Pichersky. 2015. "Biosynthesis of the Diterpenoid Lycosantalonol via Nerylneryl Diphosphate in *Solanum lycopersicum*." *PLoS One* 10: e0119302.

Muhlemann, Joëlle K., Antje Klempien, and Natalia Dudareva. 2014. "Floral Volatiles: From Biosynthesis to Function." *Plant, Cell & Environment* 37: 1936–49.

Muhlemann, Joëlle K., Hiroshi Maeda, Ching-Yun Chang, Phillip San Miguel, Ivan Baxter, Bruce Cooper, M. Ann Perera, et al. 2012. "Developmental Changes in the Metabolic Network of Snapdragon Flowers." *PLoS One* 7: e40381.

Murfitt, Lisa M., Natalia Kolosova, Craig J. Mann, and Natalia Dudareva. 2000. "Purification and Characterization of S-Adenosyl-L-Methionine:Benzoic Acid Carboxyl Methyltransferase, the Enzyme Responsible for Biosynthesis of the Volatile Ester Methyl Benzoate in Flowers of *Antirrhinum majus*." *Archives of Biochemistry and Biophysics* 382: 145–51.

Nagegowda, Dinesh A., Michael Gutensohn, Curtis G. Wilkerson, and Natalia Dudareva. 2008. "Two Nearly Identical Terpene Synthases Catalyze the Formation of Nerolidol and Linalool in Snapdragon Flowers." *The Plant Journal* 55: 224–39.

Nakamura, Ayako, Hiroshi Shimada, Tatsuru Masuda, Hiroyuki Ohta, and Ken-ichiro Takamiya. 2001. "Two Distinct Isopentenyl Diphosphate Isomerases in Cytosol and Plastid Are Differentially Induced by Environmental Stresses in Tobacco." *FEBS Letters* 506: 61–64.

Negre, Florence, Christine M. Kish, Jennifer Boatright, Beverly Underwood, Kenichi Shibuya, Conrad Wagner, David G. Clark, and Natalia Dudareva. 2003. "Regulation of Methylbenzoate Emission after Pollination in Snapdragon and Petunia Flowers." *The Plant Cell* 15: 2992–3006.

Neiland, Mary Ruth M., and Christopher C. Wilcock. 1998. "Fruit Set, Nectar Reward, and Rarity in the Orchidaceae." *American Journal of Botany* 85: 1657–71.

Nielsen, Jens K., Henrik B. Jakobsen, Palle Friis, Keld Hansen, Jørgen Møller, and Carl E. Olsen. 1995. "Asynchronous Rhythms in the Emission of Volatiles from *Hesperis matronalis* Flowers." *Phytochemistry* 38: 847–51.

Pansarin, Ludmila M., Emerson R. Pansarin, and Marlies Sazima. 2014. "Osmophore Structure and Phylogeny of *Cirrhaea* (Orchidaceae, Stanhopeinae)." *Botanical Journal of the Linnean Society* 176: 369–83.

Phillips, Michael A., John C. D'Auria, Jonathan Gershenzon, and Eran Pichersky. 2008. "The *Arabidopsis thaliana* Type I Isopentenyl Diphosphate Isomerases Are Targeted to Multiple Subcellular Compartments and Have Overlapping Functions in Isoprenoid Biosynthesis." *The Plant Cell* 20: 677–96.

Pichersky, Eran, Robert A. Raguso, Efraim Lewinsohn, and Rodney Croteau. 1994. "Floral Scent Production in *Clarkia* (Onagraceae). I. Localization and Developmental Modulation of Monoterpene Emission and Linalool Synthase Activity." *Plant Physiology* 106: 1533–40.

Pulido, Pablo, Catalina Perello, and Manuel Rodriguez-Concepcion. 2012. "New Insights into Plant Isoprenoid Metabolism." *Molecular Plant* 5: 964–67.

Qualley, Anthony V., Joshua R. Widhalm, Funmilayo Adebesin, Christine M. Kish, and Natalia Dudareva. 2012. "Completion of the Core β-Oxidative Pathway of Benzoic Acid Biosynthesis in Plants." *Proceedings of the National Academy of Sciences of the United States of America* 109: 16383–88.

Sakai, Miwa, Hiroshi Hirata, Hironori Sayama, Kazuya Sekiguchi, Hiroaki Itano, Tatsuo Asai, Hideo Dohra, Masakazu Hara, and Naoharu Watanabe. 2007. "Production of 2-Phenylethanol in Roses as the Dominant Floral Scent Compound from L-Phenylalanine by Two Key Enzymes, a PLP-Dependent Decarboxylase and a Phenylacetaldehyde Reductase." *Bioscience, Biotechnology, and Biochemistry* 71: 2408–19.

Sallaud, Christophe, Denis Rontein, Sandrine Onillon, Françoise Jabès, Philippe Duffé, Cécile Giacalone, Samuel Thoraval, et al. 2009. "A Novel Pathway for Sesquiterpene Biosynthesis from Z, Z-Farnesyl Pyrophosphate in the Wild Tomato *Solanum habrochaites*." *The Plant Cell* 21: 301–17.

Sas, Claudia, Frank Müller, Christian Kappel, Tyler V. Kent, Stephen I. Wright, Monika Hilker, and Michael Lenhard. 2016. "Repeated Inactivation of the First Committed Enzyme Underlies the Loss of Benzaldehyde Emission after the Selfing Transition in Capsella." *Current Biology* 26: 3313–19.

Schuhr, Christoph A., Tanja Radykewicz, Silvia Sagner, Christoph Latzel, Meinhart H. Zenk, Duilio Arigoni, Adelbert Bacher, Felix Rohdich, and Wolfgang Eisenreich. 2003. "Quantitative Assessment of Crosstalk between the Two Isoprenoid Biosynthesis Pathways in Plants by NMR Spectroscopy." *Phytochemistry Reviews* 2: 3–16.

Seo, Hok Seo, Jong Tae Song, Jong-Joo Cheong, Yong-Hwan Lee, Yin-Won Lee, Ingyu Hwang, Jong Seob Lee, and Yang Do Choi. 2001. "Jasmonic Acid Carboxyl Methyltransferase: A Key Enzyme for Jasmonate-Regulated Plant Responses." *Proceedings of the National Academy of Sciences of the United States of America* 98: 4788–93.

Shalit, Moshe, Inna Guterman, Hanne Volpin, Einat Bar, Tal Tamari, Naama Menda, Zach Adam, et al. 2003. "Volatile Ester Formation in Roses. Identification of an Acetyl-Coenzyme A. Geraniol/Citronellol Acetyltransferase in Developing Rose Petals." *Plant Physiology* 131: 1868–76.

Shi, Shaochuan, Guangyou Duan, Dandan Li, Jie Wu, Xintong Liu, Bo Hong, Mingfang Yi, and Zhao Zhang. 2018. "Two-Dimensional Analysis Provides Molecular Insight into Flower Scent of Lilium 'Siberia.'" *Scientific Reports* 8: 5352.

Simkin, Andrew J., Steven H. Schwartz, Michele Auldridge, Mark G. Taylor, and Harry J. Klee. 2004a. "The Tomato Carotenoid Cleavage Dioxygenase 1 Genes Contribute to the Formation of the Flavor Volatiles β-Ionone, Pseudoionone, and Geranylacetone." *The Plant Journal* 40: 882–92.

Simkin, Andrew J., Beverly A. Underwood, Michele Auldridge, Holly M. Loucas, Kenichi Shibuya, Eric Schmelz, David G. Clark, and Harry J. Klee. 2004b. "Circadian Regulation of the PhCCD1 Carotenoid Cleavage Dioxygenase Controls Emission of Beta-Ionone, a Fragrance Volatile of *Petunia* Flowers." *Plant Physiology* 136: 3504–14.

Sommer, Susanne, Klaus Severin, Bilal Camara, and Lutz Heide. 1995. "Intracellular Localization of Geranylpyrophosphate Synthase from Cell Cultures of *Lithospermum erythrorhizon*." *Phytochemistry* 38: 623–27.

Spitzer-Rimon, Ben, Moran Farhi, Boaz Albo, Alon Cna'ani, Michal Moyal Ben Zvi, Tania Masci, Orit Edelbaum, et al. 2012. "The R2R3-MYB–Like Regulatory Factor EOBI, Acting Downstream of EOBII, Regulates Scent Production by Activating *ODO1* and Structural Scent-Related Genes in *Petunia*." *The Plant Cell* 24: 5089–105.

Spitzer-Rimon, Ben, Elena Marhevka, Oren Barkai, Ira Marton, Orit Edelbaum, Tania Masci, Naveen-Kumar Prathapani, Elena Shklarman, Marianna Ovadis, and Alexander Vainstein. 2010. "*EOBII*, a Gene Encoding a Flower-Specific Regulator of Phenylpropanoid Volatiles' Biosynthesis in *Petunia*." *The Plant Cell* 22: 1961–76.

Suzuki, Hirokazu, Shin'ya Sawada, Kazufumi Watanabe, Shiro Nagae, Masa-atsu Yamaguchi, Toru Nakayama, and Tokuzo Nishino. 2004. "Identification and Characterization of a Novel Anthocyanin Malonyltransferase from Scarlet Sage (*Salvia splendens*) Flowers: An Enzyme That Is Phylogenetically Separated from Other Anthocyanin Acyltransferases." *The Plant Journal* 38: 994–1003.

Tholl, Dorothea. 2006. "Terpene Synthases and the Regulation, Diversity and Biological Roles of Terpene Metabolism." *Current Opinion in Plant Biology* 9: 297–304.

Tholl, Dorothea. 2015. "Biosynthesis and Biological Functions of Terpenoids in Plants." *Advances in Biochemical Engineering/Biotechnology* 148: 63–106.

Tholl, Dorothea, Feng Chen, Jana Petri, Jonathan Gershenzon, and Eran Pichersky. 2005. "Two Sesquiterpene Synthases Are Responsible for the Complex Mixture of Sesquiterpenes Emitted from Arabidopsis Flowers." *The Plant Journal* 42: 757–71.

Tholl, Dorothea, Christine M. Kish, Irina Orlova, Debra Sherman, Jonathan Gershenzon, Eran Pickersky, and Natalia Dudareva. 2004. "Formation of Monoterpenes in *Antirrhinum majus* and *Clarkia breweri* Flowers Involves Heterodimeric Geranyl Diphosphate Synthases." *The Plant Cell* 16: 977–92.

Tieman, Denise M., Holly M. Loucas, Joo Young Kim, David G. Clark, and Harry J. Klee. 2007. "Tomato Phenylacetaldehyde Reductases Catalyze the Last Step in the Synthesis of the Aroma Volatile 2-Phenylethanol." *Phytochemistry* 68: 2660–69.

Tollsten, Lars. 1993. "A Multivariate Approach to Post-Pollination Changes in the Floral Scent of *Platanthera bifolia* (Orchidaceae)." *Nordic Journal of Botany* 13: 495–99.

Underwood, Beverly A., Denise M. Tieman, Kenichi Shibuya, Richard J. Dexter, Holly M. Loucas, Andrew J. Simkin, Charles A. Sims, Eric A. Schmelz, Harry J. Klee, and David G. Clark. 2005. "Ethylene-Regulated Floral Volatile Synthesis in *Petunia corollas*." *Plant Physiology* 138: 255–66.

Van Moerkercke, Alex, Carlos S. Galvan-Ampudia, Julian C. Verdonk, Michel A. Haring, and Robert C. Schuurink. 2012. "Regulators of Floral Fragrance Production and Their Target Genes in Petunia Are Not Exclusively Active in the Epidermal Cells of Petals." *Journal of Experimental Botany* 63: 3157–71.

Van Moerkercke, Alex, Michel A. Haring, and Robert C. Schuurink. 2011. "The Transcription Factor EMISSION OF BENZENOIDS II Activates the MYB ODORANT1 Promoter at a MYB Binding Site Specific for Fragrant Petunias." *The Plant Journal* 67: 917–28.

Van Moerkercke, Alex, Ines Schauvinhold, Eran Pickersky, Michel A. Haring, and Robert C. Schuurink. 2009. "A Plant Thiolase Involved in Benzoic Acid Biosynthesis and Volatile Benzenoid Production." *The Plant Journal* 60: 292–302.

Vassão, Daniel G., David R. Gang, Takao Koeduka, Brenda Jackson, Eran Pickersky, Laurence B. Davin, and Norman G. Lewis. 2006. "Chavicol Formation in Sweet Basil (*Ocimum basilicum*): Cleavage of an Esterified C9 Hydroxyl Group with NAD(P)H-Dependent Reduction." *Organic and Biomolecular Chemistry* 4: 2733–44.

Verdonk, Julian C., Michel A. Haring, Arjen J. Van Tunen, and Robert C. Schuurink. 2005. "ODORANT1 Regulates Fragrance Biosynthesis in Petunia Flowers." *The Plant Cell* 17: 1612–24.

Verdonk, Julian C., C. H. Ric De Vos, Harrie A. Verhoeven, Michel A. Haring, Arjen J. Van Tunen, and Robert C. Schuurink. 2003. "Regulation of Floral Scent Production in Petunia Revealed by Targeted Metabolomics." *Phytochemistry* 62: 997–1008.

Vranová, Eva, Diana Coman, and Wilhelm Gruissem. 2013. "Network Analysis of the MVA and MEP Pathways for Isoprenoid Synthesis." *Annual Review of Plant Biology* 64: 665–700.

Wang, Jihong, Natalia Dudareva, Shyam Bhakta, Robert A. Raguso, and Eran Pickersky. 1997. "Floral Scent Production in *Clarkia breweri* (Onagraceae). II. Localization and Developmental Modulation of the Enzyme S-Adenosyl-L-Methionine:(Iso)Eugenol O-Methyltransferase and Phenylpropanoid Emission." *Plant Physiology* 114: 213-21.

Watanabe, Shuzo, Kentaro Hayashi, Kensuke Yagi, Tatsuo Asai, Hazel MacTavish, Joanne Picone, Colin Turnbull, and Naoharu Watanabe. 2002. "Biogenesis of 2-Phenylethanol in Rose Flowers: Incorporation of [^{2}H$_8$]L-Phenylalanine into 2-Phenylethanol and Its Beta-D-Glucopyranoside during the Flower Opening of *Rosa* 'Hoh-Jun' and *Rosa damascena* Mill." *Bioscience, Biotechnology, and Biochemistry* 66: 943–47.

Widhalm, Joshua R., Michael Gutensohn, Heejin Yoo, Funmilayo Adebesin, Yichun Qian, Longyun Guo, Rohit Jaini, et al. 2015. "Identification of a Plastidial Phenylalanine Exporter That Influences Flux Distribution through the Phenylalanine Biosynthetic Network." *Nature Communications* 6: 8142.

Winterhalter, Peter, and Russell Rouseff. 2001. "Carotenoid-Derived Aroma Compounds: An Introduction." In *Carotenoid-Derived Aroma Compounds*, edited by Peter Winterhalter and Russell Rouseff, 1–17. Washington, DC: American Chemical Society.

Wu, Shuiqin, Michel Schalk, Anthony Clark, R Brandon Miles, Robert Coates, and Joe Chappell. 2006. "Redirection of Cytosolic or Plastidic Isoprenoid Precursors Elevates Terpene Production in Plants." *Nature Biotechnology* 24: 1441–47.

Wu, Shuiqin, Naoharu Watanabe, Satoru Mita, Hideo Dohra, Yoshihiro Ueda, Masaaki Shibuya, and Yutaka Ebizuka. 2004. "The Key Role of O-Methyltransferase in the Biosynthesis of *Rosa chinensis* Volatile 1,3,5-Trimethoxybenzene." *Plant Physiology* 135: 95–102.

Xu, Haiyang, Björn Bohman, Darren C.J. Wong, Claudia Rodriguez-Delgado, Adrian Scaffidi, Gavin R. Flematti, Ryan D.Phillips, et al. 2017. "Complex Sexual Deception in an Orchid Is Achieved by Co-opting Two Independent Biosynthetic Pathways for Pollinator Attraction." *Current Biology* 27: 1867–77.

Zvi, Michal Moyal Ben, Elena Shklarman, Tania Masci, Haim Kalev, Thomas Debener, Sharoni Shafir, Marianna Ovadis, and Alexander Vainstein. 2012. "PAP1 Transcription Factor Enhances Production of Phenylpropanoid and Terpenoid Scent Compounds in Rose Flowers." *New Phytologist* 195: 335–45.

9 Biosynthesis and Regulation of Vegetative Plant Volatiles

Takao Koeduka, Koichi Sugimoto, and Kenji Matsui

CONTENTS

9.1 INTRODUCTION

The ecological significance of volatiles formed in plant reproductive organs, such as flowers and fruits, is well established; these volatiles are known to be involved in the attraction of pollinators and seed-dispersers. In contrast, vegetative organs have no dispersing products; therefore, the role of volatiles from vegetative organs is different from those of reproductive organs. Volatiles from vegetative organs are instead mostly involved in the defense against herbivores and pathogens. Plants use volatiles in several ways to exert constitutive and inducible defense (Chapter 13). Herbs generally have secretory trichomes filled with volatile compounds on the leaf surface to exert constitutive defense (Chapter 13). The secretory cavities, which are commonly found in citrus plant species, also provide constitutive defense to plants. Formation of isothiocyanates by cruciferous plants is an inducible defense, as the production of the active compounds, isothiocyanates, occurs only after the tissue is damaged because mixing enzymes (myrosinases) and substrates (glucosinolates) after tissue damage is prerequisite. Green leaf volatiles (GLVs) are also formed in an inducible manner. The levels of GLVs in healthy leaf tissues are generally low, but upon tissue damage, they are formed *de novo* through enzymatic reactions of lipids or fatty acids. Thus, tissue damage is a key functional event in the production of volatiles in vegetative tissues to exert inducible defenses. It has also been shown that predators that attack herbivores have acquired abilities to sense volatiles produced by damaged plants, leading to the evolution of an indirect defense system consisting of plant–herbivore–predator.

Herein, we provide an outline of the biosynthetic pathways by which volatiles are produced in vegetative tissues. We also discuss how the formation of volatiles is controlled with regard to their functions in the ecosystems.

9.2   BIOSYNTHESIS OF GLVs

9.2.1   BIOCHEMICAL PATHWAYS LEADING TO GLV FORMATION

GLVs, a group that includes six-carbon (C6) aldehydes, alcohols, and their esters, represent one of the most ubiquitous vegetative volatiles in vascular plants. GLVs are formed from membrane lipids via the lipoxygenase (LOX)-hydroperoxide lyase (HPL) pathway in oxylipin metabolism (Figure 9.1). LOX is a nonheme iron-containing enzyme that catalyzes the dioxygenation of unsaturated fatty acids and acyl groups in lipids. Common substrates of LOX in plant tissue include linoleic acid, linolenic acid, and glycerolipid. The binding site on the fatty acid that is recognized by the enzyme is a (Z,Z)-1,4-pentadiene unit. In the case of linolenic acid, LOX stereo-specifically adds molecular oxygen into its C-13 position, forming linolenic acid 13(S)-hydroperoxide (13S-HPOT). HPL utilizes the 13S-HPOT as a substrate and cleaves the C_{12}–C_{13} bond to form (Z)-3-hexenal and 12-oxo-(Z)-9-dodecenoic acid (traumatin) as its counterpart. (Z)-3-Hexenal is partly converted to (E)-2-hexenal by (Z)-3:(E)-2-hexenal isomerase in some plant species (Kunishima et al., 2016). Part of the formed (Z)-3-hexenal can also be metabolized by cinnamaldehyde and hexenal reductase (CHR) to (Z)-3-hexen-1-ol (Tanaka et al., 2018). The resulting (Z)-3-hexen-1-ol is converted to its ester form by acetyl-CoA: (Z)-3-hexen-1-ol acetyltransferase (CHAT), which belongs to the BAHD super family (D'Auria et al., 2007). In some plant species, LOX forms a 9-hydroperoxide of linolenic acid by adding molecular oxygen at the C-9 position, whereby the 9-hydroperoxide is further converted into C9 volatile compounds (as described below) (Matsui, 2006).

HPL is the key enzyme involved in GLV formation. HPL belongs to the CYP74B clade of the cytochrome P450 (CYP) oxidoreductases superfamily (Matsui et al., 1996). However, the reaction mechanism is unlike that of a typical CYP because neither molecular oxygen nor any redox partners

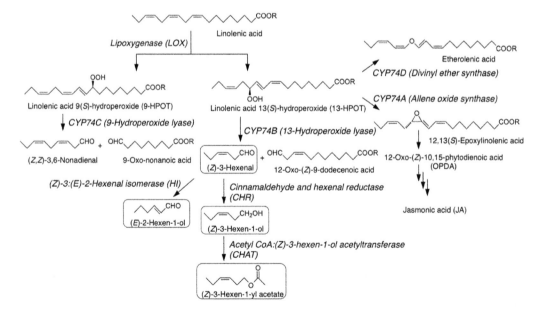

FIGURE 9.1   The GLV biosynthetic pathway in plants. Major GLVs are circled by squares. Enzymes catalyzing each reaction are italicized.

(e.g., NADPH-dependent P450 reductase) is needed for the HPL reaction. Instead, HPL uses the hydroperoxide group in the substrates as an electron donor (Grechkin and Hamberg, 2004).

In addition to HPL (CYP74B), several CYP74 enzymes have been reported to utilize fatty acid hydroperoxides as substrates. Allene oxide synthase (AOS, CYP74A) acts on the 13S-HPOT as a common substrate with HPL, and converts it into an unstable allene oxide through a similar catalytic pathway without any cofactors (Figure 9.1) (Brash, 2009). The resulting allene oxide is converted into 12-oxo-(Z)-10,15-phytodienoic acid (OPDA) by allene oxide cyclase, then further converted into jasmonic acid, a phytohormone involved in plant defense against herbivore and pathogen attacks (Wasternack and Hause, 2013). CYP74C is a type of HPL that specifically converts 9-hydroperoxides of fatty acids into (Z,Z)-3,6-nonadienal, which has a cucumber/melon-like flavor (Matsui et al., 2000). CYP74C was initially discovered in the family Cucurbitaceae, but its occurrence has since been reported in a wide variety of plant species, including rice, potato, almond, and *Lotus japonicas*, nonetheless, phylogenetic distribution of the CYP74C subfamily remains limited (Koeduka, 2018). CYP74D, known as divinyl ether synthase, also converts fatty acid hydroperoxides into divinyl ethers, which have antimicrobial activities. The distribution of CYP74D is restricted to the family Solanaceae and several other plant species including *Selaginella moellendorffii* and *Ranunculus acris* (Gorina et al., 2014, 2016; Mosblech et al., 2009).

9.2.2 EVOLUTION OF THE GENES INVOLVED IN GLV FORMATION

The CYP74 family represents a diverse group of enzymes that convert the hydroperoxides of fatty acids into numerous oxylipins, including GLVs and jasmonic acids (Figure 9.1). In addition to the wide distribution of CYP74 in higher plants, CYP74 orthologs have recently been reported in the plants belonging to the basal evolutionary lineages of land plants. In the moss *Physcomitrella patens*, three CYP74 genes were identified, two of which showed AOS-specific activity, while the third CYP74 has HPL activity that produces C9 volatiles from the 11- or 12-hydroperoxide of arachidonic acid and 9-hydroperoxide of linoleic acid (Scholz et al., 2012). In liverwort *Marchantia polymorpha*, there are two CYP74 genes, and both encode AOS. As for the green microalgae *Klebsormidium flaccidum*, only one AOS gene was found, while the genome sequences of *Chlamydomonas* and *Nostoc* did not reveal any CYP74 members (Koeduka et al., 2015). Moreover, nine putative *CYP74* genes were found in the Lycopodiophyta *Selaginella moellendorffii*, two of which showed DES activity toward 13-hydroperoxides of linolenic acid and linoleic acid. Accordingly, it can be deduced that AOS is the common ancestral activity of the CYP74 family that arose to use fatty acid hydroperoxide substrates, and that the acquisition of the CYP74 gene with HPL activity for the formation of GLV in plants occurred after the mosses have evolved. During plant evolution, it is most likely that gene duplications, followed by functional divergence between AOS and HPL or DES, occurred in the CYP74 family. Indeed, it was recently reported that the conversions of DES into AOS and AOS into HPL were caused by single point mutations (Lee et al., 2008; Scholz et al., 2012; Brühlmann et al., 2013; Toporkova et al., 2013).

9.2.3 REGULATION OF GLV FORMATION

In intact and healthy leaves, GLVs content is generally low. The quick and extensive formation of GLVs is in most cases associated with tissue damage caused either by mechanical wounding or herbivore attacks. This rapid formation is sometimes termed GLV-burst. This is also the case in *Arabidopsis* leaves, albeit the ecotype commonly used by plant biologists, Col-0, does not form GLVs due to a 10-bp deletion in the ORF of its HPL gene (Duan et al., 2005). When *Arabidopsis* leaves (ecotypes such as Ler-0, No-0, or WS) were homogenized in a mortar, GLV-burst was observed within one minute, and the amount of (Z)-3-hexenal increased to ca. 300 nmol g FW^{-1} within 5 minutes (Nakashima et al., 2013). Even though the production of the counterpart of the HPL product was expected to be equal, less than 10 nmol g FW^{-1} of traumatins (sum of free traumatin and 9-hydroxytraumatin) was detected.

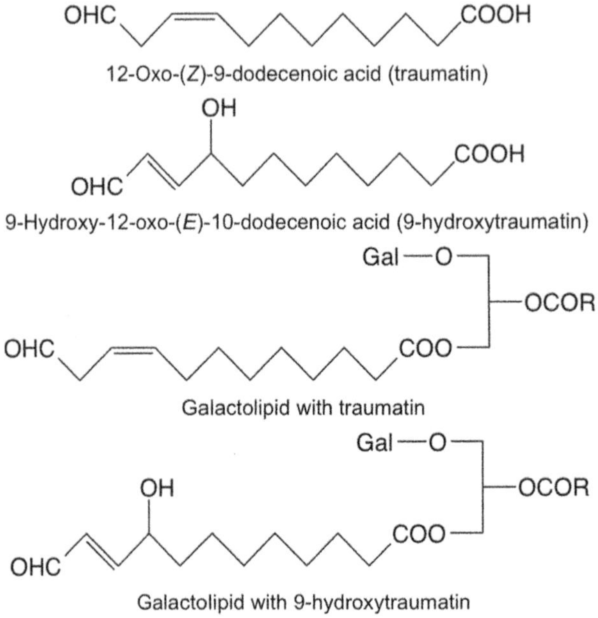

FIGURE 9.2 Structures of traumatins and galactolipid with traumatins. In galactolipid traumatins, traumatins may occur either at the *sn*-1 or *sn*-2 position. If the oxygenation and cleavage should proceed with hexadacatrienoate (C16:3), *dinor*traumatins should occur.

Instead, around 10 nmol g FW^{-1} of monogalactosyldiacylglycerol (MGDG) harboring traumatin and 9-hydroxytraumatin (Figure 9.2) were detected. Even though the total amounts of traumatins were still substantially lower than that of (Z)-3-hexenal, this observation indicates that lipid hydrolysis to release free fatty acids is not always essential for the GLV-burst. At least a portion of GLVs is formed directly from MGDG without a lipid hydrolyzing step. Because the amounts of esterified traumatins were equivalent or even higher than the amounts of free traumatins in leaves of cabbage, tobacco, tomato, and bean after GLV-burst (Nakashima et al., 2013), lipid-hydrolysis-independent-GLV-burst was likely to be a rather general mechanism among various plant species.

This raises a question regarding how the GLV-burst mechanism commences. With the lipid-hydrolysis-independent-GLV-burst, the first commitment step is performed by a LOX. Even though a LOX (EC 1.13.11.12) is classically defined as a linoleate 13-LOX and free fatty acids are its preferable substrates *in vitro*, it also shows substantial activity against lipids that contain acyl groups with (Z,Z)-1,4-pentadiene moiety. Notably, an appropriate orientation is required for the formation of lipid-water interface, as well as suitable physicochemical properties through emulsification with a given concentration of bile salt (Eskola and Laakso, 1983). We found that MGDG was also a good substrate for soybean LOX-1 when the lipids were emulsified with the appropriate concentration of deoxycholic acid (Nakashima et al., 2011).

Most plant species have multiple isoforms of LOXs, and in most studied cases, each isoform seems to have its own specific role. *Arabidopsis* has six LOXs from AtLOX1 to AtLOX6. It has been reported that AtLOX2 is involved in the JA synthesis in local tissues following tissue damage, while AtLOX6 is involved in the systemic JA synthesis (Chauvin et al., 2013). A study with a complete set of T-DNA knock-out mutants for each LOX showed AtLOX2 to be exclusively involved in GLV-burst in local tissues after mechanical wounding (Mochizuki et al., 2016). In *Arabidopsis*, AtLOX2 is constitutively expressed in leaves, even before subjecting them to any stimulus. A proteomic study indicated that AtLOX2 protein is localized in the stroma of chloroplasts, and its levels amount to 0.75% of the total stromal proteins (Peltier et al., 2006). This implies that AtLOX2 proteins have limited access to MGDGs in thylakoid membranes, despite their close proximity. Leaf tissue damage could cause the disruption of chloroplasts, which would associate with the activation of AtLOX2 and, or its translocation into the thylakoid membranes. When the leaves were damaged in the presence of the Ca^{2+}-chelating agent 1,2-bis(O-aminophenoxy)ethane-N,N,N',N'-tetraacetic acid (BAPTA), GLV-burst was extensively suppressed, and its suppression was highly correlated with the formation of MGDGs harboring hydroperoxides of linoleate as acyl groups; therefore, it was assumed that Ca^{2+} was involved somehow in the activation and, or the translocation of AtLOX2 (Mochizuki and Matsui, 2018).

A subset of animal LOXs are activated and translocated into biological membranes after binding with Ca^{2+} at a conserved flexible loop that is found in the N-terminal β-barrel domain and is termed C2-like domain or a PLAT (polycystin-1, LOX and alpha toxin) domain (Hammarberg et al., 2000). Structural modeling suggested the presence of the PLAT domain with a flexible loop on AtLOX2 (Figure 9.3). Glu and Asp residues that could account for Ca^{2+} binding were also found in the loop (Mochizuki and Matsui, 2018). It is therefore assumed that AtLOX2 remains in the latent form in the chloroplast stroma, until the Ca^{2+} influx-elicited by the disruption of chloroplast integrity causes activation and, or translocation of AtLOX2, resulting in the oxygenation of MGDGs in the thylakoid membrane, thereby initiating GLV-burst.

The effect of BAPTA addition is an indirect evidence to the Ca^{2+}-dependent activation of LOX for GLV-burst; further studies of the direct observation of Ca^{2+}-dependent activation and translocation with purified LOXs, or of the association between Ca^{2+} and LOXs, are essential. In addition, the elucidated mechanism does not exclude the possible existence of additional GLV-burst mechanisms. In fact, the effect of BAPTA was also observed in leaves of tomato and clover, but not in leaves of tobacco, rice and maize; these results imply that each plant might have multiple mechanisms that induce GLV-burst.

In comparison with damaged plant leaves, the production of GLVs in intact leaves is relatively low. However, such production is still detectable, and the temporal and spatial profiles of GLV emission seem to be properly regulated. A transient "burst" of GLVs, as well as acetaldehyde and ethanol, has been observed during light-dark transition in various plant leaves, including Grey poplar, apple, and grape (Graus et al., 2004; Giacomuzzi et al., 2017). This type of "post-illumination bursts" is also likely caused by changes in a LOX activity, and it has been postulated that a shift

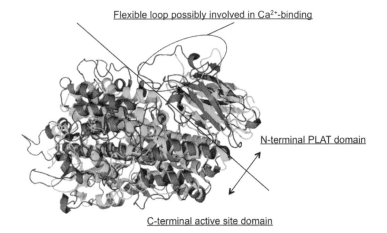

FIGURE 9.3 Model structure of AtLOX2 with the reported structure of coral 8R-LOX (2FNQ). Structure of AtLOX2 (shown with magenta) was modeled based on the soybean seed LOX-1 (3PZW) using SWISS-MODEL, and superimposed on the structure of coral 8R-LOX (2FNQ) (shown in green). N-terminal PLAT (C2) domain and C-terminal active site domain are shown. Flexible loop that might be involved in Ca^{2+}-binding is also shown. Original drawings were prepared by Dr. S. Mochizuki.

in pH values in chloroplast stroma, driven by illumination, is one such factor regulating the LOX activity responsible for the burst.

GLV formation is also regulated at the transcriptional levels. Mechanical wounding and jasmonate treatment increase the abundance of LOX and HPL transcripts (Bell et al., 1995; Mwenda et al., 2015). Both the genes are also under regulation of circadian rhythm in *Nicotiana attenuata* (Joo et al., 2018). Herbivory would continue for a while and also the damage would extend gradually into the other tissues. In this context, it would be advantageous for the plant to strengthen its ability to form GLVs in the places other than the damaged tissues. Additionally, the predators might be nocturnal or diurnal, depending on the species; therefore, GLV emission at the right time should also be advantageous to the plant (Joo et al., 2019).

GLVs are involved in the direct and indirect plant defenses through affecting the behavior of herbivores and their predators. Therefore, it is assumed that arthropods have acquired countermeasures against plant defense accountable to GLVs. The expression of *Arabidopsis* HPL is suppressed when the leaves are being attacked by *Pieris rapae*, but not by *Spodoptera exigua* (Savchenko et al., 2013). This indicates that the oral secretions of *P. rapae* (but not *S. exigua*) contain an effector that suppresses the defense response of plants that would otherwise lead to a stronger GLV emission. During infestation, the silkworm *Bombyx mori* secretes a solution containing an enzyme, fatty acid hydroperoxide dehydratase, through its spinnerets (Takai et al., 2018). The enzyme is left on the wounds created by the silkworm, enabling the enzyme to associate the fatty acid hydroperoxide, a substrate for HPL. Accordingly, GLV emission from the wounds is suppressed, which eventually makes the damaged plants less attractive to the parasitic fly (*Zenillia dolosa*). An oral secretion from the larvae of *Manduca sexta* has a hexenal isomerase activity, which decreases the $(Z)/(E)$ ratio of GLVs emitted from *Datura wrightii* and *Nicotiana attenuata* (Allmann et al., 2013; Allmann and Baldwin, 2015). The enzyme is apparently maladaptive to the herbivore, and the decrease in the $(Z)/(E)$ ratio reduces oviposition by female *M. sexta* moth, resulting in a higher foraging efficiency of predators (*Geocoris* spp.) on the larvae. The accumulating reports on insect-derived effector-like factors that interfere with plant indirect defense at the transcriptional and biochemical levels clearly show that herbivore have evolved mechanisms to control GLV production in plants.

9.3 BIOSYNTHESIS OF VOLATILE TERPENES

9.3.1 BIOCHEMICAL PATHWAY LEADING TO VOLATILE TERPENES FORMATION

Terpenes, derived from a five-carbon building block (isoprene unit), are the most structurally divergent volatile compounds. The formation of terpenes in plants proceeds *via* two alternative pathways; the cytosolic mevalonate (MVA) pathway, and the methylerythritol phosphate (MEP) pathway (Rohmer, 2003). The MVA pathway starts from acetyl-CoA to form MVA, following several reactions that synthesize isopentenyl diphosphate (IPP; C5) and dimethylallyl diphosphate (DMAPP; C5). Subsequently, the condensation of two molecules of IPP and DMAPP by prenyltransferase forms the farnesyl diphosphate (FPP; C15). In contrast, the MEP pathway, localized in the plastids, starts by the condensation of two carbons from pyruvate with glyceraldehyde-3-phosphate to form 1-deoxy-D-xylulose-5-phosphate (DXP), in a reaction catalyzed by DXP synthase (DXS) (Mandel et al., 1996; Estevez et al., 2000). Following several additional reactions, IPP and DMAPP are formed and are combined by prenyltransferase to give geranyl diphosphate (GPP; C10). GPP and FPP are precursors of monoterpene and sesquiterpene, respectively, in reactions catalyzed by terpene synthases (TPSs). To date, TPSs have been isolated and identified in many plant species, including the nonseed plants, *Selaginella* and liverwort (Li et al., 2012; Jia et al., 2016).

9.3.2 VOLATILE TERPENES IN THE SECRETARY CAVITIES OF JAPANESE PEPPER LEAVES

Members of the family Citrus (Rutaceae) are some of the most popular fruiting trees worldwide, and the trees commonly contain large volumes of monoterpenes and sesquiterpenes, in their leaves and fruit peels. These volatiles are commercially valuable and are used as flavor, aromatherapy, cleaning, and agrochemical products due to their characteristic pleasant aroma and antimicrobial properties. For example, Japanese pepper (Rutaceae, *Zanthoxylum piperitum*), a plant native to East Asia, particularly in Japan, is commonly used in Japanese cuisine as a spice and seasoning for its lemon-like aroma and strong taste, and to mask unpleasant fishy and meaty odors. Previous studies have shown that the fresh young leaves of Japanese pepper release a diverse array of volatile terpenes, together with GLVs, after being crushed, despite the absence of detectable aroma from intact leaves (Jiang and Kubota, 2001).

Based on these observations, it was hypothesized that Japanese pepper leaves may store terpenes in a specialized structures such as trichomes and oil secretory cavities, the latter are commonly observed in citrus plants (Liang et al., 2006; Voo et al., 2012; Zhou et al., 2014; Uji et al., 2015). Recently, the volatiles and the secretory cavity morphology in Japanese pepper leaves were analyzed, and the presence of secretory cavities in leaflets, forming small bulges that are responsible for the production and secretion of monoterpenes and sesquiterpenes was reported (Fujita et al., 2017).

These compounds were absent in other leaf tissues that do not have secretory cavities (Figure 9.4). Furthermore, *in situ* hybridization in Japanese pepper leaf tissues revealed that

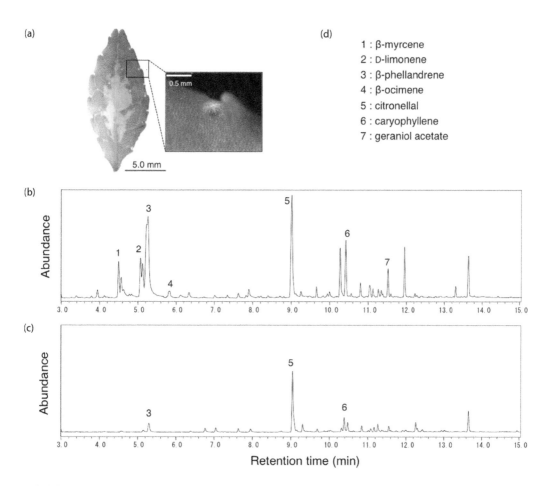

FIGURE 9.4 Photographs of the secretory cavities and GC-MS analysis in Japanese pepper leaf. (a) Secretory cavities of a Japanese pepper leaflet. GC-MS chromatogram of the volatile compounds from the secretory cavities (b) and leaf tissues without secretory cavities (c). The number of individual peaks shown in panel (b) and (c) refers to the number of compounds in panel (d).

the transcriptional localization of the TPS gene was specifically observed in the epithelial cells surrounding secretory cavities. Similar localization patterns of TPS genes have been reported for transcripts of limonene and sabinene synthase genes in rough lemon leaves (Uji et al., 2015). Thus, it is most likely that the site of terpene biosynthesis has homologous localities throughout the Rutaceae family.

9.4   BIOSYNTHESIS OF PHENYLPROPANOID VOLATILES

9.4.1   BIOCHEMICAL PATHWAYS LEADING TO PHENYLPROPANOID VOLATILES FORMATION

Phenylpropanoid volatiles contain at least one benzene ring, which is originally derived from the shikimate pathway, and can be classified into different subclasses depending on their basic structures: benzenoids with a C_6-C_1 core structure, phenylpropanoid-related C_6-C_2 compounds, and phenylpropenes having a C_6-C_3 structure. Phenylpropene volatile compounds are specifically attractive to humans, and they are key aroma components in vegetative tissues and are found in a multitude of important spices and popular culinary herbs, including clove, nutmeg, sweet basil, and fennels. For example, eugenol, chavicol, and their methyl esters are the main constituents (70%–90%) of essential oils from sweet basil (*Ocimum basilicum*) (Freitas et al., 2018). These compounds have antimicrobial, antifungal, and insect-repelling activities; therefore, humans have since ancient times used spices and herbs containing phenylpropene volatiles for seasonings in cooking and in cosmetics and aromatherapy.

Basil, strawberry, and apple, among others, are reported to produce particularly high levels of phenylpropenes (Koeduka et al., 2006; Aragüez et al., 2013; Yauk et al., 2017). The synthesis of phenylpropenes begins with phenylalanine, which is also a precursor of flavonoids, stilbenoids, curcuminoids, and lignins. First, phenylalanine is deaminated by phenylalanine ammonia lyase to biosynthesize cinnamic acid, which is converted to 4-coumaric acid by cinnamic acid 4-hydroxylase (C4H), which belongs to the cytochrome P450 family. 4-Coumaric acid is next converted to 4-coumaroyl-CoA by 4-coumarate:CoA ligase. The resulting 4-coumaroyl-CoA is sequentially reduced by cinnamoyl-CoA reductase and cinnamyl alcohol dehydrogenase to produce coumaryl alcohol. This alcohol, together with coniferyl alcohol, is the precursor in the phenylpropene biosynthetic pathway, and they also serve as monolignol alcohol intermediates in lignin biosynthesis. For phenylpropene biosynthesis, coumaryl and coniferyl alcohols are further converted to their acetyl esters by coniferyl alcohol acyltransferase (CFAT), which is a member of the benzyl alcohol *O*-acetyltransferase, anthocyanin *O*-hydroxycinnamoyltransferase, anthranilate *N*-hydroxycinnamoyl/benzoyltransferase, deacetylvindoline 4-*O*-acetyltransferase (BAHD) acyltransferase family. CFAT transfers an acetyl moiety from acetyl-CoA onto the hydroxyl group of its alcohol substrate (Dexter et al., 2007; Yauk et al., 2017). Phenylpropenes are then synthesized by the reductive elimination of the acetyl-ester moiety, by the NADPH-dependent reductases eugenol synthase (EGS) and isoeugenol synthase (IGS), giving rise to the allyl or propenyl form, respectively. EGS and IGS are members of the pinoresinol-lariciresinol reductase, isoflavone reductase, phenylcoumaran benzylic ether reductase (PIP) family. Further modifications in various plant species follow, including the methylation (see next section) and glycosylation of the benzene ring (Koeduka, 2014).

9.4.2   *O*-METHYLTRANSFERASES MODIFICATION OF THE PHENYL RING IN PHENYLPROPENES

Phenylpropenes consist of a modified benzene ring (C_6) bearing a propenyl side chain (C_3). The structural diversity of many phenylpropenes stems from variation both in the position of the propenyl double bond and in the substituents on the benzene ring. These structural differences sometimes

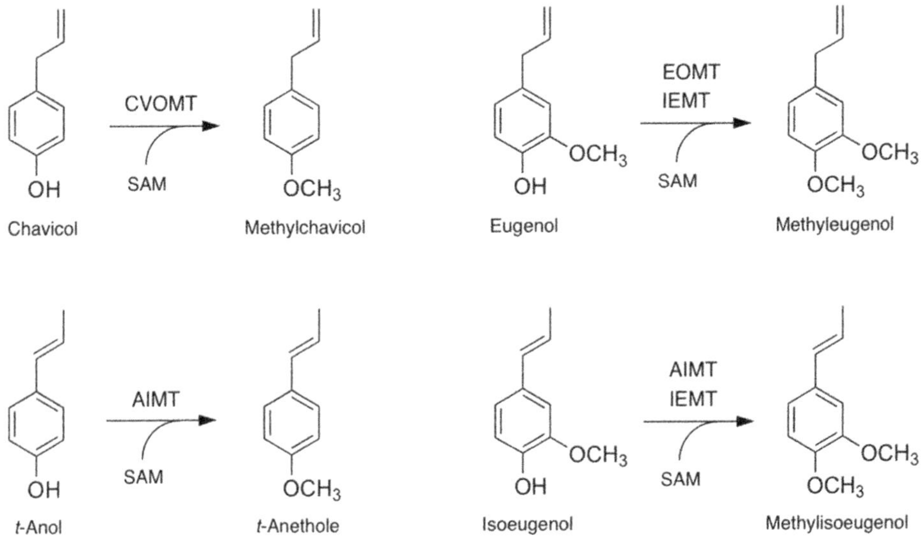

FIGURE 9.5 Biochemical reactions that produce phenylpropene volatiles by *O*-methyltransferases. CVOMT, Chavicol *O*-methyltransferase; EOMT, Eugenol *O*-methyltransferase; AIMT, *t*-Anol/Isoeugenol *O*-methyltransferase; IEMT, Isoeugenol/Eugenol *O*-methyltransferase; SAM, *S*-Adenosyl-L-methionine.

influence the aroma quality of the phenylpropenes. For example, phenylpropenes such as eugenol, isoeugenol, and chavicol have a *para*-hydroxyl group and sometimes also a *para*-methoxyl group on the benzene ring; the former group has a pungent clove-like aroma, while the latter has a soft herbaceous smell (Wang and Pichersky, 1998) (Figure 9.5).

The formation of some phenylpropenes, such as methyleugenol, methylisoeugenol, methyl-chavicol, and *t*-anethole, requires methylation of the *para*-hydroxyl group on their benzene rings. *O*-Methyltransferases are some of the enzymes that provide the chemical diversification of phen-ylpropene structures, and several *O*-methyltransferases involved in the biosynthesis of phenylpro-penes have been characterized (Figure 9.5). Chavicol *O*-methyltransferase (CVOMT) and eugenol *O*-methyltransferase (EOMT) have been characterized from *Ocimum basilicum*, and these enzymes are 90% identical at amino acid level but display significantly different substrate specificity. CVOMT is highly specific for chavicol and EOMT shows high activity toward eugenol (Gang et al., 2002). Site-directed mutagenesis to convert a specific amino acid residue in one of these enzymes to the amino acid found in the other, and vice versa, causes a corresponding shift in substrate preference, indicating that a single amino acid residue could determine their substrate preference.

In anise (*Pimpinella anisum*), a *t*-anol/isoeugenol *O*-methyltransferase (AIMT) was reported, with ten-fold higher activity for propenyl phenylpropenes such as *t*-anol and isoeugenol, com-pared to allyl phenylpropenes such as chavicol and eugenol (Koeduka et al., 2009). (Iso)Eugenol *O*-methyltransferase (IEMT) has been found in *Clarkia breweri* (Wang et al., 1997). This enzyme can use both eugenol and isoeugenol as substrates to form methyleugenol and methylisoeugenol respectively, but cannot use chavicol or *t*-anol, which do not have a methoxyl group at the *meta*-position on their benzene ring (unlike eugenol and isoeugenol). It is likely that these phenylpropene OMTs discriminate based on the double-bond position in the propenyl side chain, or the func-tional groups on the benzene ring of various potential substrates. Although the *in vivo* biochemical function of each *O*-methyltransferases remains unknown, the production of volatiles may be partly determined by the substrate/product specificities of the biosynthetic enzymes responsible for the formation of volatile compounds.

9.5 VOLATILE GLYCOSIDES, A STORAGE FORM OF VOLATILE COMPOUNDS IN VEGETATIVE TISSUES

9.5.1 GLYCOSYLTRANSFERASES FOR VOLATILE COMPOUNDS IN PLANTS

Several types of specialized metabolites, such as flavonoids and alkaloids, are glycosylated in plants for various purposes, including rendering aglycons more hydrophilic (Bowles et al., 2005), storing aglycons into specific cell compartments, e.g., vacuoles (Yazaki, 2005; Zhao and Dixon, 2010), detoxifying xenobiotic toxins (Poppenberger et al., 2003), or preventing autotoxicity of defense chemicals (Maduagwu, 1983). Plant volatiles are also glycosylated in many tissues, mainly in fruits and flowers. For example, tomato fruit produces the smoky flavor compounds eugenol, methyl salicylate, and guaiacol, and accumulates them as the diglycosides during ripening (Tikunov et al., 2010). Nonsmoky cultivars convert these diglycosides into triglycosides until the fruit have matured to prevent degradation by glycosidase (Tikunov et al., 2013). In the vegetative tissues, few studies have analyzed the endogenous volatile glycosides, but some transgenic plants ectopically expressed the volatile biosynthetic genes accumulate their volatile as glycons. For example, transgenic petunia leaves expressing the linalool synthase isolated from *Clarkia breweri* were shown to accumulate linalyl glucoside (Lücker et al., 2001), and aspen leaves expressing the eugenol biosynthetic genes isolated from petunia, were shown to produce eugenyl glycosides (Koeduka et al., 2013).

Glycosyltransferases acting on volatile compounds in leaf tissues were identified in tea plants. The products of two classes of compounds were elucidated; UGT85K11 produces volatile-monoglucosides, while UGT94P1 binds an additional sugar moiety to yield diglycosides, called primeverosides. UGT85 family is widely found in the aglycon-dependent glucosyltransferases, which does not only act on volatile compounds (e.g., monoterpenols from grape and kiwi fruits) (Yauk et al., 2014; Bönisch et al., 2014a, 2014b), but also on various other types of metabolites (e.g., diterpene) (Richman et al., 2005). UGT94 family members are categorized as glycoside-dependent glycosyltransferases, and like the previous family, their activity is not restricted to volatile glycosides (e.g., tomato fruits) (Tikunov et al., 2013) as they also act on other specialized metabolites (e.g., flavonoids and sterol) (Sawada et al., 2005; Noguchi et al., 2008; Itkin et al., 2013). Therefore, predicting the substrate of glycosyltransferases from the amino acid sequences has been challenging to date.

9.5.2 GLYCOSIDES AS AROMA/FLAVOR PRECURSORS

Due to their high vapor pressure, volatile compounds would evaporate easily following biosynthesis in the plant cells. Therefore, storage of volatile compounds is essential. The storage of such compounds requires compartmentalization into the specialized tissues, such as glandular trichomes, oil cavities, and oil glands: or conversion into nonvolatile compounds. One way to convert the volatiles into storage compounds is glycosylation, which is well analyzed in fruits and flowers. For example, dozens of glycosidically bound volatiles are accumulated in kiwi fruits (Young and Paterson, 1995; Garcia et al., 2011), tobacco flowers (Loughrin et al., 1992), rose flowers (Oka et al., 1999), grape berries (João Cabrita et al., 2006), and bananas (Aurore et al., 2011). Even though biological or ecological roles of the glycosidically bound volatiles remain to be known, volatile glycosides have a role as aroma or flavor sources; therefore, the control of the amount of glycosides could be a target for breeding, for examples, wine grape, *Vitis vinifera* (Liu et al., 2017), sweet potato, *Ipomoea batatas*, for Japanese distilled spirit shochu (Sato et al., 2018), and soybean, *Glycine max*, for soymilk and tofu (Matsui et al., 2018).

Compared to nonvegetative tissues, the impact of volatile glycosides on vegetable flavor in vegetative tissues is less studied, except in the case of tea plants. Tea leaves are widely used worldwide to produce "tea" drinks via complex manufacturing processes. During the processing called "fermentation," the volatile glycosides are cleaved to release major flavor volatiles, including geraniol, linalool, (Z)-3-hexenol, benzyl alcohol, and 2-phenylethanol (Zheng et al., 2016). The volatile glycosides isolated from tea leaves (see Section 9.5.1) are mainly formed as diglycosides, called primeverosides, vicianosides, and acuminosides (Moon et al., 1994, 1996; Guo et al., 1994; Nishikitani et al., 1996, 1999; Ma et al., 2001; Wang et al., 2000). Their characteristic structure requires specialized diglycosidase to release the volatile aglycons. Biochemical purification of primeverosidase from tea leaves clarifies the mode of action to degrade volatile diglycosides: primeverosidase splits primeverosides into free volatiles and primeverose, but does not split primeverose into glucose and xylose (Mizutani et al., 2002). Recently, the final step of diglycoside biosynthetic pathway, glycoside-dependent glycosyltransferase, was identified using modern genetic approaches (Ohgami et al., 2015). With the identification of volatile-dependent glucosyltransferases (Ohgami et al., 2015; Jing et al., 2019), whole pathways for volatile diglycoside biosynthesis and degradation were identified. Complete elucidation of these pathways provide insights into new breeding programs that facilitate the modulation of tea flavors.

Interestingly, regulation of these pathways might be involved in specialized tea production, called "oriental beauty." "Oriental beauty" is one of the famous semi-fermented tea types in Taiwan with characteristic aroma compositions. It is produced from tea leaves that were infested by the green leafhopper, *Jacobiasca formosana* (Cho et al., 2007). Leaf damage changes the gene expressions involved in volatile biosynthesis, volatile glycoside biosynthesis, and volatile glycoside degradation (Mei et al., 2017). Similar to leaf damage, treatment with jasmonate, a defense hormone in plants, also changes the tea aroma and expression of the primeverosidase gene (Shi et al., 2014). Thus, identification of metabolic regulation mechanisms is the next challenge toward the production of enhanced tea aromas.

9.5.3 THE ECOLOGICAL ROLE OF VOLATILE GLYCOSYLATION

Even though the biosynthesis and degradation of volatile glycosides are becoming better understood through the intensive biochemical and genetic studies of plants, the biological or ecological roles of these compounds remain to be described. In the case of leaf tissues, volatile compounds induced by herbivory should function to minimize plant damage; for instance, as repellents against herbivores, attractants of their natural enemies, or as chemical cues to alarm the surrounding plants of forthcoming herbivores (Arimura et al., 2009). The emission of these volatiles could be derived not only from *de novo* synthesis but also from the release of stored ones. An example of volatile storage is glycosylation; tea leaves accumulate glycosylated volatiles, e.g., phenyl ethanol, linalool, geraniol, and (Z)-3-hexenol (Moon et al., 1994, 1996; Guo et al., 1994; Nishikitani et al., 1996, 1999; Wang et al., 2000; Ma et al., 2001), and cleave these glycosides by glycosidases to release the volatiles (Mizutani et al., 2002). Such a storage and cleavage system is similar in the context of the release of toxic chemicals from their glycosides (e.g., diterpenoids in Solanaceae plants, and cyanides in Brassicaceae plants) (Poreddy et al., 2015; Shirakawa and Hara-Nishimura 2018). Compared to *de novo* biosynthesis, this system potentially contributes to a quick release of volatiles under attack.

In addition to emitter plants, the plants surrounding the emitter also produce volatile glycosides by converting the airborne volatiles from the emitter plant. For example, healthy tomato plants exposed to the volatiles emitted from others infested with *Spodoptera litura* were shown to accumulate (Z)-3-hexenyl vicianoside, the aglycon of which is (Z)-3-hexenol (Sugimoto et al., 2014). (Z)-3-Hexenyl vicianoside functions to suppress the growth of *Spodoptera litura* (Figure 9.6). Transfer of (Z)-3-hexenol between infested and healthy plants and its conversion into the defensive glycoside is a mechanism of plant–plant interactions. The recent efforts to identify glycosyltransferases that act on different volatiles (Tikunov et al., 2013; Yauk et al., 2014; Bönisch et al., 2014a, 2014b; Song et al., 2016; Louveau et al., 2011; Härtl et al., 2017; Jing et al., 2019), and the breeding of glycosyltransferase-weakened plants (Jing et al., 2019), might provide insight into the biological and ecological contribution of glycosylation-mediated plant–plant interactions by volatiles.

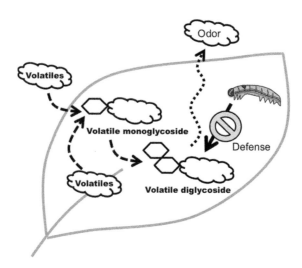

FIGURE 9.6 Summary of the biosynthesis and biological role of volatile glycoside. This cartoon summarizes the discussed topic in this section. In Section 9.5.1, we summarize the biosynthetic pathway of volatile glycosides (hatched arrow). Endogenously synthesized or exogenously exposed volatile compounds are converted into volatile glycosides by a series of glycosyltransferases. In Section 9.5.2, we introduce the utility of volatile glycosides as odor sources (dotted arrow). Some food materials accumulate the volatile glycosides and degraded them at the appropriate time for consumption. Finally, in Section 9.5.3, we discuss the possible ecological roles of volatile glycosides (solid arrow). Volatile glycosides (or volatiles from the cleaved glycosides) work as defense compounds against invaders, such as herbivores.

9.6 FINAL REMARKS

Volatiles from vegetative organs are involved in defense against biotic stressors. Therefore, they have had a significant role during plant evolution. Knowledge of the molecular mechanisms by which volatiles are formed and modulated should help elucidate the mechanisms by which plants have countered predation pressure throughout the long history of their presence on land.

In the context of agricultural systems, the findings about the function of volatiles from vegetative organs should be useful to establish eco-friendly agricultural systems by exploiting the abilities of plants to defend against pests using volatile compounds. The effects of manipulating plant abilities to form or sense vegetative volatiles should be moderate, and in most cases, farmers prefer agrochemicals, which yield clear results within a few days. However, such moderate effects that would be acquired through manipulating plants' abilities to employ vegetative volatiles for defense should nonetheless make sustainable agriculture more feasible. Even though our knowledge of the enzyme systems involved in biosynthesis of volatiles is advanced, we do not know much about how plants employ vegetative volatiles for defense. The next step would be to describe how the biosynthetic pathways are controlled, how the storage and delivery of volatile metabolites are regulated, and how the vegetative volatiles function in bringing about benefit to the producers and/or receivers. Thus, future studies on vegetative volatiles should aim to understand the mechanism of volatile functioning in ecosystems, as well as their applications to agricultural systems.

ACKNOWLEDGMENTS

This work was partly supported by JSPS KAKENHI Grant Numbers 16H03283 and 19H02887 (to KM).

REFERENCES

Allmann, S., A. Späthe, S. Bisch-Knaden, M. Kallenbach, A. Reinecke, S. Sachse, I.T. Baldwin, and B.S. Hansson. 2013. Feeding-induced rearrangement of green leaf volatiles reduces moth oviposition. *Elife* 2: e00421, doi:10.7554/eLife.00421.

Allmann, S., and I.T. Baldwin. 2015. Insects betray themselves in nature to predators by rapid isomerization of green leaf volatiles. *Science* 329: 1075–1078.

Aragüez, I., S. Osorio, T. Hoffmann, J.L. Rambla, N. Medina-Escobar, A. Granell, M.A. Botella, W. Schwab, and V. Valpuesta. 2013. Eugenol production in achenes and receptacles of strawberry fruits is catalyzed by synthases exhibiting distinct kinetics. *Plant Physiology* 163: 946–958.

Arimura, G., K. Matsui, and J. Takabayashi. 2009. Chemical and molecular ecology of herbivore-induced plant volatiles: Proximate factors and their ultimate functions. *Plant and Cell Physiology* 50: 911–923.

Aurore, G., C. Ginies, B. Ganou-parfait, C.M.G.C. Renard, and L. Fahrasmane. 2011. Comparative study of free and glycoconjugated volatile compounds of three banana cultivars from French West Indies: Cavendish, Frayssinette, and Plantain. *Food Chemistry* 129: 28–34.

Bell, E., R.A. Creelman, and J.E. Muller. 1995. A chloroplast lipoxygenase is required for wound-induced jasmonic acid accumulation in Arabidopsis. *Proceedings of National Academy of Sciences of the United States of America* 92: 8675–8679.

Bönisch, F., J. Frotscher, S. Stanitzek, E. Rühl, M. Wüst, O. Bitz, and W.A. Schwab. 2014a. UDP-glucose:monoterpenol glucosyltransferase adds to the chemical diversity of the grapevine metabolome. *Plant Physiology* 165: 561–581.

Bönisch, F., J. Frotscher, S. Stanitzek, E. Rühl, M. Wüst, O. Bitz, and W. Schwab. 2014b. Activity-based profiling of a physiologic aglycone library reveals sugar acceptor promiscuity of family 1 UDP-glucosyltransferases from grape. *Plant Physiology* 166: 23–39.

Bowles, D., J. Isayenkova, E.K. Lim, and B. Poppenberger. 2005. Glycosyltransferases: Managers of small molecules. *Current Opinion in Plant Biology* 8: 254–263.

Brash, A.R. 2009. Mechanistic aspects of CYP74 allene oxide synthases and related cytochrome P450 enzymes. *Phytochemistry* 70: 1522–1531.

Brühlmann, F., B. Bosijokovic, C. Ullmann, P. Auffray, L. Fourage, and D. Wahler. 2013. Directed evolution of a 13-hydroperoxide lyase (CYP74B) for improved process performance. *Journal of Biotechnology* 163: 339–345.

Chauvin, A., D. Caldelari, J.L. Wolfender, and E.E. Farmer. 2013. Four 13-lipoxygenases contribute rapid jasmonate synthesis in wounded *Arabidopsis thaliana* leaves: A role for lipoxygenase 6 in responses to long-distance wound signals. *New Phytologist* 197: 566–575.

Cho, J.Y., M. Mizutani, B. Shimizu, T. Kinoshita, M. Ogura, K. Tokoro, M.L. Lin, and K. Sakata. 2007. Chemical profiling and gene expression profiling during the manufacturing process of Taiwan oolong tea "Oritental Beauty." *Bioscience Biotechnology & Biochemistry* 71: 1476–1486.

D'Auria, J.C., E. Pichersky, A. Schaub, A. Hansel, and J. Gershenzon. 2007. Characterization of a BAHD acyltransferase responsible for producing the green leaf volatile (Z)-3-hexen-1-yl acetate in *Arabidopsis thaliana*. *The Plant Journal* 49: 194–207.

Dexter, R., A. Qualley, C.M. Kish, C.J. Ma, T. Koeduka, D.A. Nagegowda, N. Dudareva, E. Pichersky, and D. Clark. 2007. Characterization of a petunia acetyltransferase involved in the biosynthesis of the floral volatile isoeugenol. *The Plant Journal* 49: 265–275.

Duan, H., M.Y. Huang, K. Palacio, and M.A. Schuler. 2005. Variations in *CYP74B2* (hydroperoxide lyase) gene expression differentially affect hexenal signaling in the Columbia and Landsberg *erecta* ecotypes of *Arabidopsis*. *Plant Physiology* 139: 1529–1544.

Eskola J, and S. Laakso. 1983. Bile salt-dependent oxygenation of polyunsaturated phosphatidylcholines by soybean lipoxygenase-1. *Biochimica et Biophysica Acta* 751: 305–311.

Estevez, J.M., A. Cantero, C. Romero, H. Kawaide, L.F. Jimenez, T. Kuzuyama, H. Seto, Y. Kamiya, and P. Leon. 2000. Analysis of the expression of CLA1, a gene that encodes the 1-deoxyxylulose 5-phosphate synthase of the 2-C-methyl-D-erythritol-4-phosphate pathway in *Arabidopsis*. *Plant Physiology* 124: 95–103.

Freitas, J.V.B., E.G. Alves Filho, L.M.A. Silva, G.J. Zocolo, E.S. de Brito, and N.V. Gramosa. 2018. Chemometric analysis of NMR and GC datasets for chemotype characterization of essential oils from different species of *Ocimum*. *Talanta* 180: 329–336.

Fujita, Y., T. Koeduka, M. Aida, H. Suzuki, Y. Iijima, and K. Matsui. 2017. Biosynthesis of volatile terpenes that accumulate in the secretory cavities of young leaves of Japanese pepper (*Zanthoxylum piperitum*): Isolation and functional characterization of monoterpene and sesquiterpene synthase genes. *Plant Biotechnology* 34: 17–28.

Gang, D.R., N. Lavid, C. Zubieta, F. Chen, T. Beuerle, E. Lewinsohn, J.P. Noel, and E. Pichersky. 2002. Characterization of phenylpropene *O*-methyltransferases from sweet basil: Facile change of substrate specificity and convergent evolution within a plant *O*-methyltransferase family. *The Plant Cell* 14: 505–519.

Garcia, C.V., S.Y. Quek, R.J. Stevenson, and R.A. Winz. 2011. Characterization of the bound volatile extract from baby kiwi (*Actinidia argute*). *Journal of Agricultural and Food Chemistry* 99: 8358–8365.

Giacomuzzi, V., L. Cappellin, S. Nones, I. Khomenko, F. Biasioli, A.L. Knight, and S. Angeli. 2017. Diel rhythms in the volatile emission of apple and grape foliage. *Phytochemistry* 138: 104–115.

Gorina, S.S., Y.Y. Toporkova, L.S. Mukhtarova, E.O. Smirnova, I.R. Chechetkin, B.I. Khairutdinov, Y.V. Gogolev, and A.N. Grechkin. 2016. Oxylipin biosynthesis in spikemoss *Selaginella moellendorffii*: Molecular cloning and identification of divinyl ether synthases CYP74M1 and CYP74M3. *Biochimica et Biophysica Acta* 1861: 301–309.

Gorina, S.S., Y.Y. Toporkova, L.S. Mukhtarova, I.R. Chechetkin, B.I. Khairutdinov, Y.V. Gogolev, and A.N. Grechkin. 2014. Detection and molecular cloning of CYP74Q1 gene: Identification of *Ranunculus acris* leaf divinyl ether synthase. *Biochimica et Biophysica Acta* 1841: 1227–1233.

Graus, M., J. Schnitzler, A. Hansel, C. Cojocariu, H. Rennenberg, A. Wisthaler, and J. Kreuzwieser. 2004. Transient release of oxygenated volatile organic compounds during light-dark transitions in Grey poplar leaves. *Plant Physiology* 135: 1967–1975.

Grechkin, A.N., and M. Hamberg. 2004. The "heterolytic hydroperoxide lyase" is an isomerase producing a short-lived fatty acid hemiacetal. *Biochimica et Biophysica Acta* 1636: 47–58.

Guo, W., R. Hosoi, K. Sakata, N. Watanabe, A. Yagi, K. Ina, and S. Luo. 1994. (*S*)-Linalyl, 2-phenylethyl, and benzyl disaccharide glycosides isolated as aroma precursors from oolong tea leaves. *Bioscience, Biotechnology & Biochemistry* 58: 1532–1534.

Hammarberg, T., P. Provost, B. Persson, and O. Rådmark. 2000. The N-terminal domain of 5-lipoxygenase binds calcium and mediates calcium stimulation of enzyme activity. *Journal of Biological Chemistry* 275: 38787–38793.

Härtl, K., F.C. Huang, A.P. Giri, K. Franz-Oberdorf, J. Frotscher, Y. Shao, T. Hoffmann, and W. Schwab. 2017. Glucosylation of smoke-derived volatiles I grapevine (*Vitis vinifera*) is catalyzed by a promiscuous resveratrol/guaiacol glucosyltransferase. *Journal of Agricultural and Food Chemistry* 65: 5681–5689.

Itkin, M., U. Heinig, O. Tzfadia, A.J. Bhide, B. Shinde, P.D. Cardenas, S.E. Bocobza et al. 2013. Biosynthesis of antinutritional alkaloids in solanaceous crops is mediated by clustered genes. *Science* 341: 175–179.

Jia, Q., G. Li, T.G. Köllner, J. Fu, X. Chen, W. Xiong, B.J. Crandall-Stotler et al. 2016. Microbial-type terpene synthase genes occur widely in nonseed land plants, but not in seed plants. *Proceedings of National Academy of Sciences of the United States of America* 113: 12328–12333.

Jiang, L. and K. Kubota. 2001. Formation by mechanical stimulus of the flavor compounds in young leaves of Japanese pepper (*Xanthoxylum piperitum* DC.). *Journal of Agricultural and Food Chemistry* 49: 1353–1357.

Jing, T., N. Zhang, T. Gao, M. Zhao, J. Jin, Y. Chen, M. Xu, X. Wan, W. Schwab, and C. Song. 2019. Glucosylation of (Z)-3-hexenol informs intraspecies interactions in plants: A case study in *Camellia sinensis*. *Plant Cell & Environment* 42: 1352–1367.

João Cabrita, M., A.M. Costa Freitas, O Laureano, and R. Di Stefano. 2006. Glycosidic aroma compounds of some Portuguese grape cultivars. *Journal of the Science of Food and Agriculture* 86: 922–931.

Joo, Y., M.C. Schuman, J.K. Goldberg, A. Wissgott, S.G. Kim, and I.T. Baldwin. 2019. Herbivory elicits changes in green leaf volatile production via jasmonate signaling and the circadian clock. *Plant, Cell & Environment* 42: 972–982.

Joo, Y., M.C. Schuman, J.K. Goldberg, S.G. Kim, F. Yon, C. Brütting, and I.T. Baldwin. 2018. Herbivore-induced volatile blends with both "fast" and "slow" components provide robust indirect defence in nature. *Functional Ecology* 32; 136–149.

Koeduka, T. 2014. The phenylpropene synthase pathway and its applications in the engineering of volatile phenylpropanoids in plants. *Plant Biotechnology* 31: 401–407.

Koeduka, T. 2018. Functional evolution of biosynthetic enzymes that produce plant volatiles. *Bioscience, Biotechnology & Biochemistry* 82: 192–199.

Koeduka, T., E. Fridman, D.R. Gang, D.G. Vassão, B.L. Jackson, C.M. Kish, I. Orlova et al. 2006. Eugenol and isoeugenol, characteristic aromatic constituents of spices, are biosynthesized via reduction of a coniferyl alcohol ester. *Proceedings of National Academy of Sciences of the United States of America* 103: 10128–10133.

Koeduka, T., K. Ishizaki, C.M. Mwenda, K. Hori, Y. Sasaki-Sekimoto, H. Ohta, T. Kohchi, and K. Matsui. 2015. Biochemical characterization of allene oxide synthases from the liverwort *Marchantia polymorpha* and green microalgae *Klebsormidium flaccidum* provides insight into the evolutionary divergence of the plant CYP74 family. *Planta* 242: 1175–1186.

Koeduka, T., S. Suzuki, Y. Iijima, T. Ohnishi, H. Suzuki, B. Watanabe, D. Shibata, T. Umezawa, E. Pichersky, and J. Hiratake. 2013. Enhancement of production of eugenol and its glycosides in transgenic aspen plants via genetic engineering. *Biochemical and Biophysical Research Communications* 436: 73–78.

Koeduka, T., T.J. Baiga, J.P. Noel, and E. Pichersky. 2009. Biosynthesis of *t*-anethole in anise: Characterization of *t*-anol/isoeugenol synthase and an *O*-methyltransferase specific for a C7-C8 propenyl side chain. *Plant Physiology* 149: 384–394.

Kunishima, M., Y. Yamauchi, M. Mizutani, M. Kuse, H. Takikawa, and Y. Sugimoto. 2016. Identification of (Z)-3:(E)-2-hexenal isomerases essential to the production of the leaf aldehyde in plants. *Journal of Biological Chemistry* 291: 14023–14033.

Lee, D.S., P. Nioche, M. Hamberg, and C.S. Raman. 2008. Structural insights into the evolutionary paths of oxylipin biosynthetic enzymes. *Nature* 455: 363–368.

Li, G., T.G. Köllner, Y. Yin, Y. Jiang, H. Chen, Y. Xu, J. Gershenzon, E. Pichersky, and F. Chen. 2012. Nonseed plant *Selaginella moellendorffi* has both seed plant and microbial types of terpene synthases. *Proceedings of National Academy of Sciences of the United States of America* 109: 14711–14715.

Liang, S-J., H. Wu, X. Lun, and D-W. Lu. 2006. Secretory cavity development and its relationship with the accumulation of essential oil in fruits of *Citrus medica* L. var. *sarcodactlyis* (Noot.) Swingle. *Journal of Integrative Plant Biology* 48: 573–583.

Liu, J., X.L. Zhu, N. Ullah, and Y.S. Tao. 2017. Aroma glycosides in grapes and wine. *Journal of Food Science* 82: 248–259.

Loughrin, J.H., T.R. Hamilton-Kemp, H.R. Burton, R.A. Andresen, and D.F. Hildebrand. 1992. Glycosidically bound volatile components of *Nicotiana sylvestris* and *N. suaveolens* flowers. *Phytochemistry* 31: 1537–1540.

Louveau, T., C. Leirao, S. Green, C. Hamiaux, B. van der Rest, O. Dechy-Cabaret, R.G. Atkinson, and C. Chervin. 2011. Predicting the substrate specificity of a glycosyltransferase implicated in the production of phenolic volatiles in tomato fruit. *FEBS Journal* 278: 390–400.

Lücker, J., H.J. Bouwmeester, W. Schwab, J. Blaas, L.H.W. van der Plas, and H.A. Verhoeven. 2001. Expression of *Clarkia S*-linalool synthase in transgenic petunia plants results in the accumulation of *S*-linalyl-β-D-glucopyranoside. *The Plant Journal* 27: 315–324.

Ma, S.J., M. Mizutani, J. Hiratake, K. Hayashi, K. Yagi, N. Watanabe, and K. Sakata. 2001. Substrate specificity of β-primeverosidase, a key enzyme in aroma formation during oolong tea and black tea manufacturing. *Bioscience, Biotechnology & Biochemistry* 65: 2719–2729.

Maduagwu, E.N. 1983. Differential effects on the cyanogenic glycoside content of fermenting cassava root pulp by β-glucosidase and microbial activities. *Toxicology Letters* 15: 335–339.

Mandel M A, K.A. Feldmann, L. Herrera-Estrella, M. Rocha-Sosa, and P. León. 1996. CLA1, a novel gene required for chloroplast development, is highly conserved in evolution. *The Plant Journal* 9: 649–658.

Matsui, K. 2006. Green leaf volatiles: Hydroperoxide lyase pathway of oxylipin metabolism. *Current of Opinion in Plant Biology* 9: 274–280.

Matsui, K., C. Ujita, S. Fujimoto, J. Wilkinson, B. Hiatt, V. Knauf, T. Kajiwara, and I. Feussner. 2000. Fatty acid 9- and 13-hydroperoxide lyases from cucumber. *FEBS Letters* 481: 183–188.

Matsui, K., H. Takemoto, T. Koeduka, and T. Ohnishi. 2018. 1-Octen-3-ol is formed from its glycoside during processing of soybean [*Glycine max* (L.) Merr.] seeds. *Journal of Agricultural and Food Chemistry* 66: 7409–7416.

Matsui, K., M. Shibutani, T. Hase, and T. Kajiwara. 1996. Bell pepper fatty acid hydroperoxide lyase is a cytochrome P450 (CYP74B). *FEBS Letters* 394: 21–24.

Mei, X., X. Liu, Y. Zhou, X. Wang, L. Zeng, X. Fu, J. Li, J. Tang, F. Dong, and Z. Yang. 2017. Formation and emission of linalool in tea (*Camellia sinensis*) leaves infested by tea green leafhopper (*Empoasca (Matsumurasca) onukii* Matsuda). *Food Chemistry* 237: 356–363.

Mizutani, M., H. Nakanishi, J. Ema, S.J. Ma, E. Noguchi, M. Inohara-Ochiai, M. Fukuchi-Mizutani, M. Nakao, and K. Sakata. 2002. Cloning of β-primeverosidase from tea leaves, a key enzyme in tea aroma formation. *Plant Physiology* 130: 2164–2176.

Mochizuki, S., and K. Matsui. 2018. Green leaf volatile-burst in Arabidopsis is governed by galactolipid oxygenation by a lipoxygenase that is under control of calcium ion. *Biochemical and Biophysical Research Communications* 505: 939–944.

Mochizuki, S., K. Sugimoto, T. Koeduka, and K. Matsui. 2016. Arabidopsis lipoxygenase 2 is essential for formation of green leaf volatiles and five-carbon volatiles. *FEBS Letters* 590: 1017–1027.

Moon, J.H., N. Watanabe, K. Sakata, A. Yagi, K. Ina, and S. Luo. 1994. *trans*- and *cis*-Linalool 3,6-oxide 6-*O*-β-D-xylopyranosyl-β-D-glucopyranosides isolated as aroma precursors from leaves of oolong tea. *Bioscience, Biotechnology & Biochemistry* 58: 1742–1744.

Moon, J.H., N. Watanabe, Y. Ijima, A. Yagi, and K. Sakata. 1996. *cis*- and *trans*-Linalool 3,7-oxides and methyl salicylate glycosides and (Z)-3-hexenyl β-D-glucopyranoside as aroma precursors from tea leaves for oolong tea. *Bioscience, Biotechnology & Biochemistry* 60: 1815–1819.

Mosblech, A., I. Feussner, and I. Heilmann. 2009. Oxylipins: Structurally diverse metabolites from fatty acid oxidation. *Plant Physiology and Biochemistry* 47: 511–517.

Mwenda, C.M., A. Matsuki, K. Nishimura, T. Koeduka, and K. Matsui. 2015. Spatial expression of the Arabidopsis hydroperoxide lyase gene in controlled differently from that of the allene oxide synthase gene. *Journal of Plant Interactions* 10: 1–10.

Nakashima, A., S.H. von Reuss, H. Tasaka, M. Nomura, S. Mochizuki, Y. Iijima, Y. Aoki, D. Shibata, W. Boland, J. Takabayashi, and K. Matsui. 2013. Traumatin- and dinortraumatin-containing galactolipids in Arabidopsis: Their formation in tissue-disrupted leaves as counterparts of green leaf volatiles. *Journal of Biological Chemistry* 288: 26078–26088.

Nakashima, A., Y. Iijima, K. Aoki, D. Shibata, K. Sugimoto, J. Takabayashi, and K. Matsui. 2011. Monogalactosyl diacylglycerol is a substrate for lipoxygenase: Its implications for oxylipin formation directly from lipids. *Journal of Plant Interactions* 6: 93–97.

Nishikitani, M., D. Wang, K. Kubota, A. Kobayashi, and F. Sugawara. 1999. (Z)-3-Hexenyl and *trans*-linalool 3,7-oxide β-primeverosides isolated as aroma precursors from leaves of a green tea cultivar. *Bioscience, Biotechnology & Biochemistry* 63: 1631–1633.

Nishikitani, M., K. Kubota, A. Kobayashi, and F. Sugawara F. 1996. Geranyl 6-*O*-α-L-arabinopyranosyl-β-D-glucopyranoside isolated as an aroma precursor from leaves of a green tea cultivar. *Bioscience, Biotechnology & Biochemistry* 60: 929–931.

Noguchi, A., Y. Fukui, A. Iuchi-Okada, S. Kakutani, H. Satake, T. Iwashita, M. Nakao, T. Umezawa, and R. Ono. 2008. Sequential glucosylation of a furofuran lignin, (+)-sesaminol, by *Sesamum indicum* UGT71A9 and UGT94D1 glucosyltransferases. *The Plant Journal* 54: 415–427.

Ohgami, S., E. Ono, M. Horikawa, J. Murata, K. Totsuka, H. Toyonaga, Y. Ohba et al. 2015. Volatile glycosylation in tea plants: Sequential glycosylations for the biosynthesis of aroma β-primeverosides are catalyzed by two *Camellia sinensis* glycosyltransferases. *Plant Physiology* 168: 464–477.

Oka, N., H. Ohishi, T. Hatano, M. Hornberger, K. Sakata, and N. Watanabe. 1999. Aroma evolution during flower opening in *Rosa samascena* Mill. *Zeitschrift für Naturforschung* 54c: 889–895.

Peltier J-B, Y. Cai, Q. Sun, V. Zabrouskov, L. Giacomelli, A. Rudella, A.J. Ytterberg, H. Rutschow, and K.J. van Wijk. 2006. The oligomeric stromal proteome of *Arabidopsis thaliana* chloroplasts. *Molecular & Cellular Proteomics* 5: 114–133.

Poppenberger, B., F. Berthiller, D. Lucyshyn, T. Sieberer, R. Schuhmacher, R. Krska, K. Kuchler, J. Glössl, C. Luschnig, and G. Adam. 2003. Detoxification of the *Fusarium* Mycotoxin Deoxynivalenol by a UDP-glucosyltransferase from *Arabidopsis thaliana*. *Journal of Biological Chemistry* 278: 47905–47914.

Poreddy, S., S. Mitra, M. Schöttner, J. Chandran, B. Schneider, I.T. Baldwin, P. Kumar, and S.S. Pandit. 2015. Detoxification of hostplant's chemical defence rather than its anti-predator co-option drives β-glucosidase-mediated lepidopteran counteradaptation. *Nature Communications* 6: 8525–8537.

Richman, A., A. Swanson, T. Humphrey, R. Chapman, B. McGarvey, R. Pocs, and J. Brandle. 2005. Functional genomics uncovers three glucosyltransferases involved in the synthesis of the major sweet glucosides of *Stevia rebaudiana*. *The Plant Journal* 41: 56–67.

Rohmer, M. 2003. Mevalonate-independent methylerythritol phosphate pathway for isoprenoid biosynthesis. Elucidation and distribution. *Pure and Applied Chemistry* 75: 375–387.

Sato, Y., J. Han, H. Fukuda, S. Mikami. 2018. Enhancing monoterpene alcohols in sweet potato shochu using the diglycoside-specific β-primeverosidase. *Journal of Bioscience and Bioengineering* 125: 218–223.

Savchenko, T., I.S. Pearse, L. Ignatia, R. Karban, and K. Dehesh. 2013. Insect herbivores selectively suppress the HPL branch of the oxylipin pathway in host plants. *The Plant Journal* 73: 653–662.

Sawada, S., H. Suzuki, F. Ichimaida, M. Yamaguchi, T. Iwashita, Y. Fukui, H. Hemmi, T. Nishino, and T. Nakayama. 2005. UDP-glucuronic acid:anthocyanin glucuronosyltransferase from red daisy (*Bellis perennis*) flowers. *Journal of Biological Chemistry* 280: 899–906.

Scholz, J., F. Brodhun, E. Hornung, C. Herrfurth, M. Stumpe, A.K. Beike, B. Faltin, W. Frank, R. Reski, and I. Feussner. 2012. Biosynthesis of allene oxides in *Physcomitrella patens*. *BMC Plant Biology* 12: 228. doi:10.1186/1471-2229-12-228.

Shi, J., L. Wang, C.Y. Ma, H.P. LV, Z.M. Chen, and Z. Lin. 2014. Aroma changes of black tea prepared from methyl jasmonate treated tea plants. *Journal of Zhejiang University-Science B* 15: 313–321.

Shirakawa, M., and I. Hara-Nishimura. 2018. Specialized vabuoles of myrosin cells: Chemical defense strategy in Brassicales plants. *Plant and Cell Physiology* 59: 1309–1316.

Song, C., X. Hong, S. Zhao, J. Liu, K. Schulenburg, F.C. Huang, K. Franz-Oberdorf, and W. Schwab. 2016. Glucosylation of 4-hydroxy-2,5-dimethyl-3(2H)-furanone, the key strawberry flavor compound in strawberry fruit. *Plant Physiology* 171: 139–151.

Sugimoto, K., K. Matsui, Y. Iijima, Y. Akakabe, S. Muramoto, R. Ozawa, R. Sasaki et al. 2014. Intake and transformation to a glycoside of (Z)-3-hexenol from neighbors reveals a new mode of plant odor reception and defense. *Proceedings of National Academy of Sciences of the United States of America* 111: 7144–7149.

Takai, J., R. Ozawa, J. Takabayashi, S. Fujii, K. Arai, R.T. Ichiki, T. Koeduka et al. 2018. Silkworms suppress the release of green leaf volatiles by mulberry leaves with an enzyme from their spinnerets. *Scientific Reports* 8: 11942 doi:10.1038/s41598-018-30328-6.

Tanaka, T., A. Ikeda, K. Shiojiri, R. Ozawa, K. Shiki, N. Nagai-Kunihiro, K. Fujita et al. 2018. Identification of a hexenal reductase that modulates the composition of green leaf volatiles. *Plant Physiology* 178: 552–564.

Tikunov, Y.M., J. Molthoff, R.C.H. de Vos, J. Beekwilder, A. van Houwelingen, J.J.J. van der Hooft, M. Nijenhuis-de Vries et al. 2013. Non-smoky glycosyltransferase1 prevents the release of smoky aroma from tomato fruit. *The Plant Cell* 25: 3067–3078.

Tikunov, Y.M., R.C.H. de Vos, A.M. González Paramás, R.D. Hall, and A.G. Bovy. 2010. A role for differential glycoconjugation in the emission of phenylpropanoid volatiles from tomato fruit discovered using a metabolic data fusion approach. *Plant Physiology* 152: 55–70.

Toporkova, Y.Y., V.S. Ermilova, S.S. Gorina, L.S. Mukhtarova, E.V. Osipova, Y.V. Gogolev, and A.N. Grechkin. 2013. Structure-function relationship in the CYP74 family: Conversion of divinyl ether synthases into allene oxide synthases by site-directed mutagenesis. *FEBS Letters* 587: 2552–2258.

Uji, Y., R. Ozawa, H. Shishido, S. Taniguchi, J. Takabayashi, K. Akimitsu, and K. Gomi. 2015. Isolation of a sesquiterpene synthase expressing in specialized epithelial cells surrounding the secretory cavities in rough lemon (*Citrus jambhiri*). *Journal of Plant Physiology* 180: 67–71.

Voo, S.S., H.D. Grimes, and M.B. Lange. 2012. Assessing the biosynthetic capabilities of secretory glands in *citrus* peel. *Plant Physiology* 159: 81–94.

Wang, D., T. Yoshimura, K. Kubota, and A. Kobayashi. 2000. Analysis of glycosidically bound aroma precursors in tea leaves. 1. Qualitative and quantitative analyses of glycosides with aglycons as aroma compounds. *Journal of Agricultural and Food Chemistry* 48: 5411–5418.

Wang, J. and E. Pichersky. 1998. Characterization of S-adenosyl-L-methionine:(iso)eugenol O-methyltransferase involved in floral scent production in *Clarkia breweri*. *Archives of Biochemistry and Biophysics* 349: 153–160.

Wang, J., N. Dudareva, S. Bhakta, R.A. Raguso, and E. Pichersky. 1997. Floral scent production in *Clarkia breweri* (Onagraceae). II. Localization and developmental modulation of the enzyme S-adenosyl-L-methionine:(iso)eugenol O-methyltransferase and phenylpropanoid emission. *Plant Physiology* 114: 213–221.

Wasternack, C. and B. Hause. 2013. Jasmonates: Biosynthesis, perception, signal transduction and action in plant stress response, growth and development. *Annals of Botany* 111:1021–1058.

Yauk, Y.K., C. Ged, M.Y. Wang, A.J. Matich, L. Tessarotto, J.M. Cooney, C. Chervin, and R.G. Atkinson. 2014. Manipulation of flavor and aroma compound sequestration and release using a glycosyltransferase with specificity for terpene alcohols. *The Plant Journal* 80: 317–330.

Yauk, Y.K., E.J.F. Souleyre, A.J. Matich, X. Chen, M.Y. Wang, B. Plunkett, A.P. Dare et al. 2017. Alcohol acyltransferase 1 links two distinct volatile pathways that produce esters and phenylpropenes in apple fruit. *The Plant Journal* 91: 292–305.

Yazaki, K. 2005. Transporters of secondary metabolites. *Current Opinion in Plant Biology* 8: 301–307.

Young, H. and V.J. Paterson. 1995. Characterisation of bound flavor components in kiwifruit. *Journal of the Science of Food and Agriculture* 68: 257–260.

Zhao, J., and R.A. Dixon. 2009. The "ins" and "outs" of flavonoid transport. *Trends in Plant Science* 15: 72–80.

Zheng, X.Q., Q.S. Li, L.P. Xiang, and Y.R. Liang. 2016. Recent advances in volatiles of teas. *Molecules* 21: 338–349.

Zhou, Y.F., S.L. Mao, S.F. Li, X.L. Ni, B. Li, and W.Z. Liu. 2014. Programmed cell death: A mechanism for the lysigenous formation of secretory cavities in leaves of *Dictamnus dasycarpus*. *Plant Science* 225: 147–160.

10 Biosynthesis and Regulation of Fruit Volatiles

José L. Rambla and Antonio Granell

CONTENTS

10.1 INTRODUCTION

Volatile compounds play key roles in fruit, as they are involved in defense against pests and diseases, act as a cue to attract or repel seed dispersers, and are major components determining fruit flavor. Fruit volatile production is highly dynamic and tightly coordinated along the process of ripening, with most compounds showing dramatic changes in their biosynthesis in a short lapse of time. Ripe fruits typically produce a blend consisting of a few hundreds of volatile compounds of different chemical nature which originate from several independent metabolic pathways (Granell and Rambla 2013). In most cases these pathways are not new to the plant, and the same or similar volatiles are also produced by other plant organs. Some pathways are preferentially activated in the fruit, and some of them are expressed only in certain species. Volatiles produced by fruit appear to have played an important role in defining eating habits (i.e., frugivore vs. herbivore) and other plant–insect interactions (Dweck et al. 2015).

10.2 THE MAIN PATHWAYS RESPONSIBLE FOR THE BIOSYNTHESIS OF FRUIT VOLATILES

Unlike flower scent, which is produced and emitted by the intact flower or flower organs, many of the volatile compounds emitted by the fruit are produced after physical tissue disruption, like that produced during overripening or in the mouth when chewing the fruit. In fact, the volatile profile produced by the intact fruit is dramatically different from that of the same fruit after being cut into pieces, and even more after complete homogenization. This fact, together with the techniques used for volatile extraction and determination, is in some extent responsible for the divergence observed in the scientific literature regarding fruit volatile profiles (Rambla et al. 2015). Therefore, different approaches should be considered when analyzing fruit volatiles depending on the aim of the study. For instance, when studying frugivore attraction, focus should be set on the compounds emitted by the whole intact fruit, while when studying fruit flavor, the tissues should be homogenized to some extent, somewhat mimicking the chewing process taking place in the mouth. The same consideration applies for the identification of the enzymes and genes involved in these processes.

Fruit volatile compounds are most often produced by means of several independent metabolic pathways from different substrates previously accumulated in the fruit during its development. These substrates include several amino acids, essential fatty acids and some other healthy/beneficial compounds such as carotenoids. In fact, most of the volatile precursors have positive effects on the frugivores feeding on them. Therefore it has been proposed that fruit volatile compounds could act as sensory cues for health and nutritional value for their predators (Klee and Giovannoni 2011).

In general terms, the volatile metabolic pathways are essentially the same in most plant organs. Nevertheless, according to the specific organ function they are activated differently, each organ producing a characteristic and often dramatically different volatile profile. As in all plant organs, fruit volatile biosynthesis is a developmentally and environmentally regulated process. The volatile profile in the fruit shows ample modifications along development and is also modulated by the environmental conditions. In most species, the process of ripening implies a dramatic shift in volatile production. Below, we are going to focus on the ripe fruit, summarizing the most universal metabolic pathways leading to fruit volatile production.

10.2.1 FATTY ACID DERIVATIVES

Volatiles derived from fatty acid metabolism are among the most ubiquitous compounds. Although also emitted by the intact fruit, a burst of these compounds is produced upon tissue disruption. They are typically oxygenated linear molecules, most often aldehydes, ketones or alcohols, originated from the cleavage of fatty acids, preferentially linoleic and linolenic acids. These include the so called *green leaf volatiles* (Chapter 9), which are C_6 molecules providing *green* aromatic notes to the fruit flavor. Longer chain molecules tend to provide *fatty* notes and, in the case of some C_9 compounds, cucumber-like aroma.

Although other metabolic pathways also lead to the production of fatty acid derivatives in fruit, such as fatty acid biosynthesis (as in the case of methyl ketone synthases/acylsulfurylases) or β-oxidation and α-oxidation of preformed fatty acids, the most important and well-studied metabolic pathway involved in their biosynthesis is the lipoxygenase pathway. The initial step in this pathway is lipid hydrolysis by a lipase to release the free fatty acids (Garbowicz et al. 2018). Lipoxygenases (LOX) are fatty acid dioxygenases that catalyze the dioxygenation of polyunsaturated fatty acids with a (1Z,4Z)-pentadiene moiety, such as the C_{18} linoleic and linolenic acids, producing the corresponding hydroperoxides, which are further metabolized by hydroperoxide lyases (HPL) to produce a volatile aldehyde and an oxoacid. Depending on the position where the fatty acid is oxygenated, lipoxygenases are classified as 13-LOX or 9-LOX. The resulting hydroperoxides are cleaved by means of 13-HPL or 9-HPL respectively. Depending on the species, different activities are predominant. For instance, the most abundant compounds produced in the tomato fruit are hexanal and *(Z)*-3-hexenal, which are produced by the 13-LOX gene *TomLoxC*, while 9-LOX activity is responsible for the production of several C_9 volatiles in cucumber, providing this fruit its characteristic aroma. Most plants have several isoforms of lipoxygenases, but usually only one or a few of them are involved in volatile biosynthesis in the fruit. The aldehydes produced can be reversibly converted into alcohols by alcohol dehydrogenases (ADH). These enzymes also participate in the reduction to alcohols of the aldehydes synthesized by means of other metabolic pathways. Eventually, the alcohols produced can be further metabolized by means of alcohol acyl transferases to produce esters (Klee and Tieman 2018).

10.2.2 PHENYLPROPANOIDS AND BENZENOIDS

This group comprises a large class of volatile compounds with an aromatic ring in their chemical structure that derives from the amino acid phenylalanine. Each compound produces a different perception. For instance, eugenol and cinnamaldehyde provide *spicy* notes, phenylacetaldehyde is *pungent*, guaiacol is *smoky* or *pharmaceutical*, while 2-phenylethanol is *sweet* and *floral*. This group also includes vanillin, the world's most used flavor additive.

In general terms, most volatile compounds in the fruit are either only synthesized during ripening or show a dramatic increase when the fruit ripens. Nevertheless, this is not always the general rule in benzenoids. In some species like tomato, although some phenylalanine-derived volatiles such as 2-phenylethanol increase during ripening, several volatile phenylpropanoids such as guaiacol or methyl salicylate often decrease dramatically, thus suggesting a role of the latter compounds in mechanisms of defense against pests or pathogens, or in discouraging frugivores from feeding on the unripe fruit (Tikunov et al. 2013).

Most compounds in this group are synthesized by means of the phenylpropanoid pathway, also leading to the synthesis of lignin and many other important fruit plant metabolites such as anthocyanins. The first step in this pathway is the conversion of phenylalanine into *(E)*-cinnamic acid by means of an L-phenylalanine ammonia-lyase (PAL), from which metabolism is driven to the biosynthesis of a variety of both volatile and nonvolatile metabolites. Nevertheless, a few other volatiles are synthesized by alternate metabolic pathways, such as phenylacetaldehyde or 2-phenylethanol, which are directly produced from phenylalanine (Granell and Rambla 2013).

10.2.3 BRANCHED-CHAIN VOLATILES

These are originally small molecules originated in the metabolic pathway of branched-chain amino acids leucine, isoleucine and valine. Most of these compounds are aldehydes, alcohols or acids, and impact fruit flavor in a number of species, including tomato, banana, apple and strawberry. Most of them tend to provide the fruit *roast/burned* or *cheese-like* aromatic notes. Branched-chain alcohols can also be used as precursors for the biosynthesis of esters. In the case of banana, branched-chain esters are the compounds providing the dominant aroma notes to the fruit (Pino et al. 2017).

Their biosynthesis has not yet been completely unraveled. Branched-chain amino acid transaminases *SlBCAT1*, *SlBCAT2* and *CmBCAT1* catalyzing the conversion of the amino acids into the corresponding α-ketoacids have been identified in tomato and melon respectively. Such α-ketoacids, either obtained from amino acid catabolism or by *de novo* synthesis, have been proposed to be the most relevant precursors of the volatile compounds, which would be obtained by decarboxylation to produce aldehydes and eventually, after further modification, the corresponding alcohols, acids and esters (Gonda et al. 2010).

10.2.4 ESTERS

Biosynthesis of volatile esters is mediated by alcohol acyl transferases (AAT), which catalyze the acyl esterification of alcohols by means of an acyl-CoA thioester donor. AATs are mid-size families of enzymes from the BAHD superfamily with different tissue-specific pattern producing both volatile and nonvolatile esters. Several AATs have been identified as responsible of the production of fruit volatile esters, such as SAAT in strawberry (Aharoni et al. 2000), CmAAT1, CmAAT3 and CmAAT4 in melon (El-Sharkawy et al. 2005), VpAAT1 in mountain papaya (Balbontin et al. 2010) or MpAAT1 in apple (Dunemann et al. 2012). Methyl esters are a specific type of esters synthesized by S-adenosyl-L-methionine dependent O-methyltransferases, which make up the SABATH family of enzymes. Several genes from this family have been described such as *SAMT* in tomato, which is involved in the biosynthesis of methyl salicylate (Tieman et al. 2010), or *FaOMT* in strawberry catalyzing the methylation of furaneol into mesifurane (Zorrilla-Fontanesi et al. 2012).

The most common substrates for ester biosynthesis are fatty acid derivatives and branched-chain volatiles. Most esters provide fruity notes and sometimes even the characteristic flavor to some fruit species. Esters are predominant in the volatile profile of many fruit such as strawberry, peach, banana, or the climacteric varieties of melon. Interestingly, this ability to synthesize a large amount of diverse esters does not rely on large gene families. For instance, in wild strawberry *Fragaria vesca* only six AATs were identified after sequencing the whole genome (Shulaev et al. 2011). Nevertheless, other species lack almost completely ester compounds in their volatile profile, as in the case of tomato fruit. Interestingly, both cultivated tomato and all the red-fruited related wild species show a high esterase activity which degrade a set of (for humans) unpleasant esters which are synthesized in both the red- and green-fruited species, but which only accumulate in the green-fruited species showing a lower esterase activity (Goulet et al. 2012).

10.2.5 TERPENOIDS

Terpenoids constitute the largest class among the plant volatile compounds. Although some terpenoids are linear, most of them are cyclic molecules. Fruit terpenes can be found either as hydrocarbonated molecules or further modified, often as aldehydes, alcohols, ketones or even esters (Gonzalez-Mas et al. 2019).

These metabolites are often abundant in the vegetative tissues of many plant species, and also have an important role in the volatile profile of some fruit such as mango, different *Citrus* species, where monoterpenes are dominant or, in a lesser extent, strawberry and some white grape varieties. Nevertheless, in general terms their relevance in fruit is less important than in vegetative tissues. For instance, mono- and sesquiterpenoids are dominant in the volatile profile of tomato leaves, as they accumulate at high levels in glandular trichomes where they have a key role in plant defense against pests and pathogens (Lopez-Gresa et al. 2017). However, terpenoids represent only a tiny part of the tomato fruit volatile blend (Rambla et al. 2017).

Biosynthesis of terpenoid volatiles is produced by a mid-size family of terpene synthases by means of two alternate metabolic pathways localized in the plastids and the cytosol respectively, and they originate from the five carbon precursors isopentenyl diphosphate (IPP) and dimethylallyl diphosphate (DMAPP). These serve as substrates to prenyltransferases to produce C_{10} geranyl diphosphate (GPP) and C_{15} farnesyl diphosphate (FPP), which are the precursors of monoterpenes and sesquiterpenes respectively (Chapter 8). Terpene synthases are responsible for the production of volatile mono- and sesquiterpenes; TPSs specifically involved in the synthesis of either mono- or sesquiterpenes are the rule, but TPS enzymes with both activities have also been described (Aharoni et al. 2004). Correlation between expression levels of biosynthetic genes and volatile production suggests that terpenoid synthesis would be regulated transcriptionally, as observed for the *CsTps1* gene in orange (Sharon-Asa et al. 2003), *VvVal*, *VvGerD* and *VvTer* in grape (Lucker et al. 2004) and *FaNES1* in strawberry (Aharoni et al. 2004). It is interesting again that species producing a range of terpenes in fruit can do it with a small number of terpene synthases, i.e., just three *NES* and 4 *PINS* in wild strawberry (Shulaev et al. 2011). Similarly, in the octoploid strawberry, 89% of the biosynthesis of the terpene geranyl acetate is due to the expression of the homeologs from *F. vesca* (Edger et al. 2019).

10.2.6 APOCAROTENOIDS

Volatile apocarotenoids are a small family of linear or cyclic norisoprenoid molecules, which can therefore be considered as terpenoid compounds. Their distinctive characteristic is that they are originated from the cleavage of tetraterpenes. Apocarotenoids are not usually among the most abundant compounds in the volatile profile of fruits, but in many fruit species they have an impact on their flavor and aroma, as the human olfactory receptors are extremely sensitive to these compounds, particularly to the cyclic ones. Apocarotenoids tend to provide fruity, floral or sweet notes to the fruit, and are therefore considered as desirable compounds for consumer flavor perception (Goff and Klee 2006).

Apocarotenoids are biosynthesized by a small family of carotenoid cleavage dioxygenases (CCD), and produce either linear or cyclic compounds depending on the chemical structure of the substrate. These are promiscuous enzymes which can recognize carotenes after ζ–carotene in the carotene biosynthesis pathway, and have the ability to cleave double bonds in cyclic carotenoids in the 9–10 (9′,10′) or 7–8 (7′,8′) positions, and linear carotenoids also in the 5,6 (5′,6′) positions, thus liberating and oxygenated C_{13}, C_{10} or C_8 volatile moiety respectively (Ilg et al. 2009).

Regulation of their biosynthesis seems to be predominantly due to substrate availability, as *CCD* expression has been detected in the fruit in early developmental stages when no volatile apocarotenoids are produced at significant levels. In fact, their biosynthesis mainly takes place after chloroplast conversion into chromoplasts, where carotenoids accumulate, and after membrane degradation or tissue disruption, which favors their contact with the CCD enzymes (Vogel et al. 2010). It is interesting to note that despite their hydrolytic origin, some volatile apocarotenoids in fruit are later converted in nonvolatile forms by conjugation to sugars (Wirth et al. 2001).

10.3 BIOSYNTHESIS OF VOLATILES IN SPECIFIC FRUITS

There exist ample differences between the blend of volatiles produced in the fruit of different species, each of them harboring a particular profile that produces its distinctive aroma. One major mechanism for these differences is the transcriptional regulation of whole metabolic pathways, leading to enhanced or lowered production of a set of many compounds biosynthetically related. Nevertheless, most of the differences in the volatile profiles between species are quantitative rather than qualitative. For instance, a comparison of two apparently largely different profiles such as those of strawberry and tomato shows that almost half of the compounds identified in the ripe tomato fruit were also detected in the ripe strawberry receptacle, although many of them were only detected at very low or trace levels (Zorrilla-Fontanesi et al. 2012; Rambla et al. 2017).

An idea of the contribution specific volatile pathways makes to differentiating a selection of fruits is provided in Figure 10.1, using the volatile profiles of the world's ten most cultivated fruit species (according to FAO January 2016). The most relevant information for the main volatiles, genes and enzymes in the corresponding biosynthetic pathways for five of these fruit species are described below.

10.3.1 TOMATO

Although generally consumed as a vegetable, from a botanical point of view tomato (*Solanum lycopersicum* L.) is the fruit of the tomato plant. In fact, it is one of the most studied fruit species, as it is used in scientific research as a model to study fruit development and ripening, including fruit volatile production.

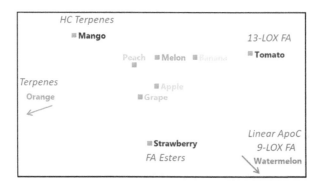

FIGURE 10.1 Distribution of the ten most cultivated crop species based on Principal Component Analysis of their ripe fruit volatile profile. Distribution based on Principal Component Analysis of major volatiles obtained from untargeted analysis of ripe fruit of the ten most cultivated crop species (own data): tomato (*Solanum lycopersicum* L.), orange (*Citrus sinensis* (L.) Osb.), strawberry (*Fragaria x ananassa* Duchesne), grape (*Vitis vinifera* L.), melon (*Cucumis melo* L.), peach (*Prunus persica* (L.) Batsch), watermelon (*Citrullus lanatus* (Thunb.) Matsum. & Nakai), apple (*Malus pumila* Miller), banana (*Musa* spp.) and mango (*Mangifera indica* L.). Discriminant volatiles or metabolic pathways are highlighted in blue: *HC Terpenes*, Hydrocarbonated terpenes; *13-LOX FA*, Fatty acid-derived C_6 volatiles from 13-lipoxygenase activity; *9-LOX FA*, Fatty acid-derived C_9 volatiles from 9-lipoxygenase activity; *Linear ApoC*, Lycopene-derived linear apocarotenoids; *FA esters*, Fatty acid-derived linear esters.

The volatile profile in tomato is dominated by C_6 fatty acid-derived *green leaf volatiles*, mainly produced from the cleavage of linoleic and linolenic acids catalyzed by the enzymes 13-lipoxygenase (TomLoxC) and hydroperoxide lyase (LeHPL). In addition to those compounds, over three hundred other volatiles have been detected in the ripe fruit (Tikunov et al. 2005), most of them in low abundance. Regarding the human perception of tomato flavor and aroma, our senses are not overly influenced by the presence of any single compound in tomato; rather up to around thirty volatiles appear to contribute to flavor. Interestingly, the most abundant volatiles are not necessarily among those with a higher impact on our perception. In fact, some of these highly produced compounds, such as hexanal, seem to have no impact on the consumer perception of flavor. On the other hand, several less abundant volatiles such as other fatty acid derivatives, phenylalanine-derived volatiles, branched-chain amino acid related compounds or volatile apocarotenoids do have an impact on tomato flavor and liking (Tieman et al. 2017).

A substantial number of genes/proteins involved in the biosynthesis of volatiles in the tomato fruit have been identified in the last two decades, thus providing highly valuable information to understand, and eventually modify, fruit volatile production both in tomato and in related species (Rambla et al. 2014a).

Fatty acid derivatives are obtained by the initial degradation of lipids to release the free fatty acids, a reaction catalyzed by lipases such the *SpLIP1* recently identified (Garbowicz et al. 2018). These fatty acids are substrates for TomLoxC and LeHPL, which are essential not only in the production of C_6 aldehydes, but also of C_5 compounds (Shen et al. 2014a). The resulting aldehydes can be converted into alcohols by means of alcohol dehydrogenase *ADH2*. Carotenoid cleavage dioxygenases *LeCCD1A* and especially *LeCCD1B* showed activity on both linear and cyclic carotenoids to produce a variety of volatile apocarotenoids with a very relaxed specificity for both substrate and cleavage site (Rambla et al. 2014a). Interestingly, it has recently been reported that lipoxygenase *TomLoxC* also plays a role in the production of volatile apocarotenoids (Gao et al. 2019), which is consistent with the previous observation of tight correlation between volatile apocarotenoids and fatty acid derivatives in the tomato fruit (Rambla et al. 2017).

Phenylalanine-derived C_6–C_2 compounds undergo initial decarboxylation catalyzed by *LeAADC1A*, *LeAADC1B* and *LeAADC2* (Tieman et al. 2006) to be further metabolized by enzymes yet to be identified to produce several volatiles and eventually leading to 2-phenylethanol by the action of phenylacetaldehyde reductases *LePAR1* and *LePAR2* (Tieman et al. 2007). The biosynthesis of phenylpropanoids methyl salicylate and guaiacol has been identified as produced by *O*-methyltransferases *SlSAMT* (Tieman et al. 2010) and *CTOMT1* (Mageroy et al. 2012) respectively.

Esters are volatiles usually detected at very low or even trace levels in the tomato fruit, unlike its green-fruited wild relatives, which accumulate at significantly higher levels. The reason for this relies on both a less efficient allele of alcohol acyltransferase *AAT1* (Goulet et al. 2015) and an increased expression of carboxylesterase *SlCXE1* resulting in ester catabolism (Goulet et al. 2012).

One of the most remarkable differences between tomato cultivars is the different pattern of volatile phenylpropanoid evolution during ripening, some of them dramatically reducing their levels while in others phenylpropanoid levels remain rather stable (Tikunov et al. 2010). This has been associated to the glycosyl transferase *NSGT1*, which adds an additional glycoside preventing the ubiquitous endogenous glycosidase activity from cleaving the newly conjugated form and release the volatile aglycon (Tikunov et al. 2013).

In addition, *loci* and candidate genes and activities associated with production of flavor volatiles have been identified by genetic approaches (Bauchet et al. 2017; Rambla et al. 2017; Tieman et al. 2017). The role of many of the fruit volatiles in relation to fruit herbivores/frugivores and in protecting or attracting seed dispersers to the fruit other than humans, together with the role of key enzymes in this process, remain mostly unknown.

10.3.2 Citrus

Oranges, mandarines and related *Citrus* species and their hybrids are among the most consumed fruit in the world, and it is therefore not surprising that a large number of studies on *Citrus* volatiles exist, including attempts to help in *Citrus* classification. The reason is that taxonomy of the species in this genus has been controversial. Recently, sequencing of the whole genome has shed light in their genetic relationship and their evolution (Wu et al. 2018). Interestingly, clustering of different *Citrus* species based on the profile of fruit rind essential oil composition revealed a high similarity to that based on the sequence of the whole genome (Gonzalez-Mas et al. 2019).

The volatile profile in *Citrus* pulp is typically dominated by the hydrocarbon monoterpene limonene, which accounts for over 90% of the quantitative volatile composition in most *Citrus* species. Nevertheless, despite its abundance, limonene contribution to flavor and aroma is modest. In fact, some minor or even trace compounds have a stronger impact on aroma, such as 1-*p*-menthene-8-thiol in grapefruit or citral in lemon (Perez-Cacho and Rouseff 2008). Over three hundred compounds have been described in *Citrus* juice, although the identification of some of them has been tentative and several of them are probably misidentified (Gonzalez-Mas et al. 2011). Besides limonene, the volatile profile of *Citrus* mostly consists of fatty acid-derived aldehydes, alcohols and ketones, and their respective esters, together with many other terpenoids, most of them monoterpenoids (Rambla et al. 2014b; Sdiri et al. 2017), although in species such as pummelo (*Citrus grandis* (L.) Osb.), sesquiterpenoid volatiles are also very abundant (Gonzalez-Mas et al. 2011).

It is not surprising that given its prevalence and importance, most of the genes identified up to date as involved with *Citrus* volatile biosynthesis are terpene synthases, the key enzymes in the terpene biosynthetic pathway. Thus, sesquiterpene synthase gene *CsTPS1* involved in the biosynthesis of valencene (Sharon-Asa et al. 2003) has been identified. A mutation in the promoter sequence that modulates its regulation (Yu et al. 2019), and also its regulation by means of the AP2/ERF transcription factor CitAP2.10 (Shen et al. 2016) have been characterized. Similarly, another monoterpene synthase, CitPS16, involved in the biosynthesis of *(E)*-geraniol and its regulation by means of the transcription factor CitERF71 (Li et al. 2017), has also been described. In addition, chromosomal regions harboring genes involved in volatile production have been recently identified in a mandarin-derived (*Citrus reticulata* Bl.) population (Yu et al. 2017).

10.3.3 STRAWBERRY

Strawberry has a particularly rich aroma which is generally highly appreciated by consumers, although major differences exist between the wild species and modern commercial octoploid hybrids (*Fragaria x ananassa*) (Urrutia et al. 2017). In fact, the generic name *Fragaria* derives from the Latin name *fragum*, which is derived from the Latin verb fragrare, "to emit sweet smell" (hence the English word "fragrant"). In the case of strawberry, the part which is consumed is not a fruit, botanically speaking, but the fleshy receptacle that holds the achenes that are the real fruits.

The volatile profile in the ripe strawberry receptacle is dominated by linear esters, which are the most abundant compounds and have a determining influence on its flavor and aroma, providing *green* and *sweet fruity* notes. Mono- and sesquiterpenes together with several fatty acid derivatives and a few lactones are also important contributors to flavor, the latter providing some *fruity peach-like* notes. Nevertheless, the most important molecules involved in the typical strawberry flavor are the furanones mesifurane and, most notably, furaneol (4-Hydroxy-2,5-dimethyl-3(2H)-furanone) (Zorrilla-Fontanesi et al. 2012). Some of the volatile compounds produced could have a role in defense, as their biosynthetic pathways have been shown to be activated by pests.

Increasing attention has been paid in the last few years to unravel the genetic control of volatile biosynthesis in strawberry, and a number of genomic regions (Zorrilla-Fontanesi et al. 2012; Urrutia et al. 2017) and genes involved in these processes have been identified. Most interesting is the finding that a small number of genes in each family of enzymes can account for the complex ester/terpenoid profiles, indicating promiscuity of individual AATs and TPSs. In some cases, more detailed studies have been conducted. Thus, several genes such as *SAAT* (Aharoni et al. 2000) and *VAAT* (Beekwilder et al. 2004) are involved in the production of esters. *FaNES1* is associated with the production of monoterpenoid linalool and sesquiterpenoid nerolidol (Aharoni et al. 2004), and *FaEGS1* and *FaEGS2* are responsible for the biosynthesis of phenylpropanoid volatile eugenol (Araguez et al. 2013). In addition, *FaEO* enone oxidoreductase, i.e., 2-methylene-furan-3-one reductase, was described as involved in the biosynthesis of furaneol, and the *O*-methyltransferase gene *FaOMT* specifies the enzyme for the conversion of furaneol into mesifurane (Lavid et al. 2002; Zorrilla-Fontanesi et al. 2012). Regarding lactone production, similarly as it was proposed in the case of peach (Sanchez et al. 2013), ω-6 fatty acid desaturase gene *FaFAD1* has been described as involved in the biosynthesis of γ–decalactone (Chambers et al. 2014; Sanchez-Sevilla et al. 2014).

10.3.4 GRAPE

Grapevine (*Vitis vinifera* L.) has historically received much attention not only because of the fruit itself, but also because of its use in making wine. The most abundant volatile compounds in the ripe fruit are C_6 fatty acid-derived aldehydes, providing the fruit with its characteristic *green* aroma, while a diversity of other compounds from different metabolic pathways are produced, with important differences depending on the particular variety (Gonzalez-Mas et al. 2009; Rambla et al. 2016).

Several genes involved in the biosynthesis of *green leaf volatiles* have been identified, such as *VvLOXA*, a 13-lypoxygenase rapidly increasing its expression in response to mechanical wounding and biotic stress (Podolyan et al. 2010), hydroperoxide lyases *VvHPL1* and *VvHPL2*, (Zhu et al. 2012) and alcohol dehydrogenase *VvADH2* (Torregrosa et al. 2008). Their relative expression, together with that of alcohol acyltransferases such as *VvAAT2*, explains most of the varietal differences observed in the production of *green leaf* aldehydes, alcohols and esters. In fact, the most limiting factor in ester biosynthesis in grape appears to be the availability of the alcohol precursor (Qian et al. 2019).

Volatile terpenoids are produced in low amounts in grapes with some relevant exceptions like some aromatic white varieties including the famous Muscat and Gewürztraminer, where they contribute to the special bouquet of the wines they produce. Indeed, grape seems to be well endowed for terpene biosynthesis as the grape genomic DNA sequence (Jaillon et al. 2007) revealed that the terpene synthase family is twice as large as compared to other plants, and more specifically shows a high diversification of monoterpene synthases involved in the production of key monoterpenoids impacting aroma (geraniol, linalool, cineole or α-terpineol), some of which have been confirmed in enzyme assays (Martin et al. 2010). Furthermore, grape genome analysis revealed that, as in plants producing high amounts of monoterpenes (like mentha or clarkia), they have heterodimeric forms of geranyl diphosphate synthases, the enzyme that produces the substrate for monoterpene synthases (Martin et al. 2012).

Biosynthesis of other relevant compounds has also been described. For instance, that of methoxypyrazines, which are compounds involved in plant defense that accumulate preferentially in unripe stages and are very potent odorants, providing the fruit intense *herbaceous* aroma notes, at least in certain varieties (Lund and Bohlmann 2006). The key gene involved in their biosynthesis is *O*-methyltranferase *VvOMT3*, which produces 2-methoxy-3-isobutylpyrazine, the major methoxypyrazine in grape, by methylation of the corresponding alcohol (Guillaumie et al. 2013). Methyl anthranilate, an important compound impacting aroma in *Vitis labrusca*, is produced by alcohol acyltransferase *AAMT*, a gene which expression occurs after the initiation of berry ripening (Wang and De Luca 2005).

Conjugation with a polar metabolite, most often glucose, is a mechanism commonly used by plants and in grapes in particular to accumulate volatile compounds in the form of nonvolatile molecules. Monoterpenol β-D-glucosyltransferase *VvGT7* has been identified as involved in the accumulation of geranyl and neryl glucosides during fruit ripening, by means of adding to these monoterpenols a glucose moiety (Bonisch et al. 2014). Most volatile alcohols are also subjected to glycosylation to some extent, which means that an important pool of volatiles are sequestered in a nonavailable way (Sarry and Gunata 2004). The understanding of the processes to cleave the volatile aglycones by means of glycosidases has received increasing attention, due to its potential for the liberation of high amounts of volatile compounds (Rambla et al. 2016).

Although significant advances have been made in the elucidation of grape volatile biosynthesis, much still remains unknown. In this sense, many metabolomic and transcriptomic integrative studies are being performed to unravel the complexity of the genetic control of the volatile network and its regulation by biotic stress and the different environmental conditions (Savoi et al. 2017).

10.3.5 Melon

Melon (*Cucumis melo* L.) is one of the cultivated species showing the highest phenotypic diversity. Variability within this species appears even in the physiology of fruit ripening, as some varieties are climacteric (a burst in ethylene biosynthesis and respiration takes place at the onset of ripening) while others are nonclimacteric. Ample diversity is also evident in the metabolome. According to their volatile profile, melon genotypes could be classified as aromatic or nonaromatic. Most aromatic varieties are climacteric, and are characterized by a high production of volatiles of around two hundred different compounds, dominated by esters derived from fatty acid or branched-chain amino acid, many of them ethyl esters. On the other hand, most nonaromatic varieties are nonclimacteric and produce significantly lower levels of most volatile compounds. In these genotypes, their volatile profile is dominated by fatty acid-derived aldehydes and alcohols, and also by acetate esters (Esteras et al. 2018).

Volatiles derived from fatty acids are highly relevant in melon, as this group includes not only linear aldehydes and alcohols, but also many of the esters. Nevertheless, although well described for other fruit species, most of the genes involved in their biosynthesis in melon have not been described yet. A 13-LOX lipoxygenase, *CmLOX18*, has been recently spotted as involved in the production of C_6 volatiles (Zhang et al. 2017) providing *green* aromatic notes to the aroma. Nevertheless, several C_9 fatty acid derivatives have been described in melon, providing characteristic cucumber-like notes to the fruit, which would be produced by a not yet identified 9-LOX enzyme.

Branched-chain volatiles are also relevant compounds, as they are found as part of many esters and also in the nonesterified form. Branched-chain amino acid transaminase gene *CmBCAAT1* has been involved in their biosynthesis (Gonda et al. 2010). Aldehydes formed from this or other metabolic pathways can be transformed into alcohols by means of alcohol dehydrogenases such as CmADH1 and CmADH2 (Manriquez et al. 2006). These alcohols can be further modified and used as substrates for ester biosynthesis. Alcohol acyl transferases *CmAAT1*, *CmAAT2*, *CmAAT3* and *CmAAT4* have been described to be expressed in the ripe fruit, although the role of some of them in ester biosynthesis has not yet been completely elucidated (El-Sharkawy et al. 2005). Interestingly, despite the absence of whole genome duplication in *C. melo*, analysis of its genomic DNA (Garcia-Mas et al. 2012) sequence revealed several gene families that expanded in melon, including those related to alcohol metabolism.

Several genes and the corresponding proteins have been described in melon in relation to the biosynthesis of aromatic volatiles that use the amino acid phenylalanine as a precursor. *CmArAT1* gene has been described as involved in the biosynthesis of several benzenoid volatiles (Gonda et al. 2010), and *CmCNL* and *CmBAMT* have been implicated in the accumulation of *(E)*-cinnamaldehyde and methyl benzoate respectively (Gonda et al. 2018). Additional genes involved in melon volatile biosynthesis have been uncovered, such as *PDC1*, which is involved in the production of pentanal, propanal and phenylacetaldehyde (Wang et al. 2019) and *CmCCD1*, involved in apocarotenoid production in the carotene-producing (yellow and orange-colored flesh) melon cultivars (Ibdah et al. 2006).

One characteristic feature of melon volatile profile as compared with other fruit species (with some exceptions like in some tropical fruits with durian as paradigmatic) (Cannon and Ho 2018) is the production of sulfur compounds, which are particularly abundant in some musky melon genotypes (Esteras et al. 2018). Human olfactory receptors tend to be extremely sensitive to these volatiles, which are sometimes perceived as unpleasant odors, especially when present at high levels, but appear to contribute to flavor and consumer preferences (Trimmer et al. 2019). L-methionine-γ-lyase CmMLG has been identified as involved in the synthesis of several sulfur and nonsulfur-containing volatiles using L-methionine as substrate (Gonda et al. 2013).

10.3.6 Peel Compounds

Although from the consumer perspective we are particularly interested in the fleshy part of the fruit, the peel also plays an important role in fruit evolution. While the pulp needs to be attractive for consumption and serve as a reward to seed dispersers, the peel has a defensive role, providing protection against pests and diseases. Consequently, pulp and peel are completely different at the cellular level and also in structure and chemical composition. This is also true in the case of volatile compounds. In melon pulp, for instance, a wide variety of esters have been identified, while most of these compounds are very rare or not present at all in the rind, as illustrated in Figure 10.2. Similarly, the volatile profiles in *Citrus* juice are completely different to those of the essential oils in the rind (Gonzalez-Mas et al. 2011, 2019). In some species including several *Citrus*, volatile compounds extracted from the fruit peel have been object of much study and have been extensively used in the food and perfume industry.

From the biological point of view, peel volatiles are key compounds involved in the plant defense strategy, their biosynthesis often induced or enhanced as a response to an external attack. Just to show a few examples, rough lemon (*Citrus jambhiri* Lush.) monoterpene synthase gene *RlemTPS2* expression was induced by microbial attack. This gene is involved in the biosynthesis of sabinene, a monoterpene involved in antifungal activity. Also, genetically engineering limonene synthase gene in orange (*Citrus sinensis* (L.) Osb.) produced a modification of the profile of monoterpenoids in the peel which resulted in modified attraction of herbivore insects (Rodriguez et al. 2011) and enhanced resistance against fungal pathogens (Rodriguez et al. 2018).

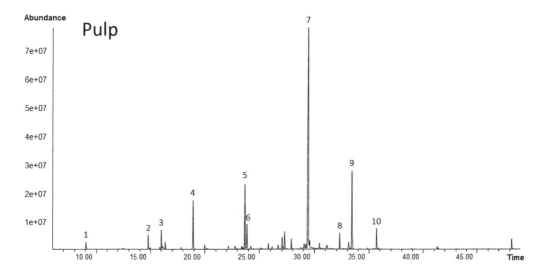

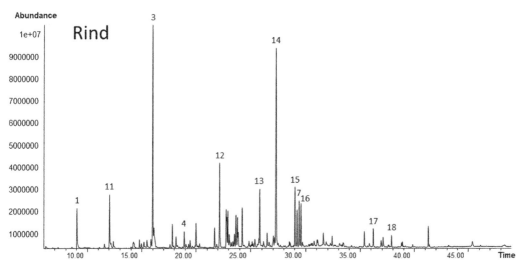

FIGURE 10.2 Differential volatile profiles in pulp and rind. Chromatograms correspond to pulp (up) and rind (down) of the same melon (*Cucumis melo* L.) cultivar. Dramatic differences are observed between their volatile profiles: pulp is dominated by esters, while rind is rich in terpenoids and fatty acid-derived aldehydes (own unpublished data). Some of the most abundant peaks have been indicated: 1, ethyl acetate; 2, 2-methylpropyl acetate; 3, hexanal; 4, 2-methylbutyl acetate; 5, *(Z)*-3-hexenyl acetate; 6, hexyl acetate; 7, benzyl acetate; 8, 2-phenylethyl acetate; 9, *(Z)*-6-nonyl acetate; 10, 3-phenylpropyl acetate; 11, pentanal; 12, *(Z)*-2-heptenal; 13, *(E)*-2-octenal; 14, *(E)*-6-nonenal; 15, *(E, Z)*-2,6-nonadienal; 16, camphor; 17, *β*-damascenone; 18, *β*-caryophyllene.

10.4   REGULATION OF VOLATILE EMISSION

Production of volatile compounds in the fruit is genetically regulated during the whole process of fruit development, each particular developmental stage producing a characteristic volatile profile. Nevertheless, the most dramatic changes occur during fruit ripening. In general terms, the ripening process takes place once the seeds have completed their development and are viable to germinate and produce descendants. This implies a shift in the physical structure and chemical composition of the fruit. When seeds are still not viable, the fruit protects the developing seeds from physical damage and also from any biotic threat which may compromise their development. Therefore, part of the metabolism is driven to the biosynthesis of compounds with the ability to defend the fruit from pests and diseases, and also to discourage frugivores from feeding on the unripe fruit (Giovannoni et al. 2017). This often includes the occurrence of specific specialized metabolites such as, among others, alkaloids and some volatile compounds. Once the seeds have completed their development and are ready for dispersal, most defensive compounds in fruit are degraded (Kozukue et al. 2004) or inactivated (Tikunov et al. 2013), while another set of metabolites are biosynthesized to transform the ripe fruit into an attractive target for seed dispersers. These often include the production of primary metabolites such as short-chain saccharides, an array of specialized metabolites providing vivid colors such as carotenoids or anthocyanins, and an ample variety of volatile compounds (Tohge et al. 2014). In fact, most of the major volatiles characteristic of the ripe fruits are often not found or detected at trace levels in the developing or unripe fruit. These changes in the volatile profile can be detected by at least some frugivore species, which are able to find fruit relying solely on olfactory cues and also have the ability to discriminate between them according to its state of ripeness (Luft et al. 2003).

As a consequence, it makes sense that ripening regulators, volatiles and associated key enzymes follow similar evolution, i.e., in the case of climacteric fruits with the increase in ethylene production (see Figure 10.3 as an illustrative case). However, it is interesting to note that despite *Citrus* fruit being classified as nonclimacteric, ethylene could modulate several aspects of fruit ripening including terpene synthases through ethylene response factors (ERF). A similar situation occurs in another typical nonclimacteric fruit, strawberry, where the ethylene response factor *FaERF#9* and the transcription factor *FaMYB98* have been involved in their regulation (Zhang et al. 2018).

A relevant aspect to be considered is that fruit is a bulky and heterogeneous organ comprised of different tissues, each of which has a slightly different and sometimes markedly particular volatile profile. This is in part due to anatomical differences, but also to the fact that the ripening process

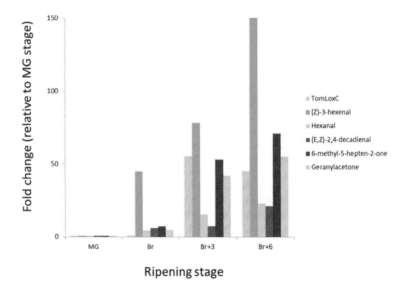

FIGURE 10.3 Parallel evolution along ripening of selected fruit volatile compounds and expression of the biosynthesis involved gene *TomLoxC*. Evolution along tomato fruit ripening of volatile fatty acid derivatives *(Z)*-3-hexenal, hexanal and *(E, Z)*-2,4-decadienal, and apocarotenoids 6-methyl-5-hepten-2-one and geranylacetone (own data), together with the expression of *TomLoxC*, a gene involved in their biosynthesis. (*TomLoxC* (*Solyc01g006540*) expression obtained from TomExpress. Toulouse. INRA.fr.) Ripening stages: MG, Mature green (unripe); Br, Breaker (onset of ripening); Br+3, 3 days after breaker; Br+6, 6 days after breaker (fully ripe). Note that *TomLoxC* gene expression increases up to 50-fold and some volatiles over 150-fold in the fully ripe fruit as compared to the unripe fruit.

does not take place in a perfectly simultaneous manner through the entire fruit, but following a tissue and positional pattern. For example, fruit ripening in tomato is initiated in the locular tissue, from where it expands to the surrounding tissues. Additionally, expression of some genes involved in volatile biosynthesis is tissue-specific, as recently described in the ripening tomato fruit for *CTOMT1* gene, involved in the biosynthesis of guaiacol and exclusively expressed in vascular tissues, and *TomLOXC*, involved in the biosynthesis of several *green leaf volatiles*, which is predominantly expressed in collenchyma cells (Giovannoni et al. 2017).

The presence of the adequate precursors is sometimes the limiting step in volatile biosynthesis. Accordingly, only the colored flesh melon varieties (those accumulating carotenoids) are able to produce significant levels of apocarotenoid volatiles (Gonda et al. 2016), and mutant watermelon and tomato varieties with an altered biosynthesis in linear and cyclic carotenoids showed a parallel modification in their apocarotenoid volatile profile (Lewinsohn et al. 2005). This is also in accordance with the frequent observation that volatile biosynthetic enzymes are often able to produce *in vitro* a wider range of products than those they actually produce *in vivo*, apparently due to the lack of availability of the adequate precursor (Beekwilder et al. 2004). Nevertheless, often the limiting step in the regulation of volatile production in the ripe fruit cannot be attributed to substrate scarcity. In fact, it has been observed that volatile compound levels in the ripe tomato fruit correlate poorly with those of their primary (Zanor et al. 2009) or specialized metabolite precursors. More likely, substrate availability rather than substrate abundance is often the limiting step in volatile biosynthesis, as biosynthetic enzymes and their substrates are often confined in different subcellular locations, and only come into contact after membrane degradation or physical tissue disruption (Rambla et al. 2015).

A relevant mechanism involved in the regulation of volatile emission is the conjugation of preformed volatiles with polar, nonvolatile moieties. It has been observed in a wide range of species such as grape (Martin et al. 2012), strawberry (Ubeda et al. 2012) or tomato (Tikunov et al. 2010) among many others, that fruit accumulates significant amounts of volatile compounds in the form of conjugates, most often as glycosyl esters. Several genes responsible of glycosylation of volatile compounds in fruit have recently been described, such as *NSGT1* in tomato (Tikunov et al. 2013), *VvGT7* in grape (Bonisch et al. 2014) or *PpUGT85A2* in peach (Wu et al. 2019). Such glycosyl conjugates are nonvolatile compounds which consequently remain in the fruit unless an enzyme with the specific glycosidase activity cleaves the aglycon away from the sugar moiety to be freely released. Conjugates tend to accumulate in subcellular compartments different from those where glycosidase enzymes are located. Therefore, conjugates would act as a reservoir of volatiles available to be suddenly released upon contact with glycosidases. This can be produced by tissue disruption such as physical damage or chewing the fruit by a frugivorous species.

From the human perspective, volatiles are major components involved in our perception of fruit flavor. Therefore, the evolution of the volatile composition after fruit harvesting has been thoroughly studied as a determinant of fruit quality. Although the most dramatic changes in fruit volatile composition take place from the onset of ripening to the fully ripe stage, the volatile profile is under continuous modification thereafter, often including an increase in ethanol fermentation in overripe fruits (Rambla et al. 2016). A common practice to slow down the ripening and postharvest evolution of the collected fruits is their storage at low temperature. Nevertheless, although this practice is very effective for fruit preservation, most often it also has an effect on some physicochemical traits, and most notably in the development of flavor volatiles. Thus, it has been described that, with few exceptions, low temperature storage generally induces reduced expression of a variety of genes involved in different metabolic pathways of volatile biosynthesis, producing an often irreversible reduction in volatile production (Shi et al. 2019), which is perceived as a negative trait by the consumer. This process has been postulated to be in part due to transient methylation in key gene promoters (Zhang et al. 2016). When temperature is low enough to produce chilling injury in the fruit, volatile depletion is even more dramatic (Liu et al. 2019).

Fruit of some species like ripe watermelon have the ability to tolerate long-term storage while maintaining rather stable physical condition and metabolomic composition. However, the volatile biosynthesis can lead to overproduction of undesirable compounds such as *(Z)*-6-nonen-1-ol, which confers on the fruit an undesirable pumpkin-like aroma. This disorder can be accelerated or attenuated depending on the postharvest conditions (Xisto et al. 2012; Mendoza-Enano et al. 2019).

Environmental conditions during plant growth and also during postharvest modulate gene expression and eventually plant metabolism, including that of the fruit volatile complement. For instance, drought is known to affect terpene biosynthesis in grape (Savoi et al. 2016), and exposure to UV irradiation enhances the levels of sesquiterpene α-farnesene and reduces the levels of the flavor-associated monoterpenoid linalool in peaches, in parallel with increased expression of terpene synthase *PpTPS1* and reduced *PpTPS2* expression (Liu et al. 2017). Other volatile biosynthetic pathways have also been described to be affected by ultraviolet radiation (Zhou et al. 2019) or direct sunlight exposure (Shen et al. 2014b). In mango, the geographical location – a proxy for environmental differences – has been shown to be responsible for variation in the volatile profile of fruit of the same variety (Singh et al. 2013).

It seems to be a general rule that despite the relatively small genetic diversity present in the domesticated pool of varieties, there is still a huge variation in the amount and profiles of volatiles, with some compounds ranging from a few fold to several orders of magnitude within the same species. The molecular genetics basis for this variation is starting to get elucidated (Tieman et al. 2017). It is also a generally admitted issue that most modern high-yielding varieties have lost the capacity to accumulate high levels of good flavor volatiles for many of the fruit crops (i.e., strawberry, peach, tomato, etc). The identification of the genes and genomic regions underlying this variation has been performed in tomato, and over one hundred genetic markers corresponding to regions with superior alleles of genes conferring good levels of flavor volatiles have been identified. The identification of many of the underlying genes is still under investigation. In some cases, variation in the promoter sequence seems to be important for the differences in accumulation of fruit volatiles. Such is the case of *TomLox C* (Gao et al. 2019), which is a key enzyme for C_6 volatile formation in tomato, and a rare allele of it seems to affect apocarotenoid production.

10.5 CONCLUSION

We are just starting to understand the diversity in fruit volatiles composition, biosynthesis and the underlying regulatory mechanisms operating in the different fruit and their role in fruit communication strategies, as they prove to be extremely variable. This diversity has a common framework that seems to be adopted by a variety of fruit species that we are now beginning to understand in some detail. With this body of knowledge we can now move to the specific volatile network in each new fruit case, as it was likely selected by nature over eons of evolution. With the availability of multiple genomes sequenced from a single species, it is now easier to identify all the actors underlying the fruit volatile pathways and their dynamics. While the molecular genetics approach is the one making faster progress and proving to be very useful in revealing alleles and isoforms of the different enzymes, we still need to conduct detailed biochemical and chemical analysis. These, together with volatile transport and mobility studies, are key issues in order to completely understand the complexity of fruit volatile pathways and how the plant regulates the volatile network in their interactions with other organisms.

ACKNOWLEDGMENTS

José L. Rambla was supported by the Spanish Ministry of Economy and Competitiveness through a "Juan de la Cierva-Formación" grant (FJCI-2016-28601). Research in AG is supported by EC-H2020 program under contracts TRADITOM (634561), TomGEM (679796), Newcotiana (760331-2) and SusCrop-ERA-NET proposal SOLNUE and by Spanish Ministry project BIO2016-78601-R.

REFERENCES

Aharoni, A., A.P. Giri, F.W.A. Verstappen, et al. 2004. Gain and loss of fruit flavor compounds produced by wild and cultivated strawberry species. *Plant Cell* 16 (11):3110–3131.

Aharoni, A., L.C.P. Keizer, H.J. Bouwmeester, et al. 2000. Identification of the SAAT gene involved in strawberry flavor biogenesis by use of DNA microarrays. *Plant Cell* 12 (5):647–661.

Araguez, I., T. Hoffmann, S. Osorio, et al. 2013. Eugenol production in achenes and receptacles of strawberry fruits is catalyzed by synthases exhibiting distinct kinetics. *Plant Physiology* 163 (2):946–958.

Balbontin, C., C. Gaete-Eastman, L. Fuentes, et al. 2010. VpAAT1, A gene encoding an alcohol Acyltransferase, is involved in ester biosynthesis during ripening of mountain papaya fruit. *Journal of Agricultural and Food Chemistry* 58 (8):5114–5121.

Bauchet, G., S. Grenier, N. Samson, et al. 2017. Identification of major loci and genomic regions controlling acid and volatile content in tomato fruit: Implications for flavor improvement. *New Phytologist* 215 (2):624–641.

Beekwilder, J., M. Alvarez-Huerta, E. Neef, F.W.A. Verstappen, H.J. Bouwmeester, and A. Aharoni. 2004. Functional characterization of enzymes forming volatile esters from strawberry and banana. *Plant Physiology* 135 (4):1865–1878.

Bonisch, F., J. Frotscher, S. Stanitzek, et al. 2014. A UDP-Glucose: Monoterpenol Glucosyltransferase adds to the chemical diversity of the grapevine metabolome. *Plant Physiology* 165 (2):561–581.

Cannon, R.J., and C.T. Ho. 2018. Volatile sulfur compounds in tropical fruits. *Journal of Food and Drug Analysis* 26 (2):445–468.

Chambers, A.H., J. Pillet, A. Plotto, J.H. Bai, V.M. Whitaker, and K.M. Folta. 2014. Identification of a strawberry flavor gene candidate using an integrated genetic-genomic-analytical chemistry approach. *BMC Genomics* 15:217.

Dunemann, F., D. Ulrich, L. Malysheva-Otto, et al. 2012. Functional allelic diversity of the apple alcohol acyl-transferase gene MdAAT1 associated with fruit ester volatile contents in apple cultivars. *Molecular Breeding* 29 (3):609–625.

Dweck, H.K.M., M. Knaden, and B.S. Hansson. 2015. Functional loss of yeast detectors parallels transition to herbivory. *Proceedings of the National Academy of Sciences of the United States of America* 112 (10):2927–2928.

Edger, P.P., T. Poorten, R. VanBuren, et al. 2019. Origin and evolution of the octoploid strawberry genome. *Nature Genetics* 51 (3):541.

El-Sharkawy, I., D. Manriquez, F.B. Flores, et al. 2005. Functional characterization of a melon alcohol acyl-transferase gene family involved in the biosynthesis of ester volatiles. Identification of the crucial role of a threonine residue for enzyme activity. *Plant Molecular Biology* 59 (2):345–362.

Esteras, C., J.L. Rambla, G. Sanchez, et al. 2018. Fruit flesh volatile and carotenoid profile analysis within the Cucumis melo L. species reveals unexploited variability for future genetic breeding. *Journal of the Science of Food and Agriculture* 98 (10):3915–3925.

Gao, L., I. Gonda, H.H. Sun, et al. 2019. The tomato pan-genome uncovers new genes and a rare allele regulating fruit flavor. *Nature Genetics* 51 (6):1044.

Garbowicz, K., Z.Y. Liu, S. Alseekh, et al. 2018. Quantitative trait loci analysis identifies a prominent gene involved in the production of fatty acid-derived flavor volatiles in tomato. *Molecular Plant* 11 (9):1147–1165.

Garcia-Mas, J., A. Benjak, W. Sanseverino, et al. 2012. The genome of melon (*Cucumis melo L.*). *Proceedings of the National Academy of Sciences of the United States of America* 109 (29):11872–11877.

Giovannoni, J., C. Nguyen, B. Ampofo, S.L. Zhong, and Z.J. Fei. 2017. The epigenome and transcriptional dynamics of fruit ripening. *Annual Review of Plant Biology*, 68:61–84.

Goff, S.A., and H.J. Klee. 2006. Plant volatile compounds: Sensory cues for health and nutritional value? *Science* 311 (5762):815–819.

Gonda, I., E. Bar, V. Portnoy, et al. 2010. Branched-chain and aromatic amino acid catabolism into aroma volatiles in *Cucumis melo* L. fruit. *Journal of Experimental Botany* 61 (4):1111–1123.

Gonda, I., R. Davidovich-Rikanati, E. Bar, et al. 2018. Differential metabolism of L-phenylalanine in the formation of aromatic volatiles in melon (*Cucumis melo* L.) fruit. *Phytochemistry* 148:122–131.

Gonda, I., S. Lev, E. Bar, et al. 2013. Catabolism of L-methionine in the formation of sulfur and other volatiles in melon (*Cucumis melo* L.) fruit. *Plant Journal* 74 (3):458–472.

Gonda, I., Y. Burger, A.A. Schaffer, et al. 2016. Biosynthesis and perception of melon aroma. *Biotechnology in Flavor Production*. Edited by D. HavkinFrenkel and N. Dudai, *2nd Edition*, Oxford, UK: Blackwell Publishing.

Gonzalez-Mas, M.C., J.L. Rambla, M.C. Alamar, A. Gutierrez, and A. Granell. 2011. Comparative analysis of the volatile fraction of fruit juice from different citrus species. *PLoS One* 6 (7): e22016.

Gonzalez-Mas, M.C., L.M. Garcia-Riano, C. Alfaro, J.L. Rambla, A.I. Padilla, and A. Gutierrez. 2009. Headspace-based techniques to identify the principal volatile compounds in red grape cultivars. *International Journal of Food Science and Technology* 44 (3):510–518.

Gonzalez-Mas, M.C., J.L. Rambla, M.P. Lopez-Gresa, M.A. Blazquez, and A. Granell. 2019. Volatile compounds in citrus essential oils: A comprehensive review. *Frontiers in Plant Science* 10: 12.

Goulet, C., M.H. Mageroy, N.B. Lam, A. Floystad, D.M. Tieman, and H.J. Klee. 2012. Role of an esterase in flavor volatile variation within the tomato clade. *Proceedings of the National Academy of Sciences of the United States of America* 109 (46):19009–19014.

Goulet, C., Y. Kamiyoshihara, N.B. Lam, et al. 2015. Divergence in the enzymatic activities of a tomato and *Solanum pennellii* alcohol acyltransferase impacts fruit volatile ester composition. *Molecular Plant* 8 (1):153–162.

Granell, A., and J.L. Rambla. 2013. Biosynthesis of volatile compounds. In *The Molecular Biology and Biochemistry of Fruit Ripening*, G B Seymour; ProQuest (Firm), Chichester, UK: Wiley-Blackwell.

Guillaumie, S., A. Ilg, S. Rety, et al. 2013. Genetic analysis of the biosynthesis of 2-methoxy-3-isobutylpyrazine, a major grape-derived aroma compound impacting wine quality. *Plant Physiology* 162 (2):604–615.

Ibdah, M., Y. Azulay, V. Portnoy, et al. 2006. Functional characterization of CmCCD1, a carotenoid cleavage dioxygenase from melon. *Phytochemistry* 67 (15):1579–1589.

Ilg, A., P. Beyer, and S. Al-Babili. 2009. Characterization of the rice carotenoid cleavage dioxygenase 1 reveals a novel route for geranial biosynthesis. *The FEBS Journal* 276 (3):736–747.

Jaillon, O., J.M. Aury, B. Noel, et al. 2007. The grapevine genome sequence suggests ancestral hexaploidization in major angiosperm phyla. *Nature* 449 (7161):463–U5.

Klee, H.J., and D.M. Tieman. 2018. The genetics of fruit flavour preferences. *Nature Reviews Genetics* 19 (6):347–356.

Klee, H.J., and J.J. Giovannoni. 2011. Genetics and control of tomato fruit ripening and quality attributes. In *Annual Review Genetics*, Vol 45. Edited by B.L. Bassler, M. Lichten and G. Schupbach, Palo Alto, CA: Annual Reviews.

Kozukue, N., J.S. Han, K.R. Lee, and M. Friedman. 2004. Dehydrotomatine and alpha-tomatine content in tomato fruits and vegetative plant tissues. *Journal of Agricultural and Food Chemistry* 52 (7):2079–2083.

Lavid, N., W. Schwab, E. Kafkas, et al. 2002. Aroma biosynthesis in strawberry: S-adenosylmethionine: Furaneol O-methyltransferase activity in ripening fruits. *Journal of Agricultural and Food Chemistry* 50 (14):4025–4030.

Lewinsohn, E., Y. Sitrit, E. Bar, et al. 2005. Not just colors: Carotenoid degradation as a link between pigmentation and aroma in tomato and watermelon fruit. *Trends in Food Science & Technology* 16 (9):407–415.

Li, X., Y.Y. Xu, S.L. Shen, et al. 2017. Transcription factor CitERF71 activates the terpene synthase gene CitTPS16 involved in the synthesis of E-geraniol in sweet orange fruit. *Journal of Experimental Botany* 68 (17):4929–4938.

Liu, B.D., W.X. Jiao, B.G. Wang, J.Y. Shen, H.D. Zhao, and W.B. Jiang. 2019. Near freezing point storage compared with conventional low temperature storage on apricot fruit flavor quality (volatile, sugar, organic acid) promotion during storage and related shelf life. *Scientia Horticulturae* 249:100–109.

Liu, H.R., X.M. Cao, X.H. Liu, et al. 2017. UV-B irradiation differentially regulates terpene synthases and terpene content of peach. *Plant Cell and Environment* 40 (10):2261–2275.

Lopez-Gresa, M.P., P. Lison, L. Campos, et al. 2017. A non-targeted metabolomics approach unravels the VOCs associated with the tomato immune response against *Pseudomonas syringae*. *Frontiers in Plant Science* 8:15.

Lucker, J., P. Bowen, and J. Bohlmann. 2004. Vitis vinifera terpenoid cyclases: Functional identification of two sesquiterpene synthase cDNAs encoding (+)-valencene synthase and (−)-germacrene D synthase and expression of mono-and sesquiterpene synthases in grapevine flowers and berries. *Phytochemistry* 65 (19):2649–2659.

Luft, S., E. Curio, and B. Tacud. 2003. The use of olfaction in the foraging behaviour of the golden-mantled flying fox, Pteropus pumilus, and the greater musky fruit bat, *Ptenochirus jagori* (Megachiroptera: Pteropodidae). *Naturwissenschaften* 90 (2):84–87.

Lund, S.T., and J. Bohlmann. 2006. The molecular basis for wine grape quality: A volatile subject. *Science* 311 (5762):804–805.

Mageroy, M.H., D.M. Tieman, A. Floystad, M.G. Taylor, and H.J. Klee. 2012. A *Solanum lycopersicum* catechol-O-methyltransferase involved in synthesis of the flavor molecule guaiacol. *Plant Journal* 69 (6):1043–1051.

Manriquez, D., I. El-Sharkawy, F.B. Flores, et al. 2006. Two highly divergent alcohol dehydrogenases of melon exhibit fruit ripening-specific expression and distinct biochemical characteristics. *Plant Molecular Biology* 61 (4–5):675–685.

Martin, D.M., A. Chiang, S.T. Lund, and J. Bohlmann. 2012. Biosynthesis of wine aroma: Transcript profiles of hydroxymethylbutenyl diphosphate reductase, geranyl diphosphate synthase, and linalool/nerolidol synthase parallel monoterpenol glycoside accumulation in Gewurztraminer grapes. *Planta* 236 (3):919–929.

Martin, D.M., S. Aubourg, M.B. Schouwey, et al. 2010. Functional annotation, genome organization and phylogeny of the grapevine (*Vitis vinifera*) terpene synthase gene family based on genome assembly, FLcDNA cloning, and enzyme assays. *BMC Plant Biology* 10: 226.

Mendoza-Enano, M.L., R. Stanley, and D. Frank. 2019. Linking consumer sensory acceptability to volatile composition for improved shelf-life: A case study of fresh-cut watermelon (*Citrullus lanatus*). *Postharvest Biology and Technology* 154:137–147.

Perez-Cacho, P.R., and R.L. Rouseff. 2008. Fresh squeezed orange juice odor: A review. *Critical Reviews in Food Science and Nutrition* 48 (7):681–695.

Pino, J.A., P. Winterhalter, and M. Castro-Benitez. 2017. Odour-active compounds in baby banana fruit (*Musa acuminata* AA Simmonds cv. Bocadillo). *International Journal of Food Properties* 20:1448–1455.

Podolyan, A., J. White, B. Jordan, and C. Winefield. 2010. Identification of the lipoxygenase gene family from Vitis vinifera and biochemical characterisation of two 13-lipoxygenases expressed in grape berries of Sauvignon Blanc. *Functional Plant Biology* 37 (8):767–784.

Qian, X., Y. Liu, G. Zhang, et al. 2019. Alcohol acyltransferase gene and ester precursors differentiate composition of volatile esters in three interspecific hybrids of *Vitis labrusca* x *V. vinifera* during berry development period. *Food Chemistry* 295:234–246.

Rambla, J.L., A. Medina, A. Fernandez-del-Carmen, et al. 2017. Identification, introgression, and validation of fruit volatile QTLs from a red-fruited wild tomato species. *Journal of Experimental Botany* 68 (3):429–442.

Rambla, J.L., A. Trapero-Mozos, G. Diretto, et al. 2016. Gene-metabolite networks of volatile metabolism in airen and tempranillo grape cultivars revealed a distinct mechanism of aroma bouquet production. *Frontiers in Plant Science* 7:1619.

Rambla, J.L., C. Alfaro, A. Medina, M. Zarzo, J. Primo, and A. Granell. 2015. Tomato fruit volatile profiles are highly dependent on sample processing and capturing methods. *Metabolomics* 11 (6):1708–1720.

Rambla, J.L., M.C. Gonzalez-Mas, C. Pons, G.P. Bernet, M.J. Asins, and A. Granell. 2014b. Fruit volatile profiles of two citrus hybrids are dramatically different from those of their parents. *Journal of Agricultural and Food Chemistry* 62 (46):11312–11322.

Rambla, J.L., Y.M. Tikunov, A.J. Monforte, A.G. Bovy, and A. Granell. 2014a. The expanded tomato fruit volatile landscape. *Journal of Experimental Botany* 65 (16):4613–4623.

Rodriguez, A., V. Kava, L. Latorre-Garcia, et al. 2018. Engineering D-limonene synthase down-regulation in orange fruit induces resistance against the fungus *Phyllosticta citricarpa* through enhanced accumulation of monoterpene alcohols and activation of defence. *Molecular Plant Pathology* 19 (9):2077–2093.

Rodriguez, A., V.S. Andres, M. Cervera, et al. 2011. Terpene down-regulation in orange reveals the role of fruit aromas in mediating interactions with insect herbivores and pathogens. *Plant Physiology* 156 (2):793–802.

Sanchez, G., M. Venegas-Caleron, J.J. Salas, A. Monforte, M.L. Badenes, and A. Granell. 2013. An integrative "omics" approach identifies new candidate genes to impact aroma volatiles in peach fruit. *BMC Genomics* 14:343.

Sanchez-Sevilla, J.F., E. Cruz-Rus, V. Valpuesta, M.A. Botella, and I. Amaya. 2014. Deciphering gammadecalactone biosynthesis in strawberry fruit using a combination of genetic mapping, RNA-Seq and eQTL analyses. *BMC Genomics* 15:218.

Sarry, J.E., and Z. Gunata. 2004. Plant and microbial glycoside hydrolases: Volatile release from glycosidic aroma precursors. *Food Chemistry* 87 (4):509–521.

Savoi, S., D.C.J. Wong, A. Degu, et al. 2017. Multi-Omics and integrated network analyses reveal new insights into the systems relationships between metabolites, structural genes, and transcriptional regulators in developing grape berries (*Vitis vinifera* L.) exposed to water deficit. *Frontiers in Plant Science* 8:1124.

Savoi, S., D.C.J. Wong, P. Arapitsas, et al. 2016. Transcriptome and metabolite profiling reveals that prolonged drought modulates the phenylpropanoid and terpenoid pathway in white grapes (*Vitis vinifera* L.). *BMC Plant Biology* 16:67.

Sdiri, S., J.L. Rambla, C. Besada, A. Granell, and A. Salvador. 2017. Changes in the volatile profile of citrus fruit submitted to postharvest degreening treatment. *Postharvest Biology and Technology* 133:48–56.

Sharon-Asa, L., M. Shalit, A. Frydman, et al. 2003. Citrus fruit flavor and aroma biosynthesis: Isolation, functional characterization, and developmental regulation of Cstps1, a key gene in the production of the sesquiterpene aroma compound valencene. *Plant Journal* 36 (5):664–674.

Shen, J.Y., D. Tieman, J.B. Jones, et al. 2014a. A 13-lipoxygenase, TomloxC, is essential for synthesis of C5 flavour volatiles in tomato. *Journal of Experimental Botany* 65 (2):419–428.

Shen, J.Y., L. Wu, H.R. Liu, et al. 2014b. Bagging treatment influences production of C-6 Aldehydes and biosynthesis-related gene expression in peach fruit skin. *Molecules* 19 (9):13461–13472.

Shen, S.L., X.R. Yin, B. Zhang, et al. 2016. CitAP2.10 activation of the terpene synthase CsTPS1 is associated with the synthesis of (+)-valencene in "Newhall" orange. *Journal of Experimental Botany* 67 (14):4105–4115.

Shi, F., X. Zhou, M.M. Yao, Q. Zhou, S.J. Ji, and Y. Wang. 2019. Low-temperature stress-induced aroma loss by regulating fatty acid metabolism pathway in "Nanguo" Pear. *Food Chemistry* 297, doi:10.1016/j.foodchem.2019.05.201.

Shulaev, V., D.J. Sargent, R.N. Crowhurst, et al. 2011. The genome of woodland strawberry (*Fragaria vesca*). *Nature Genetics* 43 (2):109–116.

Singh, Z., R.K. Singh, V.A. Sane, and P. Nath. 2013. Mango: Postharvest biology and biotechnology. *Critical Reviews in Plant Sciences* 32 (4):217–236.

Tieman, D.M., H.M. Loucas, J.Y. Kim, D.G. Clark, and H.J. Klee. 2007. Tomato phenylacetaldehyde reductases catalyze the last step in the synthesis of the aroma volatile 2-phenylethanol. *Phytochemistry* 68 (21):2660–2669.

Tieman, D.M., M. Zeigler, E.A. Schmelz, et al. 2006. Identification of loci affecting flavour volatile emissions in tomato fruits. *Journal of Experimental Botany* 57 (4):887–896.

Tieman, D., G.T. Zhu, M.F. R. Resende, et al. 2017. A chemical genetic roadmap to improved tomato flavor. *Science* 355 (6323):391–394.

Tieman, D., M. Zeigler, E. Schmelz, et al. 2010. Functional analysis of a tomato salicylic acid methyl transferase and its role in synthesis of the flavor volatile methyl salicylate. *Plant Journal* 62 (1):113–123.

Tikunov, Y.M., J. Molthoff, R.C.H. de Vos, et al. 2013. Non-smoky glycosyltransferase1 prevents the release of smoky aroma from tomato fruit. *Plant Cell* 25 (8):3067–3078.

Tikunov, Y.M., R.C.H. de Vos, A.M. G. Paramas, R.D. Hall, and A.G. Bovy. 2010. A Role for differential glycoconjugation in the emission of Phenylpropanoid volatiles from tomato fruit discovered using a metabolic data fusion approach. *Plant Physiology* 152 (1):55–70.

Tikunov, Y., A. Lommen, C.H.R. de Vos, et al. 2005. A novel approach for nontargeted data analysis for metabolomics: Large-scale profiling of tomato fruit volatiles. *Plant Physiology* 139 (3):1125–1137.

Tohge, T., S. Alseekh, and A.R. Fernie. 2014. On the regulation and function of secondary metabolism during fruit development and ripening. *Journal of Experimental Botany* 65 (16):4599–4611.

Torregrosa, L., M. Pradal, J.M. Souquet, M. Rambert, Z. Gunata, and C. Tesniere. 2008. Manipulation of VvAdh to investigate its function in grape berry development. *Plant Science* 174 (2):149–155.

Trimmer, C., A. Keller, N.R. Murphy, et al. 2019. Genetic variation across the human olfactory receptor repertoire alters odor perception. *Proceedings of the National Academy of Sciences of the United States of America* 116 (19):9475–9480.

Ubeda, C., F. San-Juan, B. Concejero, et al. 2012. Glycosidically bound aroma compounds and impact odorants of four strawberry varieties. *Journal of Agricultural and Food Chemistry* 60 (24):6095–6102.

Urrutia, M., J.L. Rambla, K.G. Alexiou, A. Granell, and A. Monfort. 2017. Genetic analysis of the wild strawberry (*Fragaria vesca*) volatile composition. *Plant Physiology and Biochemistry* 121:99–117.

Vogel, J.T., D.M. Tieman, C.A. Sims, A.Z. Odabasi, D.G. Clark, and H.J. Klee. 2010. Carotenoid content impacts flavor acceptability in tomato (*Solanum lycopersicum*). *Journal of the Science of Food and Agriculture* 90 (13):2233–2240.

Wang, J., and V. De Luca. 2005. The biosynthesis and regulation of biosynthesis of Concord grape fruit esters, including "foxy" methylanthranilate. *Plant Journal* 44 (4):606–619.

Wang, M.M., L. Zhang, K.H. Boo, E. Park, G. Drakakaki, and F. Zakharov. 2019. PDC1, a pyruvate/alpha-ketoacid decarboxylase, is involved in acetaldehyde, propanal and pentanal biosynthesis in melon (*Cucumis melo* L.) fruit. *Plant Journal* 98 (1):112–125.

Wirth, J., W.F. Guo, R. Baumes, and Z. Gunata. 2001. Volatile compounds released by enzymatic hydrolysis of glycoconjugates of leaves and grape berries from *Vitis vinifera* Muscat of Alexandria and Shiraz cultivars. *Journal of Agricultural and Food Chemistry* 49 (6):2917–2923.

Wu, B.P., X.M. Cao, H.R. Liu, et al. 2019. UDP-glucosyltransferase PpUGT85A2 controls volatile glycosylation in peach. *Journal of Experimental Botany* 70 (3):925–936.

Wu, G.A., J. Terol, V. Ibanez, et al. 2018. Genomics of the origin and evolution of citrus. *Nature* 554 (7692):311.

Xisto, A., E. Boas, E.E. Nunes, B.M. V. Boas, and M.C. Guerreiro. 2012. Volatile profile and physical, chemical, and biochemical changes in fresh cut watermelon during storage. *Ciencia E Tecnologia De Alimentos* 32 (1):173–178.

Yu, Q.B., M. Huang, H.G. Jia, et al. 2019. Deficiency of valencene in mandarin hybrids is associated with a deletion in the promoter region of the valencene synthase gene. *BMC Plant Biology* 19: 101.

Yu, Y., J.H. Bai, C.X. Chen, et al. 2017. Identification of QTLs controlling aroma volatiles using a "Fortune" x "Murcott" (*Citrus reticulata*) population. *BMC Genomics* 18:646.

Zanor, M.I., J.L. Rambla, J. Chaib, et al. 2009. Metabolic characterization of loci affecting sensory attributes in tomato allows an assessment of the influence of the levels of primary metabolites and volatile organic contents. *Journal of Experimental Botany* 60 (7):2139–2154.

Zhang, B., D.M. Tieman, C. Jiao, et al. 2016. Chilling-induced tomato flavor loss is associated with altered volatile synthesis and transient changes in DNA methylation. *Proceedings of the National Academy of Sciences of the United States of America* 113 (44):12580–12585.

Zhang, C., S.X. Cao, Y.Z. Jin, et al. 2017. Melon13-lipoxygenase CmLOX18 may be involved in C6 volatiles biosynthesis in fruit. *Scientific Reports* 7:1816.

Zhang, Y.Y., X.R. Yin, Y.W. Xiao, et al. 2018. An ethylene response factor-MYB transcription complex regulates furaneol biosynthesis by activating quinone oxidoreductase expression in strawberry. *Plant Physiology* 178 (1):189–201.

Zhou, D.D., Y. Sun, M.Y. Li, T. Zhu, and K. Tu. 2019. Postharvest hot air and UV-C treatments enhance aroma-related volatiles by simulating the lipoxygenase pathway in peaches during cold storage. *Food Chemistry* 292:294–303.

Zhu, B.Q., X.Q. Xu, Y.W. Wu, C.Q. Duan, and Q.H. Pan. 2012. Isolation and characterization of two hydroperoxide lyase genes from grape berries. *Molecular Biology Reports* 39 (7):7443–7455.

Zorrilla-Fontanesi, Y., J.L. Rambla, A. Cabeza, et al. 2012. Genetic analysis of strawberry fruit aroma and identification of O-methyltransferase FaOMT as the locus controlling natural variation in mesifurane content. *Plant Physiology* 159 (2):851–870.

11 Biosynthesis and Regulation of Belowground Signaling Molecules

Lemeng Dong and Harro Bouwmeester

CONTENTS

11.1 INTRODUCTION

A crucial process for the interaction of plants with other organisms in their environment, is the conversion of photosynthetically fixed carbon to signaling molecules, through a series of enzymatic reactions, and emitting them to the air surrounding them or the soil in which they grow. Volatile signals that are emitted above-ground can travel long distances and thus also mediate the interaction with organisms at long distances (as detailed elsewhere in this book). Belowground, however, most of the signals and interactions happen in a small volume of soil bordering the plant root, called the rhizosphere, in which plant roots, soil and the soil biota tightly interact with each other. Studies indicate that roughly 40% of photosynthetically fixed carbon is invested belowground (Jones et al. 2009). Nineteen percent of fixed carbon is invested in root biomass, and the remaining 21% is deposited into the soil. The carbon deposited into the soil is composed of compounds actively secreted by the plant root or exuded passively in a gradient-dependent manner (together they are called the root exudate), as well as dead root cells and/or sloughed-off of root border cells (Hawes et al. 2000). The composition and quantity of root exudates is affected by plant development, soil properties and abiotic and biotic environmental factors, and they are exuded into the rhizosphere as the result of the interaction network between a plant and its environment (Jones et al. 2009). Vice versa, these root exudates modify properties of the soil they are exuded into, for example change the pH and detoxify heavy metals to improve nutrient uptake and alleviate other abiotic stresses (Hinsinger 2001, Walker et al. 2003).

Although we are just beginning to understand the full complexity and biological relevance of these root exudates, in recent years quite a number of reviews and research papers were dedicated to this topic, especially in the emerging area of the plant and soil microbiome interaction (Walker et al. 2003, van Dam and Bouwmeester 2016). Plant root exudates serve as chemoattractants and chemo-repellents to attract beneficial organisms and repel harmful ones like pathogens and herbivores, respectively (Walker et al. 2003). This seems to have started of an arms race, in which other organisms hijack the root exudates as signaling molecules for host detection or use them as food source. Examples of this are the strigolactones and flavonoids that have both positive and negative roles, which will be further discussed below (Falcone Ferreyra et al. 2012, Ruyter-Spira et al. 2013). Signaling molecules so far identified in the rhizosphere mostly belong to the specialized metabolites, such as the terpenoids (including the strigolactones) and phenolics (including the flavonoids). Below, we will briefly describe the biological roles of these two classes of molecules in the rhizosphere (see more details in Chapter 18) and discuss their biosynthesis and its regulation.

11.2 TERPENOID RHIZOSPHERE SIGNALS

11.2.1 BIOLOGICAL ROLE OF RHIZOSPHERE TERPENOIDS

Terpenoids play a crucial role in the belowground communication between plants and other organisms. The volatile, low molecular weight, monoterpenoids, sesquiterpenoids and some diterpenoids contribute to long-range communication. The nonvolatile terpenoids, such as diterpenoids, sesterterpenoids, triterpenoids, tetraterpenoids and some of their breakdown products, are secreted/deposited from the epidermal cell layer of the root and are involved in communication with organisms in the rhizosphere. Terpenoids exuded by plants roots have been reported to inhibit growth of neighboring plant roots (Xu et al. 2012), repel herbivores, have antimicrobial activity (Bell et al. 1975), and attract entomopathogenic nematodes (Köllner et al. 2008). In contrast to these nonvolatile terpenoids, volatile compounds (such as for example camphor, cineole, and camphene) – detected in the soil around *Salvia leucophylla* roots – displayed allelopathy in rather high concentrations (>400 μM) (Nishida et al. 2005). These monoterpenes inhibited seed germination and seedling growth of *Brassica campestris* by inhibiting both cell-nuclear and organelle DNA synthesis in the root apical meristem.

11.2.2 BIOSYNTHESIS OF RHIZOSPHERE TERPENOIDS

Terpenoids (isoprenoids) are derived from the isomeric five-carbon building blocks (IPP) and dimethylallyl diphosphate (DMAPP). IPP and DMAPP are produced through two independent pathways in plants, the methylerythritol phosphate (MEP) pathway that is located in the plastids and the mevalonic acid (MVA) pathway in the cytosol (McGarvey and Croteau 1995). IPP and DMAPP are also available in the mitochondria where isoprenoid derived compounds such as ubiquinone and heme A are produced. The IPP in the mitochondria is imported from the cytosol and the DMAPP generated by a mitochondrially located IPP isomerase (Guirimand et al. 2012). IPP and DMAPP are condensed to form bigger (C10, C15, C20, etc.) molecules by the activity of prenyl transferases or isoprenyl diphosphate synthases. The condensation of one DMAPP molecule with one IPP molecule in a head-to-tail manner yields the C10 monoterpene precursor geranyl diphosphate (GDP) or its cisoid isomer neryl diphosphate (NDP). Interestingly, (heterologously expressed and artificially targeted) GDP synthase can use mitochondrial, cytosolic, and plastid localized IPP and DMAPP and produce GDP in the corresponding compartment (Dong et al. 2016). This GDP can also be exchanged between these compartments but not equally effective. GDP produced in the cytosol is not transferred to the plastids, while the GDP produced in the mitochondria is entirely transferred to the plastids (Dong et al. 2016). However, only 7% of GDP produced in the plastids is available in the mitochondria. The condensation of two IPP and one DMAPP units results in the C15 sesquiterpene precursor farnesyl diphosphate (2*E*,6*E*-FDP) or its isomers (2*Z*,6*Z*-FDP and 2*Z*,6*E*-FDP). The condensation of three IPP and one DMAPP units results in the C20 precursor geranylgeranyl diphosphate (GGDP), a reaction catalyzed by GGDP synthase. Just five years ago it was discovered that GGDP can combine with another IPP molecule to produce the C25 prenyl diphosphate precursor (geranylfarnesyl diphosphate), a precursor for the production of sesterterpenoids, and this step is catalyzed by a geranylfarnesyl diphosphate synthase. Interestingly, FDP and GGDP molecules can also be self-doubled (with the loss of both diphosphate groups) by enzymes beyond the family of prenyl transferases, squalene synthase (SS) and phytoene synthase (PSY) leading to the C30 triterpenoid and phytosterol precursor, squalene, and the C40 carotenoid precursor, phytoene, respectively.

After formation of the prenyl diphosphate precursors GDP, FDP and GGDP, monoterpenes, sesquiterpenes and diterpenes, respectively, are generated through the action of a large family of enzymes known as terpene synthases. Some of these terpene synthases produce a (virtually) single product, such as the monoterpene synthases geraniol synthase, the sesquiterpene synthases germacrene A synthase, and the diterpene synthases taxadiene synthase. The majority of the terpene synthases, however, produce multiple products. A multi-substrate enzyme described in one of the earliest studies on this topic is the *Mentha x piperita* β-farnesene synthase. It converts FDP to mainly (*E*)-β-farnesene (85%) and lower amounts of (Z)-β-farnesene (8%) and δ-cadinene (5%). But it can also use GDP as substrate and produce several different monoterpene products such as limonene, terpinolene, and myrcene (Crock et al. 1997).

Many of the terpenoids are direct products of terpene synthase, while others are formed through modification of the primary terpene skeleton by hydroxylation, acetylation, dehydrogenation, glycosylation and other types of reactions. Belowground produced terpenoids that (potentially) play a role in rhizosphere signaling have been investigated much less than the above-ground terpenoids, due to their more difficult accessibility. However, below we will discuss some examples, hoping to demonstrate and raise the awareness of how diverse the biosynthesis of belowground terpenoids is. Their biosynthesis largely shares the same principles as described above, yet, with some interesting particularities.

11.2.3 BIOSYNTHESIS OF VOLATILE RHIZOSPHERE TERPENOIDS

Just as above ground, volatiles produced belowground play an important role in direct and indirect plant defense against biotic stresses. An example of a direct defense belowground volatile is the monoterpene 1,8-cineole (Steeghs et al. 2004). Cineole is not present in normal or mechanically injured *A. thaliana* roots, but upon pathogen infection is rapidly produced and released. In addition to induced direct defense compounds such as 1,8-cineole, plants also constitutively produce volatiles for direct defense. The semi-volatile diterpene, rhizathalene A, for example is constitutively released from *A. thaliana* roots, and was shown to play a role in the belowground resistance toward root-feeding insects (Vaughan et al. 2013). An example of indirect defense is represented by the sesquiterpene, (*E*)-β-caryophyllene, which is released from maize (*Zea mays*) roots upon feeding by corn root worm larvae *Diabrotica virgifera* and emitted from the leaves in response to attack by lepidopteran larvae like *Spodoptera littoralis*. Belowground the induced (*E*)-β-caryophyllene attracts an entomopathogenic nematode, a natural enemy of the corn rootworm larvae, while above ground a parasitic wasp is attracted (Köllner et al. 2008). The *A. thaliana* genome contains over 32 genes potentially encoding terpene synthases (TPSs), 15 of which are expressed primarily or exclusively in the roots (Vaughan et al. 2013). Interestingly, the formation from GDP of 1,8-cineole, which is only detected in the rhizosphere and not in the roots, was catalyzed by two TPSs (*At3g25820/At3g25830*), which were exclusively expressed in *A. thaliana* roots (Chen et al. 2004). In contrast, (*E*)-β-caryophyllene synthase (TPS23) in *Z. mays* is expressed in both roots and above-ground organs (Köllner et al. 2008).

11.2.4 Volatile-Derived Nonvolatile Rhizosphere Terpenoids

Gossypol and its derivatives, that are produced in the epidermis of cotton roots (*Gossypium hivsutum* L.), are derived from a volatile sesquiterpene. Early experiments with C14 labeled precursors showed that it is produced by cyclization of FDP (Heinstein et al. 1970). This was supported by the discovery of a (+)-δ-cadinene synthase in cotton that catalyzes the first committed step in the gossypol biosynthetic pathway (Chen et al. 1995). Interestingly, a year after, Chen et al. identified a second (+)-δ-cadinene synthase from *Gossypium arboreum* which belongs to a different subfamily (80% identity with the other one) but has the same activity as the first identified one (Chen et al. 1996). In 2018, Tian et al. characterized a number of the other biosynthetic steps in this biosynthetic pathway. Three cytochrome P450s (CYP706B1, *CYP82D113* and CYP71BE79) were demonstrated to catalyze the C8, C7 and C11 oxidation of (+)-δ-cadinene, 7-hydroxy-(+)-δ-cadinene, and 8-hydroxy-7-keto-δ-cadinene, respectively. In addition, Tian et al identified an alcohol dehydrogenase and one 2-oxoglutarate/Fe(II)-dependent dioxygenase which are also involved in gossypol biosynthesis up to the intermediate 3-hydroxy-furocalamen-2-one (Tian et al. 2018). A few steps downstream from this intermediate, a specific (S-adenosyl-L-methionine) methyltransferase methylates the 6-position of desoxyhemigossypol to form desoxyhemigossypol-6-methyl ester also identified in cotton (Liu et al. 1999).

11.2.5 Nonvolatile-Derived Volatile Terpenoids

The unusual acyclic C11 homoterpene (*E*)-4, 8-dimethyl-1,3,7-nonatriene (DMNT) was initially isolated and identified from the essential oil of *Elettaria cardamomurn* (Maurer et al. 1986). But soon DMNT was found in the headspace of many plant species in response to herbivore attack (McCall et al. 1994). DMNT was shown to be an attractant for several insect species, both herbivorous as well as predators and parasites of other insects, among which *Cydia pomonella* (codling moth), *Neoseiulus womersleyi* (predatory mite), *Myllocerinus aurolineatus* (Tea weevil), *Phytoseiulus persimilis* (predatory mite), *Orseolia oryzivora* (African rice gall midge), *Cotesia marginiventris* (parasitoid wasp) and *Microplitis croceipes* (braconid wasp) (Rioja et al. 2016). In cucumber, it was shown that (*E*)-nerolidol synthase catalyzes the conversion of FDP to the sesquiterpene nerolidol, the likely first committed step of DMNT biosynthesis, based on the direct incorporation of deuterium-labeled (*E*)-nerolidol into DMNT and the close correlation between (*E*)-nerolidol synthase activity and DMNT emission after herbivore damage (Bouwmeester et al. 1999). Only 16 years later, exploiting the variation in herbivore-induced volatile formation among 26 maize inbred lines using nested association mapping and genome-wide association analysis, a P450 monooxygenase (CYP92C5) was identified that can convert nerolidol into DMNT (Richter et al. 2016). The cytochrome P450 enzyme encoded by the *A. thaliana* CYP82G1, which is induced in the *Arabidopsis* Inflorescences and leaves upon insect feeding, produces DMNT and its C16-analog (*E, E*)-4,8,12-trimethyl-1,3,7,11-tridecatetraene (TMTT) *in vitro* by the oxidative breakdown of (E)-nerolidol and (E, E)-geranyl linalool, respectively. However, in *A. thaliana* leaves CYP82G1 only functions as a TMTT synthase because of the presence of (*E, E*)-geranyllinalool but not (*E*)-nerolidol in *A. thaliana* leaves (Lee et al. 2010). Intriguingly, the roots of *A. thaliana* also emit DMNT, and emission of DMNT was increased ~seven-fold over constitutive background levels upon inoculation of detached axenically cultivated roots with soil-borne pathogen *Pythium irregulare*, but did not proceed via (*E*)-nerolidol (Lee et al. 2010). Instead, in *A. thaliana* roots DMNT is produced via degradation of a nonvolatile triterpenoid arabidiol (Sohrabi et al. 2015), representing an intriguing example of convergent evolution. The reaction is catalyzed by the Brassicaceae-specific cytochrome P450 monooxygenase CYP705A1, which clusters with the arabidiol synthase in the genome of *A. thaliana* (Sohrabi et al. 2015).

11.2.6 NONVOLATILE TERPENOIDS

Also nonvolatile compounds play important roles in the rhizosphere interaction between plants and their environment. The strigolactones, for example, have received a lot of attention in the past decade or so due to their multiple roles in plant biology. They were initially discovered as root exuded signals that induce the germination of parasitic plants such as *Striga* and *Orobanche* species (Bouwmeester et al. 2003). Under phosphate deficiency, strigolactone secretion is upregulated reportedly as symbiotic signal for arbuscular mycorrhizal fungi, which facilitate the uptake of phosphate by plants (Yoneyama et al. 2007). In 2008 and 2011, strigolatones were reported to also be endogenous signals, plant hormones, that regulate shoot branching and root development, respectively (Gomez-Roldan et al. 2008, Ruyter-Spira et al. 2011). Even though strigolactones have a hormonal function also in the shoot, they are primarily biosynthesized – though there are indications not exclusively – in the roots. Strigolactones occur widespread in the plant kingdom, and their biosynthetic pathway is a perfect example for divergent evolution. All of the so far identified strigolactones derive from the same precursor, all-*trans*-β-carotene and it seems that the initial three enzymatic steps are highly conserved. From all-*trans*-β-carotene three sequential reactions are catalyzed by β-carotene isomerase (D27) and two carotenoid cleavage dioxygenases (CCDs) 7 and 8, resulting in the formation of the already bioactive strigolactone precursor carlactone (Seto and Yamaguchi 2014). However, from carlactone over 30 different strigolactones are produced with individual plant species producing specific blends of up to about 8 different strigolactones, such as 4-deoxyorobanchol, orobanchol, 5-deoxystrigol, strigol, zealactone, and heliolactone (Wang and Bouwmeester 2018). The enzymes involved in the conversion of carlactone to all these different strigolactones are still largely unknown. A class-III cytochrome P450 monooxygenase (MAX1) was shown to catalyze the conversion of carlactone to carlactonoic acid in *A. thaliana* (Abe et al. 2014), which is further converted to methyl carlactonate by an unknown methyl transferase, and to an as yet unidentified oxidized product by LATERAL BRANCHING OXIDOREDUCTASE (LBO) (Brewer et al. 2016). In rice, two MAX1 homologs are responsible for the formation of 4-deoxyorobanchol from carlactone and orobanchol from 4-deoxyorobanchol (Zhang et al. 2014). Identification of the other enzymes involved in strigolactone structural diversification should help us to better understand the biological and evolutionary relevance of the strigolactone structural diversity in plants.

11.3 PHENOLIC RHIZOSPHERE SIGNALS

11.3.1 FUNCTION OF PHENOLIC COMPOUNDS AS SIGNALING MOLECULES

Phenolics are metabolites that have an aromatic ring (or rings) decorated with one or more hydroxyl groups. As antioxidants, pigments, auxin transport regulators, defense and signaling compounds, phenolic compounds in general are being recognized for their profound impact on plant growth, development, reproduction, UV protection, and defense (Croteau et al. 2000). An increasing number of phenolic compounds are recognized for their signaling role in the rhizosphere. In *A. thaliana*, phenolic-related compounds were shown to be positively correlated with a higher number of unique rhizosphere microbiome species compared with other groups of compounds (i.e., sugars, sugar alcohols, and amino acids). For instance, salicylic acid levels in the rhizosphere positively correlated with the presence of microbial species of the Corynebacterineae, Pseudonocardineae and Streptomycineae. This suggests that salicylic acid acts as specific substrate or signaling molecule for certain microbial species in the soil (Badri et al. 2013). The phenolic compound rosmarinic acid (RA), a caffeic acid ester widely present in the plant kingdom, presumably works as a defense compound. Upon pathogen attack RA secreted by *Ocimum basilicum* roots as part of the root exudates, was shown to be highly inhibitory against an array of rhizosphere microorganisms (Bais et al. 2002). Recently, however, a new role was discovered for RA. Bacteria monitor their own cell density using quorum sensing (QS) molecules, such as the homoserine lactones (Parsek et al. 1999). RA was demonstrated to be a homoserine-lactone mimic and can evoke several QS regulator controlled phenotypes like virulence factor biosynthesis or biofilm formation (Corral-Lugo et al. 2016).

Phenolics also play a crucial role in the interaction of legumes with symbiotic nitrogen-fixing bacteria or rhizobia. Nodules are root organelles that are developed through signal exchange between the plant roots and the rhizobia to facilitate nitrogen fixation and nodulation gene (*nod*) expression in rhizobium was shown to be essential for this process. Depending on their structure, the phenolic compounds can induce or suppress *nod* gene expression. Two flavonoids that acted as *nod* gene expression inducers were isolated from *Medicago sativa* (luteolin) (Peters et al. 1986) and *Trifolium repens* (7,4'-dihydroxyflavone) (Redmond et al. 1986). In contrast, the isoflavonoids medicarpin and coumestrol, isolated from *M. sativa* have been shown to negatively regulate *nod* gene expression in *Sinorhizobium meliloti* (Zuanazzi et al. 1998).

Just as for their role in plant-microbe interaction, phenolic compounds also play dual roles in plant–plant interaction. Parasitic weeds of the *Striga* genus, such as *Striga hermonthica* and *Striga asiatica*, constitute one of the major problems in African agriculture with yield losses up to 100% in large parts of sub-Saharan Africa (Gressel et al. 2004). The use of the legume *Desmodium uncinatum* as a "push–pull" intercrop was proposed for smallholder farmers to control *Striga* infection in maize. The strong suppressing effect of Desmodium on *Striga* infection seems to be caused by the presence of isoflavonoids in the *Desmodium* root exudate, which were demonstrated to prevent attachment of *Striga* to its host (Hooper et al. 2010). However, on the other hand, phenolic compounds, such as 2,6-dimethoxy-p-benzoquinone (DMBQ), isolated from sorghum roots were shown to induce haustorium formation in *Striga*, a process required for attachment to and penetration of its host (Estabrook and Yoder 1998).

11.3.2 Biosynthesis and Diversification of Phenolic Compounds

There are about 8000 naturally occurring plant phenolics, and phenolic compounds represent a large proportion of the metabolites present in root exudates (Baxter et al. 1998). Several classes of phenolics have been categorized according to their basic skeletons: C6 (simple phenols, benzoquinones), C6–C1 (phenolic acids and aldehydes), C6–C2 (acetophenones, phenylacetic acids), C6–C3 (hydroxycinnamic acids, coumarins, phenylpropanes, chromones, monolignols), C6–C4 (naphthoquinones), C6–C1–C6 (xanthones), C6–C2–C6 (stilbenes, anthraquinones), C6–C3–C6 (flavonoids, isoflavonoids, anthocyanins), (C6–C3–C6)2,3 (bi-, triflavonoids, proanthocyanidin dimers, trimers), (C6–C3)2 (lignans, neolignans), (C6–C3)n (lignins), (C6)n (catechol melanins, phlorotannins) and (C6–C3–C6)n (condensed tannins) (Cheynier et al. 2013). Phenolics are derived from the shikimate pathway, beginning with an aldol-type condensation of phosphoenolpyruvic acid (PEP) from the glycolysis pathway, and D-erythrose-4-phosphate, from the pentose phosphate cycle, to produce 3-deoxy-D-arabino-heptulosonic acid 7-phosphate (DAHP) (Figure 11.1). Then six more enzymatic steps result in the formation of the key branch-point compound, chorismic acid, the final product of the shikimate pathway and also the common precursor for three aromatic amino acids, tryptophan, phenylalanine, and tyrosine. Of these three amino acids, phenylalanine and tyrosine can both serve as precursor for the biosynthesis of phenolic acids, tannins, flavonoids, and isoflavonoids. Interestingly, in most vascular plants, phenylalanine is the preferred substrate for this, but the monocot enzymes can utilize both phenylalanine and tyrosine (Croteau et al. 2000). The core

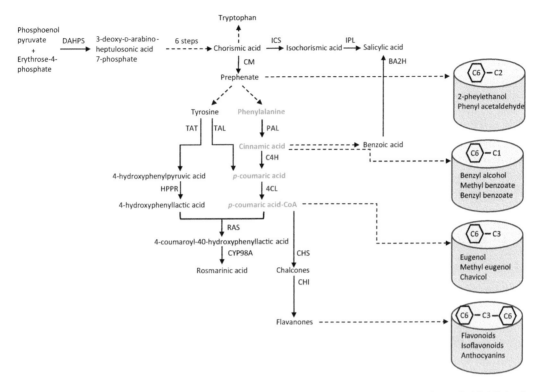

FIGURE 11.1 Biosynthesis of phenolic compounds. The core phenylpropanoid pathway is highlighted with color. DAHP synthase, 3-deoxy-D-arabino-heptulosonate 7-phosphate synthase; ICS, isochorismate synthase; IPL, isochorismate pyruvate lyase; BA2H, benzoic acid 2-hydroxylase; PAL, phenylalanine ammonialyase; C4H, cinnamate 4-hydroxylase; 4CL, 4-coumaroyl CoA-ligase; TAT, tyrosine aminotransferase; HPPR, hydroxyphenylpyruvate reductase; RAS, 4-coumaroyl CoA:40-hydroxyphenyllactic acid 4-coumaroyltransferase; CYP98A, cytochrome P450-dependent monooxygenase 98A family; CHI, chalcone isomerase; CHS, chalcone synthase.

pathway of phenolic biosynthesis from phenylalanine and tyrosine starts with the removal of the amino group by phenylalanine ammonia lyase (PAL) to form cinnamic acid, while the amino group of tyrosine can be removed by two different enzymes, tyrosine aminotransferase (or tyrosine transaminase TAT) that produces 4-hydroxyphenylpyruvic acid or tyrosine ammonia lyase that produces *p*-coumaric acid (Figure 11.1). The further transformation of phenylalanine derived cinnamic acid is catalyzed by the enzyme of the general phenylpropanoid pathway, cinnamic acid 4-hydroxylase (C4H) to form *p*-coumaric acid. In the tyrosine derived pathway, 4-hydroxyphenylpyruvic acid is further reduced to 4-hydroxyphenyllactic acid by hydroxyphenylpyruvate reductase. 4-Coumaric acid CoA-ligase (4CL) will use both phenylalanine and tyrosine derived *p*-coumaric acid and form *p*-coumaroyl-CoA (Figure 11.1).

While most of these aromatic compounds are usually nonvolatile, the volatile subset is represented by benzenoid (C6–C1), phenylpropanoid (C6–C3) and phenylpropanoid-related compounds (C6–C2). Generally, C6-C1 compounds such as benzyl alcohol, methyl benzoate, and benzyl benzoate are formed from cinnamic acid as a precursor, C6-C2 compounds such as 2-phenylethanol and phenyl acetaldehyde from phenylalanine, and C6-C3 compounds such as eugenol, methyl eugenol and chavicol from 4-coumaroyl-CoA (Dudareva et al. 2013).

All other phenolic compounds are produced either from phenylalanine or from tyrosine from the intermediary precursors from the core pathway (Figure 11.1). For example, coumarins derive from cinnamic acid, and ubiquinons, lignins, flavonoids, and anthocyanins derive from *p*-coumaric acid. *p*-Hydroxyphenylpyruvic acid is needed for the biosynthesis of tocopherols and plastoqui-nones. There are also some phenolic compounds derived from both pathways, for example, the above mentioned rosmarinic acid. *p*-Coumaroyl-CoA and *p*-hydroxyphenyllactic acid are coupled by ester formation and with the release of coenzyme A, 4-coumaroyl-4′-hydroxyphenyllactic acid is formed. The reaction is catalyzed by "rosmarinic acid synthase" (RAS; 4-coumaroylCoA:40-hydroxyphenyllactic acid 4-coumaroyltransferase). The 3- and 30-hydroxyl groups are finally intro-duced by cytochrome P450-dependent monooxygenase reactions (CYP98A) (Petersen et al. 2009) (Figure 11.1). Some phenolic compounds have several biosynthetic routes, for example, salicylic acid. Stressed *A. thaliana* synthesizes salicylic acid primarily via an isochorismate-utilizing path-way derived from chorismic acid. A distinct pathway utilizing phenylalanine derived cinnamic acid as the substrate also may contribute to salicylic acid production, although to a much lesser extent (D'Maris Amick Dempsey et al. 2011) (Figure 11.1).

11.4 REGULATION OF SIGNAL MOLECULE PRODUCTION

11.4.1 Biotic Factors

11.4.1.1 Herbivores

Plants under attack by arthropod herbivores emit volatiles from their leaves that attract natural ene-mies of the herbivores. The sesquiterpene (*E*)-β-caryophyllene was the first identified insect-induced belowground plant signal, induced upon feeding by larvae of the western corn rootworm and is highly attractive to an entomopathogenic nematode (Rasmann et al. 2005). Likewise, monoterpenes including α-pinene, camphene, β-pinene, *p*-cymene and 1,8-cineole were shown to be released from roots of *Populus trichocarpa* and *Populus nigra* after cockchafer larvae-damage and camphor was released from the roots of apple after *melolontha* larvae attack. The expression profiles of terpene synthases *PtTPS1*, *PtTPS4*, *PtTPS14*, *PtTPS16* and *PnTPS21* responsible for the production of these monoterpenes in poplar all showed significant upregulation upon root herbivory, suggesting that monoterpene emission from roots is mainly determined transcriptionally (Lackus et al. 2018).

11.4.1.2 Pathogens

Root released signaling molecules are also induced by pathogen infection. Infection of cotton hypo-cotyls by *Rhizoctonia solani* increased the concentration of gossypol like terpenoids in the root exudate (Hunter et al. 1978). Formation of the volatile homoterpene DMNT was transiently induced in a jasmonate-dependent manner upon infection by the root-rot pathogen *P. irregulare*, and two biosynthetic genes (arabidiol synthase and CYP705A1) were upregulated in the roots (Sohrabi et al. 2015). Although DMNT is especially known for its role in the attraction of natural enemies in tri-trophic interactions, in this case DMNT played a role in *Arabidopsis* resistance against *P. irregulare*.

11.4.1.3 *Arbuscular Mycorrhizal* Fungi

Both the quantity and composition of root exudates were altered by the presence of *arbuscular mycorrhizal* (AM) fungi, and these changes were plant species specific. Lendzemo et al. demon-strate that in sorghum (*Sorghum bicolor*) AM colonization results in a reduction of Striga infection, possibly through the down regulation of strigolactone production (Lendzemo et al. 2007). This is possibly supported by the fact that plants colonized by AM fungi seem to negatively regulate further mycorrhization via their root exudates (Pinior et al. 1999). The authors showed that root exudates from nonmycorrhizal cucumber plants stimulated hyphal growth, whereas root exudates from AM fungi (*Glomus intraradices* or *Glomus mosseae*) colonized cucumber plants showed no stimulation of the hyphal growth of *Gigaspora rosea* (Pinior et al. 1999). In addition to down regulation of the secretion of rhizosphere signals, mycorrhizal sorghum plant roots exuded more alcohols, alkenes, ethers, and acids but fewer linear-alkanes (Sun and Tang 2013).

11.4.1.4 Plant Growth Promoting Rhizobacteria

Plant root exudates can be influenced by volatile compounds produced by plant growth promoting rhizobacteria (PGPR). It was shown that root exudates of sorghum exposed to volatile compounds from different PGPR strains differ in terms of types, numbers, and concentrations of compounds (Hernández-Calderón et al. 2018). For example, exudates produced by plants exposed to volatile compounds from *Arthrobacter agilis* were more diverse and accumulated in higher concentrations than those of plants exposed to other bacterial strains such as *Bacillus methylotrophicus* and *Sinorhizobium meliloti* (Hernández-Calderón et al. 2018). In noninoculated and *A. agilis* inoculated plants, the root exudates were also shown to be different with one compound only found in the noninoculated exudates whereas 6 compounds were only found in exudates of plants treated with the *A. agilis* (Hernández-Calderón et al. 2018). The authors speculate that of these six compounds, citric acid is likely used as carbon source by the bacteria, ferulic acid acts as a chemo-attractant signal compound for beneficial rhizospheric bacteria, while three fatty acids (nonadecanoic, eicosanoic, and tetracosanoic acids) may possibly affect plant growth.

11.4.2 Abiotic Factors

11.4.2.1 Nutrient Deficiency

The composition of root exudates is also altered in response to abiotic signals in the rhizosphere. Phenolics, for example, are especially responsive to phosphate and nitrogen deficiency (Coronado et al. 1995, Malusà et al. 2006). Phenolic compounds are known to stimulate AM fungi hyphal growth that facilitate phosphate uptake (Abdel-Lateif, Bogusz, and Hocher 2012). Under phosphate starvation, total soluble phenolic content of bean (*Phaseolus vulgaris* L.) root exudates increased whereas the content of phenolic compounds in the root decreased (Malusà et al. 2006). The increase of phenolic exudates might be explained by an increased gene expression of *L-phenylalanine ammonia lyase* (Malusà et al. 2006). As described above, phenolic compounds secreted by legume roots are chemoattractants and nod gene inducers for the symbiotic Rhizobia. Under nitrogen deficiency, both expression of the flavonoid biosynthesis genes, *chalcone synthase* and *isoflavone reductase*, and exudation of flavonoids and isoflavonoids increased in alfalfa (*Medicago sativa* L.) roots (Coronado et al. 1995).

Terpenoid derived strigolactones are another type of signaling molecules in the rhizosphere that are induced by phosphate and nitrogen deficiency. Phosphate deficiency induces strigolactone biosynthesis in tomato, and strongly increased *Phelipanche ramosa* seed germination and hyphal branching of AM fungi compared with control plants grown with normal phosphate supply (López-Ráez et al. 2008). Mutants of striolactone signaling (*max2*) or biosynthesis mutants (*max4*) in *A. thaliana* showed reduced response to low Pi conditions, such as a lower root hair density, lower level of expression for starvation-induced genes, and higher shoot branching than the wild type suggesting that strigolactones play a role in the plant response to low phosphate conditions (Ruyter-Spira et al. 2013). Similarly, low levels of either phosphate or nitrogen stimulated strigolactone biosynthesis in rice roots (Sun et al. 2014). Strigolactone deficient mutants (d10 and d27) and strigolactone signaling mutant (d3) displayed a lost sensitivity of the root response to phosphate and nitrogen deficiency. For example, seminal root length was increased by 50% and 29% in WT plants but by only approximately 18% and 14% in the three mutants when grown under low phosphate and low nitrogen solution, respectively.

11.4.2.2 Other Abiotic Signals

The composition of root exudates is also affected by other abiotic signals in the rhizosphere. There is evidence that flavonoids play a role in resistance to aluminum toxicity and silicon induced amelioration of aluminum toxicity in maize (Kidd et al. 2001). Roots of maize plants that were exposed to aluminum and silicon exuded high levels of phenolics compounds such as catechol, catechin, and quercetin, and an aluminum-resistant variety exuded a 15-fold higher level of phenolics when pretreated with silicon than when no such pre-treatment was applied (Kidd et al. 2001). These results might be due to the metal-binding activity of many flavonoids rather than that they have a signaling role. Other abiotic factors like UV light, CO_2, and O_3 were all shown to affect root exudate composition (Formánek et al. 2014), but there is so far no evidence linking this to a signaling role of the exuded compounds.

11.5   CONCLUSIONS AND FUTURE PERSPECTIVE

In this book chapter, we described the current knowledge on the biosynthesis of a number of important classes of signaling molecules exuded by plants from their roots into the rhizosphere. Compared with above-ground signaling molecules, common and root specific biosynthetic routes were discussed and the plasticity and regulation of the biosynthesis of these signaling molecules reviewed. The biosynthesis and exudation of belowground signaling molecules are mostly regulated by abiotic and biotic factors in the soil but there is evidence that also above-ground factors can influence belowground signaling.

We are just beginning to explore the importance of the vast repertoire of belowground signaling molecules. Structure elucidation of the unknown signaling molecules of low abundance in the rhizosphere is a first challenge. Establishment of the link between signaling molecules and single or multiple organisms from the enormous diversity of soil-dwelling organisms is the next. Using the recently emerging metabolomics and metagenomics tools and systems biology approaches to unravel the relationship between signaling molecules and soil biota from our biological "big data" will likely advance our knowledge rapidly. Verifying postulated signaling relationships will rely on metabolic engineering approaches. CRISPR/Cas9 as a new approach for metabolic engineering is evolving rapidly from only gene knock-out applications to knock-in and precise gene editing with targeted nucleotide modification. In addition, metabolic engineering approaches such as virus-induced-gene-silencing and transgenic hairy roots are greatly shortening the process to discover new signaling relationships. In addition to metabolic engineering, modern breeding technique such as Targeting Induced Local Lesions IN Genomes (TILLING) and eco-TILLING to find EMS-induced or natural mutations, respectively, in genes of interest can be used for gene discovery and to study the effect of certain signaling molecules on their environment.

A better understanding of the molecular mechanisms underlying biosynthesis and regulation of belowground signals and their biological function together with advanced metabolic engineering/breeding approaches will provide the basis for a better exploitation of the hidden, belowground, part of plants and the creation of a more sustainable agriculture with less inputs of fertilizers and pesticides.

ACKNOWLEDGMENTS

The preparation of this book chapter was supported by the ERC (Advanced grant CHEMCOMRHIZO, 670211 to HJB) and the EU (Marie Curie grant NemHatch, 793795 to LD).

REFERENCES

Abdel-Lateif, K., D. Bogusz, and V. Hocher. 2012. The role of flavonoids in the establishment of plant roots endosymbioses with arbuscular mycorrhiza fungi, rhizobia and Frankia bacteria. *Plant Signaling & Behavior* 7 (6):636–641.

Abe, S., A. Sado, K. Tanaka, T. Kisugi, K. Asami, S. Ota, H. I. Kim, K. Yoneyama, X. Xie, and T. Ohnishi. 2014. Carlactone is converted to carlactonoic acid by MAX1 in Arabidopsis and its methyl ester can directly interact with AtD14 *in vitro*. *Proceedings of the National Academy of Sciences* 111 (50):18084–18089.

Badri, D.V., J.M. Chaparro, R. Zhang, Q. Shen, and J. M. Vivanco. 2013. Application of natural blends of phytochemicals derived from the root exudates of *Arabidopsis* to the soil reveal that phenolic-related compounds predominantly modulate the soil microbiome. *Journal of Biological Chemistry* 288 (7):4502–4512.

Bais, H.P., T.S. Walker, H.P. Schweizer, and J.M. Vivanco. 2002. Root specific elicitation and antimicrobial activity of rosmarinic acid in hairy root cultures of *Ocimum basilicum*. *Plant Physiology and Biochemistry* 40 (11):983–995.

Baxter, H., J.B. Harborne, and G. P. Moss. 1998. *Phytochemical Dictionary: A handbook of Bioactive Compounds from Plants*. CRC Press, Boca Raton, FL.

Bell, A.A., R.D. Stipanovic, C.R. Howell, and P.A. Fryxell. 1975. Antimicrobial terpenoids of *Gossypium*: Hemigossypol, 6-methoxyhemigossypol and 6-deoxyhemigossypol. *Phytochemistry* 14 (1):225–231.

Bouwmeester, H.J., F.W. Verstappen, M.A. Posthumus, and M. Dicke. 1999. Spider mite-induced (3*S*)-(*E*)-nerolidol synthase activity in cucumber and lima bean. The first dedicated step in acyclic C11-homoterpene biosynthesis. *Plant Physiology* 121 (1):173–180.

Bouwmeester, H.J., R. Matusova, S. Zhongkui, and M.H. Beale. 2003. Secondary metabolite signalling in host–parasitic plant interactions. *Current Opinion in Plant Biology* 6 (4):358–364.

Brewer, P.B., K.Yoneyama, F. Filardo, E. Meyers, A. Scaffidi, T. Frickey, K. Akiyama, Y. Seto, E.A. Dun, and J.E. Cremer. 2016. Lateral branching oxidoreductase acts in the final stages of strigolactone biosynthesis in *Arabidopsis*. *Proceedings of the National Academy of Sciences* 113 (22):6301–6306.

Chen, F., D.-K. Ro, J. Petri, J. Gershenzon, J. Bohlmann, E. Pichersky, and D. Tholl. 2004. Characterization of a root-specific *Arabidopsis* terpene synthase responsible for the formation of the volatile monoterpene 1, 8-cineole. *Plant Physiology* 135 (4):1956–1966.

Chen, X.-Y., M. Wang, Y. Chen, V. J. Davisson, and P. Heinstein. 1996. Cloning and heterologous expression of a second (+)-δ-cadinene synthase from *Gossypium arboreum*. *Journal of Natural Products* 59 (10):944–951.

Chen, X.-Y., Y. Chen, P. Heinstein, and V. J. Davisson. 1995. Cloning, expression, and characterization of (+)-δ-cadinene synthase: A catalyst for cotton phytoalexin biosynthesis. *Archives of Biochemistry and Biophysics* 324 (2):255–266.

Cheynier, V., G. Comte, K. M. Davies, V. Lattanzio, and S. Martens. 2013. Plant phenolics: Recent advances on their biosynthesis, genetics, and ecophysiology. *Plant Physiology and Biochemistry* 72:1–20.

Coronado, C., J.S. Zuanazzi, C. Sallaud, J.-C. Quirion, R. Esnault, H.-P. Husson, A. Kondorosi, and P. Ratet. 1995. Alfalfa root flavonoid production is nitrogen regulated. *Plant Physiology* 108 (2):533–542.

Corral-Lugo, A., A. Daddaoua, A. Ortega, M. Espinosa-Urgel, and T. Krell. 2016. Rosmarinic acid is a homoserine lactone mimic produced by plants that activates a bacterial quorum-sensing regulator. *Science Signaling* 9 (409):ra1–ra1.

Crock, J., M. Wildung, and R. Croteau. 1997. Isolation and bacterial expression of a sesquiterpene synthase cDNA clone from peppermint (*Mentha* x *piperita*, L.) that produces the aphid alarm pheromone (*E*)-β-farnesene. *Proceedings of the National Academy of Sciences* 94 (24):12833–12838.

Croteau, R., T.M. Kutchan, and N.G. Lewis. 2000. Natural products (secondary metabolites). *Biochemistry and Molecular Biology of Plants* 24:1250–1319.

D'Maris Amick Dempsey, A.C., M.C.W. Vlot, and F.K. Daniel. 2011. Salicylic acid biosynthesis and metabolism. *The Arabidopsis Book/American Society of Plant Biologists* 9:e0156.

Dong, L., E. Jongedijk, H. Bouwmeester, and A. Van Der Krol. 2016. Monoterpene biosynthesis potential of plant subcellular compartments. *New Phytologist* 209 (2):679–690.

Dudareva, N., A. Klempien, J.K. Muhlemann, and I. Kaplan. 2013. Biosynthesis, function and metabolic engineering of plant volatile organic compounds. *New Phytologist* 198 (1):16–32.

Estabrook, E.M., and J.I. Yoder. 1998. Plant-plant communications: Rhizosphere signaling between parasitic angiosperms and their hosts. *Plant Physiology* 116 (1):1–7.

Falcone Ferreyra, M.L., S. Rius, and P. Casati. 2012. Flavonoids: Biosynthesis, biological functions, and biotechnological applications. *Frontiers in Plant Science* 3:222.

Formánek, P., K. Rejšek, and V. Vranová. 2014. Effect of elevated CO_2, O_3, and UV radiation on soils. *The Scientific World Journal* 2014:730149.

Gomez-Roldan, V., S. Fermas, P.B. Brewer, V. Puech-Pagès, E.A. Dun, J.-P. Pillot, F. Letisse, R. Matusova, S. Danoun, and J.-C. Portais. 2008. Strigolactone inhibition of shoot branching. *Nature* 455 (7210):189.

Gressel, J., A. Hanafi, G. Head, W. Marasas, A.B. Obilana, J. Ochanda, T. Souissi, and G. Tzotzos. 2004. Major heretofore intractable biotic constraints to African food security that may be amenable to novel biotechnological solutions. *Crop Protection* 23 (8):661–689.

Guirimand, G., A. Guihur, M.A. Phillips, A. Oudin, G. Glévarec, C. Melin, N. Papon, M. Clastre, B. St-Pierre, and M. Rodríguez-Concepción. 2012. A single gene encodes isopentenyl diphosphate isomerase isoforms targeted to plastids, mitochondria and peroxisomes in *Catharanthus roseus*. *Plant Molecular Biology* 79 (4–5):443–459.

Hawes, M.C., U. Gunawardena, S. Miyasaka, and X. Zhao. 2000. The role of root border cells in plant defense. *Trends in Plant Science* 5 (3):128–133.

Heinstein, P., D. Herman, S. Tove, and F. Smith. 1970. Biosynthesis of gossypol incorporation of mevalonate-2-^{14}C and isoprenyl pyrophosphates. *Journal of Biological Chemistry* 245 (18):4658–4665.

Hernández-Calderón, E., M.E. Aviles-Garcia, D.Y. Castulo-Rubio, L. Macías-Rodríguez, V.M. Ramírez, G. Santoyo, J. López-Bucio, and E. Valencia-Cantero. 2018. Volatile compounds from beneficial or pathogenic bacteria differentially regulate root exudation, transcription of iron transporters, and defense signaling pathways in *Sorghum bicolor*. *Plant Molecular Biology* 96 (3):291–304.

Hinsinger, P. 2001. Bioavailability of soil inorganic P in the rhizosphere as affected by root-induced chemical changes: A review. *Plant and Soil* 237 (2):173–195.

Hooper, A. M., M.K. Tsanuo, K. Chamberlain, K. Tittcomb, J. Scholes, A. Hassanali, Z. R. Khan, and J. A. Pickett. 2010. Isoschaftoside, a *C*-glycosylflavonoid from *Desmodium uncinatum* root exudate, is an allelochemical against the development of *Striga*. *Phytochemistry* 71 (8–9):904–908.

Hunter, R., J. Halloin, J. Veech, and W. Carter. 1978. Terpenoid accumulation in hypocotyls of cotton seedlings during aging and after infection by *Rhizoctonia solani*. *Phytopathology* 68:347–350.

Jones, D.L., C. Nguyen, and R.D. Finlay. 2009. Carbon flow in the rhizosphere: Carbon trading at the soil–root interface. *Plant and Soil* 321 (1–2):5–33.

Kidd, P., M. Llugany, C. Poschenrieder, B. Gunse, and J. Barcelo. 2001. The role of root exudates in aluminium resistance and silicon – Induced amelioration of aluminium toxicity in three varieties of maize (*Zea mays* L.). *Journal of Experimental Botany* 52 (359):1339–1352.

Köllner, T.G., M. Held, C. Lenk, I. Hiltpold, T.C. Turlings, J. Gershenzon, and J. Degenhardt. 2008. A maize (*E*)-β-caryophyllene synthase implicated in indirect defense responses against herbivores is not expressed in most American maize varieties. *The Plant Cell* 20 (2):482–494.

Lackus, N.D., S. Lackner, J. Gershenzon, S.B. Unsicker, and T.G. Köllner. 2018. The occurrence and formation of monoterpenes in herbivore-damaged poplar roots. *Scientific Reports* 8 (1):17936.

Lee, S., S. Badieyan, D.R. Bevan, M. Herde, C. Gatz, and D. Tholl. 2010. Herbivore-induced and floral homoterpene volatiles are biosynthesized by a single P450 enzyme (CYP82G1) in *Arabidopsis*. *Proceedings of the National Academy of Sciences* 107 (49):21205–21210.

Lendzemo, V.W., T.W. Kuyper, R. Matusova, H.J. Bouwmeester, and A.V. Ast. 2007. Colonization by arbuscular mycorrhizal fungi of sorghum leads to reduced germination and subsequent attachment and emergence of *Striga hermonthica*. *Plant Signaling & Behavior* 2 (1):58–62.

Liu, J., C.R. Benedict, R.D. Stipanovic, and A.A. Bell. 1999. Purification and characterization of S-adenosyl-l-methionine: Desoxyhemigossypol-6-*O*-methyltransferase from cotton plants. An enzyme capable of methylating the defense terpenoids of cotton. *Plant Physiology* 121 (3):1017–1024.

López–Ráez, J.A., T. Charnikhova, V. Gómez-Roldán, R. Matusova, W. Kohlen, R. De Vos, F. Verstappen, V. Puech-Pages, G. Bécard, and P. Mulder. 2008. Tomato strigolactones are derived from carotenoids and their biosynthesis is promoted by phosphate starvation. *New Phytologist* 178 (4):863–874.

Malusà, E., M.A. Russo, C. Mozzetti, and A. Belligno. 2006. Modification of secondary metabolism and flavonoid biosynthesis under phosphate deficiency in bean roots. *Journal of Plant Nutrition* 29 (2):245–258.

Maurer, B., A. Hauser, and J.-C. Froidevaux. 1986. (*E*)-4, 8-dimethyl-1, 3, 7-nonatriene and (*E, E*)-4, 8, 12-trimethyl-1, 3, 7, 11-tridecatetraene, two unusual hydrocarbons from cardamom oil. *Tetrahedron Letters* 27 (19):2111–2112.

McCall, P.J., T.C. Turlings, J. Loughrin, A.T. Proveaux, and J.H. Tumlinson. 1994. Herbivore-induced volatile emissions from cotton (*Gossypium hirsutum* L.) seedlings. *Journal of Chemical Ecology* 20 (12):3039–3050.

McGarvey, D.J., and R. Croteau. 1995. Terpenoid metabolism. *The Plant Cell* 7 (7):1015.

Nishida, N., S. Tamotsu, N. Nagata, C. Saito, and A. Sakai. 2005. Allelopathic effects of volatile monoterpenoids produced by *Salvia leucophylla*: Inhibition of cell proliferation and DNA synthesis in the root apical meristem of *Brassica campestris* seedlings. *Journal of Chemical Ecology* 31 (5):1187–1203.

Parsek, M.R., D.L. Val, B.L. Hanzelka, J.E. Cronan, and E. Greenberg. 1999. Acyl homoserine-lactone quorum-sensing signal generation. *Proceedings of the National Academy of Sciences* 96 (8):4360–4365.

Peters, N.K., J.W. Frost, and S.R. Long. 1986. A plant flavone, luteolin, induces expression of *Rhizobium meliloti* nodulation genes. *Science* 233 (4767):977–980.

Petersen, M., Y. Abdullah, J. Benner, D. Eberle, K. Gehlen, S. Hücherig, V. Janiak, K. H. Kim, M. Sander, and C. Weitzel. 2009. Evolution of rosmarinic acid biosynthesis. *Phytochemistry* 70 (15–16):1663–1679.

Pinior, A., U. Wyss, Y. Piché, and H. Vierheilig. 1999. Plants colonized by AM fungi regulate further root colonization by AM fungi through altered root exudation. *Canadian Journal of Botany* 77 (6):891–897.

Rasmann, S., T.G. Köllner, J. Degenhardt, I. Hiltpold, S. Toepfer, U. Kuhlmann, J. Gershenzon, and T.C. Turlings. 2005. Recruitment of entomopathogenic nematodes by insect-damaged maize roots. *Nature* 434 (7034):732.

Redmond, J.W., M. Batley, M.A. Djordjevic, R.W. Innes, P.L. Kuempel, and B.G. Rolfe. 1986. Flavones induce expression of nodulation genes in *Rhizobium*. *Nature* 323 (6089):632.

Richter, A., C. Schaff, Z. Zhang, A.E. Lipka, F. Tian, T. G. Köllner, C. Schnee, S. Preiß, S. Irmisch, and G. Jander. 2016. Characterization of biosynthetic pathways for the production of the volatile homoterpenes DMNT and TMTT in *Zea mays*. *The Plant Cell* 28 (10):2651–2665.

Rioja, T., R. Ceballos, L. Holuigue, and R. Vargas. 2016. Different population densities and continuous feeding by *Oligonychus yothersi* (McGregor)(Acari: Tetranychidae) affect the emissions of herbivore-induced plant volatiles on avocado (*Persea americana* Mill. cv. Hass) shoots under semi-field conditions. *International Journal of Acarology* 42 (6):310–318.

Ruyter-Spira, C., S. Al-Babili, S. Van Der Krol, and H. Bouwmeester. 2013. The biology of strigolactones. *Trends in Plant Science* 18 (2):72–83.

Ruyter-Spira, C., W. Kohlen, T. Charnikhova, A. van Zeijl, L. van Bezouwen, N. de Ruijter, C. Cardoso, J. A. Lopez-Raez, R. Matusova, and R. Bours. 2011. Physiological effects of the synthetic strigolactone analog GR24 on root system architecture in *Arabidopsis*: Another belowground role for strigolactones? *Plant Physiology* 155 (2):721–734.

Seto, Y., and S. Yamaguchi. 2014. Strigolactone biosynthesis and perception. *Current Opinion in Plant Biology* 21:1–6.

Sohrabi, R., J.-H. Huh, S. Badieyan, L. H. Rakotondraibe, D. J. Kliebenstein, P. Sobrado, and D. Tholl. 2015. In planta variation of volatile biosynthesis: An alternative biosynthetic route to the formation of the pathogen-induced volatile homoterpene DMNT via triterpene degradation in *Arabidopsis* roots. *The Plant Cell* 27 (3):874–890.

Steeghs, M., H.P. Bais, J. de Gouw, P. Goldan, W. Kuster, M. Northway, R. Fall, and J. M. Vivanco. 2004. Proton-transfer-reaction mass spectrometry as a new tool for real time analysis of root-secreted volatile organic compounds in *Arabidopsis*. *Plant Physiology* 135 (1):47–58.

Sun, H., J. Tao, S. Liu, S. Huang, S. Chen, X. Xie, K. Yoneyama, Y. Zhang, and G. Xu. 2014. Strigolactones are involved in phosphate-and nitrate-deficiency-induced root development and auxin transport in rice. *Journal of Experimental Botany* 65 (22):6735–6746.

Sun, X.-G., and M. Tang. 2013. Effect of arbuscular mycorrhizal fungi inoculation on root traits and root volatile organic compound emissions of *Sorghum bicolor*. *South African Journal of Botany* 88:373–379.

Tian, X., J.-X. Ruan, J.-Q. Huang, C.-Q. Yang, X. Fang, Z.-W. Chen, H. Hong, L.-J. Wang, Y.-B. Mao, and S. Lu. 2018. Characterization of gossypol biosynthetic pathway. *Proceedings of the National Academy of Sciences* 115 (23):E5410–E5418.

van Dam, N. M., and H. J. Bouwmeester. 2016. Metabolomics in the rhizosphere: Tapping into belowground chemical communication. *Trends in Plant Science* 21 (3):256–265.

Vaughan, M. M., Q. Wang, F. X. Webster, D. Kiemle, Y. J. Hong, D. J. Tantillo, R. M. Coates, A. T. Wray, W. Askew, and C. O'Donnell. 2013. Formation of the unusual semivolatile diterpene rhizathalene by the Arabidopsis class I terpene synthase TPS08 in the root stele is involved in defense against belowground herbivory. *The Plant Cell* 25 (3):1108–1125.

Walker, T. S., H.P. Bais, E. Grotewold, and J. M. Vivanco. 2003. Root exudation and rhizosphere biology. *Plant Physiology* 132 (1):44–51.

Wang, Y., and H. J. Bouwmeester. 2018. Structural diversity in the strigolactones. *Journal of Experimental Botany* 69 (9):2219–2230.

Xu, M., R. Galhano, P. Wiemann, E. Bueno, M. Tiernan, W. Wu, I. M. Chung, J. Gershenzon, B. Tudzynski, and A. Sesma. 2012. Genetic evidence for natural product-mediated plant–plant allelopathy in rice (*Oryza sativa*). *New Phytologist* 193 (3):570–575.

Yoneyama, K., X. Xie, D. Kusumoto, H. Sekimoto, Y. Sugimoto, Y. Takeuchi, and K. Yoneyama. 2007. Nitrogen deficiency as well as phosphorus deficiency in sorghum promotes the production and exudation of 5-deoxystrigol, the host recognition signal for arbuscular mycorrhizal fungi and root parasites. *Planta* 227 (1):125–132.

Zhang, Y., A.D. Van Dijk, A. Scaffidi, G.R. Flematti, M. Hofmann, T. Charnikhova, F. Verstappen, J. Hepworth, S. Van Der Krol, and O. Leyser. 2014. Rice cytochrome P450 MAX1 homologs catalyze distinct steps in strigolactone biosynthesis. *Nature Chemical Biology* 10 (12):1028.

Zuanazzi, J.A.S., P.H. Clergeot, J.-C. Quirion, H.-P. Husson, A. Kondorosi, and P. Ratet. 1998. Production of *Sinorhizobium meliloti* nod gene activator and repressor flavonoids from *Medicago sativa* roots. *Molecular Plant-Microbe Interactions* 11 (8):784–794.

12 Evolution of Scent Genes

Sylvie Baudino, Philippe Hugueney, and Jean-Claude Caissard

CONTENTS

12.1 INTRODUCTION

Elucidation of biosynthesis pathways of volatile organic compounds (VOCs) in plants is a very active field of research. Since the initial isolation of the limonene synthase and the linalool synthase genes from *Mentha spicata* and *Clarkia breweri* (Colby et al., 1993; Dudareva et al., 1996), dozens of genes involved in the biosynthesis of terpenes, phenylpropanoids/benzenoids and fatty acid derivatives have been characterized in plants belonging to various families. Although most work has been done on angiosperms and conifers, the genetic basis of VOC diversity found in more basal plant lineages is also starting to be investigated (Jia et al., 2018). Until recently, most studies were focused on identifying the genes and enzymes involved in the biosynthesis of aroma compounds. Early works used model plants such as *C. breweri*, *Antirrhinum majus*, *Arabidopsis thaliana* and *Petunia* x *hybrida* or plants of agronomical or horticultural interests such as *Solanum lycopersicum* and *Rosa* x *hybrida*. During the past ten years, as more and more pathways were elucidated, the central question of the evolutionary origin of this chemical diversity arose. In a growing number of cases, impact of gene and/or allele evolution on specialized metabolism has been analyzed in detail. In this chapter, we will describe some of these molecular mechanisms by which a plant can acquire or loose the capacity to synthesize specific VOCs. As illustrations, examples will be selected from different categories of VOCs in various plant families, with an emphasis on scent and aroma compounds from plants of particular agronomical or horticultural interests.

12.2 THE MOTORS OF NATURAL SELECTION

12.2.1 Friends and Foes Interacting with Plants

Plants are sessile organisms, which have, even more than other organisms, exploited chemical diversity to cope with their changing environments. The metabolites that a plant is able to synthesize have crucial roles for its survival in challenging conditions, due to both biotic and abiotic factors. VOCs are particularly important for interactions with pollinators and for defense against enemies. It is generally admitted that plant defense was the initial function of VOCs and that scent emission for the attraction of pollinators evolved later (Raguso, 2009). Defense-related VOCs may have direct effects through toxic properties or act as anti-digestive and anti-nutritive compounds. They may also attract enemies of the herbivores in the now well-known indirect defense phenomena. Accordingly, any mutation of a gene leading to the production of an advantageous volatile compound, or eliminating a disadvantageous compound, increases the plant fitness and thus can be quickly selected.

Flower scent is a strong signal directing pollinators to flowers that can deliver a reward like nectar or pollen. Sometimes the reward for insects is a place used as a nest that favors mating and protection. Scent can also be deterrent or toxic to undesirable visitors like ants, caterpillars, or even bacteria (Muhlemann et al., 2014). Scent is thus considered as a major fitness component, however, due to the complexity of the volatile profiles of many flowers, detailed analysis of the role of individual compounds is difficult.

Community-wide patterns of flower scent signals have rarely been explored. Nevertheless, scent advertisement in a plant community has been shown to be inversely proportional to the population size of pollinators: more scent is emitted by species blooming early in the season, when pollinators are scarce and the emission tends to decrease when more pollinators are present (Filella et al., 2013). The dynamics of seasonal emission of scent leading to the structuration of the plant–pollinator network is thus quite complex. Although it is generally assumed that floral volatiles provide fitness benefits for plants, VOCs may impact both mutualists that pollinate flowers and herbivores that damage them (Boachon et al., 2015). By silencing both benzylacetone biosynthesis and nectar production in *N. attenuata* by RNAi approaches, the costs and benefits of these traits for the plant could be evaluated, in relationships with two pollinators and one herbivore (Kessler et al., 2015). This study concluded that both herbivores and pollinators shaped the evolution of floral traits and that it made little sense to study the effects of the ones without considering the effects of the others.

Besides the well described plant–insect interactions, more recent works highlight the influence of plant VOCs on the microbial communities present at the surface of plant tissues (reviewed in Junker and Tholl, 2013). Microbial VOCs have also been shown to influence plant growth or reproduction (Piechulla et al., 2017). The diversity of both beneficial and pathogenic organisms interacting with plants makes it complicated to understand the forces that drive the evolution of VOC biosynthesis and diversity. Furthermore, evolution may occur on genes that directly synthesize VOCs, but also on transcription factors and regulators that are linked to development cycle, to rhythmicity, or to response to environmental, biotic or abiotic factors.

12.2.2 Molecular Mechanisms Leading to Biochemical Diversity

The term of plant specialized metabolism was chosen to reflect that each species (or genus or family…) is capable of synthetizing its own bouquet of metabolites, which is supposed to be well adapted to its biotic and abiotic environment (Pichersky and Lewinsohn, 2011). This implies that each plant has acquired the necessary set of genes and enzymes to produce the said compounds. And indeed, genes involved in the biosynthesis of specialized metabolites, including VOCs, are often restricted and specific to the lineages that produce these compounds. There are various mechanisms by which such genes evolve.

It is generally held that a great deal of plant gene diversity evolved through gene duplication. These duplications include whole-genome duplications, linked to polyploidization events, that are common in many plant lineages, as well as local duplications due to errors during meiosis or unequal crossing over for example. Since Ohno's famous book "evolution by gene duplication" (Ohno, 1970), many theoretical models have been proposed to explain what happens after the duplication event (Figure 12.1) (Innan and Kondrashov, 2010; Panchy et al., 2016). The differences between these models are based on the modes of selection (neutral, relaxed, purifying or positive), on the types of mutations (gain-of-function Figure 12.1a, loss-of-function Figure 12.1b, or no mutation), on the fact that duplicate genes immediately increase the fitness (e.g., concerted evolution Figure 12.1c, or positive dosage) or increase fitness after a series of mutations, and on the ways of functional evolution (neofunctionalization Figure 12.1d, subfunctionalization Figure 12.1e, subneofunctionalization, or pseudogenization Figure 12.1f). In neofunctionalization (Figure 12.1d), one of the duplicated copy accumulates mutations, which eventually lead to a new function, while the other copy retains the original function. However, in this case it is highly probable that the randomly-accumulated mutations are deleterious, leading to nonfunctionalization of the gene, or pseudogenization (Figure 12.1f). To circumvent this weakness of the theory, another model was proposed, in which each duplicated copy was able to perform a part of the initial function(s) of the ancestral gene. This model is broadly referred to as subfunctionalization (Force et al., 1999) (Figure 12.1e). When applied to metabolism, this view is supported by the fact that many enzymes are promiscuous and have the ability to catalyze more than one reaction. Mutations in regulatory elements of the duplicated copies may also lead to divergent patterns of expression. Other types of models, in particular hybrid ones that include both subfunctionalization and neofunctionalization, have been proposed. In the subneofunctionalization model, a short period of subfunctionalization precedes a longer neofunctionalization period (He and Zhang, 2005).

It is noticeable that concerning specialized metabolism, authors only described neofunctionalization or subfunctionalization in terms of an extended definition of these concepts, probably because the link between the theoretical models and the experimental observations is not obvious. Unfortunately, there are still too few descriptions of neo-, sub-, or subneofunctionalization scenarios for genes involved in VOC biosynthesis.

Conversely, such scenarios are well-documented for some nonvolatile specialized compounds (Moghe and Last, 2015). For example, capsaicin biosynthesis in pepper involves capsaicin synthase, which evolved in the genome of *Capsicum* species through multiple gene duplications in the acyltransferase gene family. Furthermore, evolution of this new function was accompanied by the evolution of a fruit-specific expression pattern, that did not occur in other Solanaceae. This allowed the authors to propose a true neofunctionalization scenario giving rise to capsaicin biosynthesis (Kim et al., 2014).

Duplicated genes often revert back to a single copy and neo-, sub-, or subneofunctionalization models correspond to rare events (Moghe and Last, 2015). Most often, duplicated copies accumulate mutation over time and evolve to pseudogenes without functions (Figure 12.1f). Nevertheless, microRNAs, transcription binding sites and Pfam domains (Punta et al., 2012) could originate from pseudogenes. For example, significant differences were found in the number of Pfam domains in the pseudogenes of terpene synthases (TPSs) and P450 oxidases in the genomes of several plant species (Xie et al., 2019). Such creation of new regulation elements could give rise to a "rewiring" of the pre-existing regulatory mechanism and/or de novo evolution of some genes. For example, a duplicated TPS pseudogene may act as a new *cis*-acting element in front of the older copy or give rise to a new noncoding RNA (Figure 12.1g).

Apart from duplication and divergence, the evolution of an entirely new gene, without a functional ancestor, was first seen as highly improbable, but has been shown to occur in many organisms, including plants (Andersson et al., 2015). Finally, transposable elements (TE) are a source of diversification and creation of genes by providing regulatory elements and/or exons (Krasileva, 2019). For example, TE may have participated to the evolution of nicotine

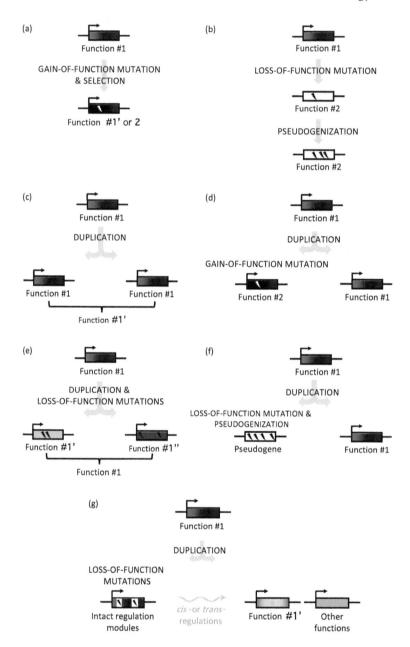

FIGURE 12.1 Mechanisms of gene functionalization. (a) gain-of-function mutation (GOF) changes the ancestral function #1 giving a modified function #1', or a novel function #2; (b) loss-of-function mutation (LOF) suppresses the ancestral function, but is beneficial and selected. Thus, the absence of function #1 can be considered as a function #2, even if the gene becomes a pseudogene; (c) mutations on duplicated copies are eliminated during evolution because two copies of the gene, giving rise to the enhanced function #1', are beneficial. This can be considered as concerted evolution; (d) GOF on one duplicated copy gives a new beneficial function #2: this is a neofunctionalization process. The other copy keeps the function #1; (e) LOF on both copies splits the ancestral function #1 in two subfunctions #1' and #1'', but the "sum" of the functions is the function #1: this is a subfunctionalization process. Therefore, it is sometimes considered that the ancestral gene had two functions. The subfunctionalization event can be followed by a neofunctionalization event of one copy in a subneofunctionalization process; (f) LOF on one copy gives rise to a pseudogene, but the ancestral function #1 is kept by the other copy; (g) the pseudogenization process is not complete and some regulatory modules remain intact (transcription factor binding sites in the promoter, expression of a noncoding RNA…). This can lead to regulation of the other copy, giving rise to a function #1', or to regulation of other genes.

biosynthesis in the genus *Nicotiana*. Indeed, many transcription factor binding sites present in upstream regions of nicotine biosynthesis genes are likely derived from TE insertions (Xu et al., 2017).

To summarize, novel functions may arise from duplications or de novo creations by different mechanisms. Whatever the evolution mechanisms involved, they can either directly affect the activity of the enzymes in biosynthetic pathways or modify the expression patterns of genes encoding these enzymes. These processes may create chemodiversity following divergent evolution, or lead to the production of the same volatiles by convergent evolution. Finally, single loss-of-function or gain-of-function mutations may result in substantial modifications of VOC profiles.

12.3   DIVERGENT EVOLUTION: THE CREATION OF CHEMODIVERSITY

As described above, various mechanisms, among which duplication and divergence are predominant, lead to new enzymes with new functions in plants. Lineage-specific new enzymes have the potential to synthesize novel compounds, ultimately leading to the great diversity of plant-derived VOCs present in nature. Moreover, modification of expression patterns may lead to new products, due to the access to novel substrates. To illustrate these processes, we chose to describe in details a few examples of compounds from different chemical families.

12.3.1   EVOLUTION OF TEA SCENT IN ROSE

The rose is considered the queen of ornamental flowers and is widely used as a garden plant and for the cut flower market. The extraordinary diversity of modern rose varieties has been obtained, by intensive breeding, from a small number of progenitors including both Chinese species, such as *R. chinensis*, and European species such as *R. gallica*, the former contributing recurrent flowering and the latter cold and disease resistance characters. In addition, the European and Chinese progenitors each contributed distinctive scent characteristics. The major scent components provided by European roses include 2-phenylethanol (2PE) and a number of monoterpenes (see Section 12.4.2). Chinese roses, on the other hand, produce phenolic methyl ethers such as 3,5-dimethoxytoluene (DMT), and hence contributed these "tea scent" components to modern rose varieties. Rose petal transcriptome analyses led to the identification of two orcinol *O*-methyltransferases, OOMT1 and OOMT2, that catalyze the successive methylations of the orcinol substrate to produce DMT (Lavid et al., 2002; Scalliet et al., 2002). OOMT1 and OOMT2 are very similar enzymes (96.5% identity), but they exhibit different substrate specificities, that are consistent with their operating sequentially in DMT biosynthesis. These substrate specificities were shown to depend mostly on a single amino acid polymorphism in the phenolic substrate binding site of OOMTs (Scalliet et al., 2008). An analysis of the *OOMT* gene family in the genus *Rosa* indicated that the genomes of all rose species contained *OOMT2* genes. However, only Chinese roses possessed *OOMT1* genes in addition to *OOMT2* genes. Hence, the Chinese-rose-specific *OOMT1* genes most probably evolved, by duplication and divergence, from an *OOMT2*-like gene that has homologues in the genomes of all extant roses. As OOMT2 is capable of performing the first step of the pathway, although with less efficiency than OOMT1, the evolutionary mechanism involved is probably a case of subfunctionalization. The emergence of the *OOMT1* gene may have been a critical step in the evolution of scent production in Chinese roses, maybe in relationship with specific pollinators in their wild habitat. This gene and the "tea scent," originally limited to Chinese wild rose species, were thus transmitted during the breeding of modern roses and are found today in thousands of cultivars present in gardens all over the world. Besides roses, the importance of *O*-methyltransferase enzymes (OMTs) as a source of VOC diversity has been documented in many plant species, since the pioneering analysis of the difference in substrate preference between the closely related chavicol-OMT and eugenol-OMT from sweet basil (Gang et al., 2002).

12.3.2   ENZYME PROMISCUITY AND GENE CLUSTERS: THE EXAMPLE OF TERPENE SYNTHASES

TPSs are key enzymes in plants, which catalyze the formation of the terpenoid backbone. Even though the primary products of TPS activity may be further metabolized, the diversity of terpenoids found in plants can largely be attributed to the activity of TPSs. TPSs have been the subject of multiple studies, which led to a better understanding of their biochemical properties and environmental regulations (reviewed in Chen, Tholl, et al., 2011). One of the characteristic features of TPSs is their promiscuity, which means that they are broad-specificity proteins (Kreis and Munkert, 2019). Indeed, a single TPS may synthesize multiple products from the same substrate, as for example the TPS35 from *M. truncatula* (Garms et al., 2010), which catalyze the formation of 27 different sesquiterpenes. Conversely, other enzymes only produce one compound, such as the geraniol synthase of sweet basil (Iijima et al., 2004). Few amino acid changes may change drastically the products of a TPS, leading to a large diversity of terpene profiles in one species or different accessions. For example, amino acid changes at five positively selected sites in the *Oryza* TPS1 protein from rice are responsible for the diverse sesquiterpene blends produced by different *Oryza* species in response to insect damages (Chen et al., 2014). Moreover, recent advances in functional characterization of TPSs, with the use of prenyl diphosphates with different chain length or different (*E/Z*) configurations, led to the conclusion that many TPSs can accept more than one substrate (Pazouki and Niinemets, 2016). Typical examples are the linalool/nerolidol synthases, which can accept both geranyl diphosphate (GPP) and farnesyl diphosphate (FPP) substrates, leading to formation of the monoterpene linalool or the sesquiterpene nerolidol (Aharoni et al., 2004).

The genomic organization of *TPS* genes has been investigated in several plant species. These genes are frequently organized in clusters of tandemly duplicated genes, as observed in tomato (Falara et al., 2011) and grapevine (Martin et al., 2010). A survey of 13 angiosperm genomes showed that *TPS* genes were significantly more enriched in tandem duplicates compared to genome-wide average, across most of the analyzed species (Hofberger et al., 2015). This suggests that this gene family has experienced rapid evolution through gene duplication, which has led to divergence in enzyme functions resulting in a wide diversity of terpenes. Cultivated tomato, *S. lycopersicum*, and its related wild species, *S. habrochaites* and *S. pimpinelli*, synthesize a range of terpenes in their glandular trichomes, which are implicated in defense against insects. The terpene chemical diversity in *S. habrochaites* could be attributed to TPSs related to TPS20, previously characterized in *S. lycopersicum* (Gonzales-Vigil et al., 2012). These TPS20-related enzymes, unlike most TPSs in other plants, use (Z)-prenyl diphosphates, like neryl diphosphate and (2Z, 6Z)-farnesyl diphosphate, as substrates and belong to TPS-e/f subfamily. They are clustered on chromosome 8 of the *S. lycopersicum* genome, together with other genes involved in terpene biosynthesis. Combined biochemical, genomic and phylogenetic analyses on the tomato species led the authors to propose a complex evolutionary scenario (Matsuba et al., 2013). Briefly, in all *Solanum* species analyzed, the monoterpenes synthases genes in the cluster evolved from a diterpene synthase by duplication and divergence. In *S. habrochaites*, a new sesquiterpene synthase gene was created following another duplication event involving a gene encoding a monoterpene synthase. This enzyme evolved to used (2Z, 6Z)-farnesyl diphosphate. This has led to an organization in which monoterpene and sesquiterpene synthases are clustered on the same locus and the prenyltransferases that provide their substrates are also part of the same gene cluster. This clustering of *TPS* genes with other genes in the terpenoid pathway seems quite common in angiosperms (Boutanaev et al., 2015). There are many hypotheses that could explain the clustering of genes in a metabolic pathway (Nützmann et al., 2016). One of these is the fact that it may facilitate coexpression of the genes in the cluster. The genomes of some species harbor much more compact *TPS* gene families. For instance, the genome of cultivated apple (*Malus domestica*) contains only ten potentially functional *TPS* genes and a larger number of *TPS* pseudogenes. The genome duplication events that have shaped the organization of the apple genome appears to have significantly impacted the diversity of the *TPS* genes in this species (Nieuwenhuizen et al., 2013). In addition, the reduced number of TPSs in a modern cultivar, compared to heritage varieties, suggests that breeding strategies may have reduced the range of terpenes in the apple fruit.

12.4 CONVERGENT EVOLUTION

Convergent evolution is an independent evolution resulting in similar traits in different organisms. For example, the biosynthesis of the same compound in different plant species may be catalyzed by phylogenetically unrelated enzymes. There are many examples of convergent evolution in the plant specialized metabolism. Phylogenetic analysis of caffeine N-methyltransferases (NMTs) in *Coffea canephora* showed that these genes expanded through tandem duplications, independently of NMTs from *Theobroma cacao* and *Camellia sinensis*, indicating a convergent evolution of caffeine biosynthesis in different plant lineages (Denoeud et al., 2014). Similarly, for decades it was believed that nearly all VOCs were produced through a limited number of pathways shared by all plants. However, there is more and more evidence that different pathways leading to similar or identical compounds have evolved independently in plants. These alternative pathways often occur through convergent evolution (Sun et al., 2016).

12.4.1 FUNCTIONAL CONVERGENCE: DISTINCT ENZYME LINEAGES PRODUCING SIMILAR COMPOUNDS

The monoterpene alcohol linalool is present in the floral scent of many species. Its diverse ecological functions have recently been reviewed (Raguso, 2016a). This compound may be both an attractant for pollinators (in orchids for example) and a repellant for various herbivores. Emission of linalool by *N. attenuata* leaves has been proposed to attract predators of the generalist herbivore *Manduca sexta* (He et al., 2019). When present in nectar, it could contribute to avoid microbe proliferation. Two enantiomeric forms of this molecule exist: (*R*)-(−)-linalool and (*S*)-(+)-linalool. Both forms have distinct biochemical properties and impacts on insects. For instance, female moths oviposit more on *Datura wrightii* plants emitting (*S*)-(+)-linalool over control plants, while plants emitting (*R*)-(−)-linalool are less preferred than control plants (Reisenman et al., 2010). Since the first characterization of *C. breweri* linalool synthase (Dudareva et al., 1996), TPSs able to catalyze the formation of linalool from GPP have been functionally characterized in many plant species, including *A. thaliana* (Ginglinger et al., 2013), *F. x ananassa* (Aharoni et al., 2004), and *Rosa* genus (Magnard et al., 2018). Phylogenetic analysis indicates that these genes belong to four TPS subfamilies (Figure 12.2). For example, in *R. chinensis*, three genes encoding potentially functional TPSs are expressed in petals, *RcLINS*, *RcLIN-NERS1*, and *RcLIN-NERS2*. The *LIN-NERS1/2* genes are clustered on chromosome 5 of the rose genome (Raymond et al., 2018) whereas *RcLINS* gene is on chromosome 2. RcLINS is responsible for the small amounts of (*R*)-(−)-linalool present in rose scent (Magnard et al., 2018). *RcLIN-NERS1*, and *RcLIN-NERS2* genes, although weakly expressed, are probably not active in planta as neither nerolidol, nor (*S*)-(+)-linalool are present in rose petals. A comparison of roses with strawberries, show that these two members of the Rosaceae family have evolved different enzymes to produce linalool. *F. x ananassa* uses the (*S*)-(+)-linalool /nerolidol synthase FaNES1 (Aharoni et al., 2004), whereas the orthologous RcLIN-NERS2 is ineffective in rose flowers. Conversely, the small quantity of linalool produced in rose flowers derives from the activity of the RcLINS (*R*)-(−)-linalool synthase, which is unable to produce nerolidol. More generally, linalool or linalool/nerolidol synthase activities are well correlated with their belonging to a specific TPS subfamily (Figure 12.2). Monofunctional linalool synthases cluster into the TPS-b subfamily. Like RcLINS, most of these b class enzymes with linalool synthase activity generally produce (*R*)-(−)-linalool. The bifunctional linalool/nerolidol synthases, including RcLIN-NERS1 and RcLIN-NERS2 belong to the TPS-g subgroup together with some linalool/nerolidol/geranyllinalool

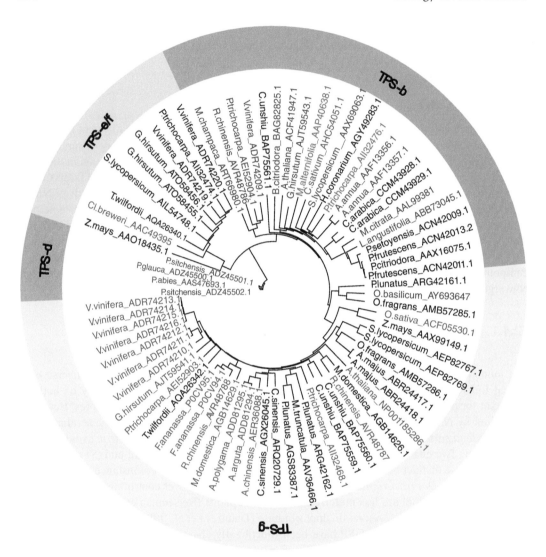

FIGURE 12.2 Diversity of linalool synthases characterized from different plant species. Unrooted Neighbor Joining tree depicting the classification of linalool synthases and linalool/nerolidol synthases from plants into terpene synthase (TPS) subfamilies. Linalool/nerolidol/geranyllinalool synthases were also included. Protein sequences were aligned with ClustalW and tree was constructed using Geneious software. Linalool isomers are indicated by colors of the GenBank accession letters. Black letters, unidentified linalool isomer; blue letters (*R*)-(−)-linalool synthase; orange letters (*R/S*)-linalool synthase (racemic); pink letters (*S*)-(+)-linalool synthase.

synthases. The linalool isomer produced by synthases belonging to this group is usually (*S*)-(+)-linalool, like RcLIN-NERS1 and 2. Although not all studies report the linalool enantiomer that is produced by the different species, the general trend is that (*S*)-(+)- and (*R*)-(−)-linalool synthases have evolved independently, probably under the selection pressures exerted by both herbivores and pollinators.

Another example of convergent evolution is the biosynthesis of phenylpropenes. These compounds are phenylpropanoids derived from phenylalanine, which participate in the unique aroma of many fruits, herbs and spices. In flowers, they attract pollinators and may be used as defense compounds against fungi and bacteria. Many plants synthesize volatile phenylpropenes in their floral and vegetative organs as a defense against insect pests and herbivores. For example, the stamens of roses emit eugenol (Yan et al., 2018). The last step of the biosynthesis pathway to eugenol and isoeugenol is performed by the NADPH-dependent reductases eugenol synthase and isoeugenol synthase, which catalyze the elimination of an acyl moiety from coniferyl acetate (reviewed in Koeduka, 2014). Phenylpropene synthases belong to the PIP family of reductases and are distributed in two distinct protein lineages in *C. breweri* and *P.* x *hybrida* (Koeduka et al., 2008), suggesting independent evolutionary origins. These independent origins have since been supported by studies in a number of other plants such as *F.* x *ananassa* and *Daucus carota* (Aragüez et al., 2013; Yahyaa et al., 2019).

12.4.2 PATHWAY CONVERGENCE: DIFFERENT PATHWAYS PRODUCING THE SAME COMPOUNDS

Monoterpenes are well known as components of floral scent and also as constituents of essential oils of aromatics plants that are widely used in the cosmetic, perfume, food, and pharmaceutical industries. Because of their high economic relevance, monoterpene biosynthesis has been extensively studied. In rose, acyclic monoterpenic alcohols, such as geraniol, nerol and β-citronellol and their aldehyde and acetate derivatives are especially abundant. In most plants studied so far, geraniol is synthesized in plastids by geraniol synthase (GES), belonging to the TPS family. GES genes have been cloned and functionally characterized in many plant species, including *O. basilicum* (Iijima et al., 2004) and *Catharanthus roseus* (Simkin et al., 2013). In roses, no such enzyme has been isolated and it has been recently shown that rose flowers use an alternative route to produce geraniol.

By comparing the volatile profiles and differential gene expression of scented and unscented rose cultivars, an unexpected enzyme – a Nudix hydrolase, RhNUDX1 – was found to be responsible for the formation of geraniol (Magnard et al., 2015). RhNUDX1 uses GPP as substrate, as classical TPSs, but hydrolyzes one phosphate resulting in geranyl monophosphate (Figure 12.3). Thus, a

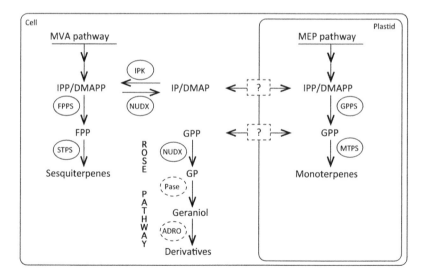

FIGURE 12.3 Biosynthesis of monoterpenes and sesquiterpenes in plants. Roles of NUDX proteins in terpene biosynthesis have been shown in *A. thaliana* (Henry et al., 2018) and *Rosa* sp. (Magnard et al., 2015). ADRO, acetyltransferases, dehydrogenases, reductases, and oxidases; FPP, farnesyl diphosphate; FPPS, farnesyl diphosphate synthase; GP, geranyl phosphate; GPP, geranyl diphosphate; DMAP, dimethylallyl diphosphate; DMAPP, dimethylallyl diphosphate; IP, isopentenyl phosphate; IPK, isopentenyl phosphate kinase; IPP, isopentenyl diphosphate; MEP, 2-C-methyl-D-erythritol 4-phosphate; MTPS, monoterpene synthases; MVA, mevalonic acid; NUDX, nudix hydrolase; Pase, phosphatase; STPS, sesquiterpene synthases; ?, unknown transporters; One arrow, one enzymatic step; Two arrows, several enzymatic steps; Solid circle, functionally characterized enzymes; Dashed circle, one or several hypothetical enzymes; Dashed box with opposite arrows, hypothetical transporters.

further dephosphorylation reaction catalyzed by an unidentified phosphatase present in rose petals is required to form geraniol. This process differs from terpene synthase-catalyzed reactions, where a diphosphate group is removed in one reaction. In addition, the RhNUDX1 enzyme has been shown to be located in the cytosol, raising the question of the origin of its GPP substrate, which may be either transported from plastids or generated in the cytosol.

Nudix hydrolases belong to a superfamily of pyrophosphatases found in animals, plants, bacteria and archaea. They generally catalyze the hydrolysis of nucleoside diphosphates linked to other moieties (X) and contain the Nudix domain G(X5)E(X7)REUXEEXX. For example, the *E. coli* mutator protein MutT eliminates harmful oxidized nucleotides, such as 8-oxo-7,8-dihydrodeoxy-guanosine triphosphate (8-oxo-dGTP), which may be misincorporated in DNA during replication (Maki and Sekiguchi, 1992). In plants, Nudix hydrolases are poorly characterized, although up to 29 genes have been identified in the *A. thaliana* genome (Kraszewska, 2008). In *A. thaliana*, some Nudix hydrolases have been linked to specific functions *in vivo*. For example, some act as regulators of plant defense responses (Fonseca and Dong, 2014). RhNUDX1 shows the closest similarity to AtNUDX1 from *A. thaliana*.

Although AtNUDX1 actual function is still controversial, this protein has been proposed to have a similar function to that of the MutT protein from *E. coli* (Jemth et al., 2019). However, recent studies indicate that AtNUDX1 is involved in the regulation of terpene synthesis. Indeed, AtNUDX1, in combination with another Nudix hydrolase, catalyzed the dephosphorylation of isopentenyl diphosphate (IPP), which is the common precursor of both mono- and sesquiterpenes, to isopentenyl phosphate (IP). IP is transformed to IPP by isopentenyl phosphate kinase (IPK) (Henry et al., 2015). The ratio of IPP to IP, modulated by activities of Nudix hydrolases and IPK, is supposed to regulate terpenoid biosynthesis (Henry et al., 2018). These results suggest that NUDX1 proteins may have been recruited to perform various functions in different plants species. In rose, one striking feature of *RhNUDX1* is its very high level of expression in petals, whereas it is barely expressed in other plant organs. Upregulation of the expression in petals might thus have been an important step for the evolution of *RhNUDX1* function in scent production. Moreover, this gene is present in tandem repeated copies in the genome of *R. chinensis* cv. "Old Blush" (Hibrand Saint-Oyant et al., 2018; Raymond et al., 2018). The repeated duplication of the gene may therefore have been selected possibly leading to higher transcription levels. Monoterpenes are present in the scent of many flowers. The discovery that very different pathways involving TPSs or Nudix hydrolases evolved independently to produce monoterpenes suggests that they may be important for plant fitness, as pollination or defense-related compounds. The Nudix pathway to monoterpene biosynthesis has so far only been characterized in the *Rosa* genus. The questions of its evolutionary origin and of its presence in other members of the Rosaceae family still remain.

There are other examples of compounds formed by different ways in different species.

Some volatiles derived from phenylalanine (Phe), for example phenylacetaldehyde (PAld) and 2PE, are important scent compounds in numerous flowers such as roses and *P. x hybrida*. 2PE is also a component of plant defenses (Günther et al., 2019). The genes and enzymes responsible for the biosynthesis of these phenylalanine-derived compounds have been identified and functionally characterized in a number of plant species. Interestingly, plants have evolved different pathways to convert Phe to PAld (Figure 12.4). In *P. x hybrida* and *R. x hybrida*, a bifunctional phenylacetaldehyde synthase (PAAS) can catalyze both decarboxylation and oxidative deamination reactions and convert Phe to PAld (Kaminaga et al., 2006; Farhi et al., 2010) (Figure 12.4, route 1). Genetic analysis of 2PE production in a segregating rose population has recently shown that this trait depends on specific *PAAS* alleles (Roccia et al., 2019). Aromatic aldehyde synthases (AASs) with the same catalytic properties as PAAS are also active in flowers or leaves of *A. thaliana*, depending on the ecotype (Gutensohn et al., 2011).

In *S. lycopersicum* fruits, a similar pathway proceeds in two steps, Phe being first converted to 2-phenylethylamine by an aromatic amino acid decarboxylase (AADC) and then subjected to oxidation by an unidentified monoamine oxidase (Tieman et al., 2006) (Figure 12.4, route 2). AADCs

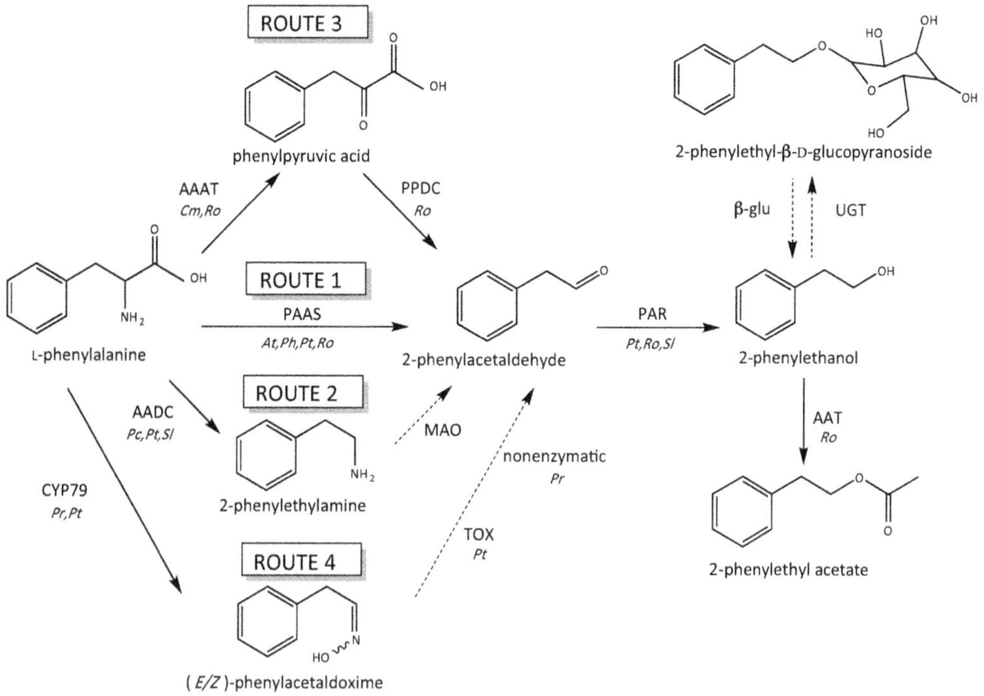

FIGURE 12.4 Biosynthesis of 2-phenylethanol and derivatives in plants. AAAT, aromatic amino acid aminotransferase; AADC, L-aromatic amino acid decarboxylase; AAT, alcohol acetyltransferase; β-glu, β-glucosidase; CYP79, cytochrome P450 family 79 enzyme; MAO, monoamine oxidase; PAAS, phenylacetaldehyde synthase; PAR, phenylacetaldehyde reductase; PPDC, phenylpyruvic acid decarboxylase; UGT, UDP-glycosyltransferase; TOX, transoximase. *At, Arabidopsis thaliana* (Gutensohn et al., 2011); *Cm, Cucumis melo* (Gonda et al., 2010); *Sl, Solanum lycopersicum* (Tieman et al., 2006, 2007); *Pc, Populus x canescens* (Günther et al., 2019); *Ph, Petunia x hybrida* (Kaminaga et al., 2006), *Pr, Plumeria rubra* (Dhandapani et al., 2019); *Pt, Populus trichocarpa* (Günther et al., 2019; Irmisch et al., 2014); *Ro, Rosa* sp. (different rose species and cultivars were used) (Chen, Kobayashi, et al., 2011; Farhi et al., 2010; Hirata et al., 2012, 2016; Kaminaga et al., 2006; Roccia et al., 2019; Shalit et al., 2003). Solid lines, well established reactions with characterized enzymes; Dashed lines, hypothetical reactions.

and AASs are members of the same aromatic amino acid decarboxylase (AAAD) family. One particular amino acid, present in a catalytic loop of these enzymes, determines whether they act as monofunctional AADCs (decarboxylation) or bifunctional AASs (decarboxylation-deamination). A mutation of this amino acid is sufficient to switch enzyme activities (Torrens-Spence et al., 2013). In addition, another pathway leading to PAld biosynthesis from Phe also exists in rose (Figure 12.4, route 3). In this pathway, Phe is first converted into phenylpyruvic acid by an aromatic amino acid aminotransferase (AAAT3), and then undergoes decarboxylation to form PAld by a phenylpyruvic acid decarboxylase (PPDC) (Hirata et al., 2012, 2016) (Figure 12.4). Similarly, in the fruits of *Cucumis melo*, Phe is also converted by an amino acid transaminase (ArAT1) to phenylpyruvic acid (Gonda et al., 2010). Experiments with transgenic petunia plants producing high levels of phenyl-pyruvate suggest that such an aminotransferase pathway could also be active in *P.* x *hybrida* petals (Oliva et al., 2017).

Recently, the metabolism of Phe was studied in *Populus trichocarpa* (Günther et al., 2019). In this species, separate pathways contribute to the formation of 2PE and its glycosylated derivative (2-phenylethyl-β-D-glucopyranoside), which are released by leaves upon herbivory. PtAAS1 controls the emission of 2PE (Figure 12.4, route 1), whereas PtAADC1 is involved in the formation of 2-phenylethylamine which can be further metabolized into 2-phenylethyl-β-D-glucopyranoside (Figure 12.4, route 2). *PtAAS1* and *PtAADC1* sequences are very similar and are clustered on the same chromosome, which suggests that they may have evolved from a common ancestor by gene duplication and neofunctionalization.

Lastly, in *P. trichocarpa* and *Plumeria rubra*, 2PE biosynthesis proceeds via a fourth route (Figure 12.4, route 4), which involves the conversion of Phe into (*E/Z*)-phenylacetaldoxime by cytochrome P450s from the CYP79 family (Irmisch et al., 2014; Dhandapani et al., 2019). In all species, the conversion from PAld to 2PE is supposed to be catalyzed by a phenylacetaldehyde reductase (PAR), characterized in rose, poplar and tomato (Tieman et al., 2007; Chen, Kobayashi, et al., 2011; Günther et al., 2019). 2PE may be further converted into ester derivatives, as shown in rose (Shalit et al., 2003). Biosynthesis of 2PE is thus another example of multiple routes used to produce the same compound, which is probably of high interest for pollination and/or defense in plants. Interestingly in rose, the pathway which uses AAAT3 and PPDC becomes active only in summer under high-temperature conditions, in contrast to the pathway using RhPAAS, which is active throughout the year, regardless of temperature, and produces a nearly constant amount of 2PE. The seasonally induced pathway may have originated in wild rose species as a heat adaptation when roses spread to low latitudes and/or low altitudes. Alternatively, this pathway may have been selected during cultivation to produce blossoms in summer (Hirata et al., 2016).

12.5 LOSS- OR GAIN-OF-FUNCTION MUTATIONS AS A SOURCE OF VOC DIVERSITY

Several examples in wild and cultivated plant species show that local mutations in a single gene may result in loss- or gain-of-function alleles that may greatly impact VOC profiles.

12.5.1 Loss of Floral Scent Production during Pollinator-Mediated Speciation

As described in other chapters of this book, scent is one of the major components that mediate the interaction of flowers with their pollinators (Gervasi and Schiestl, 2017). There are still too few examples to draw general conclusions but we present here two examples, in which modifications in flower scent profiles led to changes in pollinator communities.

The *Petunia* genus was used as a model to investigate the genetic basis of gain and loss of scent during the transition from bee to moth and from moth to hummingbird pollination (Klahre et al., 2011; Amrad et al., 2016). *P. inflata*, which is supposed to be ancestral to other species, has purple flowers, is pollinated by bees and emits the benzenoid compound benzaldehyde, in small amounts. Conversely, *P. axillaris* has white flowers and is nocturnally pollinated by hawkmoths. It emits large amounts of benzenoids such as methylbenzoate, benzylalcohol and benzaldehyde. Finally, the hummingbird-pollinated flowers of *P. exserta* are red and scentless. Here we will focus on the differences found between *P. axillaris* and *P. exserta*. At first, two QTLs linked to the capacity to produce benzenoid volatiles were detected in an F1 population derived from an interspecific cross between *P. axillaris* and *P. exserta* (Klahre et al., 2011). One of these QTLs, present on chromosome VII, could be associated to the MYB transcription factor ODO1, which regulates several enzymes of the phenylpropanoids/benzenoid pathway (Verdonk et al., 2005). *ODO1* gene is highly expressed in *P. axillaris*, which emits higher amounts of volatiles than *P. exserta*. A combination of genetic and molecular approaches identified the gene underlying the QTL on chromosome II as cinnamate-Co-A ligase (*CNL1*), which conjugates cinnamic acid to coenzyme A (Amrad et al., 2016). The lower activity found in *P. exserta*, compared to *P. axillaris*, was attributed to multiple defects in the gene, including a nonsense mutation in the fourth exon, which is predicted to shorten the protein by 171 amino acids. These works concluded that functional inactivation in CNL1, together with the decrease expression of the MYB transcription factor ODO1, accounted for the loss of scent during the shift from hawkmoth to hummingbird pollination.

Another study reported that the loss of floral scent during the transition from outcrossing to selfing in the genus *Capsella* was also caused by loss of CNL function (Sas et al., 2016). Indeed, the insect-pollinated *C. grandiflora* emits benzaldehyde as a major constituent of its floral scent, whereas it has been lost in the selfing *C. rubella*. By comparing CNL sequences in 207 accessions of the two species, they found that it was inactivated twice by independent mutations in *C. rubella*.

These studies on *Petunia* and *Capsella* suggest that loss-of-function of *CNL* genes has repeatedly occurred during evolutionary transition to scentless flowers, in two phylogenetically unrelated genera. The selective forces driving scent losses during evolution of the pollination syndromes are not known. One of the possible explanations is that the plants must both attract pollinators and avoid attracting the florivores. In some ecological contexts, it may be advantageous for plants to reduce scent emissions, in order to minimize the cost of florivory (Raguso, 2016b).

12.5.2 Loss- and Gain-of-Function Mutations Impacting Scents and Aromas in Cultivated Plant Species

Gain or loss of scent and aroma attributes has frequently, and sometimes unintentionally, been selected during plant domestication. One of the first molecular analysis of such phenomenon characterized the fruit aroma of cultivated *F.* x *ananassa* in comparison to those of its wild relative *F. vesca*. The poor aroma of cultivated strawberries could be attributed to a two base-pair insertion causing a frameshift in a *TPS* gene, which was responsible for the biosynthesis of bouquet of monoterpenes containing α-pinene and β-myrcene in wild strawberries (Aharoni et al., 2004).

Rice (*O. sativa*) is a crucial food source for the majority of the world's population. Fragrance in the grain is highly valued by the consumers and has been the subject of intensive researches. Although rice fragrance may involve several compounds, the major contributor is 2-acetyl-1-pyrroline (2AP), which possesses a popcorn-like aroma. Early analysis of the genetic basis of rice flavor led to the identification of a single locus on chromosome 8 (*FGR*) associated with fragrance. This locus encodes a betaine aldehyde dehydrogenase (BADH2), which catalyzes the oxidation of γ-aminobutaraldehyde, which is itself derived from putrescine, to γ-aminobutyric acid. However, the recessive allele *badh2.1* encodes an inactive protein, due to an 8 bp deletion in exon 7 of the gene. γ-aminobutaraldehyde thus accumulates in the grain and is transformed to 2AP, through yet unknown steps, leading to the typical fragrant phenotype (Chen et al., 2008). *O. sativa* consists of two major varietal groups, *Indica* and *Japonica*, that have been recognized in China since ancient times.

In-depth survey of the genetic of the *BADH2* gene in a large panel of genetically and geographically diverse rice germplasm showed that one particular *badh2.1* allele is predominant in all fragrant rice varieties today, such as Jasmine and Basmati varieties (Kovach et al., 2009). Sequencing of the *BADH2* gene in a panel of 242 *O. sativa* accessions, revealed a single origin of the *badh2.1* allele within the *Japonica* varietal group and showed that it was later introgressed in the *Indica* varietal group, presumably by breeders. This work also suggested that the *badh2.1* allele, associated with fragrance within a small set of varieties from China, was selected as a de novo mutation in *O. sativa* after domestication from its wild progenitor, presumably after the divergence of the *Japonica* populations. All these studies show that the *BADH2* gene was selected during rice domestication for its effect on the plant sensorial quality. Interestingly, in other plants producing 2AP, the genes and mechanisms involved were found to be similar. For example, in *Glycine max*, genetic mapping of a QTL for fragrance linked a *GmBADH2* allele to 2AP biosynthesis (Juwattanasomran et al., 2011). These examples illustrate the fact that a loss-of-function mutation may lead to the synthesis of a new volatile compound, and it is striking to see that selection of fragrant cultivars in different plant species resulted in the selection of mutations in orthologous genes.

In a few cases, the fascinating aromas that characterize specific cultivars involve gain-of-function or enhanced-function mutations. For instance, the typical floral flavor of Muscat and Gewurztraminer grape varieties is greatly appreciated both in wines and in table grapes. Muscat flavor is due to high levels of monoterpenols (geraniol, linalool and nerol) accumulated in grapes. Genetic analyses of monoterpenol contents in Muscat and Gewurztraminer varieties revealed the co-localization of a gene encoding 1-deoxy-D-xylulose 5-phosphate synthase (*VvDXS*) with a major QTL positioned on chromosome 5 (Battilana et al., 2009; Duchêne et al., 2009). Furthermore, an association study showed that a single nucleotide polymorphism (SNP) responsible for the substitution of a lysine (K284) with an asparagine at position 284 (N284) of the VvDXS protein was significantly associated with Muscat-flavored varieties (Emanuelli et al., 2010). Expression of the Muscat-specific *VvDXS* allele in tobacco confirmed that the N284 mutation was sufficient to drive a massive accumulation of monoterpenes in the corresponding transgenic lines, compared to lines expressing the regular K284 allele. Kinetic analyses showed that the N284 mutation impacted the catalytic efficiency of the VvDXS enzyme, VvDXS N284 being twice as efficient as VvDXS K284 (Battilana et al., 2011). This represents a remarkable example of a natural enzyme variant with enhanced catalytic properties, causing terpene over-accumulation in the much-appreciated Muscat-flavored grapevine varieties.

12.6   CONCLUSION

The capacity to produce scent is a key component of plant fitness, together with other vegetative and floral traits. Although the picture is far from complete, in the past decade, many pathways that lead to the biosynthesis of terpenoids, phenylpropanoids/benzenoid compounds and fatty acid derivatives were deciphered. The knowledge of enzymes and regulator genes of VOC biosynthesis has permitted to addressed the crucial question of the evolutionary origin of these pathways. On the one hand, many plant species have acquired the capacity to produce highly diversified blends of volatiles through different mechanisms leading to divergent evolution and pathway plasticity, such as the expansion of large gene families following gene duplication, catalytic promiscuity or easy change in substrate specificity within enzyme families. On the other hand, some ubiquitous volatiles such as geraniol, linalool or 2PE are present in flower scent or vegetative VOCs of many plant species. Recent studies have shown that such compounds may be produced by unrelated enzymes, or even totally different pathways, which evolved independently in different plant species. Evolutionary convergence toward the biosynthesis of a limited number of ubiquitous volatiles suggests that these compounds may be especially important for plant fitness, as attractant for pollinators or for defense against herbivores or pathogens. Very recent studies analyzing the roles of VOCs in different ecological situations suggest that the response of a specialist herbivore to a single compound like linalool may be very complex, depending on enantiomer, plant genotype, and environmental conditions (He et al., 2019). Many more studies will undoubtedly aim at deciphering the ecological functions of plants volatiles in the future.

ACKNOWLEDGMENTS

The authors' laboratory work is supported by ANR Rosascent, GDR 3658 MediatEC, Fondation UJM, IDEX Lyon/Saint-Etienne and PEPS CNRS.

REFERENCES

Aharoni, A., A.P. Giri, F.W.A. Verstappen, et al. 2004. Gain and loss of fruit flavor compounds produced by wild and cultivated strawberry species. *Plant Cell* 16:3110–31.
Amrad, A., M. Moser, T. Mandel, et al. 2016. Gain and loss of floral scent production through changes in structural genes during pollinator-mediated speciation. *Curr. Biol.* 26:3303–12.
Andersson, D.I., J. Jerlström-Hultqvist, and J. Näsvall. 2015. Evolution of new functions de novo and from preexisting genes. *Cold. Spring Harb. Perspect. Biol.* 7:a017996.

Aragüez, I., S. Osorio, T. Hoffmann, et al. 2013. Eugenol production in achenes and receptacles of strawberry fruits is catalyzed by synthases exhibiting distinct kinetics. *Plant Physiol.* 163:946–58.

Battilana, J., F. Emanuelli, G. Gambino, et al. 2011. Functional effect of grapevine 1-deoxy-D-xylulose 5-phosphate synthase substitution K284N on Muscat flavour formation. *J. Exp. Bot.* 62:5497–508.

Battilana, J., L. Costantini, F. Emanuelli, et al. 2009. The 1-deoxy-D-xylulose 5-phosphate synthase gene co-localizes with a major QTL affecting monoterpene content in grapevine. *Theor. Appl. Genet.* 118:653–69.

Boachon, B., R.R. Junker, L. Miesch, et al. 2015. CYP76C1 (Cytochrome P450)-mediated linalool metabolism and the formation of volatile and soluble linalool oxides in Arabidopsis flowers: A strategy for defense against floral antagonists. *Plant Cell* 27:2972–90.

Boutanaev, A.M., T. Moses, J. Zi, et al. 2015. Investigation of terpene diversification across multiple sequenced plant genomes. *Proc. Natl. Acad. Sci. USA* 112:E81–8.

Chen, F., D. Tholl, J. Bohlmann, and E. Pichersky. 2011. The family of terpene synthases in plants: A mid-size family of genes for specialized metabolism that is highly diversified throughout the kingdom. *Plant J.* 66:212–29.

Chen, H., G. Li, T.G. Köllner, Q. Jia, J. Gershenzon, and F. Chen. 2014. Positive darwinian selection is a driving force for the diversification of terpenoid biosynthesis in the genus *Oryza. BMC Plant Biol.* 14:239.

Chen, S., Y. Yang, W. Shi, et al. 2008. *Badh2*, encoding betaine aldehyde dehydrogenase, inhibits the biosynthesis of 2-acetyl-1-pyrroline, a major component in rice fragrance. *Plant Cell* 20:1850–61.

Chen, X.-M., H. Kobayashi, M. Sakai, et al. 2011. Functional characterization of rose phenylacetaldehyde reductase (PAR), an enzyme involved in the biosynthesis of the scent compound 2-phenylethanol. *J. Plant Physiol.* 168:88–95.

Colby, S.M., W.R. Alonso, E.J. Katahira, D.J. McGarvey, and R. Croteau. 1993. 4S-limonene synthase from the oil glands of spearmint (*Mentha spicata*): cDNA isolation, characterization, and bacterial expression of the catalytically active monoterpene cyclase. *J. Biol. Chem.* 268:23016–24.

Denoeud, F., L. Carretero-Paulet, A. Dereeper, et al. 2014. The coffee genome provides insight into the convergent evolution of caffeine biosynthesis. *Science* 345:1181–4.

Dhandapani, S., J. Jin, V. Sridhar, N.-H. Chua, and I.-C. Jang. 2019. CYP79D73 participates in biosynthesis of floral scent compound 2-phenylethanol in *Plumeria rubra. Plant Physiol.* 180:171–84.

Duchêne, E., G. Butterlin, P. Claudel, V. Dumas, N. Jaegli, and D. Merdinoglu. 2009. A grapevine (*Vitis vinifera* L.) deoxy-D-xylulose synthase gene colocates with a major quantitative trait loci for terpenol content. *Theor. Appl. Genet.* 118:541–52.

Dudareva, N., L. Cseke, V.M. Blanc, and E. Pichersky. 1996. Evolution of floral scent in *Clarkia*: Novel patterns of *S*-linalool synthase gene expression in the *C. breweri* flower. *Plant Cell* 8:1137–48.

Emanuelli, F., J. Battilana, L. Costantini, et al. 2010. A candidate gene association study on Muscat flavor in grapevine (*Vitis vinifera* L.). *BMC Plant Biol.* 10:241.

Falara, V., T.A. Akhtar, T.T. Nguyen, et al. 2011. The tomato terpene synthase gene family. *Plant Physiol.* 157:770–89.

Farhi, M., O. Lavie, T. Masci, et al. 2010. Identification of rose phenylacetaldehyde synthase by functional complementation in yeast. *Plant Mol. Biol.* 72:235–45.

Filella, I., C. Primante, J. Llusià, et al. 2013. Floral advertisement scent in a changing plant-pollinators market. *Sci. Rep.* 3:3434.

Fonseca, J.P., and X. Dong. 2014. Functional characterization of a Nudix hydrolase *AtNUDX8* upon pathogen attack indicates a positive role in plant immune responses. *PLoS One* 9:e114119.

Force, A., M. Lynch, F.B. Pickett, A. Amores, Y.L. Yan, and J. Postlethwait. 1999. Preservation of duplicate genes by complementary, degenerative mutations. *Genetics* 151:1531–45.

Gang, D.R., N. Lavid, C. Zubieta, et al. 2002. Characterization of phenylpropene *O*-methyltransferases from sweet basil: Facile change of substrate specificity and convergent evolution within a plant *O*-methyltransferase family. *Plant Cell* 14:505–19.

Garms, S., T.G. Köllner, and W. Boland. 2010. A multiproduct terpene synthase from *Medicago truncatula* generates cadalane sesquiterpenes via two different mechanisms. *J. Org. Chem.* 75:5590–600.

Gervasi, D.D., and F.P. Schiestl. 2017. Real-time divergent evolution in plants driven by pollinators. *Nat. Commun.* 8:14691.

Ginglinger, J.F., B. Boachon, R. Höfer, et al. 2013. Gene coexpression analysis reveals complex metabolism of the monoterpene alcohol linalool in *Arabidopsis* flowers. *Plant Cell* 25:4640–57.

Gonda, I., E. Bar, V. Portnoy, et al. 2010. Branched-chain and aromatic amino acid catabolism into aroma volatiles in *Cucumis melo* L. fruit. *J. Exp. Bot.* 61:1111–23.

Gonzales-Vigil, E., D.E. Hufnagel, J. Kim, R.L. Last, and C.S. Barry. 2012. Evolution of TPS20-related terpene synthases influences chemical diversity in the glandular trichomes of the wild tomato relative *Solanum habrochaites. Plant J.* 71:921–35.

Günther, J., N.D. Lackus, A. Schmidt, et al. 2019. Separate pathways contribute to the herbivore-induced formation of 2-phenylethanol in poplar. *Plant Physiol.* 180:767–82.

Gutensohn, M., A. Klempien, Y. Kaminaga, et al. 2011. Role of aromatic aldehyde synthase in wounding/herbivory response and flower scent production in different Arabidopsis ecotypes. *Plant J.* 66:591–602.

He, J., R.A. Fandino, R. Halitschke, et al. 2019. An unbiased approach elucidates variation in (*S*)-(+)-linalool, a context-specific mediator of a tri-trophic interaction in wild tobacco. *Proc. Natl. Acad. Sci. USA* 116:14651–60.

He, X., and J. Zhang. 2005. Rapid subfunctionalization accompanied by prolonged and substantial neofunctionalization in duplicate gene evolution. *Genetics* 169:1157–64.

Henry, L.K., M. Gutensohn, S.T. Thomas, J.P. Noel, and N. Dudareva. 2015. Orthologs of the archaeal isopentenyl phosphate kinase regulate terpenoid production in plants. *Proc. Natl. Acad. Sci. USA* 112:10050–55.

Henry, L.K., S.T. Thomas, J.R. Widhalm, et al. 2018. Contribution of isopentenyl phosphate to plant terpenoid metabolism. *Nat. Plants* 4:721–29.

Hibrand Saint-Oyant, L., T. Ruttink, L. Hamama, et al. 2018. A high-quality genome sequence of *Rosa chinensis* to elucidate ornamental traits. *Nat. Plants* 4:473–84.

Hirata, H., T. Ohnishi, H. Ishida, et al. 2012. Functional characterization of aromatic amino acid aminotransferase involved in 2-phenylethanol biosynthesis in isolated rose petal protoplasts. *J. Plant Physiol.* 169:444–51.

Hirata, H., T. Ohnishi, K. Tomida, et al. 2016. Seasonal induction of alternative principal pathway for rose flower scent. *Sci. Rep.* 6:20234.

Hofberger, J.A., A.M. Ramirez, E. Bergh, et al. 2015. Large-scale evolutionary analysis of genes and supergene clusters from terpenoid modular pathways provides insights into metabolic diversification in flowering plants. *PLoS One* 10:e0128808.

Iijima, Y., D.R. Gang, E. Fridman, E. Lewinsohn, and E. Pichersky. 2004. Characterization of geraniol synthase from the peltate glands of sweet basil. *Plant Physiol.* 134:370–79.

Innan, H., and F. Kondrashov. 2010. The evolution of gene duplications: Classifying and distinguishing between models. *Nat. Rev. Genet.* 11:97–108.

Irmisch, S., A. Clavijo McCormick, J. Günther, et al. 2014. Herbivore-induced poplar cytochrome P450 enzymes of the CYP71 family convert aldoximes to nitriles which repel a generalist caterpillar. *Plant J.* 80:1095–107.

Jemth, A.-S., E. Scaletti, M. Carter, T. Helleday, and P. Stenmark. 2019. Crystal structure and substrate specificity of the 8-oxo-dGTP hydrolase NUDT1 from *Arabidopsis thaliana*. *Biochemistry* 58:887–99.

Jia, Q., T.G. Köllner, J. Gershenzon, and F. Chen. 2018. MTPSLs: New terpene synthases in nonseed plants. *Trends Plant Sci.* 23:121–28.

Junker, R.R., and D. Tholl. 2013. Volatile organic compound mediated interactions at the plant-microbe interface. *J. Chem. Ecol.* 39:810–25.

Juwattanasomran, R., P. Somta, S. Chankaew, et al. 2011. A SNP in *GmBADH2* gene associates with fragrance in vegetable soybean variety "Kaori" and SNAP marker development for the fragrance. *Theor. Appl. Genet.* 122:533–41.

Kaminaga, Y., J. Schnepp, G. Peel, et al. 2006. Plant phenylacetaldehyde synthase is a bifunctional homotetrameric enzyme that catalyzes phenylalanine decarboxylation and oxidation. *J. Biol. Chem.* 281:23357–66.

Kessler, D., M. Kallenbach, C. Diezel, E. Rothe, M. Murdock, and I.T. Baldwin. 2015. How scent and nectar influence floral antagonists and mutualists. *eLife* 4:e07641.

Kim, S., M. Park, S.-I. Yeom, et al. 2014. Genome sequence of the hot pepper provides insights into the evolution of pungency in *Capsicum* species. *Nat. Genet.* 46:270–78.

Klahre, U., A. Gurba, K. Hermann, et al. 2011. Pollinator choice in *Petunia* depends on two major genetic Loci for floral scent production. *Curr. Biol.* 21:730–39.

Koeduka, T. 2014. The phenylpropene synthase pathway and its applications in the engineering of volatile phenylpropanoids in plants. *Plant Biotechnol.* 31:401–07.

Koeduka, T., G.V. Louie, I. Orlova, et al. 2008. The multiple phenylpropene synthases in both *Clarkia breweri* and *Petunia hybrida* represent two distinct protein lineages. *Plant J.* 54:362–74.

Kovach, M.J., M.N. Calingacion, M.A. Fitzgerald, and S.R. McCouch. 2009. The origin and evolution of fragrance in rice (*Oryza sativa* L.). *Proc. Natl. Acad. Sci. USA* 106:14444–49.

Krasileva, K.V. 2019. The role of transposable elements and DNA damage repair mechanisms in gene duplications and gene fusions in plant genomes. *Curr. Opin. Plant Biol.* 48:18–25.

Kraszewska, E. 2008. The plant Nudix hydrolase family. *Acta Biochim. Pol.* 55:663–71.

Kreis, W., and J. Munkert. 2019. Exploiting enzyme promiscuity to shape plant specialized metabolism. *J. Exp. Bot.* 70:1435–45.

Lavid, N., J. Wang, M. Shalit, et al. 2002. *O*-methyltransferases involved in the biosynthesis of volatile pheno-lic derivatives in rose petals. *Plant Physiol.* 129:1899–907.

Magnard, J.-L., A. Rius Bony, F. Bettini, et al. 2018. Linalool and linalool nerolidol synthases in roses, several genes for little scent. *Plant Physiol. Biochem.* 127:74–87.

Magnard, J.-L., A. Roccia, J.-C. Caissard, et al. 2015. Biosynthesis of monoterpene scent compounds in roses. *Science* 349:81–83.

Maki, H., and M. Sekiguchi. 1992. MutT protein specifically hydrolyses a potent mutagenic substrate for DNA synthesis. *Nature* 355:273–75.

Martin, D.M., S. Aubourg, M.B. Schouwey, et al. 2010. Functional annotation, genome organization and phylogeny of the grapevine (*Vitis vinifera*) terpene synthase gene family based on genome assembly, FLcDNA cloning, and enzyme assays. *BMC Plant Biol.* 10:226.

Matsuba, Y., T.T. Nguyen, K. Wiegert, et al. 2013. Evolution of a complex locus for terpene biosynthesis in *Solanum*. *Plant Cell* 25:2022–36.

Moghe, G.D., and R.L. Last. 2015. Something old, Something new: Conserved enzymes and the evolution of novelty in plant specialized metabolism. *Plant Physiol.* 169:1512–23.

Muhlemann, J.K., A. Klempien, and N. Dudareva. 2014. Floral volatiles: From biosynthesis to function. *Plant Cell Environ.* 37:1936–49.

Nieuwenhuizen, N.J., S.A. Green, X. Chen, et al. 2013. Functional genomics reveals that a compact terpene synthase gene family can account for terpene volatile production in apple. *Plant Physiol.* 161:787–804.

Nützmann, H.-W., A. Huang, and A. Osbourn. 2016. Plant metabolic clusters: From genetics to genomics. *New Phytol.* 211:771–89.

Ohno, S. 1970. *Evolution by Gene Duplication*. Berlin, Germany: Springer.

Oliva, M., E. Bar, R. Ovadia, et al. 2017. Phenylpyruvate contributes to the synthesis of fragrant benzenoid-phenylpropanoids in *Petunia* x *hybrida* flowers. *Front. Plant Sci.* 8:769.

Panchy, N., M. Lehti-Shiu, and S.-H. Shiu. 2016. Evolution of gene duplication in plants. *Plant Physiol.* 171:2294–316.

Pazouki, L., and U. Niinemets. 2016. Multi-substrate terpene synthases: Their occurrence and physiological significance. *Front. Plant Sci.* 7:1019.

Pichersky, E., and E. Lewinsohn. 2011. Convergent evolution in plant specialized metabolism. *Annu. Rev. Plant Biol.* 62:549–66.

Piechulla, B., M.C. Lemfack, and M. Kai. 2017. Effects of discrete bioactive microbial volatiles on plants and fungi. *Plant Cell Environ.* 40:2042–67.

Punta, M., P.C. Coggill, R.Y. Eberhardt, et al. 2012. The Pfam protein families database. *Nucleic Acids Res.* 40:D290–301.

Raguso, R.A. 2009. Floral scent in a whole-plant context: Moving beyond pollinator attraction. *Funct. Ecol.* 23:837–40.

Raguso, R.A. 2016a. More lessons from linalool: Insights gained from a ubiquitous floral volatile. *Curr. Opin. Plant Biol.* 32:31–36.

Raguso, R.A. 2016b. Plant evolution: Repeated loss of floral scent: A path of least resistance? *Curr. Biol.* 26:R1282–R85.

Raymond, O., J. Gouzy, J. Just, et al. 2018. The *Rosa* genome provides new insights into the domestication of modern roses. *Nat. Genet.* 50:772–77.

Reisenman, C.E., J.A. Riffell, E.A. Bernays, and J.G. Hildebrand. 2010. Antagonistic effects of floral scent in an insect-plant interaction. *Proc. R. Soc. B* 277:2371–79.

Roccia, A., L. Hibrand-Saint Oyant, E. Cavel, et al. 2019. Biosynthesis of 2-phenylethanol in rose petals is linked to the expression of one allele of *RhPAAS*. *Plant Physiol.* 179:1064–79.

Sas, C., F. Müller, C. Kappel, et al. 2016. Repeated inactivation of the first committed enzyme underlies the loss of benzaldehyde emission after the selfing transition in *Capsella*. *Curr. Biol.* 26:3313–19.

Scalliet, G., F. Piola, C.J. Douady, et al. 2008. Scent evolution in Chinese roses. *Proc. Natl. Acad. Sci. USA* 105:5927–32.

Scalliet, G., N. Journot, F. Jullien, et al. 2002. Biosynthesis of the major scent components 3,5-dimethoxytolu-ene and 1,3,5-trimethoxybenzene by novel rose *O*-methyltransferases. *FEBS Lett.* 523:113–18.

Shalit, M., I. Guterman, H. Volpin, et al. 2003. Volatile ester formation in roses: Identification of an acetyl-coenzyme A. geraniol/citronellol acetyltransferase in developing rose petals. *Plant Physiol.* 131:1868–76.

Simkin, A.J., K. Miettinen, P. Claudel, et al. 2013. Characterization of the plastidial geraniol synthase from Madagascar periwinkle which initiates the monoterpenoid branch of the alkaloid pathway in internal phloem associated parenchyma. *Phytochemistry* 85:36–43.

Sun, P., R.C. Schuurink, J.-C. Caissard, P. Hugueney, and S. Baudino. 2016. My way: Noncanonical biosyn-thesis pathways for plant volatiles. *Trends Plant Sci.* 21:884–94.

Tieman, D.M., H.M. Loucas, J.Y. Kim, D.G. Clark, and H.J. Klee. 2007. Tomato phenylacetaldehyde reductases catalyze the last step in the synthesis of the aroma volatile 2-phenylethanol. *Phytochemistry* 68:2660–69.

Tieman, D., M. Taylor, N. Schauer, A.R. Fernie, A.D. Hanson, and H.J. Klee. 2006. Tomato aromatic amino acid decarboxylases participate in synthesis of the flavor volatiles 2-phenylethanol and 2-phenylacetaldehyde. *Proc. Natl. Acad. Sci. USA* 103:8287–92.

Torrens-Spence, M.P., P. Liu, H. Ding, K. Harich, G. Gillaspy, and J. Li. 2013. Biochemical evaluation of the decarboxylation and decarboxylation-deamination activities of plant aromatic amino acid decarboxylases. *J. Biol. Chem.* 288:2376–87.

Verdonk, J.C., M.A. Haring, A.J. van Tunen, and R.C. Schuurink. 2005. *ODORANT1* regulates fragrance biosynthesis in petunia flowers. *Plant Cell* 17:1612–24.

Xie, J., Y. Li, X. Liu, et al. 2019. Evolutionary origins of pseudogenes and their association with regulatory sequences in plants. *Plant Cell* 31:563–78.

Xu, S., T. Brockmöller, A. Navarro-Quezada, et al. 2017. Wild tobacco genomes reveal the evolution of nicotine biosynthesis. *Proc. Natl. Acad. Sci. USA* 114:6133–38.

Yahyaa, M., A. Berim, B. Nawade, M. Ibdah, N. Dudareva, and M. Ibdah. 2019. Biosynthesis of methyleugenol and methylisoeugenol in *Daucus carota* leaves: Characterization of eugenol/isoeugenol synthase and *O*-Methyltransferase. *Phytochemistry* 159:179–89.

Yan, H., S. Baudino, J.-C. Caissard, et al. 2018. Functional characterization of the eugenol synthase gene (*RcEGS1*) in rose. *Plant Physiol. Biochem.* 129:21–26.

13 Volatiles in Glands

Eran Pichersky

CONTENTS

13.1 INTRODUCTION

Volatile compounds, by their volatile nature, play a role in the interactions between the plant and its biotic environment. As detailed elsewhere in the book, they constitute emitted signals that attract pollinators, seed dispersers, and other beneficial organisms in the atmosphere and rhizosphere. Given the function of volatiles as signals, it is not surprising that plants evolved to have specific anatomical structures and types of cells at or near the surface of the plant body that are devoted to the production, storage and emission of volatiles (Fahn, 1988).

However, the location of cells that biosynthesize and store volatile compounds need not be right at the outer surface of the plant body, if appropriate anatomical features and cellular and molecular mechanisms also exist to convey the volatiles to the surface. But a common problem for the plant, regardless of where the biosynthesis of volatiles occurs, is that many of these compounds are also toxic to the plant itself. Indeed, the value of many volatile compounds to the plant is in their toxicity and thus in their direct contribution to defense, rather than in their volatility. This toxicity often stems from the same set of physio-chemical traits – being small (<300 d) and lipophilic molecules with high vapor pressure – that make them volatile as well (Pichersky et al., 2006). Such molecules tend to have low solubility in aqueous solutions, and therefore they end up in the membranes, and sometimes in "oil bodies," of the cells that produce them. High concentrations of such compounds in membranes could disrupt membrane structure and interfere with the stability of membrane proteins. Indeed, the nonspecific nature of the toxicity of many of them could be due to this phenomenon (Rajkowska et al., 2016).

While some plant cells may have adapted specific mechanisms to counteract this toxic effect (see below), a more general mechanism appears to be first their transport toward the outer membrane of the cell, and then either active secretion or passive diffusion through the outer membrane (Adebesin et al., 2017). When the cell producing such toxic compounds is part of the epidermis, or protrudes from it (such as a glandular trichome cell), the emitted volatiles could be released directly into the atmosphere. But for biosynthetic cells embedded deeply inside the plant body, volatiles escaping from their outer membrane could cause toxicity to neighboring cells. In such cases, and even in some cases of biosynthetic epidermal and glandular trichomes, plants have evolved mechanisms to contain and store volatile compounds outside the cell's outer membrane. Indeed, such mechanisms have evolved not only to protect the plant from the toxicity of volatiles (and other specialized compounds that are toxic but are not volatiles), but also to be able to deliver such compounds under specific situations – for example, when the plant is being damaged by an herbivore.

In the following sections, I describe various specific cells and simple anatomical structures, consisting of a few cells, including glandular trichones, idioblasts, secretory cavities, secretory ducts, and laticifers, that have evolved in plants to synthesize, store and emit volatiles. These structures could be collectively called "glands." The botanical term "gland" is borrowed from the animal field, where it means an organ that specializes in the synthesis of a specific compound (or several compounds) that are then secreted out of the organ to be used internally or externally. This definition suits these plant structures well. The primary goals of the following sections are to describe the volatile contents and particular physiological functions of plant glands, but I will also explore their structures and examine how such structures operate to effect release of volatiles to nonself organisms while protecting the plant from self-toxicity. Other regions inside the plant that are noted for the presence of volatiles but are more anatomically complex, such as so-called "scent glands" (osmophores), are also briefly examined, although it is argued here that these entities, as described in the literature, actually do not fit well the definition of glands.

13.2   GLANDULAR TRICHOMES

13.2.1   Distribution of Volatile-Containing Glandular Trichomes in the Plant Kingdom

Structures that protrude from the epidermis layer of plants are often referred to as "hairs," or trichomes, and they appear to perform a variety of functions (Schilmiller et al., 2008; Tissier, 2012; Lange, 2015). A subset of trichomes are defined as "glandular," because they contain one or more cells that actively synthesize and store specialized compounds, some of which may be volatile (Tissier, 2012). Glandular trichomes come in many shapes and levels of complexity even in the same plant (Tissier, 2012; Lange, 2015), from a single cell to more elaborate structures. Their biochemical repertoire is also extremely diverse (Schilmiller et al., 2008), and includes different classes of compounds, some of which are volatile and some are not. Some glandular trichomes continuously secret their compounds, and others store them extracellularly and these compounds are released into the open only when the trichomes are physically impacted. However, with the exception of some glandular trichomes in ferns that produce nonvolatile farinose waxes, reports of volatile compounds detected in glandular trichomes are restricted to flowering plants so far, and mostly to eudicotyledenous plants at that (Tissier, 2012).

Moreover, even among the eudicots, the vast majority of reports of volatile-producing glandular trichomes come from just a few families, including the Lamiaceae, Solanaceae, Asteraceae, and Cannabaceae (Tissier, 2012) (Figure 13.1). And the vast majority of these volatile chemicals are terpenoids, with benzenoids constituting a second but much smaller class, and a smattering of fatty acid derivatives, amino acid derivatives, and polyketides rounding up the rest. In the following sections, the occurrence of volatiles in glandular trichomes will be described using a phylogenetic reference frame.

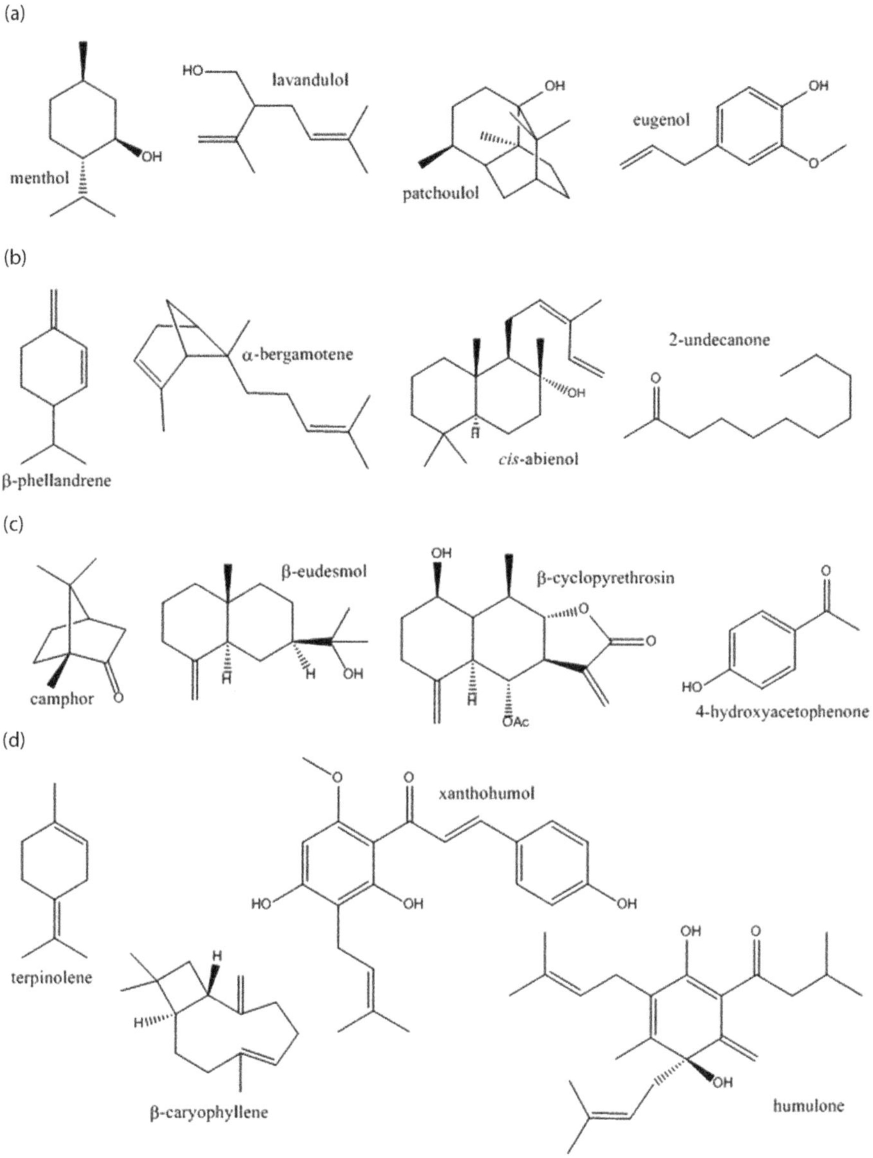

(a)

menthol

lavandulol

patchoulol

eugenol

(b)

β-phellandrene

α-bergamotene

cis-abienol

2-undecanone

(c)

camphor

β-eudesmol

β-cyclopyrethrosin

4-hydroxyacetophenone

(d)

terpinolene

β-caryophyllene

xanthohumol

humulone

FIGURE 13.1 Examples of volatiles in glandular trichomes. (a) Lamiaceae, (b) Solanaceae, (c) Asteraceae, and (d) Cannabaceae.

13.2.1.1 Lamiaceae

This family, with approximately 7,000 species, contains many well known scented plants, such as basil, mint, rosemary, sage, oregano, thyme, lavender, and perilla. There are many reports of the occurrence of various types of glandular trichomes in Lamiaceae species. However, most reports on the presence of volatiles in trichomes in this family, as indeed in many other families described below, follow indirect routes to placing the volatiles in the trichomes. Typically, the presence of trichomes on specific organs is first noted, sometimes followed by histochemical staining of the trichomes using chemical stains that are presumed to identify specific classes of chemicals. Then, volatiles are extracted from the whole organ (e.g., leaf, stem, or flower) and chemically analyzed by gas chromatography-mass spectrometry (GC-MS). The volatiles identified are then assumed to come from the trichomes. This is probably a good assumption for volatiles found in large amounts that, because of their toxicity, are unlikely to be present in any structures other than glandular trichomes (where they are typically stored in the space between the outer membrane of the cell and the cuticle). However, minor volatile compounds identified in such investigations are just as likely to come from cells other than those making up the glandular trichomes. This potential shortcoming being noted, it is nevertheless becoming more common to isolate trichomes and directly analyze their volatiles by GC-MS, or to directly sample volatiles from trichomes with micro-sampling techniques without first separating the trichomes from the rest of the plant (Gang et al., 2001; Falara et al., 2008; McDowell et al., 2011).

With these caveats, the wealth of data on trichome volatiles in Lamiaceae points to the dominance of terpenes in these structures. In some species, volatile monoterpenes predominate, while in others, sesquiterpenes constitute the majority. For example, mint (*Mentha spicata*) trichomes contain high levels of the commercially important monoterpene menthol, while the trichomes of *Pogostemon cablin* produce mostly sesquiterpenes, including patchoulol, a sesquiterpene alcohol highly valued for perfume (Figure 13.1) (Maffei et al., 1986; Deguerry et al., 2006). In addition to the "standard" monoterpenes and sesquiterpenes that occur in the glandular trichomes throughout this plant family, some species in the *Lavandula* genus also make some unusual, or "irregular," monoterpenes in their trichomes (Figure 13.1). These irregular monoterpenes include lavandulol, which is derived from the condensation of two DMAPP molecules, and lavandulyl acetate (Demissie et al., 2013).

Species in *Ocimum*, another genus in the Lamiaceae family, are noted for the presence of phenylpropenes in their trichomes, such as eugenol, methyleugenol, chavichol, and methylchavichol, in addition to mono- and sesquiterpenes (Gang et al., 2001, 2002; Raina et al., 2018). In some cultivated basil (*O. basilicum*) lines that are highly bred, phenylpropenes could constitute up to 75% of total volatiles (Iijima et al., 2004). It should be borne in mind that Lamiaceae trichomes often contain high amounts of nonvolatile compounds as well, much of it consisting of chemicals that belong to the phenylpropanoid class (from which the phenylpropones are derived), such as rosmarinic acid, caffeic acid, etc. (Gang et al., 2002).

Volatiles belonging to other classes of chemicals that have sometimes been reported for Lamiaceae species include straight-chain alkanes, aldehydes, and alcohols, but their low concentrations makes their assignment to the trichome less secure. However, Maggi et al. (2010) reported that the plant known as bastard balm (*Melittis melissophyllum* subsp. *melissophyllum*), a perennial herb growing in Italian woodlands, contains high levels of 1-octen-3-ol, up to 70% of the total essential oil (most of the rest are terpenoids). While a direct proof that this compound is present in the trichomes was not presented, the authors did show that the aerial surfaces of this plant are covered by several types of glandular trichomes, making it likely that 1-octen-3-ol is indeed present in the trichomes.

13.2.1.2   Solanaceae

The Solanaceae (nightshade) family, with about 2,700 species, is considerably smaller than the Lamiaceae but appears to contain as much volatile chemical diversity in the various types of glandular trichomes that its species possess. Within the family, investigations of the chemistry of glandular trichomes have so far been essentially restricted to the large *Solanum* genus, which includes the cultivated tomato (*S. lycopersicum*), and the much smaller *Nicotinana* genus, which includes the cultivated tobacco (*N. tabacum*),

The recently enlarged genus *Solanum* now contains about half of the species in the family, but most of the work on volatiles has concentrated on the cultivated tomato and closely related species. Both cultivated and wild tomato plants contain multiple types of glandular trichomes on their leaves, stems, developing fruits, and sometimes on the flowers (Figure 13.2). The glandular trichomes with a single biosynthetically active cell at the top of the stock (designated as type 1 and type 4, depending on the length of the stock) appear to produce mostly toxic, sticky, nonvolatile compounds such as acylated sugars (Schilmiller et al., 2010; McDowell et al., 2011). On the other

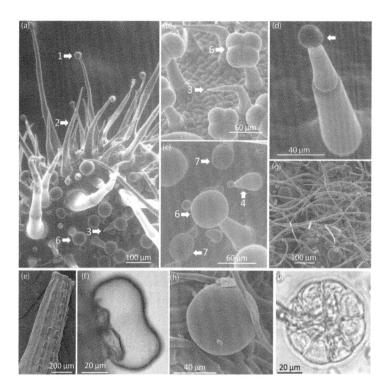

FIGURE 13.2 Examples of plant glandular trichome diversity. (a) Scanning electron microscopy (SEM) image of a young wild tomato (*Solanum habrochaites*, Solanaceae, accession LA2409) leaf. This picture illustrates the complexity of the leaf trichome landscape. At least five trichome types, numbered according to Luckwill (1943) can be observed: Type 1, tall glandular trichomes with a single secretory cell; Type 2, tall non-glandular trichomes; Type 3, short hooked non-glandular trichome; Type 6, glandular trichomes with four head cells; Type 7, short glandular trichomes. (b) Detail of a tomato leaf (*Solanum lycopersicum* LA 4024) showing Type 6 and Type 3 trichomes. (c) Detail of a wild tomato leaf (*S. habrochaites* LA 1777) showing Type 4, Type 6 and Type 7 trichomes. In *S. habrochaites*, the four glandular cells of Type 6 trichomes are enveloped in a peri-cellular cuticle, whereas in the cultivated tomato the four glandular cells can be distinctly seen (see b). (d) Capitate trichome of *Nicotiana sylvestris* (Solanaceae), similar to Type 4 trichomes of tomato. A droplet of diterpenoid-rich exudate can be seen on the side of the glandular head (white arrow). (e) Scanning electron microscopy (SEM) image of a floret from *Tanacetum parthenium* (Asteraceae) showing files of glandular trichomes. (f) A close-up view of the same type of glandular trichome in light microscopy showing the large subcuticular cavity where hydrophobic volatile compounds are stored, similar to the peltate trichomes of the Lamiaceae. (g) Scanning electron microscopy (SEM) image of a *Salvia officinalis* (Lamiaceae) leaf, showing peltate trichomes in a network of hairs. (h, i) Detailed view of a peltate trichome of *S. officinalis*, showing the envelope delimiting the subcuticular storage space (h), and the plate of eight glandular cells (i). (Reproduced, with permission, from Tissier, A., *Plant J.*, 70, 51–68, 2012.)

hand, glandular trichome type 6, which consists of four cells forming a sphere on top of a long stock, appears to specialize in producing volatiles. The type 6 glandular trichomes of various *Solanum* species, such as *S. lycopersicum, S. pimpinellifolium, S. pennellii, and S. habroichaites*, produce and store a large repertoire of monoterpenes and sesquiterpenes (Sallaud et al., 2009; Falara et al., 2011; McDowell et al., 2011). Even within the same species, type 6 glands found on one organ, such as leaves, may have a somewhat different composition of terpenoid volatiles than the type 6 glands found on stems (Falara et al., 2011). In some cases, glandular terpenes are extensively oxidized and are turned into less-volatile, sticky acids, as happens to a large portion of the sesquiterpene bergamotene and santalene in the type 6 glands of *S. habrochaites* (Sallaud et al., 2009).

Solanum type 6 glands exhibit a unique biochemical innovation in how they synthesize some of their mono- and sesquiterpenes. While some *Solanum* terpene synthases (TPSs) present in type 6 glands use the canonical substrates (GPP for monoterpene synthases and *e,e*-FPP for sesquiterpene synthases), others use neryl diphosphate (NPP), the *cis*-isomer of GPP, or *z,z*-FPP, respectively (Schilmiller et al., 2009, Gonzales-Vigil et al., 2012). The *cis*-prenyldiphosphate substrates are produced in the plastids only. Not only are the majority of monoterpenes in type 6 glands derived from NPP rather than GPP (Schilmiller et al., 2009; Gonzales-Vigil et al., 2012), in certain *S. habrochaites* accession the majority of the sesquiterpenes produced in these glands are made from or *z,z*-FPP rather than *e,e*-FPP, and, furthermore, are made in the plastids of the trichome rather than in the cytosol, where sesquiterpenes are usually synthesized (Sallaud et al., 2009). Interestingly, there is no evidence that the use of *cis*-prenyldiphosphate substrates allows for different type of monoterepenes or sequiterpenes to be made, as many of the same monoterpenes and sesquiterpenes produced from *cis*-prenyldiphosphates (Figure 13.1) are produced in other plant species from the corresponding *trans*-prenyldiphosphates. Thus, the advantage for the evolution of this alternative pathway for making mono- and sesquiterpenes is not clear.

Another innovation found in type 6 glands in *Solanum* is the synthesis of a class of volatiles called methylketones. Several S. American tomato species synthesize methylketones, mostly 2-undecanone and 2-tridecanone, but the concentrations of these compounds are particularly high in several accessions of *S. habrochaites* (Antonious, 2001; Fridman et al., 2005). In these accessions, the synthesis of terpenes in these glands is correspondingly low (Ben-Israel., 2009). The biosynthesis of methylketones in *Solanum* has been shown to involve two enzymes, the first being and esterase that interferes with fatty acid biosynthesis by hydrolyzing the carbon-sulfur bond in the 3-ketoacyl-ACP intermediate to produce a C11 or a C13 3-ketoacid, and a second enzyme that brings about decarboxylation (Fridman et al., 2005; Ben-Israel., 2009; Yu et al., 2010). 2-Undecanone and 2-tridecanone are somewhat volatile, but their role is believed to be in defense against insect herbivores, to which these methylketones are toxic (Williams et al., 1980). Recently, 2-undecanone was detected in the sticky exudate of *S. sarrachoides*, an African species in the *Solanum nigrum* complex (Murungi et al., 2016).

Volatiles are important flavor compounds in smoking tobacco. Many of them are formed during the leaf-curing process (which often involves heat as well as catabolism by microbial and plant enzymes) and through pyrolysis during smoking, and so they do not constitute plant volatiles. However, extensive research on the origin of tobacco flavor has shown that many of these volatile/ flavor compounds are derived from diterpenes produced in the glandular trichomes on leaves of *N. tabacum*, and these diterpenes are nominally volatile by themselves. Foremost among them is the labdane-type diterpene *z*-abienol (Sallaud et al., 2012) (Figure 13.1), and another group of semi-volatile diterpenes produced by these glands are the macrocyclic cembratrien-diols (Guo and Wagner, 1995). Some tobacco varieties also produce cembratrien-ols (the precursors of the cembratrien-diols) which are more volatile, and actually appear to be more toxic to aphids (Wang et al., 2001).

13.2.1.3 Asteraceae

With more than 30,000 species, the Asteraceae is one of the two largest families in the angiosperms (the other is Orchidaceae). Many Asteraceae species have been reported to have glandular trichomes (Figure 13.2). In many cases, sesquiterpene lactones (Figure 13.1), which are only minimally volatiles, have been found in these trichomes (Lopes et al., 2013; Ramirez et al., 2012, 2013). However, when examined more closely, these glands are also found to hold more volatile terpenes. For example, the glands of *Artemesia annua*, in addition to the malaria drug sesquiterpene lactone artemisinin, also contain a large number of volatile mono- and sesquiterpenes, with the monoterpenes pinene or camphor dominating, depending on the plant accessions examined (Tellez et al., 1999; Yadav et al., 2014). *Stevia rebaudiana* glandular trichomes, the source of the diterpene glycosides that serve as natural sweeteners, also synthesize and store a variety of volatile mono- and sesquiterpenes (Woelwer-Rieck et al., 2014). The trichomes of the Mexican medicinal plant *Montanoa tomentosa* contain the two diterpene acids which are believed to be responsible for its medicinal properties (including inducting childbirth), but also a number of monoterpenes and sesquiterpenes, with valencene, β-eudesmol (Figure 13.1), and geranyl acetate being the most abundant (Robles-Zepeda et al., 2009). And Piazza et al. (2018) report that the capitate glands of the Argentinian species *Flourensia campestris* contain a large number of volatile mono- and sesquiterpenes, as well as small amounts of two polyketide acetopenones.

13.2.1.4 Cannabaceae

Two species in particular have been extensively studied in this family in respect to volatiles in glands, *Cannabis sativa* (the marijuana plant) and *Humulus lupulus* (hop). Both are diecious plants, and the bracts of the female inflorescences in both species are particularly covered with a dense array of glandular trichomes. In *C. sativa*, these glands contain the psychoactive chemical tetrahydrocannabinol (THC), which is a nonvolatile compound made from a backbone of a prenylated (geranylated) polyketide. However, in addition to THC these glands also contain high amounts of volatile monoterpenes (mostly α-pinene, β-pinene, myrcene, limonene, β-ocimene, and terpinolene) and sesquiterpenes (mostly β-caryophyllene) (Booth et al., 2017). In the glands of the closely related hop plants, these glands contain high levels of the monoterpene myrcene and the sesquiterpenes β-caryophyllene and α-humulene (Wang et al., 2008). Hop glands also contain prenylated (and methylated) polyketides such as humulone and xanthohumol (Figure 13.1) which are somewhat volatile, a property that allows them to contribute to beer flavor when added during the beer making process (Nagel et al., 2008).

13.2.2 Summary for Trichome Section

The origin and evolutionary trajectories of glandular trichomes are still very murky (Tissier, 2012; Lange, 2015). Therefore, it is still not possible to determine with certainty if morphologically similar glands in different taxa are monophyletic. It follows that when considering the volatile repertoire of glandular trichomes from different lineages, it is not possible to argue that similar biochemical profiles are due to inheritance of particular genetic programs, even if this is in fact the case.

The most obvious conclusion from the foregoing description of glandular trichome volatiles in angiosperms is that they overwhelming belong to the terpenoid class, and more specifically consist of mostly monoterpenes and sesquiterpenes. This is not surprising since there are scores of basic monoterpene skeletons and hundreds or even thousands of sesquiterpene skeletons, all produced by enzymes that belong to the ubiquitous terpene synthase family, of which all vascular plant species possess multiple copies (Chen et al., 2011). The presence of a bevy of enzymes that oxidize, methylate, acylate, or otherwise modify these skeletons leads to an even greater diversity of volatile terpenes (Pichersky et al., 2006). The fact that the precursors of terpenes, IPP and DMAPP, are also present in all cells since they also serve as precursors of many general metabolites (Pichersky and Raguso, 2016) also helps in establishing the terpene-producing pathways in glandular trichomes. Therefore, the predominance of volatile terpenes in glandular trichomes of flowering plants could equally likely have come from repeated, independent evolution (i.e., convergence) as from a common ancestral origin.

In addition to volatile terpenes, other volatiles do show up in glandular trichomes of various species. Phenylpropenes such as eugenol constitute one such class. Another class of volatile compounds are polyketides, whose skeleton could be made directly from acetyl groups (as in the case of acetophenones) or from phenylpropanoid acids such coumaric acid (as in the case of xanthohumol), and which are often modified by oxidation/reduction, acylation, methylation, or prenylation. Yet another group consists of compounds derived from fatty acids, such as methylketones and mid-chain alcohols and aldehydes.

An important observation is that a glandular trichome could contain volatile compounds from more than one class, as do basil glands in some accessions (having both terpenes and phenylpropenes) (Iijima et al., 2004) or tomato glands (having both terpenes and methylketones) (Fridman et al., 2005). Moreover, each class of compounds could be represented in the same gland by both volatile and nonvolatiles compounds. While in some cases volatiles from more than one class are present in large amounts, more often one class of compounds dominates, with this class most often being the terpenes. Finally, it has been generally observed that the composition of volatiles in glands is usually determined by the genotype of the plant and by long-term environmental factors, such as seasonal variations, but not by short-term environmental cues such as herbivore damage and temperature fluctuations. It appears that glandular trichomes represent an investment of resources by the plant in a constitutive, pre-formed, first line of defense that buys the plant time to induce additional defenses as the need is perceived.

13.3 IDIOBLASTS

13.3.1 DEFINITION OF IDIOBLASTS

An idioblast is a cell that appears markedly different from neighboring cells. Often such cells appear to have distinct biochemistry, for example by showing oil bodies under various microscopic examinations or by staining distinctly with chemical reagents that are diagnostic for specific classes of chemicals. While by definition idioblasts can occur anywhere in the plant body, in practice a single specialized compound-producing cell that occurs as part of the epidermis (and often protruding from it) would typically be classified as a glandular trichome (see Section 13.2). Therefore, idioblasts will be considered here as cells found below the epidermis.

While many idioblasts are implicated in synthesis and storage of volatile compounds, their position inside the plant body, surrounded by other cells, makes it difficult to directly demonstrate the specific presence of volatiles in idioblasts; extraction of chemicals out of them presents serious technical difficulties. Nevertheless, there have been a few highly successful and informative investigations of volatile-containing idioblasts.

13.3.2 OIL CELLS IN MAGNOLIALES AND LAURALES

Many angiosperm plants have been reported to contain "oil cells." A particularly large numbers of such reports involve plants in the orders Magnoliales and Laurales. The oil cells occur in many different organs and tissues, and they are typically differentiated from other parenchymal cells surrounding them by their bigger size and their thick, suberized cell wall. Staining generally indicates the presence of lipophilic material (hence the name "oil cells"), although not necessary of volatile nature. Additional evidence, however, sometimes suggests the presence of volatile compounds, although direct evidence is typically lacking. Thus, in 1676 Leeuwenhoek already noticed, by examining the bark of *Cinnamomum* (cinnamon, most likely, *C. verum* but could also have been *C. cassia*; family Lauraceae) under his microscope, that it contains oil cells that had a natural yellow/reddish color, and attributed the flavor of cinnamon to the material in these cells (Baas and Gregory, 1985). This seems a reasonable assumption, as the main flavor compound in cinnamon bark is cinnamyl aldehyde (Figure 13.3), a compound with a yellow tinge, and many of its oxidized derivatives are reddish. A more recent investigation used a chemical reagent supposedly diagnostic of the class of sesquiterpene lactones to show the presence of such compounds in the oil cells of *Liriodendron tulipifera* (Magnolicaeae) (Mariani et al., 1989).

13.3.3 CITRAL-CONTAINING IDIOBLASTS IN LEMONGRASS

The leaves of lemongrass (*Cymbopogon citratus*, Poaceae) contain high levels of the monoterpene aldehyde citral. Citral, which is a mixture of the tautomers geranial and neral (Figure 13.3), is the chemical that gives the plant its distinct lemon smell when the leaves are crushed. In an effort to pinpoint the location of citral in the leaves, Lewinsohn et al. (1998) stained cross-sections of

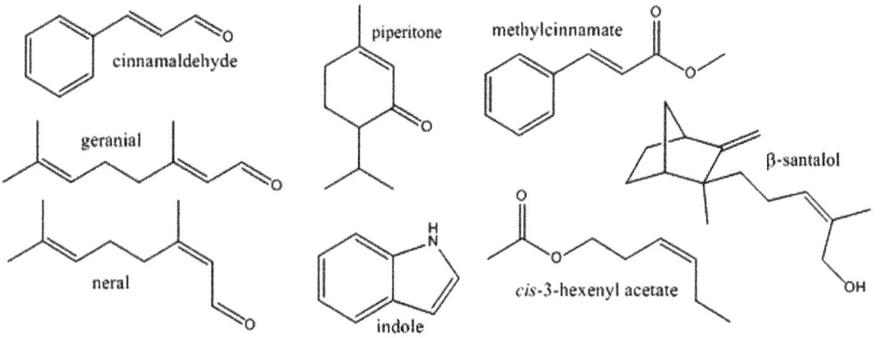

FIGURE 13.3 Examples of volatiles in idioblasts, secretory cavities, secretory ducts, and heartwood.

lemongrass leaves, which are devoid of trichomes, with the histochemical stain known as Schiff's Reagent. The staining identified idioblasts in the adaxial side of the leaf mesophyll, and between vascular bundles, that stained purple-red. The same color was observed when the Schiff's Reagent was reacted with citral *in vitro*, while no other aldehydes tested, including other terpene aldehydes, gave the same color. While still not a direct proof for the presence of citral in the idioblasts, the absence of similar staining elsewhere in the plant strongly suggests that citral is indeed stored, and possibly synthesized, in these idioblasts. The authors further showed, by staining with other specific reagents, that the cell wall of these idioblasts is thick with lignin and suberin, likely protecting the rest of the plant from the toxicity of citral.

13.3.4 SUMMARY OF IDIOBLAST SECTION

It is likely that idioblasts with volatile compounds are much more prevalent in the various organs of plants throughout the plant kingdom, but that the current lack of techniques to identify volatiles in single cells, particularly those embedded deep inside, is a serious impediment in proving this hypothesis. Therefore, it is incumbent upon us to develop such techniques.

13.4 SECRETORY CAVITIES

Some plants have internal structures called subdermal secretory cavities that appear to be an elaboration of the idioblast. A single idioblast cell produces volatile and other specialized chemicals and stores them inside the area of the cell defined by the outer membrane, or outside this area but within the space surrounded by the cuticle. A subdermal secretory cavity, on the other hand, consists of multiple cells, arranged in a sphere, that produce such compounds and secrete them into an inner space at the center of the sphere (this inner space, which is called "lumen," properly constitutes the secretory cavity, but typically researchers refer to the whole structure – the lumen and the cells surrounding it – as the "secretory cavity") (Lange, 2015). The secretory cavity is typically present right under the epidermal layer, and sometimes causes a visible bulge on the surface of the leaf (or other parts of the plant), particularly when the lumen fills up with chemicals (Russin et al., 1992).

Knowledge of synthesis and storage of volatile compounds in subdermal secretory cavities has benefited from the development of methods to isolate such structures apart from the rest of the plant. This has been accomplished by selective digestion of cell walls with commercially available enzymes followed by mechanically teasing out the structures out of the leaves or petals, then collecting them by filtration via meshes with defined pore sizes (Russin et al., 1992; Goodger et al., 2010). Alternatively, laser-dissection and pressure-catapulting have also been used to isolated secretory glands (Voo et al., 2012). Once isolated, volatiles can be extracted and analyzed, as can RNA.

Subdermal secretory cavities that contain mostly indole but also traces of two monoterpenes, piperitone and piperitenone (Figure 13.3), were reported from marigold (*Tagetes erecta*, Asteraceae) petals (Russin et al., 1988). The bulk of the reports on secretory cavities and their volatile contents, however, comes from two groups of plants – eucalyptus species (Myrtaceae) and citrus and related species (Rutaceae). Leaves of various eucalyptus species have secretory cavities whose lumen contains several volatile monoterpenes and sesquiterpenes, in addition to nonvolatile material (Goodger et al., 2010). Moreover, the cells lining up the lumen were shown to express at least one monoterpene synthase gene, indicating that the synthesis of these essential oils likely takes place in the secretory cavity structure (Goodger et al., 2010).

Using laser-dissection, Voo et al. (2012) isolated secretory cavities from the peel of young grapefruit fruits and showed that they synthesize and store a variety of mono- and sesquiterpenes as well as other, nonvolatile specialized and primary metabolites. And the secretory cavities on the fruit skin of another plant in the Rutaceae family, Japanese pepper (*Zanthoxylum piperitum*), were showed to contain a large number of mono- and sesquiterpenes as well as methyl cinnamate and (Z)-3-hexenyl acetate (Figure 13.3) (Fujita et al., 2017).

13.5   SECRETORY DUCTS AND LATICIFERS

The secretory duct and the laticifer appear to be further elaborations, but in different directions, on the idioblast and the subdermal secretory cavity. Rather than a sphere as a secretory cavity is, a secretory duct is a long tube made up of nonlignified epithelial cells and an empty center (lumen). Secretory ducts are prominent in gymnosperms but are also prevalent in angiosperms and to some extent in other vascular plants (Pickard, 2008; Lange, 2015). Such ducts are associated with defense and may constitute an extensive network; when the plant organ is damaged, the material, stored in the lumen under pressure, bursts forth out of the wound.

When the liquid in the lumen is lipophilic, the structures are often called "resin ducts." The epithelial cells of gymnosperm resin ducts synthesize mono-, sesqui- and diterpenes (and a bit of triterpenes) and secret these compounds into the lumen. The ducts are present along the vasculature of the needles and throughout the secondary xylem of the trunk and stems, and when the tissue is damaged and a section of the secretory duct is exposed to the outside, the terpene mixture oozes out (Lange, 2015). Many of the terpenes in this mixture are toxic to microorganisms and insect. After the clear terpene mixture comes out of the plant, the volatile mono- and sesquiterpenes evaporate. In the absence of the solvent, the nonvolatile diterpenes now appear as white solid, and over time they harden, sealing the wound and often physically trapping and even suffocating the insects that might have caused the damage (Trapp and Croteau, 2001; Langenheim, 2003). If ingested by the herbivore, the toxic effects of these chemicals might also cause impairment or death. Furthermore, the volatilized terpenes waft away and might be detected by nearby carnivorous or parasitic insects that then approach the wound site and kill or parasitize the offending insects (Pichersky and Gershenzon, 2002). Gymnosperms also have terpene-containing resin "blisters," which may simply be very short resin ducts or more like secretory cavities (Lange, 2015).

Secretory ducts associated with the vasculature are also found in the dicotyledonous angiosperms, notably in the tropical tree family Burseraceae (Zapata et al., 2013; Souza et al., 2016). In many Burseraceae species, the lumen of the ducts is filled with clear liquid containing mostly mono- and sesquiterpenes. The liquid accumulates under high pressure, so that when the physical integrity of the duct is disrupted, for example by a herbivore, the liquid bursts forth and sprays the offender (Becerra, 1994). Clear, turpentine resin ducts are found in many other plant species in addition to the Burseraceae (Langenheim, 2003). For example, *Pistacia terebinthus* and related *Pistacia* species (family Anacardiaceae) are well known for their resin ducts whose resin, containing triterpenes dissolved in mono- and sesquiterpenes, has been tapped and used as medicine and a flavorant for thousands of years (Papageorgiou et al., 1999; Swaidis et al., 2000; Assimopoulou and Papageorgiou, 2005; Rand et al., 2014).

A laticifer consists of an elongated cell, or multiple elongated cells connected to each other, that synthesize and store specialized metabolites (Pickard, 2008; Lange, 2015). When chemicals that are not soluble in the aqueous cytosol are also produced in the laticifer cells, the resulting emulsion is called latex (Agrawal and Konno, 2009). Latificers are also associated with the vasculature, and when the plant is injured the liquid solution containing various chemicals within the laticifer cell bursts out, similarly to what occurs with secretory ducts. It is generally assumed that the compounds that laticifers produce and store are toxic or otherwise damaging to animals that injure the plant (Agrawal and Konno, 2009). However, because storage of the chemicals occurs inside the living laticifer cell (unlike the situation in secretory ducts with their lumen), concentrations of chemicals such as mono- and sesquiterpenes that are also toxic to plant cells cannot be high. For this reason, laticifers are not major contribution for volatile production and storage in plants.

13.6   SCENT GLANDS (OSMOPHORES) AND HEARTWOOD

In addition to well-defined single cells or small structures that synthesize and emit volatiles, over the years researchers have described other, anatomically less distinct "areas" in the plant body that have been observed to contain volatiles. First among them is the "scent gland," also known as an "osmophore." The osmophore concept was championed by the eminent German botanist Stefan Vogel in the second half of the 20th century, but it goes back to the 19th century (Raguso, 2017). Since capillary gas chromatography-mass spectrometry analysis was not yet available for the majority of his studies, Vogel and his followers generally identified scent glands, usually found on flowers, by using stains that are said to detect specific classes of lypophilic volatile oils. Putting aside the lack of identification of specific volatiles, the major shortcoming of the osmophore concept has been its independence of scale. Thus, a histochemically identified "scent gland" could be an individual cell (e.g., a glandular trichome), a cell cluster (such as a multicellular glandular trichome or cuticular cells), larger and heterogeneous sections of the flower, or even entire appendages (e.g., banner petals) (Pridgeon and Stern, 1983; Stern et al., 1986, 1987).

Similarly to scent glands, there are other parts within the plant body that actively synthesize and store volatile compounds but are either anatomically ill-defined or simply cover an indeterminate area. One such noteworthy part is heartwood. Heartwood is defined as the inner, nonliving section of a tree trunk that is generally distinguished from the outer sapwood by its dark color. The dark color of heartwood comes from the specialized compounds, such as (mostly nonvolatile) phenolic compounds as well as volatile and nonvolatile terpenes. These compounds act as preservatives, deterring insects, fungi and microorganisms and thus prevent wood decay. The heartwood of some plant species is particularly odorous. For example, the angiospermous sandalwood trees (genus *Santalum*, family Santalaceae) are well known for their scented heartwood, which is due mostly to volatile sesquiterpenes such as *cis*-santalol (Figure 13.3) (Celedon et al., 2016).

Heartwood develops from the inside out. It has been reported that the parenchymal cells in a "transition zone" between the sapwood and the heartwood areas synthesize these specialized compounds, eventually leading to their own death and converting themselves and surrounding cells into heartwood, as the transition zone moves outward (Celedon and Bohlmann, 2018). However, recent work with *Santalum album* showed that its heartwood area still contains at least some live cells that are active in the synthesis of sesquiterpenes and that the final step in the production of *cis*-santalol and related sesquiterpene alcohols occurs exclusively in heartwood (Celedon et al., 2016).

13.7   CONCLUSION

Volatile compounds serve many functions in plants, not all dependent on their volatility. The physical properties that make them volatile – low molecular mass, strong lipophilicity – also make it likely that these compounds exert toxic effect on living organisms, including the plants that produce them.

Thus, whether these compounds have a role as air-borne signal molecules or as defense compounds, plants have evolved specialized structures, known collectively as glands, that can produce and store high levels of such compounds without damage to the rest of the plant. Such structures include glandular trichomes that protrude from the epidermal layer and store the noxious compounds outside the cell outer membrane, in a space that is bound by the cuticle. Inside the plant body, single-cell "idioblasts" synthesize such compounds and protect the rest of the plant by a heavy suberized and lignified cell wall. Secretory cavities and secretory ducts secret such compounds into an internal "lumen," from which the compounds escape only when the area is physically damaged. In addition, some anatomically less well-defined internal areas of the plant, such as heartwood and "scent glands," contain some as yet uncharacterized cells that produce volatile compounds.

ACKNOWLEDGMENT

I thank Dr. Alain Tissier for the permission to reprint the pictures of glandular trichomes from various plant species. Work in my laboratory supported by the National Science Foundation grants IOS-PGRP-1546617 and CBET 1565355.

REFERENCES

Adebesin, F., J.R. Widhalm, B. Boachon, et al. 2017. Emission of volatile organic compounds from petunia flowers is facilitated by an ABC transporter. *Science* 356:1386–1388.

Agrawal, A.A., and K. Konno 2009. Latex: A model for understanding mechanisms, ecology, and evolution of plant defense against herbivory. *Annu. Rev. Ecol. Evol. Syst.* 40:311–313.

Antonious, G.F. 2001. Production and quantification of methylketones in wild tomato accession. *J. Environ. Sci. Health B* 36:835–848.

Assimopoulou, A.N., and V.P. Papageorgiou. 2005. GC-MS analysis of penta- and tetra-cyclic triterpenes from resins of *Pistacia* species: Part II. *Pistacia terebinthus* var. *chia. Biomed. Chromatography.* 19:586–605.

Baas, P., and M. Gregory. 1985. A survey of oil cells in the dicotyledons with comments on their replacement by and joint occurrence with mucilage cells. *Israel J. Bot.* 34:167–186.

Becerra, J.X. 1994. Squirt-gun defense in Bursera and the chrysomelid counterploy. *Ecology* 75: 1991–96.

Ben-Israel, I., G. Yu, M.B. Austin, et al. 2009. Multiple biochemical and morphological factors underlie the production of methylketones in tomato trichomes. *Plant Physiol.* 151:1952–1964.

Booth, J.K., J.E. Page, and J. Bohlmann. 2017. Terpene synthases from *Cannabis sativa. PLoS One* 12:e0173911.

Celedon, J.M., A. Chiang, M.M. S. Yuen, et al. 2016. Heartwood specific transcriptome and metabolite signatures of tropical sandalwood (*Santalum album*) reveal the final step of (Z)-santalol fragrance biosynthesis. *Plant J.* 86:289–299.

Celedon, J.M., and J. Bohlmann. 2018. An extended model of heartwood secondary metabolism informed by functional genomics. *Tree Physiol.* 38:311–319.

Chen, F., D. Tholl, J. Bohlmann, and E. Pichersky. 2011. The family of terpene synthase in plants: A mid-size family that is highly diversified throughout the kingdom. *Plant J.* 66:212–229.

Deguerry, F., L. Pastore, S. Wu, A. Clark, J. Chappell, and M. Schalk. 2006. The diverse sesquiterpene profile of patchouli, *Pogostemon cablin*, is correlated with a limited number of sesquiterpene synthases. *Arch. Biochem. Biophys.* 454:123–136.

Demissie, Z., L.A. Erland, M.R. Rheault, et al. 2013. The biosynthetic origin of irregular monoterpenes in Lavandula: Isolation and biochemical characterization of a novel *cis*-prenyl diphosphate synthase gene, lavandulyl diphosphate synthase. *J. Biol. Chem.* 288:6333–6341.

Fahn, A. 1988. Secretory tissues in vascular plants. *New Phytol.* 108:229–257.

Falara, V., T. Akhtar, T.T. H. Nguyen, et al. 2011. The tomato (*Solanum lycopersicum*) terpene synthase gene family. *Plant Physiol.* 157:770–789.

Falara, V., V. Fotopoulos, T. Margaritis, et al. 2008. Transcriptome analysis approaches for the isolation of trichome-specific genes from the medicinal plant *Cistus creticus* subsp. *creticus. Plant Mol. Biol.* 68:633–651.

Fridman, E., J. Wang, Y. Iijima, et al. 2005. Metabolic, genomic, and biochemical analyses of glandular trichomes from the wild tomato species *Lycopersicon hirsutum* identify a key enzyme in the biosynthesis of methylketones. *Plant Cell* 17:1252–1267.

Fujita, Y., T. Koeduka, M. Aida, H. Suzuki, Y. Iijima, and K. Matsui. 2017. Biosynthesis of volatile terpenes that accumulate in the secretory cavities of young leaves of Japanese pepper (*Zanthoxylum piperitum*): Isolation and functional characterization of monoterpene and sesquiterpene synthase genes. *Plant Biochem.* 34:17–28.

Gang, D.R., J. Wang, N. Dudareva, et al. 2001. An investigation of the storage and biosynthesis of phenylpropenes in sweet basil (*Ocimum basilicum* L.). *Plant Physiol.* 125:539–555.

Gang, D.R., N. Lavid, C. Zubieta, et al. 2002. Characterization of phenylpropene *O*-methyltransferases from sweet basil: Facile change of substrate specificity and convergent evolution within a plant OMT family. *Plant Cell* 14:505–519.

Gang, D.R., T. Beuerle, P. Ullmann, D. Werck-Reichhart, and E. Pichersky. 2002. Differential production of *meta* hydroxylated phenylpropanoids in sweet basil (*Ocimum basilicum* L.) peltate glandular trichomes and leaves is controlled by the activities of specific acyltransferases and hydroxylases. *Plant Physiol.* 130:1563–1544.

Gonzales-Vigil, E., D.E. Hufnagel, J. Kim, and R. Last. 2012. Evolution of TPS20-related terpene synthases influences chemical diversity in the glandular trichomes of the wild tomato relative *Solanum habrochaites*. *Plant J.* 71:921–935.

Goodger, J.Q. D., A.M. Heskes, M.C. Mitchell, D.J. Kink, E.H. Neilson, and I.E. Woodrow. 2010. Isolation of intact sub-dermal secretory cavities from Eucalyptus. *Plant Methods* 6:20.

Guo, Z., and G.J. Wagner. 1995. Biosynthesis of cembratrienols in cell-free extracts from trichomes of *Nicotiana tabacum*. *Plant Science* 110:1–10..

Iijima, Y., R. Davidovich-Rikanati, E. Fridman, D.R. Gang, E. Bar, E. Lewinsohn, and E. Pichersky. 2004. The biochemical and molecular basis for the divergent pattern in the biosynthesis of terpene and phenylpropenes in the peltate glands of three cultivars of sweet basil. *Plant Physiol.* 136:3724–3736.

Lange, B.M. 2015. The evolution of plant secretory structures and emergence of terpenoid chemical diversity. *Annu. Rev. Plant Biol.* 66:139–59.

Langenheim, J.H. 2003. *Plant Resins: Chemistry, Evolution, Ecology, and Ethnobotany*. Timber Press, Portland, OR.

Lewinsohn, E., N. Dudai, Y. Tadmor, et al. 1998. Histochemical localization of citral accumulation in lemongrass leaves (*Cymbopogon citratus* (DC.) Stapf., Poaceae). *Ann. Bot.* 81:35–39.

Lopes, A.A., E.S. Pina, D.B. Silva DB, et al. 2013. A biosynthetic pathway of sesquiterpene lactones in *Smallanthus sonchifolius* and their localization in leaf tissues by MALDI imaging. *Chem. Comm.* 49: 9989–9991.

Maffei M, A. Codignola, and M. Fieschi. 1986. Essential oils from *Mentha spicata* L. (Spearmint) cultivated in Italy. *Flavour Fragrance J.* 1:105–109.

Maggi, F., F. Papa, G. Cristalli, G. Sagratini, S. Vittori, and C. Giuliani. 2010. Histochemical localization of secretion and composition of the essential oil in *Melittis melissophyllum* L. subsp. *melissophyllum* from central Italy. *Flavour Fragrance J.* 25:63–70.

Mariani, P., E.M. Cappelletti, D. Campoccia, B. Baldan. 1989. Oil cell ultrastructure and development in *Liriodendron tulipifera* L. *Bot. Gaz.* 150:391–396.

McDowell, R.T., J. Kapteyn, A. Schmidt, et al. 2011. Comparative functional genomic analysis of *Solanum* glandular trichome types. *Plant Physiol.* 155:524–539.

Murungi, L.K., H. Kirwa, D. Sallifu, and B. Torto. 2016. Opposing roles of foliar and glandular trichome volatile components in cultivated nightshade interaction with a specialist herbivore. *PLoS One* E0160383.

Nagel, J., L.K. Culley, Y. Lu Y, et al. 2008. EST analysis of hop glandular trichomes identifies an O-methyltransferase that catalyzes the biosynthesis of xanthohumol. *Plant Cell* 20:186–200.

Papageorgiou, V.P., A.N. Assimopoulou, and N. Yannovits-Argiriadis. 1999. Chemical composition of the essential oil of Chios turpentine. *J. Ess. Oil Res.* 11:367–368.

Piazza, L.A., D. Lopez, M.P. Silva, et al. 2018. Volatiles and non-volatiles in *Flourensia campestris* Griseb: (Asteraceae), how much do capitate glandular trichomes matter? *Chem. Biodiversity* 15: e1700511.

Pichersky, E., and J. Gershenzon. 2002. The formation and function of plant volatiles: Perfumes for pollinator attraction and defense. *Curr. Op. Plant Biol.* 5:237–243.

Pichersky, E., and R.A. Raguso. 2016. Why do plants make so many terpenoid compounds? *New Phytol.* 220:692–702.

Pichersky, E., J.P. Noel, and N. Dudareva. 2006. Biosynthesis of plant volatiles: Nature's diversity and ingenuity. *Science* 311:808–811.

Pickard, W.F. 2008. Laticifers and secretory ducts: Two other tube systems in plants. *New Phytol.* 177:877–888.

Pridgeon, A.M., and W.L. Stern. 1983. Ultrastructure of osmophores in *Restrepia* (Orchidaceae). *Am. J. of Bot.* 70:1233–1243.

Raguso, R.A., and G. Gottsberger. 2017. An ode to osmophores: Stefan Vogel's seminal contributions to the study of scent. *Flora* 232:150–152.

Raina, A.P., and R.C. Misra. 2018. Chemo-divergence in essential oil composition among germplasm collection of five *Ocimum* species from eastern coastal plains of India. *J. Ess. Oil Res.* 30:47–55.

Rajkowska, K., A. Nowak, A. Kunicka-Styczynska, and A. Siadura, 2016. Biological effects of various chemically characterized essential oils: Investigation of the mode of action against *Candida albicans* and HeLa cell. *RSC Advances* 6:97199–97207.

Ramirez, A.M., G. Stoopen, T.R. Menzel, R. Golss, et al. 2012. Bidirectional secretions from glandular trichomes of pyrethrum enable immunization of seedlings. *Plant Cell* 24:4252–4265.

Ramirez, A.M., N. Saillard, T. Yang, M.C. R. Franssen, H.J. Bouwmeester, and M.A. Jongsma. 2013. Biosynthesis of sesquiterpene lactones in pyrethrum (*Tanacetum cinerariifolium*). *PLoS One* 8:e65030.

Rand, K., E. Bar, M. Ben-Ari, E. Lewinsohn, and M. Inbar. 2014. The mono- and sesquiterpene content of aphid-induced galls on *Pistacia palaestina* is not a simple reflection of their composition in intact leaves. *J. Chem. Ecol.* 40:632–642.

Robles-Zepeda, R.E., E. Lozoya-Gloria, M.G. Lopez, M.L. Villareal, E. Ramirez-Chavez, and J. Molina-Torres. 2009. *Montanoa tomentosa* glandular trichomes containing kaurenoic acids chemical profile and distribution. *Fitoterapia* 80: 12–17.

Russin, W.A., T.F. Uchytil, and R.D. Durbin. 1992. Isolation of structurally intact secretory cavities from leaves of African marigold, *Tagetes erecta* L. (Asteraceae). *Plant Sci.* 85:115–119.

Russin, W.A., T.F. Uchytil, G. Feistner, and R.D. Durbin. 1988. Developmental changes in content of foliar secretory cavities of *Tagetes erecta* (Asteraceae). *Am. J. Bot.* 75:1787–1793.

Sallaud, C., C. Giacalone, R. Töpfer, et al. 2012. Characterization of two genes for the biosynthesis of the labdane diterpene Z-abienol in tobacco (*Nicotiana tabacum*) glandular trichomes. *Plant J.* 72:1–17.

Sallaud, C., D. Rontein, S. Onillon, et al. 2009. A novel pathway for sesquiterpene biosynthesis from Z, Z-farnesyl pyrophosphate in the wild tomato *Solanum habrochaites*. *Plant Cell* 21: 301–317.

Schilmiller, A.L., I. Schauvinhold, M. Larson, et al. 2009. Monoterpenes in the glandular trichomes of tomato are synthesized via a neryl diphosphate intermediate rather than geranyl diphosphate. *Proc. Natl. Acad. Sci. USA* 106:10865–10870.

Schilmiller, A.L., R.L. Last, and E. Pichersky. 2008. Harnessing plant trichome biochemistry for the production of useful compounds. *Plant J.* 54:702–711.

Schilmiller, A., F. Shi, J. Kim, et al. 2010. Mass spectrometry screening reveals widespread diversity in trichome specialized metabolites of tomato chromosomal substitution lines. *Plant J.* 62:391–403.

Souza, L.R., F.G. Trindade, R.A. de Oliveira, L.C. B. Costa, V.M. Gomes, M. Da Cunha. 2016. Histochemical characterization of secretory ducts and essential oil analysis of *Protium* species (Burseraceae). *J. Ess. Oil Res.* 28:166–171.

Stern, W.L., K.J. Curry, and A.M. Pridgeon. 1987. Osmophores of *Stanhopea* (Orchidaceae). *Am. J. of Bot.* 74:1323–1331.

Stern, W.L., K.J. Curry, and W.M. Whitten. 1986. Staining fragrance glands in orchid flowers. *Bull. Torrey Bot. Club* 113:288–297.

Swaidis, T., S. Dafnis, and E. Weryzko-Chmielewska. 2000. Distribution, development and structure of resin ducts in *Pistacia lentiscus* var *chia* Duhamel. *Flora* 195:83–94.

Tellez, M.R., C. Canel, A.M. Rimando, and S.O. Duke. 1999. Differential accumulation of isoprenoids in glanded and glandless *Artemisia annua* L. *Phytochem.* 52: 1035–1040.

Tissier, A. 2012. Glandular trichomes: What comes after expressed sequence tags? *Plant J.* 70:51–68.

Trapp, S., and R. Croteau. 2001. Defensive resin biosynthesis in conifers. *Annu. Rev. Plant Physiol. Plant Mol. Biol.* 52:689–724.

Voo, S.S., H.D. Grimes, B.M. Lange. 2012. Assessing the biosynthetic capabilities of secretory glands in citrus peel. *Plant Physiol.* 159:81–94.

Wang, E.M., R. Wang, J. DeParasis, J.H. Loughrin, S.S. Gan, and G.J. Wagner. 2001. Suppression of a P450 hydroxylase gene in plant trichome glands enhances natural-product-based aphid resistance. *Nat. Biotech.* 19:371–374.

Wang, G., L. Tian, N. Aziz, et al. 2008. Terpene biosynthesis in glandular trichomes of hop. *Plant Physiol.* 148:1254–1266.

Williams, W.G., G.G. Kennedy, R.T. Yamamoto, J.D. Thacker, and J. Bordner. 1980. 2-Tridecanone: A naturally occurring insecticide from the wild tomato species *Lycopersicon hirsutum f. glabratum*. *Science* 207;888–889.

Woelwer-Rieck, U., B. May, C. Lankes, and M. Wuest. 2014. Methylerythritol and mevalonate pathway contributions to the biosynthesis of mono-, sesqui- and diterpenes in glandular trichomes and leaves of *Stevia rebaudiana* Bertoni. *J. Agri. Food Chem.* 62: 2428–2435.

Yadav, R.K., R.S. Sangwan, F. Sabir, A.K. Srivastava, and N.S. Sangwan. 2014. Effect of prolonged water stress on specialized secondary metabolites, peltate glandular trichomes, and pathway gene expression in *Artemisia annua* L. *Plant Physiol. Biochem.* 74:70–83.

Yu, G., T.T. H. Nguyen, Y. Guo, et al. 2010. The enzymatic functions of the wild tomato *Solanum habrochaites glabratum* methylketone synthases 1 and 2. *Plant Physiol.* 154:67–77.

Zapata, F., and P.V. A. Fine. 2013. Diversification of the monoterpene synthase gene family (TPSb) in *Protium*, a highly diverse genus of tropical trees. *Mol. Phyl. Evol.* 68:432–442.

14 Emission and Perception of Plant Volatiles

Itay Maoz, Pulu Sun, Michel A. Haring, Robert C. Schuurink, and Natalia Dudareva

CONTENTS

14.1 INTRODUCTION

Plants direct up to 10% of photosynthetically fixed carbon to the synthesis of thousands of different volatile organic compounds (VOCs) (Staudt and Bertin 1998), which are released from every plant tissue. These VOCs have important biological functions, including above- and belowground plant defense (Penuelas and Llusia 2004), attraction of pollinators, plant–plant signaling, and responses to biotic stresses (Kessler and Baldwin 2001; Pichersky and Gershenzon 2002; Loreto and Schnitzler 2010). Besides playing important roles in plant fitness, plant VOCs constitute one of the main sources of reactive carbon molecules entering the atmosphere, thus significantly contributing to atmospheric chemistry and climate (Misztal et al. 2015). It is estimated that plants release between 500 and 1100 Tg of carbon to the atmosphere annually with isoprene accounting for ~44% of the total emission (Fehsenfeld et al. 1992; Guenther et al. 1995). Global warming over the past 30 years has increased plant VOC emission by approximately 10%, and a further 2°C–3°C rise in the mean global temperature, which is predicted to occur this century, may increase atmospheric VOC levels by an additional 30%–45% (Penuelas and Llusia 2003). Released VOCs are often subjected to spontaneous oxidation and degradation by reacting with OH, O_3, and NO_3 radicals leading to formation of aerosols, which are important for sunlight scattering or serve as condensation centers for the production of water droplets (Andreae and Crutzen 2014). Elevated ozone levels reduce the distribution distance of VOCs from their emitting sources thus drastically altering the chemical ecology of plant-animal-microbe interactions (McFrederick, Kathilankal, and Fuentes 2008). Moreover, while higher production and emission of VOCs under elevated temperatures might enhance plant–pollinator interactions, global warming may change flowering time and fruit development and thus negatively impact the relationships between plants and pollinating insects by altering the overlap between flower phenology and insect availability for pollination (Sherry et al. 2007). To date, little is known about how VOCs move within and between cells and are subsequently released into the environment.

VOCs are structurally diverse and based on their biosynthetic origin can be divided into several groups: terpenoids (mono- and sesquiterpenes), phenylpropanoids and benzenoids, amino acids derivatives, and fatty acids derivatives (Chapter 8; Dudareva, Pichersky, and Gershenzon 2004). They are generally lipophilic low-molecular-weight (~100–200 Da) compounds with high vapor pressures at ambient temperature. To be released from cells, VOCs have to move from their site of biosynthesis through the cytosol, then cross the plasma membrane (PM), hydrophilic cell wall, and often the outer waxy cuticle layer. Thus, VOC emission will primarily depend on the physicochemical properties of the compound itself (e.g., octanol-water partition coefficient, diffusivity, volatility), the barrier it goes through (e.g., membrane lipid composition, cell wall porosity, cuticle composition, etc.), and environmental conditions (e.g., temperature, relative humidity, etc.). Until recently it was accepted that the emission of VOCs occurs by a passive diffusion, however, recent studies challenge this hypothesis (Widhalm et al. 2015) as in the absence of active emission mechanism VOCs would accumulate to toxic level jeopardizing cell viability (Adebesin et al. 2017). To decipher the biological mechanisms involved in VOC emission, we have to know more about the cellular and subcellular localization of VOC biosynthesis, about physicochemical properties of volatiles and structural parameters of cellular barriers as well as the plant's transport mechanisms.

14.2   LOCALIZATION AND TYPE OF CELLS INVOLVED IN THE EMISSION OF VOCs

In the last decades, the discovery of genes encoding enzymes catalyzing the VOCs' formation (Chapter 8; Dudareva, Pichersky, and Gershenzon 2004) enabled not only to improve our understanding of the regulation of VOCs production but also to discover the sites of their biosynthesis within the plant tissue. It has been shown that the biosynthesis of VOCs in flowers occurs almost exclusively in epidermal cells, which are in closest proximity to the atmosphere (Dudareva et al. 1996; Scalliet et al. 2006). Indeed, snapdragon S-adenosyl-L-methionine:benzoic acid carboxyl methyltransferase (BAMT) that catalyzes methylbenzoate formation (Kolosova et al. 2001) and rose orcinol O-methyltransferase responsible for 3,5-dimethoxytoluene production (Scalliet et al. 2006), both display uniform distribution along the petal cross sections with highest expression in the cells of adaxial epidermal cell layer, and lower levels in the cell of abaxial epidermis. Interestingly, the adaxial epidermal cell layer of some flowers has a unique conical-shape presumably to enhance emission (Kolosova et al. 2001; Bergougnoux et al. 2007).

In leaves, and stems, epidermal cells can develop into specialized structures, leaf-hairs or trichomes, that are involved in the biosynthesis, storage and release of VOCs (McCaskill, Gershenzon, and Croteau 1992; Turner, Gershenzon, and Croteau 2000; Schilmiller, Last, and Pichersky 2008). Glandular trichomes on aerial organs are widespread in the plant kingdom and can be found in 30% of all vascular plants (Glas et al. 2012). These structures are biosynthetically active and produce volatiles of all major classes, but the blends are often dominated by terpenoids and phenylpropanoids (Dai et al. 2010). Despite the fact that trichomes accumulate high levels of VOCs, they may not actively emit volatiles but release their content upon mechanical disruption (Iijima et al. 2004).

Vegetative VOC production is not limited to trichomes and can also occur in the inner leaf tissues. Indeed, in gray poplar (*Populus* x *canescens*) expression of isoprene synthase, catalyzing isoprene formation was shown to be highest in palisade parenchyma and significantly decreasing within the spongy mesophyll (Cinege et al. 2009). While VOCs produced in leaves could be released via stomata (Niinemets and Reichstein 2003; Hüve et al. 2007), until now, there is no direct or conclusive evidence supporting this hypothesis. Regardless of stomata involvement, at the subcellular level, vegetative VOCs still have to move from the site of their biosynthesis through the cytosol, the PM, and a hydrophilic layer of the cell wall to be released either to the intercellular air spaces connected to stomata or exit the cell through the cuticle.

Volatiles produced in roots are emitted to rhizosphere as part of belowground defense mechanism (Rasmann et al. 2005; Robert et al. 2012). Analysis of VOCs production revealed that it occurs in many types of root cells, but not in the root tip. Expression of 1,8-cineole synthase, for example, was found primarily in the epidermis, cortex, and stele of mature primary and lateral roots, but not in the root meristem or the elongation zone (Chen et al. 2004). In addition, roots of many plant species have a unique type of cells, border cells, which form a boundary between the root and the rhizosphere (Baetz and Martinoia 2014) and were shown to accumulate and secrete specialized metabolites, including VOCs (e.g., hexanal) (Watson et al. 2015).

14.3 EMISSION MECHANISM OF VOCs: FROM PASSIVE TO ACTIVE

The physicochemical properties of VOCs enable them to be released into the environment, but cannot alone explain their relative abundances in emitted volatile blends. Comparison of two monoterpenes in *Citrus medica* flowers, revealed that limonene (with a lower boiling point, 176°C) was accumulated in the tissue, whereas linalool (with a higher boiling point, 197°C) was the most abundantly emitted compound (Altenburger and Matile 1990). Similar results were obtained for peppermint where two compounds of relatively low volatility, menthofuran and pulegone, were more abundant in the emitted mixture than in the total internal pool (Gershenzon, McConkey, and Croteau 2000). To date, many examples exist where VOC emission cannot be explained by a simple concentration-dependent diffusion mechanism. For example, analysis of monoterpenes in *Clarkia breweri* flowers during development revealed that linalool internal pools peaked and declined paralleling emission, whereas linalool oxide internal pools peaked one day after the peak of emission (Pichersky et al. 1994). A relatively large internal pool of methylbenzoate was also found in snapdragon flowers late in flower development when emission was reaching its minimal levels (Goodwin et al. 2003). Taken together, the lack of correlation between VOC emission and compound internal pools within plant tissues strongly suggests that emission of volatile compounds is not merely a matter of biosynthesis and evaporation, but rather a biologically controlled process.

To be emitted from a cell to the atmosphere, VOCs need to move through multi-tier barriers, including the (i) cytosol and/or, in some cases, organelles such as mitochondria and chloroplasts, (ii) PM, (iii) cell wall and (iv) cuticle (if present). Thus, physicochemical properties of each barrier will also provide resistance to the overall emission process. In addition, there is a large variation in water and membrane solubility of plant volatiles, which varies by more than six orders of magnitude (Niinemets and Reichstein 2003). As nonpolar compounds, VOCs will favorably partition into membranes, making diffusion into aqueous compartments slow.

For many years the default assumption was that VOCs simply diffuse out of cell, which may be true for small molecules like isoprene. Thus, using Fick's first law, calculations were performed to determine the concentrations of VOCs at each cellular barrier required to achieve experimentally measured VOC emission fluxes in snapdragon flowers, assuming that VOC emission is solely driven by diffusion (Widhalm et al. 2015). These calculations revealed that the concentrations of methylbenzoate and nerolidol in the PM would have to be 120 and 48 mM, respectively, to sustain their reported emission rates. These extremely high concentrations would be detrimental to membrane integrity and function (Sikkema, DeBont, and Poolman 1995) and even lead to a leakage of organellar/cellular content (Chowhan et al. 2013; Mendanha and Alonso 2015). Moreover, due to their lipophilic nature VOCs could simply partition from the cytosol into any cellular membranes. Lack of membrane selectivity and directionality of VOC movement would have no control over the potential toxic effects of VOCs freely moving around the cell. This led to the conclusion that active biological mechanisms have to exist to keep VOC concentrations in membranes below toxic levels and to achieve VOC emission rates (Widhalm et al. 2015).

14.4   TRANSPORT OF VOCs FROM SITES OF BIOSYNTHESIS TO THE ATMOSPHERE

While transport of water soluble metabolites has been studied systematically (Takano, Noguchi, and Yasumori 2002; Guo et al. 2014), only few studies were devoted, so far, to the investigation of biological mechanisms involved in the emission of lipophilic VOCs (Crouzet et al. 2013; Wang et al. 2016; Adebesin et al. 2017). Since the mode of emission remains largely unknown, several potential active mechanisms involved in shuttling VOCs from undamaged cells were proposed (Widhalm et al. 2015), based on analogy to intracellular trafficking of other hydrophobic compounds, like cuticular lipids (Bird et al. 2007) and diterpenes (Jasinski et al. 2001; Crouzet et al. 2013).

Plant VOCs are synthesized in diverse subcellular compartments. Direct experimental evidences for the localization of enzymes responsible for the final steps in VOC biosynthesis were obtained only for a limited number of proteins (Kolosova et al. 2001; Scalliet et al. 2006). However, based on the primary structure of biosynthetic enzymes involved in the final steps of VOCs formation, it is predicted that many of phenylpropanoids/benzenoids are synthesized in the cytosol (Pichersky, Noel, and Dudareva 2006). Sesquiterpenes are also produced in the cytosol, from farnesyl diphosphate precursor formed via the mevalonic acid pathway. In contrast, monoterpenes are produced in plastids and must cross the plastid membranes in order to be emitted (McGarvey and Croteau 1995).

14.4.1   THE INVOLVEMENT OF TRANSPORTERS AND POTENTIAL BIOLOGICAL CARRIERS IN VOC EMISSION

The potential contribution of an interorganellar membrane bilayer fusion mechanism, hemifusion, to the exchange of lipophilic metabolites between plastids and ER has been recently demonstrated (Mehrshahi et al. 2013). VOCs will favorably partition into internal membranes, therefore hemifusion between subcellular organelles (plastids-ER and ER-PM) could support emission of VOCs (Hawes, Kiviniemi, and Kriechbaumer 2015). So far, there is no direct evidence for the involvement of hemifusion in VOCs trafficking (Figure 14.1a). Transport proteins could also facilitate movement of volatile monoterpenes out of plastids to the cytosol, but have not yet been identified.

Alternatively, delivery of VOCs to the PM could involve secretory vesicle trafficking processes associated with ER, Golgi, *trans*-Golgi network (TGN), and/or vacuole (Figure 14.1a). In plants vesicular transport is thought to play a role in the movement of cytotoxic phytochemicals (Weston, Ryan, and Watt 2012), hormones (Geldner et al. 2001), anti-microbial compounds (Kwon, Bednarek, and Schulze-Lefert 2008), cuticular wax components (McFarlane et al. 2014) and even has been proposed to contribute to VOC emission (Bergougnoux et al. 2007; Widhalm et al. 2015). Secretory vesicles can deliver VOCs partitioned in their lipid bilayers and release them to the PM via exocytosis (fusion) mitigated by the SNARE (soluble *N*-ethylmaleimide-sensitive factor receptors) protein complexes. SNAREs are attachment factors residing on the surface of vesicles (v-SNAREs) and target membranes (t-SNAREs) that are responsible for the docking of vesicles to membranes in the cell (Filippini et al. 2001). At present, there is no direct evidence for the involvement of ER, Golgi apparatus or secretory vesicles in VOC emission. Recently, expression of *VAMP72* gene, encoding a v-SNARE, was RNAi down-regulated in *N. benthamiana* leaves ectopically expressing *caryophyllene synthase* or *linalool synthase* (Ting et al. 2015). Unexpectedly, enhanced emission of caryophyllene and linalool was observed, which was attributed to an increase in stability of proteins in the terpenoid biosynthetic pathways. Overall, the results were unable to resolve the role of vesicles in volatile transport. Transport of VOCs to the PM could also be mediated by a nonvesicular mechanism, which involves soluble carrier proteins with hydrophobic pockets capable of escorting lipophilic compounds (Samuels and McFarlane 2012). A lipid-based mechanism was suggested to take part in the trafficking of VOCs from the site of biosynthesis to the PM and/or through the cell wall (Widhalm et al. 2015).

The hydrophobic PM separates two hydrophilic matrices, i.e., cytosol and cell wall, and would accumulate VOCs to toxic levels in the absence of an active transport across the membrane. Thus, to overcome this effect on PM integrity, the involvement of transmembrane transporter(s) was proposed (Widhalm et al. 2015) (Figure 14.1a). Indeed, an *ABC* (ATP-binding cassette) transporter, *PhABCG1*, was recently shown to be directly involved in the transport of phenylpropanoid and benzenoid volatiles across the PM in petunia flowers. Moreover, RNAi downregulation of *PhABCG1* expression resulted in decreased emission of VOC, which accumulated in the PM to toxic levels (Adebesin et al. 2017). While the *ABCG1* transporter was unable to transport terpenoids compounds, pleiotropic drug resistance (PDR) transporters, also members of the ABC transmembrane transporter family, are capable to transport sesquiterpenes and diterpenes. AaPDR3 was shown to

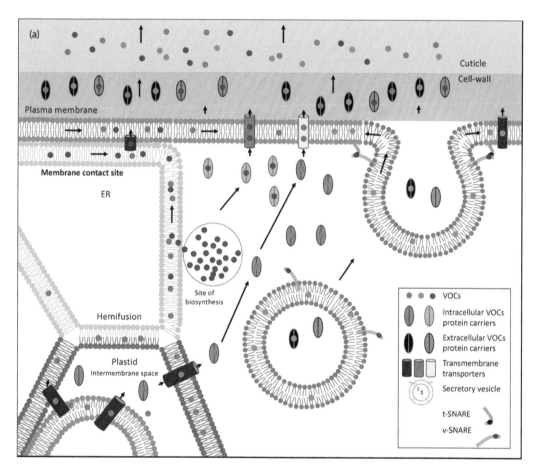

FIGURE 14.1 Proposed models for biological mechanisms involved in the emission or uptake of VOCs in plant cells. (a) VOCs originating from different metabolic pathways are represented by different colors (blue, green and pink). Trafficking of VOCs can be mediated via hemifusion or direct membrane contact sites (supported or independent of transporters). Secretory vesicles (blue) may also be involved in transport of soluble protein carriers and/or VOCs out of the cell. Extracellular (black and gray) and intracellular (orange and green) VOCs protein carriers, such as lipid transfer proteins (LTPs), can transfer VOCs from the cytosol to the plasma membrane (PM) and/or from the PM to the cuticle through the cell wall. Different colors of the VOCs protein carriers represent specific binding affinity to VOCs (if exist). The VOCs protein carriers may also facilitate transport of VOCs within organelles (orange). Transmembrane transporters, such as ABC transporters (blue and light green), localized in the PM, can also be involved in shuttling of VOCs across the PM to the cell wall. Other transporters can mitigate VOCs movement between organelles and the cytosol (purple). Movement of VOCs through the cuticle VOCs is likely to occur by diffusion.   *(Continued)*

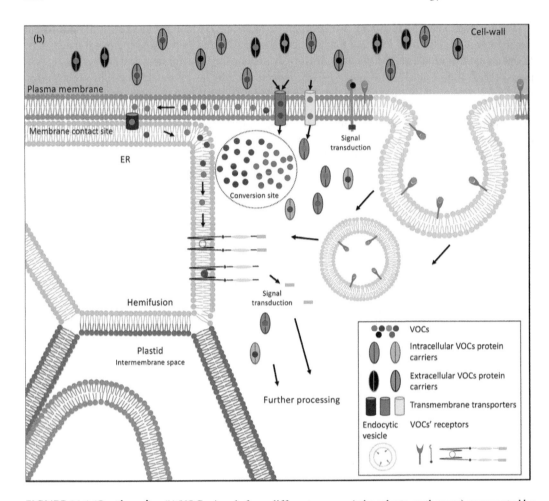

FIGURE 14.1 (Continued) (b) VOCs signals from different sources (other plants, pathogens) represented by different colored circles (blue, red, green, pink, black and gray) can be taken up initiating signal transduction, and/or further metabolized, stored or re-emitted by the plant. VOCs might be perceived by PM receptor or intracellular receptors. Further metabolism might occur in the cytosol or in targeted organelle. VOC trafficking across the cell wall could be mediated by protein carriers (black and gray), such as LTPs. Transmembrane transporters can facilitate movement of VOCs across the PM. ABC transmembrane transporters (blue and light green), localized in the PM can have a dual-function as importers and exporters, supporting VOCs uptake. Hemifusion or direct membrane contact sites may also be involved in trafficking VOCs signals (supported or independent of transporters) (purple). Endocytic vesicles (yellow) may be involved in recycling of VOC receptors, which bind VOCs and are responsible for VOC signal perception. Intracellular VOCs protein carriers, such as LTPs, can also mitigate VOCs trafficking through the cytosol (orange and green). Additional mechanisms for the emission or uptake/perception of VOCs and cannot be excluded.

transport β-caryophyllene in leaves of *Artemisia annua* (Fu et al. 2017), NbPDR1 and NbPDR2 transport the sesquiterpene caspidiol (Shibata et al. 2016; Pierman et al. 2017) and NpPDR1 transports the diterpene scalerol in tobacco (Jasinski et al. 2001). Recently it was demonstrated that in prokaryotes the specificity of ABC transporters can depend on extra-cytoplasmic accessory substrate-binding proteins (SBP) (DeBoer et al. 2019). SBPs (i) are associated with ABC transporters, (ii) are able to deliver compounds, including metal ions, sugars, peptides, and amino acids, and (iii) can have different conformations upon substrate binding thus providing transport specificity. It can be envisioned that a similar mechanism might exist in plants.

Subsequently, to be emitted VOCs must pass the hydrophilic cell wall layer. In general, VOCs have a relatively high partition coefficient, $Log(K_{o/w})$ usually greater than 1, e.g., $K_{o/w\ [methylbenzoate]} = 10^{2.1}$ (Berthod and Carda-Broch 2004). $K_{o/w}$ defines the ratio of compound concentration in the octanol phase (hydrophobic) to its equilibrium aqueous (hydrophilic) phase concentration. To prevent repartitioning of VOCs back to the PM, their movement across the cell wall must be mediated by lipid transfer proteins (LTPs) or other carrier proteins. LTPs are small abundant proteins with molecular mass ranging between 6 and 10 kDa (Edqvist et al. 2018). Often LTPs contain an N-terminus secretory signal peptide directing them to the extracellular layer (Edqvist et al. 2018). In plants, LTPs were shown to be involved in the transport of wax lipids as well as sesqui- and diterpenes from the PM, across the cell wall, to the cuticle (Cameron, Teece, and Smart 2006; DeBono et al. 2009; Choi et al. 2012; Wang et al. 2016). Given that LTPs contain a hydrophobic cavity and generally lack substrate specificity it is possible that they are also involved in VOC transport (Figure 14.1a).

14.4.2 DOES CUTICLE COMPOSITION PLAY A ROLE IN THE VOC EMISSION?

The cuticle is the final interface between the plant cells and the atmosphere. Passage of VOCs through the cuticle solely relies on diffusion, and therefore highly dependent on the physicochemical properties of the compound itself and the composition of the cuticle and its thickness (Goodwin et al. 2003; Jeffree 2006; Bernard and Joubès 2013). As the cuticle is lipophilic matrix, VOCs typically have high cuticle-water partition coefficients, $K_{c/w}$, where $K_{c/w}$, defines the ratio of metabolite concentrations, at equilibrium, in the cuticle relative to aqueous phase. Indeed, $K_{c/w}$ for limonene was determined to be $10^{4.43}$ in leaves of *Citrus aurantium* (Schmid, Steinbrecher, and Ziegler 1992).

The plant cuticle comprises a hydrophobic layer composed of cutin, suberin, and a layer of wax(es). The main monomers in cutin consist of C_{16} and C_{18} ω-hydroxy fatty acids and generally show high similarity between the plant organs and plant species (Jetter 2006). Suberin contains less ω-hydroxy fatty acids relative to cutin, but was shown to have higher levels of glycerol and phenols (Pollard et al. 2008). Cuticular wax is a mixture of alkanes, aldehydes, primary and secondary alcohols, ketones and esters, all derived from very-long-chain fatty acids (ranging between C_{22} and C_{34}) together with cyclic compounds such as triterpenoids and sterols (Goodwin et al. 2003; Jeffree 2006; Pollard et al. 2008; DeBono et al. 2009). Wax compounds in the cuticular matrix self-assemble into a multiphase system of crystalline and amorphous regions (Goodwin and Jenks 2005; Ensikat et al. 2006), and their relative amounts and arrangements govern the diffusivity of solutes and likely VOCs (Merk, Blume, and Riederer 1998). There have been multiple reports correlating high levels of wax (higher crystallinity) with lower cuticle permeability (Sieber et al. 2000; Bourdenx et al. 2011); however, this effect on VOC emission has not been established (Jetter 2006).

While alkanes primarily make up vegetative cuticle waxes, petal waxes contain relatively higher levels of primary alcohols (Buschhaus, Hager, and Jetter 2015), hydroxy esters and branched alkanes (Goodwin et al. 2003). It is possible that such a shift in cuticle composition was evolutionary favorable to release VOCs without requiring high internal concentrations, which would be toxic. Although the effect of cuticle composition on VOC emission was not studied, recent modeling of VOC efflux showed that the cuticle imposes the largest resistance to emission. Moreover, in the presence of biologically mediated mechanisms at the other cellular barriers, the effect of the cuticle on emission was predicted to be even higher (Widhalm et al. 2015).

14.5 PERCEPTION OF VOLATILE SIGNALS IN PLANT–PLANT COMMUNICATION

Living organisms can communicate with each other mainly through four sensing ways: sound (e.g., languages), movement (e.g., hand gestures), visual cues (e.g., color) and molecule perception (e.g., volatiles). As light is essential for plants, they have many photoreceptors for detecting it, but they cannot "see," or perceive sounds and movements, as most animals. Therefore, communication of plants with its environment is heavily dependent on perception of molecules, volatile or nonvolatile. Communication is not only about transmission of signals to the environment but also about uptake and perception of these signals. As plants cannot escape from their enemies by running away, chemicals have become part of their defense mechanism to fight against herbivores and pathogens. Indeed, green leaf volatiles (GLVs), consisting of C_6-aldehydes, alcohols and esters are chemical "weapons" rapidly released upon herbivory and pathogen attack (Scala et al. 2013). In addition, GLVs can be also exploited by neighboring plants as warning signals to prepare themselves in advance – priming – in order to activate defenses to respond faster and stronger against subsequent herbivore and pathogen attacks (Frost et al. 2008). Similar to GLVs, other larger volatile compounds, such as indole (C_8) (Erb et al. 2015), β-ocimene (C_{10}) (Cascone et al. 2015) and *cis*-jasmone (C_{11}) (Oluwafemi et al. 2013), also contribute to plant defense and priming effects. Apart from biotic stresses, abiotic stress-induced volatile blends can also trigger responses in neighboring plants. For example, the stomatal conductance of *Vicia faba* decreases upon exposure to volatiles released from salinity-stressed plants (Caparrotta et al. 2018). Even under seemingly stress-free conditions, a plant has to compete with surrounding plants for resources such as light, nutrients and water, and VOCs are often used as chemical tools for its survival. Indeed, *Arabidopsis* seeds exposed to the headspace of snapdragon flowers or methyl benzoate alone display inhibition of root growth and seed germination rate, respectively (Horiuchi et al. 2007). Monoterpenes released from invasive plants (*Artemisia vulgaris*) also exhibit allelopathic activity and inhibit the growth of neighboring plants (e.g., *Solidoago canadensis*) in order to expand their territory through soil (Barney et al. 2009). Conversely, some emitted VOCs are exploited by parasitic plants to search for host plants. For instance, the parasitic plant *Cuscuta pentagona* was shown to grow specifically toward neighboring tomato plants (*Lycopersicon esculentum*) or toward extracted tomato-plant volatiles (mainly monoterpenes) (Runyon, Mescher, and De Moraes 2006). Taken together, these findings indicate that plants can respond to volatiles adjusting their growth, development and defense.

However, it is still largely unknown how volatiles are perceived by plants and generate cellular responses. Small volatile molecules, such as CO_2, were once believed to diffuse into plant cells because they are permeable through plasma membranes. A recent study indicates that CO_2 diffuses into leaves through the stomata thus entering the intercellular air space. As diffusion of CO_2 in the liquid ($D = 0.0016$ mm^2/s) is 10^4 slower than that in the air ($D = 16$ mm^2/s) (Uskoković 2012), carbonic anhydrases convert CO_2 into HCO_3^- to positively drive the CO_2 from gas phase into liquid phase to further enhance CO_2 diffusion efficiency (Tiwari et al. 2005). Once CO_2 reaches the cell surface, cooporins (previously known as aquaporins, membrane proteins that are permeable to water and CO_2) facilitate its transport across the PM and chloroplast envelope to reach the photosynthetic active site (Terashima and Ono 2002). Hence, the uptake of CO_2 from the atmosphere is not based only on diffusion. Alternatively, the perception of ethylene as a plant hormone is only required in trace amounts to trigger downstream signal cascades. Therefore, while ethylene has similar air to liquid coefficient as CO_2 (Huq and Wood 1968; Elliott and Watts 1972), in this case, passive diffusion into plant cells could be sufficient for ethylene.

As for bigger volatile molecules, there are several studies experimentally demonstrating uptake of VOCs by plants. The involvement of VOCs in plant inter-organ aerial transport via natural fumigation was recently discovered in petunia flowers (Boachon et al. 2019). Using a combination of biochemical and reverse genetic approaches, it was demonstrated that sesquiterpenes emitted from the tubes accumulate in stigma of flower buds. In addition, and unexpectedly, stigmas not exposed to tube-produced terpenes were smaller and had altered microbiome profile, suggesting that the sesquiterpenes play a role in reproductive organ development and defense (Boachon et al. 2019).

There are some examples when plants not only take up VOCs but also subject them to further modifications before exploiting their biochemical activities. For example, (Z)-3-hexenol released from tomato plants infested with the common cutworm (CCW; *Spodoptera litura*) was perceived by uninfested neighboring plants and converted to (Z)-3-hexenyl vicianoside (HexVic), a glycoside with defensive activity that lowers CCW survival rates (Sugimoto et al. 2014). It was also demonstrated that airborne (*E*)-nerolidol is metabolized in *Achyranthes bidentata* leaves and emitted as homoterpene (3*E*)-4,8-dimethyl-1,3,7-nonatriene (DMNT), a defensive VOC involved in plant-insect interaction (Tamogami et al. 2011). However, in latter case additional environmental signals were required like exposure to methyljasmonate or herbivory. In the receiver plant, airborne methyljasmonate can also be converted to jasmonic acid and jasmonoyl isoleucine, leading to VOC emissions and induction of *de novo* jasmonate production (McGale et al. 2018). Similarly, methylsalicylate (MeSA) is perceived and metabolized to its acid form by a salicylic acid-binding protein 2 (SABP2) for systemic acquired resistance signaling in tobacco (Park et al. 2007) and in *Arabidopsis* (Vlot et al. 2008). While abovementioned examples include plant responses to uptake of individual VOC, a mixture of five wound-induced volatile compounds at specific concentrations was required to induce the pyrethrin biosynthesis in *Chrysanthemum cinerariaefolium* receiver plants (Kikuta et al. 2011).

In some cases, modification(s) of VOCs occurs to prevent their toxicity for plant tissues. *Arabidopsis* plants exposed to U-^{13}C-labeled 1-hexanal or (*Z*)-3-hexenal converted them into their alcohols and esters in leaves and released them back to the atmosphere (Matsui et al. 2012). The C_6 aldehydes were also shown to be more toxic to plant tissue than their respective alcohols and esters, as measured by chlorophyll fluorescence parameter [F_M–F_0]/F_M (where F_0 and F_M are initial and maximum fluorescence levels, respectively, and the ratio reflects the maximal photochemical efficiency of photosystem II), an indicator of cell deterioration (Matsui et al. 2012). VOCs can be also subjected to metabolic processes such as glycosylation, glutathionylation, oxidation and reduction to prevent their toxicity and enable continuous uptake (Matsui 2016).

Altogether, these results clearly demonstrate that plants perceive VOCs and harness the biochemical activity of compounds as part of a preemptive defense mechanism. It was proposed that a passive transport mechanism may facilitate the uptake of VOCs, similar to CO_2 diffusion into intercellular air space through stomata (Matsui et al., 2012). However, there is no evidence directly supporting such passive VOC transport. Moreover, the involvement of biological players in reception of volatiles in plants cannot be excluded.

14.6 POTENTIAL MOLECULAR MECHANISMS INVOLVED IN PERCEPTION OF VOCs IN PLANTS

In nature, perception of volatiles is known in mammals as well as in nonvertebrate organisms. Perception occurs by the transmembrane olfactory receptors that belong to G-protein-coupled receptor family and are localized on the olfactory sensory neurons (Firestein 2001). In insects, perception mechanism also includes ligand-gated channels (Wicher et al. 2008). As the number of volatiles is greater than the number of unique receptors, there two types of receptors: generalists with broad binding specificity to volatiles and specialists that are more selective and are associated with pheromone perception (Bohbot and Dickens 2012). Independent of sensorial mechanism, the binding of a volatile to a receptor initiates a signal cascade leading to odor perception. However, perception of volatiles in plants and subsequent signal transduction is still unclear, while there are evidences for the presence of volatile receptors. For example, in the major ethylene signaling pathway hormone is perceived by ethylene receptors (ETRs), transmembrane proteins located on ER membranes, that bind to ethylene with high affinity (Lacey et al., 2014) to trigger downstream signaling. A PM-localized ethylene receptor *NTHK1* was also identified in tobacco, suggesting the existence of an additional pathway for ethylene signaling (Xie et al. 2003).

Ion channels could be potential candidates for volatile perception as well, since they show relatively low ligand specificity. Indeed, CO_2 is perceived via the slow anion channel-associated 1 (SLAC1), the anion channel located in the PM of guard cells to regulate stomatal movement (Zhang et al. 2018). The gamma-aminobutyric acid (GABA) has also been reported to be perceived by an anion channel – aluminum-activated malate transporter (ALMT) in wheat (*Triticum aestivum*) (Ramesh et al. 2015). Moreover, glutamate receptor-like (GLRs) channels, ligand-gated cation channels in plants, have been shown to accept broad range of amino acids for diverse biological processes (Tapken et al. 2013). Another reason to study the potential involvement of ion channels in volatile perception is that they are abundant in guard cells. As guard cells are located in the epidermis of leaves and regulate gas exchange, they have the earliest contact with volatiles. Since VOCs are highly diluted in the air, guard cells may be involved in concentrating volatiles, enabling efficient signal transduction. Whether stomata and volatile receptors serve as part of plant sensing mechanism still remain to be investigated. Alternatively, similar to hormone responses, VOCs can directly bind to proteins and regulate gene expression. Indeed, TOPLESS-like protein, a transcriptional co-repressor with VOC-binding activity, was recently shown to be involved in intracellular caryophyllene perception and signaling in tobacco (Nagashima et al. 2019).

Even though not much is known about the involvement of transporters in the uptake and perception of plant volatiles, transporters have been shown to be involved in the emission of VOCs or secretion of nonvolatiles (Kamimoto et al. 2012; Bernard and Joubès 2013; Adebesin et al. 2017). As ABC transporters can function as both importers and exporters (Berntsson et al. 2010), it is possible that they are also involved in the uptake of VOCs (Figure 14.1b), thus mitigating their perception in the plant cell. ABC transporters that act as importers were previously demonstrated in plants for nonvolatile compounds. Indeed, AtABCg40/AtPDR12 acts as an abscisic acid (ABA) importer (Kang et al. 2010, 2011) while AtABCB21 appears to work as an auxin importer/exporter (Kamimoto et al. 2012). Since ABC transporters require adenosine triphosphate (ATP), the existence of extracellular ATP pools were reported in plants and suggested to be involved in plant stress responses (Kim, Sivaguru, and Stacey 2006).

Another two families of transporters found in plants, multidrug and toxin extrusion (MATE) transporters and nitrate/peptide transporters (NPF), should also be considered as VOC transporters, since they are capable to transport various small molecules. MATE transporters were found to facilitate (i) cytosolic accumulation of salicylic acid upon stresses in *A. thaliana* (Serrano et al. 2013), (ii) transport of flavonoids glycosides and glycoside malonates in *Medicago truncatula* (Zhao et al. 2011), (iii) efflux of ABA in *Arabidopsis* (Zhang et al. 2014), (iv) sequestration of nicotine in *Nicotiana tabacum* (Shitan et al. 2014) and (v) export of hydroxycinnamic acid amides in potato (Dobritzsch et al. 2016). NPFs have been shown to be involved in transport of various compounds such as nitrate, peptides, amino acids, glucosinolates and ABA (Léran et al. 2014). Therefore, ABC/MATE/NPF can be potential candidates for VOCs importers (Figure 14.1b).

As mentioned above, VOCs entering the plant cell are often targets for further modifications. Thus, endocytosis could be a potential trafficking mechanism for preventing VOCs from undesired modifications as well as directing and escorting them to a targeted organelle(s) for further metabolism. Therefore, it is possible that VOCs receptors localized on the PM, are involved in initiating endocytosis or signal transduction (Figure 14.1b).

Previous studies demonstrated the involvement of endocytosis in multiple cellular processes including stress resistance, nutrient uptake, signaling and response to pathogens (Di Rubbo et al. 2013; Bitterlich et al. 2014; Baral et al. 2015; Fan et al. 2015). Plants contain two endocytic pathways: (i) clathrin-mediated and (ii) membrane microdomain-associated routes (Fan et al. 2015). Clathrin-mediated major endocytic pathway relies on the involvement of the vesicle coat protein clathrin (Gadeyne et al. 2014), while the alternative microdomain-associated endocytic route was suggested to be involved in the regulation of signal transduction initiated by receptors of small molecules such as sugars (Bitterlich et al. 2014) and salts (Baral et al. 2015).

To date, endocytosis-mediated recognition of microbial flagellin protein by PM-localized flagellin sensing 2 receptor was shown in *Arabidopsis* demonstrating the existence of a plant-pathogenic sensorial mechanism (Robatzek, Chinchilla, and Boller 2006). Additional PM receptors were identified as endocytic cargos providing a mechanism to regulate diverse plant responses. Examples include *brassinosteroid insensitive 1* the brassinosteroid receptor (Di Rubbo et al. 2013), and the ethylene-inducing xylanase receptor 2, a leucine-rich repeat receptor-like protein that induces plant defense in response to the fungal protein ethylene-inducing xylanase (Bar and Avni 2009). Overall, it is clear that receptor-mediated endocytosis is, at least partially, involved in the regulation of plant sensorial mechanism, however neither evidence for the involvement of endocytosis in formation of vesicles trafficking VOCs nor the existence of VOCs receptors on the PM were demonstrated (Figure 14.1b). To date, the perception/uptake of VOCs by plant cells remain largely unknown and are awaiting their turn.

14.7 CONCLUSIONS

In this chapter we summarized the current knowledge about the emission mechanism of VOCs from plants to the atmosphere. VOCs emitted by plants serve major roles from an ecological and environmental perspectives and therefore it is rather surprising how little we still know about the mechanism(s) of emission. To be emitted from the plants, VOCs have to pass multi-barriers with distinct properties. If considering the resistance of each of these barriers and the physicochemical properties of VOCs, a simple diffusion process cannot explain the observed emission rates. While evidences for the active transport of VOCs across the PM were obtained, how VOCs are transported from the sites of their biosynthesis to the PM and through the cell wall still remain largely unknown. Therefore, more work is required to decipher the end-to-end mechanism involved in the emission of VOCs. Plants not only emit VOCs, but are also targets of released compounds as a part of plant–plant communication. They are continuously exposed to volatiles from the atmosphere and can differentiate and respond to specific cues. There is enough evidence to date showing that plants able to take up VOCs from their surroundings, convert and release them back to the atmosphere or utilize them as part of the defense mechanism. However, whether this mechanism is merely based on diffusion or alternatively on a biological mechanism including receptors and transporters will be groundbreaking for our understanding of plant interactions in the ecosystem.

ACKNOWLEDGMENTS

This work was supported by grant IOS-1655438 from the National Science Foundation and by the USDA National Institute of Food and Agriculture Hatch project 177845 to ND.

REFERENCES

Adebesin, Funmilayo, Joshua R. Widhalm, Benoît Boachon, François Lefèvre, Baptiste Pierman, Joseph H. Lynch, Iftekhar Alam, et al. 2017. "Emission of Volatile Organic Compounds from *Petunia* Flowers Is Facilitated by an ABC Transporter." *Science* 356 (6345): 1386–1388.
Altenburger, Rolf, and Philippe Matile. 1990. "Further Observations on Rhythmic Emission of Fragrance in Flowers." *Planta* 180 (2): 194–197.
Andreae, Meinrat O., and Paul J. Crutzen. 2014. "Atmospheric Aerosols: Biogeochemical Sources and in Atmospheric Chemistry Role." *Science* 276 (5315): 1052–1058.
Baetz, Ulrike, and Enrico Martinoia. 2014. "Root Exudates: The Hidden Part of Plant Defense." *Trends in Plant Science* 19 (2): 90–98.
Bar, Maya, and Adi Avni. 2009. "EHD2 Inhibits Ligand-Induced Endocytosis and Signaling of the Leucine-Rich Repeat Receptor-like Protein LeEix2." *Plant Journal* 59 (4): 600–611.
Baral, Anirban, Niloufer G. Irani, Masaru Fujimoto, Akihiko Nakano, Satyajit Mayor, and M.K. Mathew. 2015. "Salt-Induced Remodeling of Spatially Restricted Clathrin-Independent Endocytic Pathways in Arabidopsis Root." *The Plant Cell* 27 (4): 1297–1315.
Barney, Jacob N, Jed P Sparks, Jim Greenberg, Thomas H Whitlow, and Alex Guenther. 2009. "Biogenic Volatile Organic Compounds from an Invasive Species: Impacts on Plant–Plant Interactions." *Plant Ecology* 203 (2): 195–205.
Bergougnoux, Véronique, Jean Claude Caissard, Frédéric Jullien, Jean Louis Magnard, Gabriel Scalliet, J. Mark Cock, Philippe Hugueney, and Sylvie Baudino. 2007. "Both the Adaxial and Abaxial Epidermal Layers of the Rose Petal Emit Volatile Scent Compounds." *Planta* 226 (4): 853–66.

Bernard, Amélie, and Jérôme Joubès. 2013. "Arabidopsis Cuticular Waxes: Advances in Synthesis, Export and Regulation." *Progress in Lipid Research* 52 (1): 110–129.

Berntsson, Ronnie P.-A., Sander H.J. Smits, Lutz Schmitt, Dirk-Jan Slotboom, and Bert Poolman. 2010. "A Structural Classification of Substrate-Binding Proteins." *FEBS Letters* 584: 2606–2617.

Berthod, Alain, and Samuel Carda-Broch. 2004. "Determination of Liquid-Liquid Partition Coefficients by Separation Methods." *Journal of Chromatography A* 1037 (1–2): 3–14.

Bird, David, Fred Beisson, Alexandra Brigham, John Shin, Stephen Greer, Reinhard Jetter, Ljerka Kunst, Xuemin Wu, Alexander Yephremov, and Lacey Samuels. 2007. "Characterization of Arabidopsis ABCG11/WBC11, an ATP Binding Cassette (ABC) Transporter That Is Required for Cuticular Lipid Secretion." *Plant Journal* 52 (3): 485–498.

Bitterlich, Michael, Undine Krügel, Katja Boldt-Burisch, Philipp Franken, and Christina Kühn. 2014. "Interaction of Brassinosteroid Functions and Sucrose Transporter SlSUT2 Regulate the Formation of Arbuscular Mycorrhiza." *Plant Signaling and Behavior* 9 (10): e970426.

Boachon, Benoît, Joseph H. Lynch, Shaunak Ray, Jing Yuan, Kristian Mark P. Caldo, Robert R. Junker, Sharon A. Kessler, John A. Morgan, and Natalia Dudareva. 2019. "Natural Fumigation as a Mechanism for Volatile Transport between Flower Organs." *Nature Chemical Biology* 15: 583–588.

Bohbot, Jonathan D., and Joseph C. Dickens. 2012. "Selectivity of Odorant Receptors in Insects." *Frontiers in Cellular Neuroscience* 6: 2010–2013.

Bourdenx, Brice, Amélie Bernard, Frédéric Domergue, Stéphanie Pascal, Amandine Léger, Dominique Roby, Marjorie Pervent, et al. 2011. "Overexpression of Arabidopsis *ECERIFERUM1* Promotes Wax Very-Long-Chain Alkane Biosynthesis and Influences Plant Response to Biotic and Abiotic Stresses." *Plant Physiology* 156 (1): 29–45.

Buschhaus, Christopher, Dana Hager, and Reinhard Jetter. 2015. "Wax Layers on *Cosmos bipinnatus* Petals Contribute Unequally to Total Petal Water Resistance." *Plant Physiology* 167 (1): 80–88.

Cameron, Kimberly D., Mark A. Teece, and Lawrence B. Smart. 2006. "Increased Accumulation of Cuticular Wax and Expression of Lipid Transfer Protein in Response to Periodic Drying Events in Leaves of Tree Tobacco." *Plant Physiology* 140 (1): 176–183.

Caparrotta, Stefania, Sara Boni, Cosimo Taiti, Emily Palm, Stefano Mancuso, and Camilla Pandolfi. 2018. "Induction of Priming by Salt Stress in Neighboring Plants." *Environmental and Experimental Botany* 147: 261–270.

Cascone, Pasquale, Luigi Iodice, Massimo E Maffei, Simone Bossi, Gen-ichiro Arimura, and Emilio Guerrieri. 2015. "Tobacco Overexpressing β-Ocimene Induces Direct and Indirect Responses against Aphids in Receiver Tomato Plants." *Journal of Plant Physiology* 173: 28–32.

Chen, Feng, Dae-Kyun Ro, Jana Petri, Jonathan Gershenzon, Jörg Bohlmann, Eran Pichersky, and Dorothea Tholl. 2004. "Characterization of a Root-Specific Arabidopsis Terpene Synthase Responsible for the Formation of the Volatile Monoterpene 1, 8-Cineole." *Plant Physiology* 135 (4): 1956–1966.

Choi, Yong Eui, Soon Lim, Hyun Jung Kim, Jung Yeon Han, Mi Hyun Lee, Yanyan Yang, Ji Ah Kim, and Yun Soo Kim. 2012. "Tobacco NtLTP1, a Glandular-Specific Lipid Transfer Protein, Is Required for Lipid Secretion from Glandular Trichomes." *Plant Journal* 70 (3): 480–491.

Chowhan, Nadia, Harminder Pal Singh, Daizy R. Batish, Shalinder Kaur, Nitina Ahuja, and Ravinder K. Kohli. 2013. "β-Pinene Inhibited Germination and Early Growth Involves Membrane Peroxidation." *Protoplasma* 250 (3): 691–700.

Cinege, Gyöngyi, Sandrine Louis, Robert Hänsch, and Jörg Peter Schnitzler. 2009. "Regulation of Isoprene Synthase Promoter by Environmental and Internal Factors." *Plant Molecular Biology* 69 (5): 593–604.

Crouzet, Jérôme, Julien Roland, Emmanuel Peeters, Tomasz Trombik, Eric Ducos, Joseph Nader, and Marc Boutry. 2013. "NtPDR1, a Plasma Membrane ABC Transporter from *Nicotiana tabacum*, Is Involved in Diterpene Transport." *Plant Molecular Biology* 82 (1–2): 181–192.

Dai, Xinbin, Guodong Wang, Dong Sik Yang, Yuhong Tang, Pierre Broun, M. David Marks, Lloyd W. Sumner, Richard A. Dixon, and Patrick Xuechun Zhao. 2010. "TrichOME: A Comparative Omics Database for Plant Trichomes." *Plant Physiology* 152 (1): 44–54.

DeBoer, Marijn, Giorgos Gouridis, Ruslan Vietrov, Stephanie L. Begg, Florence Husada Gea K. Schuurman-Wolters, Nikolaos Eleftheriadis, Bert Poolman, Christopher A. McDevitt, and Thorben Cordes. 2019. "Conformational and Dynamical Plasticity in Substrate-Binding Proteins Underlies Selective Transport in ABC Importers." *eLife* 8: e44652.

DeBono, Allan, Trevor H. Yeats, Jocelyn KC Rose, David Bird, Reinhard Jetter, Ljerka Kunst, and Lacey Samuels. 2009. "Arabidopsis LTPG Is a Glycosylphosphatidylinositol-Anchored Lipid Transfer Protein Required for Export of Lipids to the Plant Surface." *The Plant Cell* 21 (4): 1230–1238.

Di Rubbo, Simone, Niloufer G. Irani, Soo Youn Kim, Zheng-Yi Xu, Astrid Gadeyne, Wim Dejonghe, Isabelle Vanhoutte, et al. 2013. "The Clathrin Adaptor Complex AP-2 Mediates Endocytosis of Brassinosteroid Insensitive 1 in Arabidopsis." *The Plant Cell* 25 (8): 2986–2997.

Dobritzsch, Melanie, Tilo Lübken, Lennart Eschen-Lippold, Karin Gorzolka, Elke Blum, Andreas Matern, Sylvestre Marillonnet, Christoph Böttcher, Birgit Dräger, and Sabine Rosahl. 2016. "MATE Transporter-Dependent Export of Hydroxycinnamic Acid Amides." *The Plant Cell* 28 (2): 583–596.

Dudareva, Natalia, Eran Pichersky, and Jonathan Gershenzon. 2004. "Biochemistry of Plant Volatiles." *Plant Physiology* 135 (4): 1893–1902.

Dudareva, Natalia, Leland Cseke, Victoria M. Blanc, and Eran Pichersky. 1996. "Evolution of Floral Scent in Clarkia: Novel Patterns of *S*-Linalool Synthase Gene Expression in the *C. breweri* Flower." *The Plant Cell* 8 (7): 1137–1148.

Edqvist, Johan, Kristina Blomqvist, Jeroen Nieuwland, and Tiina A. Salminen. 2018. "Plant Lipid Transfer Proteins: Are We Finally Closing in on the Roles of These Enigmatic Proteins?" *Journal of Lipid Research* 59 (8): 1374–1382.

Elliott, Robert W., and Harry Watts. 1972. "Diffusion of Some Hydrocarbons in Air: A Regularity in the Diffusion Coefficients of a Homologous Series." *Canadian Journal of Chemistry* 50: 31–34.

Ensikat, Hans Jürgen, Markus Boese, Werner Mader, Wilhelm Barthlott, and Kerstin Koch. 2006. "Crystallinity of Plant Epicuticular Waxes: Electron and X-Ray Diffraction Studies." *Chemistry and Physics of Lipids* 144 (1): 45–59.

Erb, Matthias, Nathalie Veyrat, Christelle A. M. Robert, Hao Xu, Monika Frey, Jurriaan Ton, and Ted C. J. Turlings. 2015. "Indole Is an Essential Herbivore-Induced Volatile Priming Signal in Maize." *Nature Communications* 6: 6273.

Fan, Lusheng, Ruili Li, Jianwei Pan, Zhaojun Ding, and Jinxing Lin. 2015. "Endocytosis and Its Regulation in Plants." *Trends in Plant Science* 20 (6): 388–397.

Fehsenfeld, Fred, Jack Calvert, Ray Fall, Paul Goldan, Alex B. Guenther, C. Nicholas Hewitt, Brian Lamb, et al. 1992. "Emissions of Volatile Organic Compounds from Vegetation and the Implications for Atmospheric Chemistry." *Global Biogeochemical Cycles* 6 (4): 389–430.

Filippini, Francesco, T. Valeria Rossi, Hierry Galli, Alberta Budillon, Michele D'Urso, and Maurizio D'Esposito. 2001. "Longins: A New Evolutionary Conserved VAMP Family Sharing a Novel SNARE Domain." *Trends in Biochemical Sciences* 26 (7): 407–409.

Firestein, Stuart. 2001. "How the Olfactory System Makes Sense of Scents." *Nature* 413: 211–218.

Frost, Christopher J., Mark C. Mescher, Christopher Dervinis, John M. Davis, John E. Carlson, and Consuelo M. De Moraes. 2008. "Priming Defense Genes and Metabolites in Hybrid Poplar by the Green Leaf Volatile *Cis*-3-Hexenyl Acetate." *New Phytologist* 180 (3): 722–734.

Fu, Xueqing, Pu Shi, Qian He, Qian Shen, Yueli Tang, Qifang Pan, Yanan Ma, et al. 2017. "AaPDR3, a PDR Transporter 3, Is Involved in Sesquiterpene β-Caryophyllene Transport in *Artemisia annua*." *Frontiers in Plant Science* 8: 723.

Gadeyne, Astrid, Clara Sánchez-Rodríguez, Steffen Vanneste, Simone Di Rubbo, Henrik Zauber, Kevin Vanneste, Jelle Van Leene, et al. 2014. "The TPLATE Adaptor Complex Drives Clathrin-Mediated Endocytosis in Plants." *Cell* 156 (4): 691–704.

Geldner, Niko, Jiří Friml, Ork-Dieter Stierhof, Gerd Jürgens, and Klaus Palme. 2001. "Auxin Transport Inhibitors Block PIN1 Cycling and Vesicle Trafficking." *Nature* 413: 425–428.

Gershenzon, Jonathan, Marie E. McConkey, and Rodney B. Croteau. 2000. "Regulation of Monoterpene Accumulation in Leaves of Peppermint." *Plant Physiology* 122 (1): 205–213.

Glas, Joris J., Bernardus C.J. J Schimmel, Juan M. Alba, Rocío Escobar-Bravo, Robert C. Schuurink, and Merijn R. Kant. 2012. "Plant Glandular Trichomes as Targets for Breeding or Engineering of Resistance to Herbivores." *International Journal of Molecular Sciences* 13 (12): 17077–17103.

Goodwin, S Mark, and Matthew A Jenks. 2005. "Plant Cuticle Function as a Barrier to Water Loss." In *Plant Abiotic Stress*, edited by Matthew A. Jenks and Paul M. Hasegawa, pp. 14–36. Blackwell Publishing Pondicherry, Pondicherry, India.

Goodwin, S. Mark, Natalia Kolosova, Christine M. Kish, Karl V. Wood, Natalia Dudareva, and Matthew A. Jenks. 2003. "Cuticle Characteristics and Volatile Emissions of Petals in *Antirrhinum majus*." *Physiologia Plantarum* 117 (3): 435–443.

Guenther, Alex, C. Nicholas Hewitt, Erickson David, Ray Fall, Geron Chris, Graedel Tom, Harley Peter, et al. 1995. "A Global Model of Natural Volatile Organic Compound Emissions." *Journal of Geophysical Research* 100 (94): 8873–8892.

Guo, Woei-Jiun, Reka Nagy, Hsin-Yi Chen, Stefanie Pfrunder, Ya-Chi Yu, Diana Santelia, Wolf B. Frommer, and Enrico Martinoia. 2014. "SWEET17, a Facilitative Transporter, Mediates Fructose Transport across the Tonoplast of Arabidopsis Roots and Leaves." *Plant Physiology* 164 (2): 777–789.

Hawes, Chris, Petra Kiviniemi, and Verena Kriechbaumer. 2015. "The Endoplasmic Reticulum: A Dynamic and Well-Connected Organelle." *Journal of Integrative Plant Biology* 57 (1): 50–62.

Horiuchi, Jun-ichiro, Dayakar V Badri, Bruce A Kimball, Florence Negre, Natalia Dudareva, Mark W Paschke, and Jorge M Vivanco. 2007. "The Floral Volatile, Methyl Benzoate, from Snapdragon (*Antirrhinum majus*) Triggers Phytotoxic Effects in *Arabidopsis thaliana*." *Planta* 226 (1): 1–10.

Huq, Aminul, and Trevor Wood. 1968. "Diffusion Coefficient of Ethylene Gas in Water." *Journal of Chemical and Engineering Data* 13 (2): 256–259.

Hüve, K., M. M. Christ, E. Kleist, R. Uerlings, Ülo Niinemets, A. Walter, and J. Wildt. 2007. "Simultaneous Growth and Emission Measurements Demonstrate an Interactive Control of Methanol Release by Leaf Expansion and Stomata." *Journal of Experimental Botany* 58 (7): 1783–1793.

Iijima, Yoko, Rachel Davidovich-Rikanati, Eyal Fridman, David R. Gang, Einat Bar, Efraim Lewinsohn, and Eran Pichersky. 2004. "The Biochemical and Molecular Basis for the Divergent Patterns in the Biosynthesis of Terpenes and Phenylpropenes in the Peltate Glands of Three Cultivars of Basil." *Plant Physiology* 136 (3): 3724–3736.

Jasinski, Michal, Yvan Stukkens, Herve Degand, Benedicte Purnelle, Jacqueline Marchand-Brynaert, and Marc Boutry. 2001. "A Plant Plasma Membrane ATP Binding Cassette-Type Transporter Is Involved in Antifungal Terpenoid Secretion." *The Plant Cell* 13 (5): 1095–1107.

Jeffree, Christopher E. 2006. "The Fine Structure of the Plant Cuticle." In *Annual Plant Reviews. Biology of the Plant Cuticle*, edited by Markus Riederer and Caroline Müller, vol. 23, pp. 11–125, Blackwell, Oxford.

Jetter, Reinhard. 2006. "Examination of the Processes Involved in the Emission of Scent Volatiles from Flowers." In *Biology of Floral Scent*, edited by Natalia Dudareva and Eran Pichersky, pp. 125–146. CRC Press, Boca Raton, FL.

Kamimoto, Yoshihisa, Kazuyoshi Terasaka, Masafumi Hamamoto, Kojiro Takanashi, Shoju Fukuda, Nobukazu Shitan, Akifumi Sugiyama, et al. 2012. "Arabidopsis ABCB21 Is a Facultative Auxin Importer/Exporter Regulated by Cytoplasmic Auxin Concentration." *Plant and Cell Physiology* 53 (12): 2090–2100.

Kang, Joohyun, Jae-Ung Hwang, Miyoung Lee, Yu-Young Kim, Sarah M. Assmann, Enrico Martinoia, and Youngsook Lee. 2010. "PDR-Type ABC Transporter Mediates Cellular Uptake of the Phytohormone Abscisic Acid." *Proceedings of the National Academy of Sciences* 107 (5): 2355–2360.

Kang, Joohyun, Jiyoung Park, Hyunju Choi, Bo Burla, Tobias Kretzschmar, Youngsook Lee, and Enrico Martinoia. 2011. "Plant ABC Transporters." *The Arabidopsis Book* 9: e0153.

Kessler, André, and Ian T. Baldwin. 2001. "Defensive Function of Herbivore-Induced Plant Volatile Emissions in Nature." *Science* 291 (5511): 2141–2144.

Kikuta, Yukio, Hirokazu Ueda, Koji Nakayama, Yoshio Katsuda, Rika Ozawa, Junji Takabayashi, Akikazu Hatanaka, and Kazuhiko Matsuda. 2011. "Specific Regulation of Pyrethrin Biosynthesis in *Chrysanthemum cinerariaefolium* by a Blend of Volatiles Emitted from Artificially Damaged Conspecific Plants." *Plant and Cell Physiology* 52 (3): 588–596.

Kim, Sung-Yong, Mayandi Sivaguru, and Gary Stacey. 2006. "Extracellular ATP in Plants. Visualization, Localization, and Analysis of Physiological Significance in Growth and Signaling." *Plant Physiology* 142 (3): 984–992.

Kolosova, Natalia, Debra Sherman, Dale Karlson, and Natalia Dudareva. 2001. "Cellular and Subcellular Localization of *S*-Adenosyl-L-Methionine: Benzoic Acid Carboxyl Methyltransferase, the Enzyme Responsible for Biosynthesis of the Volatile Ester Methylbenzoate in Snapdragon Flowers." *Plant Physiology* 126 (3): 956–964.

Kwon, Chian, Pawel Bednarek, and Paul Schulze-Lefert. 2008. "Secretory Pathways in Plant Immune Responses." *Plant Physiology* 147 (4): 1575–1583.

Lacey, Randy F., and Brad M. Binder. 2014. "How Plants Sense Ethylene Gas: The Ethylene Receptors." *Journal of Inorganic Biochemistry* 133: 58–62.

Léran, Sophie, Kranthi Varala, Jean-Christophe Boyer, Maurizio Chiurazzi, Nigel Crawford, Françoise Daniel-Vedele, Laure David, et al. 2014. "A Unified Nomenclature of *Nitrate Transporter 1/Peptide Transporter* Family Members in Plants." *Trends in Plant Science* 19 (1): 5–9.

Loreto, Francesco, and Jörg Peter Schnitzler. 2010. "Abiotic Stresses and Induced BVOCs." *Trends in Plant Science* 15 (3): 154–166.

Matsui, Kenji, Kohichi Sugimoto, Jun'ichi Mano, Rika Ozawa, and Junji Takabayashi. 2012. "Differential Metabolisms of Green Leaf Volatiles in Injured and Intact Parts of a Wounded Leaf Meet Distinct Ecophysiological Requirements." *Plos One* 7 (4): e36433.

Matsui, Kenji. 2016. "A Portion of Plant Airborne Communication Is Endorsed by Uptake and Metabolism of Volatile Organic Compounds." *Current Opinion in Plant Biology* 32: 24–30.

McCaskill, David, Jonathan Gershenzon, and Rodney Croteau. 1992. "Morphology and Monoterpene Biosynthetic Capabilities of Secretory Cell Clusters Isolated from Glandular Trichomes of Peppermint (*Mentha piperita* L.)." *Planta* 187 (4): 445–454.

McFarlane, Heather E., Yoichiro Watanabe, Weili Yang, Yan Huang, John Ohlrogge, and Lacey A. Samuels. 2014. "Golgi- and *trans*-Golgi Network-Mediated Vesicle Trafficking Is Required for Wax Secretion from Epidermal Cells." *Plant Physiology* 164 (3): 1250–1260.

McFrederick, Quinn S., James C. Kathilankal, and Jose D. Fuentes. 2008. "Air Pollution Modifies Floral Scent Trails." *Atmospheric Environment* 42 (10): 2336–2348.

McGale, Erica, Ian T. Baldwin, Emmanuel Gaquerel, Celia Diezel, Meredith C. Schuman, Stefan Meldau, and Sara Greenfield. 2018. "The Active Jasmonate JA-Ile Regulates a Specific Subset of Plant Jasmonate-Mediated Resistance to Herbivores in Nature." *Frontiers in Plant Science* 9: 787.

McGarvey, Douglas J., and Rodney Croteau. 1995. "Terpenoid Metabolism." *The Plant Cell* 7 (7): 1015–1026.

Mehrshahi, Payam, Giovanni Stefano, Federica Andaloro, Joshua Michael Brandizzi, John E. Froehlich, and Dean DellaPenna. 2013. "Transorganellar Complementation Redefines the Biochemical Continuity of Endoplasmic Reticulum and Chloroplasts." *Proceedings of the National Academy of Sciences* 110 (29): 12126–12131.

Mendanha, Sebastião Antonio, and Antonio Alonso. 2015. "Effects of Terpenes on Fluidity and Lipid Extraction in Phospholipid Membranes." *Biophysical Chemistry* 198: 45–54.

Merk, Susanne, Alfred Blume, and Markus Riederer. 1998. "Phase Behaviour and Crystallinity of Plant Cuticular Waxes Studied by Fourier Transform Infrared Spectroscopy." *Planta* 204 (1): 44–53.

Misztal, Pawel Konrad, C. Nicholas Hewitt, Jurgen Wildt, James D. Blande, Allyson S.D. Eller, Silvano Fares, Drew R. Gentner, et al. 2015. "Atmospheric Benzenoid Emissions from Plants Rival Those from Fossil Fuels." *Scientific Reports* 5: 12064.

Nagashima, Ayumi, Takumi Higaki, Takao Koeduka, Ken Ishigami, Satoko Hosokawa, Hidenori Watanabe, Kenji Matsui, et al. 2019. "Transcriptional Regulators Involved in Responses to Volatile Organic Compounds in Plants." *Journal of Biological Chemistry* 294 (7): 2256–2266.

Niinemets, Ülo, and Markus Reichstein. 2003. "Controls on the Emission of Plant Volatiles through Stomata: Differential Sensitivity of Emission Rates to Stomatal Closure Explained." *Journal of Geophysical Research* 108 (D7): 4211.

Oluwafemi, Sunday, Sarah Y. Dewhirst, Nathalie Veyrat, Stephen Powers, Toby J. A. Bruce, John C. Caulfield, John A. Pickett, and Michael A. Birkett. 2013. "Priming of Production in Maize of Volatile Organic Defence Compounds by the Natural Plant Activator *cis*-Jasmone." *Plos One* 8 (6): e62299.

Park, Sang-Wook, Evans Kaimoyo, Dhirendra Kumar, Stephen Mosher, and Daniel F. Klessig. 2007. "Methyl Salicylate Is a Critical Mobile Signal for Plant Systemic Acquired Resistance." *Science* 318 (5847): 113–116.

Penuelas, Josep, and Llusia Llusia. 2003. "BVOCs: Plant Defense against Climate Warming?" *Trends in Plant Science* 8 (3): 105–109.

Penuelas, Josep, and Llusia Llusia. 2004. "Plant VOC Emissions: Making Use of the Unavoidable." *Trends in Ecology and Evolution* 32 (2): 402–404.

Pichersky, Eran, and Jonathan Gershenzon. 2002. "The Formation and Function of Plant Volatiles: Perfumes for Pollinator Attraction and Defense." *Current Opinion in Plant Biology* 5 (3): 237–243.

Pichersky, Eran, Joseph P. Noel, and Natalia Dudareva. 2006. "Biosynthesis of Plant Volatiles: Nature's Diversity and Ingenuity." *Science* 311 (5762): 808–811.

Pichersky, Eran, Robert A. Raguso, Efraim Lewinsohn, and Rodney Croteau. 1994. "Floral Scent Production in *Clarkia* (Onagraceae): 1. Localization and Developmental Modulation of Monoterpene Emission and Linalool Synthase Activity." *Plant Physiology* 106 (4): 1533–1540.

Pierman, Baptiste, Frédéric Toussaint, Aurélie Bertin, Daniel Lévy, Nicolas Smargiasso, Edwin De Pauw, and Marc Boutry. 2017. "Activity of the Purified Plant ABC Transporter NtPDR1 Is Stimulated by Diterpenes and Sesquiterpenes Involved in Constitutive and Induced Defenses." *Journal of Biological Chemistry* 292 (47): 19491–19502.

Pollard, Mike, Fred Beisson, Yonghua Li, and John B Ohlrogge. 2008. "Building Lipid Barriers: Biosynthesis of Cutin and Suberin." *Trends in Plant Science* 13 (5): 236–246.

Ramesh, Sunita A, Stephen D Tyerman, Bo Xu, Jayakumar Bose, Satwinder Kaur, Vanessa Conn, Patricia Domingos, et al. 2015. "GABA Signalling Modulates Plant Growth by Directly Regulating the Activity of Plant-Specific Anion Transporters." *Nature Communications* 6: 7879.

Rasmann, Sergio, Tobias G. Köllner, Jörg Degenhardt, Ivan Hiltpold, Stefan Toepfer, Ulrich Kuhlmann, Jonathan Gershenzon, and Ted C.J. Turlings. 2005. "Recruitment of Entomopathogenic Nematodes by Insect-Damaged Maize Roots." *Nature* 434: 732–737.

Robatzek, Silke, Delphine Chinchilla, and Thomas Boller. 2006. "Ligand-Induced Endocytosis of the Pattern Recognition Receptor FLS2 in Arabidopsis." *Genes & Development* 20 (5): 537–542.

Robert, Christelle A.M., Matthias Erb, Marianne Duployer, Claudia Zwahlen, Gwladys R. Doyen, and Ted C.J. Turlings. 2012. "Herbivore-Induced Plant Volatiles Mediate Host Selection by a Root Herbivore." *New Phytologist* 194 (4): 1061–1069.

Runyon, Justin B., Mark C. Mescher, and Consuelo M. De Moraes. 2006. "Volatile Chemical Cues Guide Host Location and Host Selection by Parasitic Plants." *Science* 313 (5795): 1964–1967.

Samuels, Lacey, and Heather E. McFarlane. 2012. "Plant Cell Wall Secretion and Lipid Traffic at Membrane Contact Sites of the Cell Cortex." *Protoplasma* 249 (1): 19–23.

Scala, Alessandra, Silke Allmann, Rossana Mirabella, A. Michel Haring, and C. Robert Schuurink. 2013. "Green Leaf Volatiles: A Plant's Multifunctional Weapon against Herbivores and Pathogens." *International Journal of Molecular Sciences* 14 (9): 17781–17811.

Scalliet, Gabriel, Claire Lionnet, Mickaël Le Bechec, Laurence Dutron, Jean-Louis Magnard, Sylvie Baudino, Véronique Bergougnoux, et al. 2006. "Role of Petal-Specific Orcinol O-Methyltransferases in the Evolution of Rose Scent." *Plant Physiology* 140 (1): 18–29.

Schilmiller, Anthony L., Robert L. Last, and Eran Pichersky. 2008. "Harnessing Plant Trichome Biochemistry for the Production of Useful Compounds." *Plant Journal* 24: 702–711.

Schmid, Christine, Rainer Steinbrecher, and Hubert Ziegler. 1992. "Partition Coefficients of Plant Cuticles for Monoterpenes." *Trees* 6 (1): 32–36.

Serrano, Mario, Bangjun Wang, Bibek Aryal, Christophe Garcion, Eliane Abou-Mansour, Silvia Heck, Markus Geisler, Felix Mauch, Christiane Nawrath, and Jean-Pierre Métraux. 2013. "Export of Salicylic Acid from the Chloroplast Requires the Multidrug and Toxin Extrusion-Like Transporter EDS5." *Plant Physiology* 162 (4): 1815–1821.

Sherry, Rebecca A, Xuhui Zhou, Shiliang Gu, John A Arnone III, David S Schimel, Paul S Verburg, Linda L Wallace, and Yiqi Luo. 2007. "Divergence of Reproductive Phenology under Climate Warming." *Proceedings of the National Academy of Sciences* 104 (1): 198–202.

Shibata, Yusuke, Makoto Ojika, Akifumi Sugiyama, Kazufumi Yazaki, David A. Jones, Kazuhito Kawakita, and Daigo Takemotoa. 2016. "The Full-Size ABCG Transporters Nb-ABCG1 and Nb-ABCG2 Function in Pre- and Postinvasion Defense against *Phytophthora infestans* in *Nicotiana benthamiana*." *The Plant Cell* 28 (5): 1163–1681.

Shitan, Nobukazu, Shota Minami, Masahiko Morita, Minaho Hayashida, Shingo Ito, Kojiro Takanashi, Hiroshi Omote, et al. 2014. "Involvement of the Leaf-Specific Multidrug and Toxic Compound Extrusion (MATE) Transporter Nt-JAT2 in Vacuolar Sequestration of Nicotine in *Nicotiana tabacum*." *PLoS ONE* 9 (9): e108789.

Sieber, Patrick, Martine Schorderet, Ulrich Ryser, Antony Buchala, Pappachan Kolattukudy, Jean-Pierre Métraux, and Christiane Nawrath. 2000. "Transgenic Arabidopsis Plants Expressing a Fungal Cutinase Show Alterations in the Structure and Properties of the Cuticle and Postgenital Organ Fusions." *The Plant Cell* 12 (5): 721–737.

Sikkema, Jan, Jan A.M. DeBont, and Bert Poolman. 1995. "Mechanisms of Membrane Toxicity of Hydrocarbons." *Microbiological Reviews* 59 (2): 201–222.

Staudt, M., and N. Bertin. 1998. "Light and Temperature Dependence of the Emission of Cyclic and Acyclic Monoterpenes from Holm Oak (*Quercus ilex* L.) Leaves." *Plant, Cell and Environment* 21 (4): 385–395.

Sugimoto, Koichi, Kenji Matsui, Yoko Iijima, Yoshihiko Akakabe, Shoko Muramoto, Rika Ozawa, Masayoshi Uefune, et al. 2014. "Intake and Transformation to a Glycoside of (Z)-3-Hexenol from Infested Neighbors Reveals a Mode of Plant Odor Reception and Defense." *Proceedings of the National Academy of Sciences* 111 (19): 7144–7149.

Takano, Junpei, Kyotaro Noguchi, and Miho Yasumori. 2002. "Arabidopsis Boron Transporter for Xylem Loading." *Nature* 420: 337–340.

Tamogami, Shigeru, Yukiko Takahashi, Makoto Abe, Koji Noge, Randeep Rakwal, and Ganesh Kumar Agrawal. 2011. "Conversion of Airborne Nerolidol to DMNT Emission Requires Additional Signals in *Achyranthes bidentata*." *FEBS Letters* 585 (12): 1807–1813.

Tapken, Daniel, Uta Anschütz, Lai-Hua Liu, Thomas Huelsken, Guiscard Seebohm, Dirk Becker, and Michael Hollmann. 2013. "A Plant Homolog of Animal Glutamate Receptors Is an Ion Channel Gated by Multiple Hydrophobic Amino Acids." *Science Signaling* 6 (279): 1–10.

Terashima, Ichiro, and Kiyomi Ono. 2002. "Effects of $HgCl_2$ on CO_2 Dependence of Leaf Photosynthesis: Evidence Indicating Involvement of Aquaporins in CO_2 Diffusion across the Plasma Membrane." *Plant and Cell Physiology* 43 (1): 70–78.

Ting, Hieng Ming, Thierry L. Delatte, Pim Kolkman, Johana C. Misas-Villamil, Renier A.L. L Van Der Hoorn, Harro J. Bouwmeester, and Alexander R. Van Der Krol. 2015. "SNARE-RNAi Results in Higher Terpene Emission from Ectopically Expressed Caryophyllene Synthase in *Nicotiana benthamiana*." *Molecular Plant* 8 (3): 454–466.

Tiwari, A., P. Kumar, S. Singh, and S. A. Ansari. 2005. "Carbonic Anhydrase in Relation to Higher Plants." *Photosynthetica* 43 (1): 1–11.

Turner, Glenn W., Jonathan Gershenzon, and Rodney B. Croteau. 2000. "Development of Peltate Glandular Trichomes of Peppermint." *Plant Physiology* 124 (2): 665–680.

Uskoković, Vuk. 2012. "Dynamic Light Scattering Based Microelectrophoresis: Main Prospects and Limitations." *Journal of Dispersion Science and Technology* 33 (12): 1762–1786.

Vlot, Anna Corina, Po-Pu Liu, Robin K Cameron, Sang-Wook Park, Yue Yang, Dhirendra Kumar, Fasong Zhou, et al. 2008. "Identification of Likely Orthologs of Tobacco Salicylic Acid-Binding Protein 2 and Their Role in Systemic Acquired Resistance in *Arabidopsis thaliana*." *Plant Journal* 56 (3): 445–456.

Wang, Bo, Arman Beyraghdar Kashkooli, Adrienne Sallets, Hieng Ming Ting, Norbert C.A. de Ruijter, Linda Olofsson, Peter Brodelius, et al. 2016. "Transient Production of Artemisinin in *Nicotiana benthamiana* Is Boosted by a Specific Lipid Transfer Protein from *A. Annua*." *Metabolic Engineering* 38: 159–169.

Watson, Bonnie S., Mohamed F. Bedair, Ewa Urbanczyk-Wochniak, David V. Huhman, Dong Sik Yang, Stacy N. Allen, Wensheng Li, Yuhong Tang, and Lloyd W. Sumner. 2015. "Integrated Metabolomics and Transcriptomics Reveal Enhanced Specialized Metabolism in *Medicago truncatula* Root Border Cells." *Plant Physiology* 167 (4): 1699–1716.

Weston, Leslie A., Peter R. Ryan, and Michelle Watt. 2012. "Mechanisms for Cellular Transport and Release of Allelochemicals from Plant Roots into the Rhizosphere." *Journal of Experimental Botany* 63 (9): 3445–3454.

Wicher, Dieter, Ronny Schäfer, René Bauernfeind, Marcus C. Stensmyr, Regine Heller, Stefan H. Heinemann, and Bill S. Hansson. 2008. "Drosophila Odorant Receptors Are Both Ligand-Gated and Cyclic-Nucleotide-Activated Cation Channels." *Nature* 452: 1007–1011.

Widhalm, Joshua R., Rohit Jaini, John A. Morgan, and Natalia Dudareva. 2015. "Rethinking How Volatiles Are Released from Plant Cells." *Trends in Plant Science* 20 (9): 545–50.

Xie, Can, Jin Song Zhang, Hua Lin Zhou, Jian Li, Zhi Gang Zhang, Dao Wen Wang, and Shou Yi Chen. 2003. "Serine/Threonine Kinase Activity in the Putative Histidine Kinase-Like Ethylene Receptor NTHK1 from Tobacco." *Plant Journal* 33 (2): 385–393.

Zhang, Haiwen, Huifen Zhu, Yajun Pan, Yuexuan Yu, Sheng Luan, and Legong Li. 2014. "A DTX/MATE-Type Transporter Facilitates Abscisic Acid Efflux and Modulates ABA Sensitivity and Drought Tolerance in Arabidopsis." *Molecular Plant* 7 (10): 1522–1532.

Zhang, Jingbo, Nuo Wang, Yinglong Miao, Felix Hauser, J Andrew McCammon, Wouter-Jan Rappel, and Julian I Schroeder. 2018. "Identification of SLAC1 Anion Channel Residues Required for CO_2/ Bicarbonate Sensing and Regulation of Stomatal Movements." *Proceedings of the National Academy of Sciences* 115 (44): 11129–11137.

Zhao, Jian, David Huhman, Gail Shadle, Xian-Zhi He, Lloyd W Sumner, Yuhong Tang, and Richard A Dixon. 2011. "MATE2 Mediates Vacuolar Sequestration of Flavonoid Glycosides and Glycoside Malonates in *Medicago truncatula*." *The Plant Cell* 23 (4): 1536–1555.

Section III

Volatiles in Plant–Plant, Plant–Insect and Plant–Microbial Interactions

15 Floral Volatiles for Pollinator Attraction and Speciation in Sexually Deceptive Orchids

*Rod Peakall, Darren C. J. Wong, Björn Bohman,
Gavin R. Flematti, and Eran Pichersky*

CONTENTS

15.1 INTRODUCTION

15.1.1 What Is Sexual Deception?

Pollination by sexual deception is a worldwide phenomenon, which has evolved independently multiple times. Sexually deceptive plants secure insect pollination by chemical and physical mimicry of the female of their pollinator. Several hundred cases are now known in the Orchidaceae, with many more cases likely to be discovered with future studies (Gaskett, 2011, Phillips et al., 2014, Bohman et al., 2016a). Single cases are also known in the Asteraceae (Ellis and Johnson, 2010) and Iridaceae (Vereecken et al., 2012), suggesting this pollination strategy may be more widespread than currently known (Table 15.1).

In the sexually deceptive orchids investigated so far, semiochemicals are of paramount importance for pollinator attraction (Bohman et al., 2016a). Furthermore, as a by-product of the mimicry of specific insect sex pheromones, the pollination of these orchids is highly specialized: typically an orchid species depends on just a single pollinator species in a given plant population (Paulus and Gack, 1990, Ayasse et al., 2011, Gaskett, 2011, Phillips et al., 2014, Bohman et al., 2016a, Paulus, 2018). Thus, the remarkable pollination strategy of sexual deception ranks as one of the most highly pollinator-specific specialized pollination strategies, yet is also one of the most widespread.

Here we first provide an updated summary of sexually deceptive orchid genera and their pollinators. Next, we build on our recent reviews of the chemistry and biosynthesis of sexual deception (Bohman et al., 2016a, Wong et al., 2017b). Despite the short timeframe since these publications, there have been some exciting semiochemical and biosynthesis discoveries that we highlight. Finally, we briefly consider the empirical evidence supporting the hypothesis of pollinator-driven ecological speciation as a major mode of evolution in sexually deceptive orchids.

15.2 THE TAXONOMIC BREADTH AND GEOGRAPHIC DISTRIBUTION OF SEXUALLY DECEPTIVE PLANTS AND THEIR POLLINATORS

15.2.1 Plant Diversity

In Table 15.1 we summarize the taxonomic breadth of sexually deceptive plants and their pollinators. The orchid examples are drawn from two of the largest of the five subfamilies. Within the subfamily Orchidoideae, 16 genera drawn from nine subtribes and three tribes, are involved. Their geographic distribution includes Europe and the Mediterranean region, Australia and New Zealand, South Africa and South America. Two major centers of species diversity are found in this group: European *Ophrys* with upper estimates of ~150 species (Table 15.1), and multiple genera of Australian orchids from within the tribe Diurideae including *Arthrochilus*, *Caladenia*, *Chiloglottis*, *Cryptostylis*, *Drakaea* and *Paracaleana*, among others, where >230 species are sexually deceptive. However, the total number could be much higher, with new evidence suggesting that sexual deception is also widespread in the large Australasian genus *Pterostylis* from the tribe Cranichideae (Reiter et al., 2019) (Table 15.1).

The six reported cases in the subfamily Epidendroideae span four genera representing four tribes. With one recently discovered exception in Asia, these examples are drawn from Central and South America (Table 15.1). Despite the small number of confirmed cases, it is possible that many additional orchid species will prove to be sexually deceptive in this subfamily, particularly in the large genus *Lepanthes* with more than 1000 species currently known (Martel et al., 2019, Reiter et al., 2019).

15.2.2 Pollinator Diversity

The taxonomic diversity of the male pollinators exploited by sexually deceptive plants is perhaps even wider than the plant diversity (Table 15.1). Three orders of insects are exploited: Hymenoptera, Diptera and Coleoptera. Among Hymenoptera, 14 different families representing six superfamilies are involved. Within European sexually deceptive plant genera, bees

TABLE 15.1

Sexually Deceptive (SD) Plant Genera and Their Taxonomy, Distribution, Pollinator Type and Pollinator Taxonomy

Genera/Species	Plant Taxonomy[1] Subfamily: Tribe: Subtribe	Floral Chemistry	Male Pollinator	Pollinator Taxonomy[2,3] Order: Superfamily: Family	#Spp A[1]	#Spp B[4]	#SD[5]	#PSD	References
Orchidaceae-European genera									
Ophrys	Orchidoideae: Orchideae: Orchidinae		Bees and apoid wasps, scoliid wasp, beetles	See below	34	149		>100	Gaskett (2011), Paulus (2018)
O. sphegodes		C	*Andrena nigroaenea* bee	Hymenoptera: Apoidea: Andrenidae					Paulus (2018), Schiestl et al. (1999)
O. exaltata		C	*Colletes cunicularius* bee	Hymenoptera: Apoidea: Colletidae					Paulus (2018), Mant et al. (2005a)
O. chestermanii		P	*Bombus vestalis* bumblebee	Hymenoptera: Apoidea: Apidae					Paulus (2018), Gögler et al. (2011)
O. leochroma		P	*Eucera kullenbergi* bee	Hymenoptera: Apoidea: Anthrophoridae					Paulus (2018), Cuervo et al. (2017)
O. atlantica		U	*Megachile parietina* bee (Formerly *Chalicodoma perietina*)	Hymenoptera: Apoidea: Megachilididae					Paulus (2018)
O. insectifera		U	*Argogorytes mystaceus* digger wasp	Hymenoptera: Apoidea: Crabronidae					Paulus (2018), Borg-Karlson, (1990)
O. speculum		C	*Dasyscolia ciliata* wasp (Formerly *Campsoscolia ciliata*)	Hymenoptera: Scolioidea: Scoliidae					Paulus (2018), Ayasse et al. (2003)
O. scolopax		P	*Phylloperta horticola* beetles, also *Eucera* bees	Coleoptera: Scarabaeoidea: Scarabaeidae					Paulus (2018), Borg-Karlson, (1990)
Serapias lingua	Orchidoideae: Orchideae: Orchidinae		Apid bee	Hymenoptera: Apoidea: Apidae	13	36	1	1	Vereecken et al. (2012)
S. lingua		U	*Ceratina cucurbitina*						Vereecken et al. (2012)

(Continued)

TABLE 15.1 (Continued)
Sexually Deceptive (SD) Plant Genera and Their Taxonomy, Distribution, Pollinator Type and Pollinator Taxonomy

Genera/Species	Plant Taxonomy[1] Subfamily: Tribe: Subtribe	Floral Chemistry	Male Pollinator	Pollinator Taxonomy[2,3] Order: Superfamily: Family	#Spp A[1]	#Spp B[4]	#SD[5]	#PSD	References
Orchis	Orchidoideae: Orchideae: Orchidinae		Halictid bee	Hymenoptera: Apoidea: Halictidae	21	61	1	1	Bino et al. (1982)
O. galilea		U	*Halictus marginatus*						Bino et al. (1982)
Orchidaceae-Asian genera									
Luisia	Epidendroideae: Vandeae: Aeridinae		Chafer beetle	Coleoptera: Scarabaeoidea: Scarabaeidae	39	40	1	1	Arakaki et al. (2016)
L. teres		U	*Protaetia pryeri*						Arakaki et al. (2016)
Cryptostylis[a]	Orchidoideae: Diurideae: Cryptostylidinae		Ichneumonid wasp	Hymenoptera: Ichneumonoidea: Ichneumonidae	18	18	?	?	Gaskett (2011)
Orchidaceae-Australian genera									
Caladenia	Orchidoideae: Diurideae: Caladeniinae		Thynnine wasps	Hymenoptera: Thynnoidea: Thynnidae	267	285	88	>150	Gaskett (2011), Phillips et al. (2009)
C. crebra		C	*Campylothynnus flavipictus*						Bohman et al. (2017)
C. plicata		C	*Zeleboria sp.*						Xu et al (2017)
Chiloglottis	Orchidoideae: Diurideae: Drakaeinae		Thynnine wasps	Hymenoptera: Thynnoidea: Thynnidae	23	24	29	>30	Gaskett (2011),
C. trapeziformis		C	*Neozeleboria cryptoides*						Peakall et al. (2010), Schiestl et al. (2003)
Drakaea	Orchidoideae: Diurideae: Drakaeinae		Thynnine wasps	Hymenoptera: Thynnoidea: Thynnidae	10	10	10	10	Gaskett (2011)

(Continued)

TABLE 15.1 (Continued)
Sexually Deceptive (SD) Plant Genera and Their Taxonomy, Distribution, Pollinator Type and Pollinator Taxonomy

Genera/Species	Plant Taxonomy[1] Subfamily: Tribe: Subtribe	Floral Chemistry	Male Pollinator	Pollinator Taxonomy[2,3] Order: Superfamily: Family	#Spp A[1]	#Spp B[4]	#SD[5]	#PSD	References
D. glyptodon		C	Zaspilothynnus trilobatus						Peakall (1990), Bohman et al. (2014)
Paracaleana	Orchidoideae: Diurideae: Drakaeinae		Thynnine wasps	Hymenoptera: Thynnoidea Thynnidae	13	13	3	>10	Gaskett (2011)
P. minor		P	Thynnoturneria sp.						
Arthrochilus	Orchidoideae: Diurideae: Drakaeinae		Thynnine wasps	Hymenoptera: Thynnoidea Thynnidae	15	15	3	>10	Gaskett (2011)
A. huntianus		P	Arthrothynnus huntianus						
Spiculaea	Orchidoideae: Diurideae: Drakaeinae		Thynnine wasps	Hymenoptera: Thynnoidea Thynnidae	1	1	1	1	Gaskett (2011)
S. ciliata		U	Thynnoturneria sp.						
Caleana	Orchidoideae: Diurideae: Drakaeinae		Pergid sawfly	Hymenoptera: Tenthredinoidea: Pergidae	1	1	1	1	Gaskett (2011)
C. major		U	Lophyrotoma sp.						
Leporella	Orchidoideae: Diurideae: Megastylidinae		Formicid ant	Hymenoptera: Formicoidea: Formicidae:	1	1	1	1	Gaskett (2011)
L. fimbriata		U	Myrmecia urens						
Calochilus	Orchidoideae: Diurideae: Thelymitrinae		Scoliid wasps	Hymenoptera: Scolioidea: Scoliidae	27	27	2	>20	Gaskett (2011)

(Continued)

TABLE 15.1 (Continued)
Sexually Deceptive (SD) Plant Genera and Their Taxonomy, Distribution, Pollinator Type and Pollinator Taxonomy

Genera/Species	Plant Taxonomy[1] Subfamily: Tribe: Subtribe	Floral Chemistry	Male Pollinator	Pollinator Taxonomy[2,3] Order: Superfamily: Family	#Spp A[1]	#Spp B[4]	#SD[5]	#PSD	References
C. campestris		U	Radumeris tasmaniensis (Formerly Campsomeris tasmaniensis)						Gaskett (2011)
Cryptostylis[b]	Orchidoideae: Diurideae: Cryptostylidinae		Ichneumonid wasp	Hymenoptera: Ichneumonoidea: Ichneumonidae:	5	5	5	5	Gaskett (2011)
C. ovata		P	Lissopimpla excelsa						Gaskett (2011)
Pterostylis	Orchidoideae: Cranichideae: Pterostylidinae		Mycetophilid and Keroplatid fungus gnats	See below	211	215	5	>150	Phillips et al. (2014), Reiter et al. (2019)
P. sanguinea		U	Mycomyia sp.	Diptera: Sciaroidea: Mycetophilidae					Phillips et al. (2014)
P. basaltica		U	Xenoplatyura conformis	Diptera: Sciaroidea: Keroplatidae					Reiter et al. (2019)
Orchidaceae-South African genera									
Disa	Orchidoideae: Orchideae: Disinae		Sphecid wasp	Hymenoptera: Apoidea: Sphecidae	182	185	1		Gaskett (2011)
D. atricapilla		U	Podalonia canescens						Steiner et al. (1994)
Orchidaceae-Central and South American genera									
Maxillaria	Epidendroideae: Malaxideae: Maxillariinae		Meliponine bees	See below	658		2	?	Gaskett (2011)
M. ringens (Formerly Mormolyca ringens)		U	Nannotrigona testaceicornis & Scaptotrigona sp.	Hymenoptera: Apoidea: Apidae		26 in Morm.	1	?	Singer et al. (2004), Flach et al. (2006)

(Continued)

TABLE 15.1 (Continued)
Sexually Deceptive (SD) Plant Genera and Their Taxonomy, Distribution, Pollinator Type and Pollinator Taxonomy

Genera/Species	Plant Taxonomy[1] Subfamily: Tribe: Subtribe	Floral Chemistry	Male Pollinator	Pollinator Taxonomy[2,3] Order: Superfamily: Family	#Spp A[1]	#Spp B[4]	#SD[5]	#PSD	References
M. obtusum (Formerly Trigonidium obtusum)		U	Plebeia droryana	Hymenoptera: Apoidea: Apidae		13 in Trig.	1	?	Singer (2002)
Bipinnula	Orchidoideae: Cranichideae: Chloraeinae	U	Scoliid wasp	Hymenoptera: Scolioidea: Scoliidae	11	10	1	?	Ciotek et al. (2006)
B. penicillata (Formerly Geoblasta penicillata)		U	Campsomeris bistrimaculata						Ciotek et al. (2006)
Lepanthes	Epidendroideae: Epidendreae: Pleurothallidinae		Sciarid fungus gnats	Diptera: Sciaroidea: Sciaridae	1085	1087	2	?	Gaskett (2011)
L. glicensteinii		U	Bradysia floribunda						Blanco and Barboza (2005)
Telipogon	Epidendroideae: Cymbidieae: Oncidiinae		Tachinid flies	Diptera: Oestroidea: Tachinidae	205	206	1	?	Martel et al. (2019)
T. peruvianus		P	Eudejeania sp.						Martel et al. (2019)
Nonorchids – Europe									
Iris	Iridaceae		Apid bee	Hymenoptera: Apoidea: Apidae Xylocopa valga		362	1	?	Arakaki et al. (2016)
I. paradoxa		U							Arakaki et al. (2016)

(Continued)

TABLE 15.1 (Continued)

Sexually Deceptive (SD) Plant Genera and Their Taxonomy, Distribution, Pollinator Type and Pollinator Taxonomy

Plant Taxonomy[1] Subfamily: Tribe: Genera/Species, Subtribe	Floral Chemistry	Male Pollinator	Pollinator Taxonomy[2,3] Order: Superfamily: Family	#Spp A[1]	#Spp B[4]	#SD[5]	#PSD	References
Nonorchids – South Africa								
Gorteria / Asteraceae/ Compositae		Bombyliid fly	Diptera: Asiloidea: Bombyliidae:		3	1	1	Ellis and Johnson (2010)
G. diffusa	U	*Megalpalpus capensis*						Ellis and Johnson (2010)

Note: An example species with known pollinator is listed for each genus, with further examples provided when different types of pollinator are exploited, or when pollinator attraction involves different semiochemicals. Whether the floral chemistry is confirmed (C), pending (P), or unknown (U), two estimates of the number of species per plant genus [#Spp A, #Spp B], number of confirmed SD cases [#SD], and predicted number of SD cases [#PSD] are shown.

a *Cryptostylis* is listed under Asian and Australian genera since its distribution extends from Australia into Asia (Pridgeon et al., 2001). Several Asian species exhibit similar floral morphology to the Australian sexually deceptive species, suggesting some will also be sexually deceptive.

b All five Australian species share the one species of male ichneumonid wasp pollinator. *C. subulata* also occurs in New Zealand (Gaskett, 2011).

1. Chase et al., 2015
2. Ševčík et al., 2016
3. Peters et al., 2017
4. The Plant List, 2013
5. Gaskett, 2011

and apoid wasps dominate as pollinators, being drawn from across seven families of Apoidea. One exception is *Ophrys speculum*, which is pollinated by a scoliid wasp (Scoliidae). Within Australian sexually deceptive orchids, wasps from the family Thynnidae, are widely exploited as pollinators. However, male ichneumonid wasps, scoliid wasps, a sawfly and a winged ant are pollinators drawn from the Ichneumonidae, Scoliidae, Pergidae and Formicidae, respectively. Male sphecid wasps (Apoidea) are pollinators of two South African *Disa* orchids (Table 15.1).

Diptera, including three families of fungus gnats (Sciaridae, Mycetophilidae and Keroplatidae) are also exploited by sexually deceptive *Lepanthes* and *Pterostylis* orchids. A bombyliid fly is the pollinator of the *Gorteria* daisy. Male beetles (Coleoptera) are occasional pollinators of some *Ophrys*, and the pollinator of the Japanese *Luisia teres* orchid (Table 15.1).

15.3   SEMIOCHEMICALS INVOLVED IN SEXUAL DECEPTION

15.3.1   ALKANES AND ALKENES IN *OPHRYS SPHEGODES* AND *O. EXALTATA*

In their revolutionary study of *Ophrys sphegodes*, pollinated by male *Andrena nigroaena* bees, Schiestl et al. used GC-Electroantennography Detection (GC-EAD) to identify 14 electrophysiologically active alkanes and alkenes in common between cuticular extracts of female bees and the orchid (Schiestl et al., 1999, 2000). Subsequent field bioassays with seven unbranched (Z)-9, (Z)-11 and (Z)-12 alkenes (**1a–1g**; C25, C27, C29), and seven alkanes (**2a–2g**; C21-27), in two blends matching the proportions found in the female and orchid, respectively, both elicited attempted copulation (Schiestl et al., 1999). Field bioassays with the seven alkanes only, elicited some approaches, while the seven alkenes on their own elicited attempted copulation, but at lower rates than the combined blend (Schiestl et al., 2000) (Figure 15.1a).

In *O. exaltata*, which is pollinated by male *Colletes cunnicularius* bees, Mant et al. (2005a) detected 24 GC-EAD active alkanes and alkenes in common between the cuticular extracts of female bees and the orchid labellum. Field bioassays with 12 synthetic candidates, consisting of seven unbranched (Z)-7 (**1h–1j**; C21, C23, C25), (Z)-9 (**1k, 1a, 1b, 1c**; C23, C25, C27, C29) and (Z)-8-(Z)-20 (**1l**; C31) alkenes, and four alkanes (**2a, 2c, 2e, 2g**; C21, C23, C25, C27), showed strong sexual attraction. In additional subtractive bioassays, a mixture of just three (Z)-7 (**1h–1j**; C21, C23, C25) alkenes were sufficient to induce attempted copulation (Figure 15.1b).

Mant et al. (2005a) also investigated the role of linalool (**9a**) (Figure 15.1b), a terpene of which the (S)-(+)-enantiomer was previously shown to be attractive to patrolling *C. cunnicularius* males (Borg-Karlson et al., 2003). Noting that racemic linalool induced less "close attraction" compared to the (S)-(+)-enantiomer in this earlier study, Mant et al (2005a) also found racemic linalool rarely induced attempted copulation. However, in blends with the hydrocarbons, "contacts" increased, indicating linalool acts as a long-range attractant, while the alkanes/alkenes are essential to induce sexual behavior.

Alkanes and alkenes are reported across other *Andrena* bee pollinated *Ophrys* orchids (e.g., Stökl et al., 2007), and in the bee pollinated *Maxillaria ringens* (formerly *Mormolyca ringens*) (Flach et al., 2006) and *Serapias lingua* (Vereecken et al., 2012), and in the nonorchid *Iris paradoxa* (Vereecken et al., 2012). However, bioassay confirmations of the function of these compounds are presently lacking.

15.3.2   (ω-1)-HYDROXY CARBOXYLIC ACIDS IN *OPHRYS SPECULUM*

Ayasse et al. (2003) identified ten GC-EAD active compounds including: unusual (ω-1)-hydroxy- and (ω-1)-keto carboxylic acids (**3d–3f**); aliphatic aldehydes (**4g–4i**; C11, C13, (Z)-9-C18); and fatty acid esters (**6b–6c**), that were shared between *Ophrys speculum* and the female of its scoliid wasp pollinator *Dasyscolia ciliata* (Figure 15.1c). While most compounds exhibited different relative proportions between orchids and females, a virtually identical enantiomeric composition ($R{:}S = 6{:}4$) of 7-hydroxyoctanoic acid (**3d**) and 9-hydroxydecanoic acid (**3f**) was found between the plant and insect.

Bioassays with all ten synthetic compounds, matching the blend ratios in the wasp and orchid, respectively, yielded equivalent strong sexual responses and high rates of attempted copulation, compared to odorless female wasp dummies. Bioassays with only aldehydes and esters yielded much weaker sexual responses than the full blend, while (ω-1)-hydroxy acids ($R{:}S = 6{:}4$) only, secured similar levels of attraction as the full blend, but induced fewer attempted copulations. A racemic mixture of these acids was barely attractive, and elicited no attempted copulation, confirming a critical role of the (ω-1)-hydroxy acids at the exact enantiomeric composition found in nature (Ayasse et al., 2003).

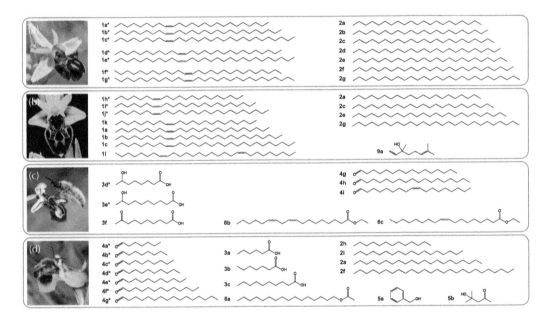

FIGURE 15.1 The chemistry of sexual deception in four European *Ophrys* orchids *Semiochemicals shown to be essential for sexual attraction of the pollinator during field bioassays. For continuity with Bohman et al. (2016a), we retain a similar chemical grouping and numbering system. (a) *Ophrys sphegodes* and the GC-EAD active (Z)-9, (Z)-11, and (Z)-12 alkenes (**1a–1g**; C25, C27, C29) and the alkanes (**2a–2g**; C21–C27) tested in field bioassays. The alkenes alone (**1a–1g**) were sufficient to secure full sexual behavior, but at lower rates than for all compounds (From Schiestl, F.P. et al., *Nature*, 399, 421–422, 1999; Schiestl, F.P. et al., *J. Comp. Physiol. A-Sens. Neural Behav. Physiol.*, 186, 567–574, 2000.) (Photo courtesy of Jorun Tharaldsen.) (b) *Ophrys exaltata* and the GC-EAD active (Z)-7 (**1h–1j**; C21, C23, C25) and (Z)-9 (**1k, 1a, 1b, 1c**; C23, C25, C27, C29) alkenes, (Z)-8-(Z)-20-hentriacontadiene (**1l**), four alkanes (**2a, 2c, 2e, 2g**; C21, C23, C25, C27), and linalool (**9a**) tested in field bioassays. The three (Z)-7 alkenes alone (**1h–1j**) were sufficient to secure attempted copulation, but at lower rates than for all test compounds. (From Mant J. et al., *J. Chem. Ecol.*, 31, 1765–1787, 2005a). (Photo courtesy of Rod Peakall.) (c) *Ophrys speculum* with its male scoliid wasp pollinator, *Dasyscolia ciliata*, and the GC-EAD active (ω-1)-hydroxy- and (ω-1)-keto carboxylic acids (**3d–3f**), aldehydes (**4g–4i**; C11, C13, (Z)-9-C16), ethyl esters (**6b, 6c**) tested in field bioassays. The 7-hydroxyoctanoic acid (**3d**) and 9-hydroxydecanoic acid (**3f**) ($R{:}S = 6{:}4$) were sufficient to secure attempted copulation. (From Ayasse M. et al, *Proc Royal Soc of Lond B-Biol Sci* 270, 517–522, 2003.) (Photo courtesy of Manfred Ayasse.) (d) *Ophrys leochroma* with its *Eucera kullenbergi* male bee pollinator and the GC-EAD active aldehydes (**4a–4g**; C7-C11, C13, C16), carboxylic acids (**3a–3c**; C6, C7, C10), benzyl alcohol (**5a**), 4-hydroxy-4-methyl-2-pentanone (**5b**), and hexadecyl acetate (**6a**) tested in field bioassays. Bioassays indicate missing compounds. However, aldehydes (**4a–4g**) are nonetheless sexual attractants (From Cuervo, M. et al., *J. Chem. Ecol.*, 43, 469–479, 2017.) (Photo courtesy of Manfred Ayasse.)

15.3.3 ALIPHATIC ALCOHOLS AND ALDEHYDES IN *OPHRYS LEOCHROMA* AND *O. SCOLOPAX*

Cuervo et al. (2017) found 20 shared GC-EAD active peaks between *Ophrys leochroma* and attractive females of its *Eucera kullenbergi* male bee pollinator. The GC-EAD active compounds included four alkanes (**2h**, **2i**, **2a**, **2f**; C13, C18, C21, C26), seven straight chain aliphatic aldehydes (**4a–4g**; C7-C11, C13, C16), three carboxylic acids (**3a-3c**; C6, C7, C10), benzyl alcohol (**5a**), 4-hydroxy-4-methyl-2-pentanone (**5b**), hexadecyl acetate (**6a**) and several unidentified compounds (Figure 15.1d). Two synthetic blends of 14 compounds matching the female bee, and 17 compounds matching the orchid flower, respectively, both elicited some attempted copulation. Blends of the four alkanes only, and the seven aldehydes only, elicited similar behavior. However, all tested blends showed weaker sexual responses compared with flower or female extracts, indicating missing compounds. Nonetheless, a key finding was that not only hydrocarbons but also aldehydes (**4a–4g**) were partially sexually active (Cuervo et al., 2017).

In some bee pollinated *Ophrys*, sexually attracted beetles can be legitimate secondary pollinators. In field bioassays, Borg-Karlson (1989) tested the responses of male *Phyllopertha horticola* beetles to extracts of orchids and beetles, and synthetic blends of alcohols and carboxylic acids. A blend containing the aliphatic alcohols: 1-hexanol, *cis*-3-hexen-1-ol, 1-octanol and 1-hexadecanol at ratios similar to those found within *O. scolopax* labellum extracts, elicited the strongest sexual attraction, including lands and attempted copulation at the dummy.

15.3.4 CYCLOHEXANEDIONES (CHILOGLOTTONES) IN *CHILOGLOTTIS*

Working with the Australian *Chiloglottis trapeziformis*, Schiestl et al. (2003) found one GC-EAD active compound in common between the orchid and the female of the thynnine wasp pollinator, *Neozeleboria cryptoides*. The new-to-science compound was identified as 2-ethyl-5-propylcyclohexan-1,3-dione (chiloglottone 1 (**7a**) Figure 15.2a). Field bioassays with the compound elicited full sexual behavior in the male pollinator, with rates of attempted copulation comparable to that observed at the flowers (Figure 15.2d). Soon after, five additional chiloglottones were found across *Chiloglottis* orchids and the allied genera *Arthrochilus* and *Paracaleana* (Peakall et al., 2010).

Presently, *Chiloglottis* is chemically the most well-characterized genus of sexually deceptive orchid with the sexual attractants identified and confirmed by bioassays for more than 10 orchid taxa, representing approximately one third of the estimated number of species in the genus (Schiestl et al., 2003, Schiestl and Peakall, 2005, Franke et al., 2009, Poldy et al., 2009, 2012, Peakall et al., 2010, Peakall and Whitehead, 2014).

Chiloglottone 1 (**7a**), first discovered in *C. trapeziformis*, is also used by at least three other *Chiloglottis* species to attract their different pollinators (Peakall et al., 2010). However, of these taxa sharing chiloglottone 1, only the distantly related *C. trapeziformis* and *C. valida* exhibit some overlap in distribution and flowering times. Their semiochemical sharing accounts for the rare, but sterile, hybrid *C. × pescottiana*, representing the only known case of hybridization in the genus (Peakall et al., 1997, Schiestl and Peakall, 2005).

Other examples, each involving only a single chiloglottone are also known: Chiloglottone 3 (**7c**, Figure 15.2b) is used by three allopatric sister orchid taxa to attract four different pollinators, with *C.* aff. *jeanesii* illustrating a case of geographic pollinator replacement involving nonsister taxa (Whitehead and Peakall, 2014). Chiloglottone 4 (**7d**, Figure 15.2c) is also used by three allopatric sister taxa to attract three sister *Neozeleboria* species. In *C.* aff. *valida* a 1:10 blend of chiloglottones 1 and 2 (**7a**, **7b**, Figure 15.2e), attracts two *Neozeleboria* pollinators in another example of pollinator geographic replacement (Peakall et al., 2010, Peakall and Whitehead, 2014). In *C. trilabra*, a 1:1 blend of chiloglottones 1 and 3 (**7a**, **7c**, Figure 15.2f) secures pollinator attraction. Five of the six known chiloglottones are GC-EAD active in *Paracaleana minor* (Figure 15.3g), including **7f**, not yet known elsewhere (Peakall et al., 2010).

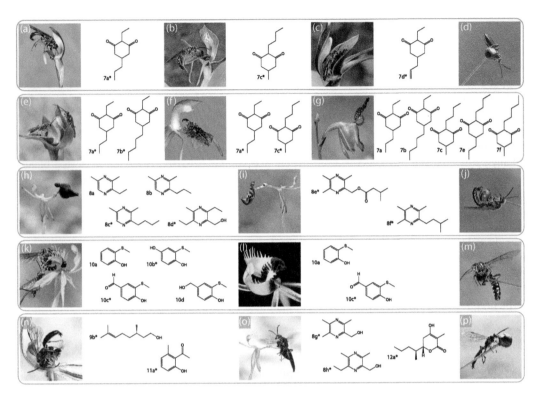

FIGURE 15.2 The chemistry of sexual deception in Australian *Chiloglottis*, *Drakaea* and *Caladenia* orchids *Semiochemicals shown to be essential for sexual attraction of the pollinator during field bioassays. For continuity with Bohman et al. (2016a), we retain a similar chemical grouping and numbering system. (Photos courtesy of Rod Peakall.) (a) *Chiloglottis trapeziformis* with its male thynnine wasp pollinator, *Neozeleboria cryptoides*, and 2-ethyl-5-propylcyclohexan-1,3-dione (**7a**, chiloglottone 1), the single compound active in GC-EAD and field bioassays, and confirmed sex pheromone of the female of the pollinator. (From Schiestl, F.P. et al., *Science*, 302, 437–438, 2003; Franke, S. et al., *Proc. Nat. Acad. Sci. USA*, 106, 8877–8882, 2009.) (b) *C.* aff. *jeanesii* with its male thynnine wasp pollinator, *N.* sp. (impatiens2) and 2-butyl-5-methylcyclohexan-1,3-dione (**7c**, chiloglottone 3) the single compound active in GC-EAD and field bioassays. (From Franke, S. et al., *Proc. Nat. Acad. Sci. USA*, 106, 8877–8882, 2009; Peakall, R. et al., *New Phytol.*, 188, 437–450, 2010.) (c) *C. turfosa* with its male thynnine wasp pollinator, *N.* sp. (carinicollis4) and 5-allyl-2-ethylcyclohexan-1,3-dione (**7d**, chiloglottone 4) the single compound active in GC-EAD and field bioassays. (From Franke, S. et al., *Proc. Nat. Acad. Sci. USA*, 106, 8877–8882, 2009; Peakall, R. et al., *New Phytol.*, 188, 437–450, 2010; Poldy, J. et al., *Europ. J. Org. Chem.*, 2012, 5818–5827, 2012.) (d) A male thynnine wasp pollinator, *N. cryptoides*, sexually attracted to a bioassay bead spiked with 2-ethyl-5-propylcyclohexan-1,3-dione (**7a**, chiloglottone 1). (From Peakall, R. et al., *New Phytol.*, 188, 437–450, 2010.) (e) *Chiloglottis* aff. *valida* with male thynnine wasp pollinator, *Neozeleboria* sp. (impatiens4), and 2-ethyl-5-propylcyclohexan-1,3-dione (**7a**, chiloglottone 1) and 5-pentylcyclohexan-1,3-dione (**7b**, chiloglottone 2), the two compounds active in GC-EAD and field bioassays at optimal blend ratio of 1:10. (From Peakall, R. et al., *New Phytol.*, 188, 437–450, 2010; Peakall, R. and Whitehead, M. R., *Ann. Botany*, 113, 341–355, 2014.) (f) *Chiloglottis trilabra* and 2-ethyl-5-propylcyclohexan-1,3-dione (**7a**, chiloglottone 1) and 2-butyl-5-methylcyclohexan-1,3-dione (**7c**, chiloglottone 3), the two compounds active in GC-EAD and field bioassays at optimal blend ratio of 1:1. (From Peakall, R. et al., *New Phytol.*, 188, 437–450, 2010.) (g) *Paracaleana minor* and its five GC-EAD active chiloglottones (**7a-7c**, **7e** and **7f**) including 2-butyl-5-propylcyclohexan-1,3-dione (**7e**, chiloglottone 5), also found in *C. sphyrnoides*, and 2-hexyl-5-methylcyclohexan-1,3-dione (**7f**, chiloglottone 6) presently only known in *P. minor*. (From Peakall, R. et al., *New Phytol.*, 188, 437–450, 2010.) (h) *Drakaea glyptodon* and 2-ethyl-3,5-dimethylpyrazine (**8a**), 2-propyl-3,5-dimethylpyrazine (**8b**), 2-butyl-3,5-dimethylpyrazine (**8c**) and 2-hydroxymethyl-3,5-diethyl-6-methylpyrazine (**8d**), the active compounds in GC-EAD and field bioassays and the confirmed female sex pheromone blend of the

(*Continued*)

15.3.5 Pyrazines in *Drakaea*

Despite being closely related to *Chiloglottis* orchids, *Drakaea glyptodon* sexually lures its thynnine wasp pollinator, *Zaspilothynnus trilobatus*, by a combination of alkylpyrazines (**8a–8c**) and an hydroxymethylpyrazine (**8d**) rather than chiloglottones (Bohman et al., 2014) (Figure 15.5h). Subtractive bioassays revealed that only 2-butyl-3,5-dimethylpyrazine (**8c**) and 2-hydroxymethyl-3,5-diethyl-6-methylpyrazine (**8d**), both also present in the female wasp, are required to elicit attraction and attempted copulation in this system. A blend of **8c**:**8d** at a 3:1 ratio on simplistic female dummies, secured copulation rates similar to that of orchid flowers (Figure 15.2j). Bioassays with structural analogs to the hydroxymethylpyrazine **8d** showed that all structurally related compounds tested were attractive, with one compound eliciting similarly strong sexual behavior as the natural sex pheromone (Bohman et al., 2016b).

The taxon *Drakaea livida* consists of several different chemotypes, each attracting different thynnine wasp pollinators (Bohman and Peakall, 2014). Two of the chemotypes contain pyrazines related to those found in *D. glyptodon* (Bohman et al., 2012a, 2012b, Bohman and Peakall, 2014) (Figure 15.2i). In one chemotype attracting a *Catocheilus* wasp, there are five GC-EAD active pyrazines. Bioassays with this pollinator confirmed that two of the five compounds: (3,5,6-trimethylpyrazin-2-yl)methyl 3-methylbutanoate (**8e**) and 2-(3-methylbutyl)-3,5,6-trimethylpyrazine (**8f**), elicit strong attraction, lands and attempted copulation. In *Drakaea micrantha*, a previously undescribed β-hydroxylactone, 4-hydroxy-3-methyl-6*S*-(pentan-2*S*-yl)-5,6-dihydro-2*H*-pyran-2-one (drakolide) **12a**, in combination with two hydroxymethylpyrazines (**8g, 8h**), act as the pollinator attractants (Bohman et al., 2020) (Figure 15.2o).

15.3.6 (Methylthio)phenols in *Caladenia crebra* and *C. attingens*

In *Caladenia crebra*, pollinated by male *Campylothynnus flavopictus* thynnine wasps, field bioassays with dissected floral parts indicated that the glandular sepal tips are the source of the sexual attractant. Subsequently, high resolution mass spectrometry (GC-HRMS) screening of solvent extracts, in combination with careful interpretation of the molecular ion isotope ratios, revealed a

---

FIGURE 15.2 (Continued) thynnine wasp pollinator, *Zaspilothynnus trilobatus*. (From Bohman B., et al. 2014. *New Phytol* 203, 939–952, 2014. (i) *Drakaea livida* and (3,5,6-trimethylpyrazin-2-yl)methyl 3-methylbutanoate (**8e**) and 2-(3-methylbutyl)-3,5,6-trimethylpyrazine (**8f**), active compounds in field bioassays of a chemotype which sexually attracts a male *Catocheilus* thynnine wasp as the pollinator. (From Bohman, B. and Peakall, R., *Insects*, 5, 474–487, 2014.) (j) A male thynnine wasp, *Zaspilothynnus trilobatus*, attempting to mate with a bioassay bead spiked with 2-butyl-3,5-dimethylpyrazine (**8c**) and 2-hydroxymethyl-3,5-diethyl-6-methylpyrazine (**8d**), representing the key constituents of the sex pheromone of the female, and the semiochemical blend of the orchid *D. glyptodon*. (From Bohman B., et al. 2014. *New Phytol* 203, 939–952, 2014. (k) *Caladenia crebra* with male thynnine wasp pollinator, *Campylothynnus flavopictus*, 2-(methylthio) phenol (**10a**), 2-(methylthio)benzene-1,4-diol (**10b**), 4-hydroxy-3-(methylthio)benzaldehyde (**10c**) and 4-(hydroxymethyl)-2-(methylthio)phenol (**10d**), the field bioassay active constituents of the sex pheromone of the female of the pollinator, and the semiochemical blend of the orchid. (From Bohman, B. et al., *Angewandte Chemie Int. Ed.*, 56, 8455–8458, 2017.) (l) *Caladenia attingens* and floral volatiles 2-(methylthio)phenol (**10a**) and 4-hydroxy-3-(methylthio)benzaldehyde (**10c**). In field bioassays, compound **10c** secures strong sexual attraction of males of the thynnine wasp, *Campylothynnus* (sp. A), one of two known species of wasp pollinators of this orchid. (From Bohman, B. et al., *Fitoterapia*, 126, 78–82, 2018b.) (m) A male of the thynnine wasp, *Campylothynnus* (sp. A), sexually attracted to a bioassay bead spiked with 4-hydroxy-3-(methylthio) benzaldehyde (**10c**). (From Bohman, B. et al., *Fitoterapia*, 126, 78–82, 2018b.) (n) *Caladenia plicata* with its male thynnine wasp pollinator, *Zeleboria* sp., and (*S*)-β-citronellol (**9b**) and 2-hydroxy-6-methylacetophenone (**11a**), the active compounds in field bioassays and the confirmed female sex pheromone of the pollinator. (From Xu, H. et al., *Curr. Biol.*, 27, 1867–1877, 2017.) (o) *Drakaea micrantha* with its male thynnine wasp pollinator, *Zeleboria* sp. A, and 4-hydroxy-3-methyl-6*S*-(pentan-2*S*-yl)-5,6-dihydro-2*H*- pyran-2-one (drakolide) (**12a**), 2-hydroxymethyl-3,5,6-trimethylpyrazine (**8g**) and 2-hydroxymethyl-3,5- dimethyl-6-ethylpyrazine (**8h**) (From Bohman, B et al., *Angewandte Chemie Int. Ed.*, 132, 1140–1144, 2020.) (p) A male of the undescribed thynnine wasp, *Zeleboria* sp. C, sexually attracted to a bioassay bead spiked with (*S*)-β-citronellol (**9b**) and 2-hydroxy-6-methylacetophenone (**11a**). (From Xu, H. et al., *Curr. Biol.*, 27, 1867–1877, 2017.)

series of unusual sulfurous compounds in common between the floral sepal tips and females of the pollinator, suggesting these as candidate semiochemicals. In total, four (methylthio)phenols (**10a–10d**) were identified. (Figure 15.2k). In the field, a combination of additive and subtractive bioassays revealed that only two (methylthio)phenols, compounds **10b** and **10c** (at a 10:1 blend ratio), are required to achieve strong sexual attraction, including frequent attempted copulations at simplistic female dummies, in this system. Compounds **10a** and **10d** appear to be unnecessary for eliciting strong sexual behavior, but not inhibitory in blends with **10b** and **10c** (Bohman et al., 2017).

In further field bioassays, Bohman et al. (2018b) also found that male *Campylothynnus* (sp. A) thynnine wasps are strongly attracted to the (methylthio)phenols **10b** and **10c**, which elicited very high attempted copulation rates (88% at 1:1). Compound **10c** on its own also elicited strong attraction and high attempted copulation rates (60%) (Figure 15.2l and m). This wasp species is a pollinator of northern populations of *Caladenia attingens* subsp. *attingens* (a distant relative of *C. crebra*), in which chemical analysis revealed that two (methylthio)phenols, **10a** and **10c**, are present in the glandular sepal and petal tips, and in the labellum (Figure 16.2l). Thus, (methylthio)phenols are likely to be more widely used as orchid semiochemicals than presently known.

15.3.7 CITRONELLOL AND AN ACETOPHENONE DERIVATIVE IN *CALADENIA PLICATA*

An unusual combination of the monoterpene (S)-β-citronellol (**9b**), and a new floral volatile compound, 2-hydroxy-6-methylacetophenone (**11a**), secure pollination in *Caladenia plicata*. Both compounds were also found in head extracts of the female of the pollinator, an undescribed *Zeleboria* thynnine wasp species (Figure 15.2n & o). Field bioassays revealed neither compound **9b** nor **11a** was active on its own. However, a 20:1 blend was weakly attractive, while a 1:20 blend achieved equivalent wasp attraction, but significantly fewer lands and attempted copulations, compared to an "optimal" 1:4 (**9b:11a**) blend (Xu et al., 2017).

Additional field bioassays revealed that a blend where (S)-β-citronellol was replaced with its enantiomer, (R)-β-citronellol (in combination with **11a** at a 1:4 ratio), was barely attractive. Furthermore, none of the nine possible alternative hydroxy-methylacetophenone regioisomers could replace the natural 2-hydroxy-6-methylacetophenone **11a** (Bohman et al., 2018a). Given their structural similarity, the complete absence of any activity with these alternative isomers was surprising, and in contrast to what was found in similar experiments with hydroxymethylpyrazines for the *Zaspilothynnus trilobatus* thynnine wasp pollinator (see Section 15.3.2 and Bohman et al., 2016b).

15.3.8 (S)-2-(TETRAHYDROFURAN-2-YL)ACETIC ACID AND ESTERS IN *CRYPTOSTYLIS OVATA*

In an unusual case of pollinator sharing among sexually deceptive orchids, all five known Australian *Cryptostylis* species share the same ichnemonid wasp pollinator, *Lissopimpla excelsa* (Gaskett 2011). Focusing on *C. ovata*, Bohman et al. (2019) used an iterative bioassay-guided fractionation approach to isolate a previously unknown semiochemical: (S)-2-(tetrahydrofuran-2-yl)acetic acid, and its ester derivatives (see also Chapter 3). While strongly attractive to the male pollinator, bioassays with this semiochemical could not replicate the high rates of attempted copulation observed at *Cryptostylis* flowers, indicating additional compounds may be required for a full sexual response.

15.4 BIOSYNTHESIS OF SEMIOCHEMICALS INVOLVED IN SEXUAL DECEPTION

15.4.1 ALKANES AND ALKENES

Schlüter and Schiestl (2008) predicted that in *Ophrys*, the biosynthesis of alkenes should follow the same biosynthetic pathway as alkanes (reviewed in Samuels et al., 2008, Bohman et al., 2016a), with the exception of an additional desaturation step, potentially achieved by stearoyl-acyl carrier protein desaturases (SAD), prior to the long chain elongation step (Figure 15.3a). In subsequent work, three

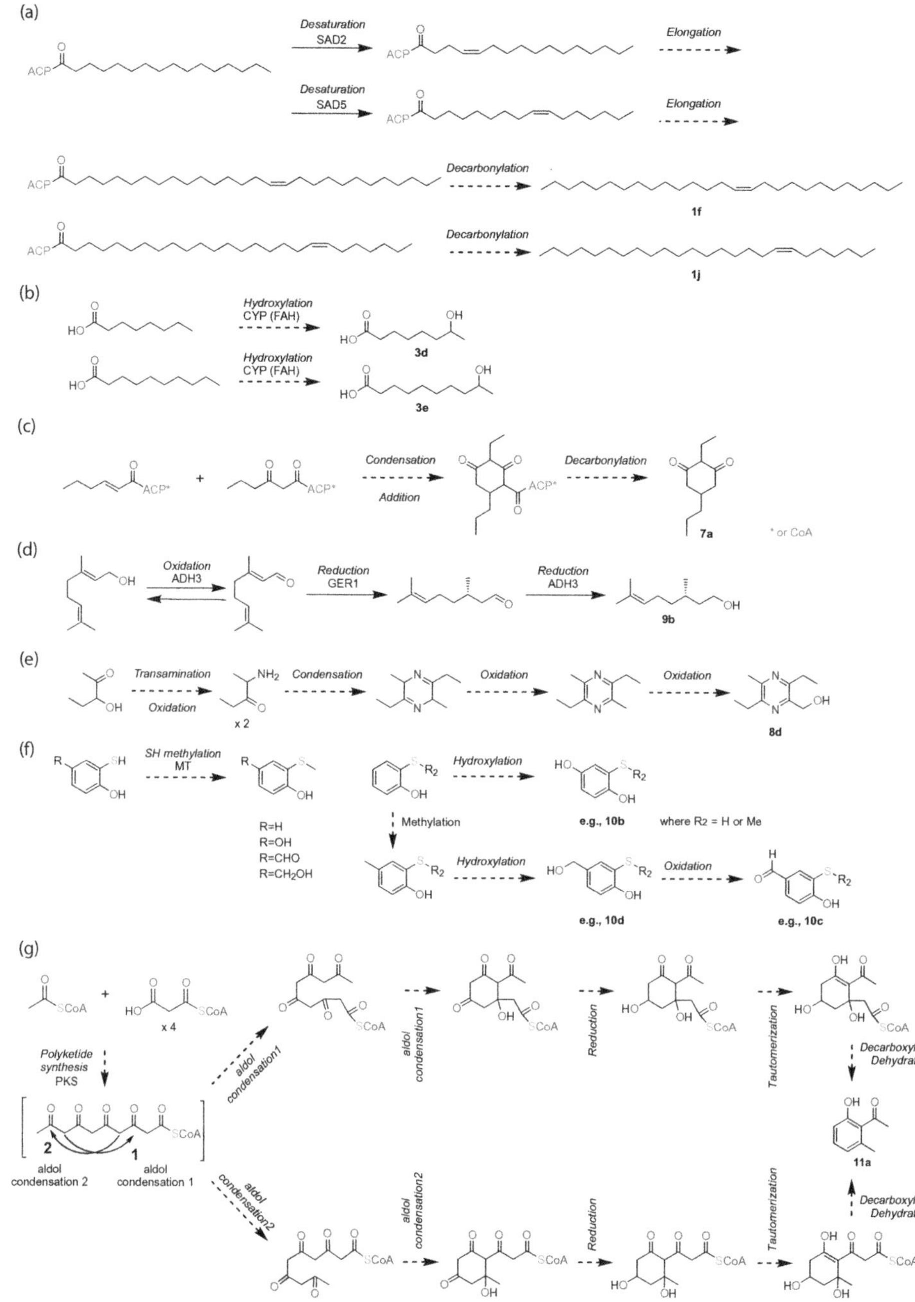

FIGURE 15.3 Confirmed and proposed biosynthetic pathways for representative groups of semiochemicals involved in the sexual attraction of pollinators in some European and Australian sexually deceptive orchids. (a) Proposed biosynthesis of (Z)-12-heptacosene (**1f**) in *Ophrys exaltata* and (Z)-7-pentacosene (**1j**) in *O. sphegodes* via 16:0-ACP precursors. (b) Proposed biosynthesis of chiloglottone 1 (**7a**) in *Chiloglottis trapeziformis* via activated fatty acid biosynthetic/degradation pathway intermediates. (c) Proposed biosynthesis of 7-hydroxyoctanoic acid and 9-hydroxydecanoic acid (**3d, 3e**) in *Ophrys speculum* via fatty acid precursors. (d) Confirmed biosynthesis of (S)-β-citronellol (**9b**) in *Caladenia plicata* via geraniol. (e) Proposed biosynthesis of 2-hydroxymethyl-3,5-diethyl-6-methylpyrazine (**8d**) in *Drakaea glyptodon* via α-hydroxy carbonyl intermediates. (f) Proposed biosynthesis of (methylthio)phenols (**10a–10d**) in *Caladenia crebra*. (g) Proposed biosynthesis of 2-hydroxy-6-methylacetophenone (**11a**) in *Caladenia plicata* via one acetyl-CoA and four malonyl-CoA starter units.

putative SAD genes (*SAD1-SAD3*) were isolated (Schlüter et al., 2011). Transgenic expression and *in vitro* enzyme assays demonstrated that SAD2 is a functional desaturase capable of introducing 18:1 Δ9 and 16:1 Δ4 fatty acid intermediates, from which it is hypothesized that (Z)-9 alkenes and (Z)-12 alkenes are synthesized. Three additional putative SAD genes (*SAD4-6*) were also identified from an *O. sphegodes* transcriptome (Sedeek et al., 2013).

In *O. sphegodes* and *O. exaltata*, *SAD1* and *SAD2* expression levels significantly correlated with (Z)-9 and (Z)-12-alkene production, while high *SAD5* expression was correlated with the (Z)-7-alkene production unique to *O. exaltata* (Xu et al., 2012) (Figure 15.3a). The patterns of gene expression in two natural putative F1 hybrids suggest *cis*-regulation of *SAD1/2*, since expression matched an allele expected from *O. sphegodes*, but not *O. exaltata*. By contrast, for *SAD5*, the two putative hybrids failed to express an allele from either parent, suggesting *trans*-regulation (Xu et al., 2012, Xu and Schlüter, 2015).

In vitro enzyme-activity studies further showed that the putative housekeeping desaturase, SAD3, catalyzes the general reactions of stearate to oleate (18:0-ACP to 18:1 Δ9-ACP), and palmitate to palmitoleate (16:0-ACP to 16:1 Δ9-ACP), whereas SAD5 is a specialized 16:0 Δ9-ACP enzyme (Sedeek et al., 2016). Subsequent elongation of a 16:1 Δ9-ACP to 26:1 Δ19-coenzyme A precursors, followed by decarbonylation, would yield the (Z)-7 alkene (25:1 Δ7) that characterizes *O. exaltata* (Figure 15.3a).

15.4.2   (ω-1)-Hydroxy Carboxylic Acids

Some cytochrome P450 (*CYP*) plant genes are known to be involved in the biosynthesis of hydroxy fatty acids where they catalyze the hydroxylation of the terminal methyl (ω position) and/or the oxidation of sub-terminal or in-chain (ω-*n*) carbon positions (Pinot and Beisson, 2011). Enzymatic assays with fatty acids of different chain lengths (e.g., C10–C18) have established that specificity and the position of attack vary among genes. However, only *Vicia sativa* CYP94A1 (Tijet et al., 1998) and *Helianthus tuberosus* CYP81B1 (Cabello-Hurtado et al., 1998) have been found to hydroxylate decanoic acid at the ω-1 position to form 9-hydroxydecanoic acid. We hypothesize that in *Ophrys speculum*, a similar *CYP* gene, with specific ω-1 activity, could be involved in the formation of the 7-hydroxyoctanoic and 9-hydroxydecanoic acids (Figure 15.3b).

15.4.3   Cyclohexanediones (chiloglottones)

Franke et al. (2009) proposed that chiloglottone 1 could be formed by the fusion of two intermediates in the fatty acid biosynthetic pathway (Figure 15.3c), with one precursor being an activated 3-ketohexanyl compound and the other being an activated 2-hexenyl compound. Both these acyl intermediates occur in the fatty acid biosynthetic pathway in the plastids (as 3-ketohexanoyl-ACP and 2-hexenoyl-ACP) as well as in the fatty acid degradation pathway in the peroxisomes (as CoA derivatives) (Heldt and Piechulla, 2004). Precursors of other chiloglottones are likely to be similar fatty acid intermediates with varying chain lengths.

Support for this hypothesis of fatty acids as precursors of chiloglottone production continues to grow with recent studies. For example, in *Chiloglottis trapeziformis*, the production of chiloglottone 1 occurs in the calli on the labellum of open flowers under continuous UV-B light (Falara et al., 2013). Therefore, in a first step toward testing the hypothesis of fatty acid (FA) based biosynthesis, Wong et al. (2017a) applied differential gene expression analysis across active and nonactive tissue types and UV-B treatments. This work revealed strong coordinated induction of entire FA biosynthesis and β-oxidation pathways.

Wong et al. (2018) further showed that upregulation of fatty acid biosynthesis and β-oxidation genes occurs over the transition from early buds through to flowers, correlating with the ramping of chiloglottone production. Enzyme inhibition experiments targeting ketoacyl-ACP synthase activity significantly inhibited chiloglottone biosynthesis by up to 88%. Wong et al. (2019) next investigated the evolutionary history of gene duplication and selection in β-ketoacyl-ACP synthase 1 (*KASI*) gene lineages. This revealed a duplicate *KASI-2B* clade unique to the Orchidaceae. Selection analysis further indicated that *KASI-2B* may have evolved novel enzymatic functions along a two-tiered evolutionary path – strong positive selection for a new function shared with all other orchids, followed by relaxed purifying selection unique to *Chiloglottis*. Thus, it is possible that the relaxed selection from the core function of the widespread plant *KASI* genes has enabled the evolution of short chain length acyl-ACP substrate specificities, predicted to be crucial for chiloglottone biosynthesis (Wong et al., 2019).

15.4.4 CITRONELLOL

Despite long interest in solving the biosynthesis of the monoterpene β-citronellol, the pathway had remained unknown. Remarkably, the breakthrough was made by Xu et al. (2017) in the nonmodel Australian sexually deceptive orchid *Caladenia plicata*. This orchid exclusively produces its pollinator attracting semiochemicals (*S*)-β-citronellol and 2-hydroxy-6-methylacetophenone in the sepal tips. Therefore, to identify the genes and enzymes involved, *de novo* transcriptome assembly and differential expression analysis were performed between active (sepal tips) and nonactive tissues. A candidate geraniol synthase gene was identified based on its differential expression and its homology to the terpene synthase (TPS) gene family, with the corresponding protein demonstrated to catalyze the formation of geraniol from geranyl diphosphate (Xu et al., 2017). One highly expressed alcohol dehydrogenase (*Cp*ADH3) and one tissue-specific double-bond reductase (*CpGER1*) were also identified as candidate genes. Contrary to previous predictions of a one-step conversion of geraniol to β-citronellol (e.g., Schwab and Wüst, 2015), subsequent biochemical assays for *Cp*ADH3 and *Cp*GER1 revealed that β-citronellol biosynthesis from geraniol proceeds in three steps, beginning with the oxidation of geraniol to geranial by *Cp*ADH3, enantioselective reduction of geranial to (*S*)-β-citronellal by *Cp*GER1, and a further reduction of (*S*)-β-citronellal to (*S*)-β-citronellol by *Cp*ADH3 (Xu et al., 2017; Figure 15.3d).

15.4.5 HYPOTHESES ON THE BIOSYNTHESIS OF SOME OTHER SEMIOCHEMICALS

15.4.5.1 Pyrazines

While pyrazines impart characteristic flavors to many fruits (Dunlevy et al., 2013), they are rare floral volatiles (Bohman et al., 2014), and even the biosynthesis of the basic pyrazine ring has not yet been elucidated in plants (or animals). However, in myxobacteria, feeding experiments with isotope-labeled precursors indicate that the pyrazine ring of dialkyl-substituted pyrazines is formed by dimerization and oxidation of α-amino aldehydes derived from amino acids (Nawrath et al., 2010). In the bacterium *Corynebacterium glutamicum*, simple alkylated pyrazines (up to four substituents, usually methyl or ethyl) are derived from α-hydroxy carbonyls (also originating from amino acids) via transamination and oxidation to the corresponding α-amino ketones, followed by dimerization and oxidation (Dickschat et al., 2010). Knocking down the genes for keto-acid reductoisomerase and acetolactate synthase, stopped pyrazine biosynthesis in *C. glutamicum*. In both bacterial systems, dimerization of the α-amino carbonyls is thought to be enzyme-controlled, however candidate enzymes are currently unknown (Dickschat et al., 2010, Nawrath et al., 2010). Other substituted and oxygenated pyrazines are also likely formed via α-amino carbonyls (Nawrath et al., 2010). Therefore, in *Drakaea* orchids a pathway starting from known α-hydroxy carbonyls could lead to the alkyl pyrazines, with further functionalization (e.g., oxidation and esterification) to give the oxygenated analogs (Figure 15.3e).

15.4.5.2 (Methylthio)phenols

(Methylthio)phenols are only known from one example in bacteria (Brock et al., 2014), and recently as floral volatiles in *Caladenia* orchids (Bohman et al., 2017, 2018b). Hence, their biosynthesis remains to be investigated. Three distinct classes of plant methyl transferases (MT) are known to be able to catalyze the methylation of thiols (Attieh et al., 2002, Coiner et al., 2006, Zhao et al., 2012). We predict that the biosynthesis of (methylthio)phenols in *Caladenia* orchids may involve enzymes related to type I plant *O*-methyltransferase or members of the SABATH gene family. Under this hypothesis, the biosynthesis of 2-(methylthio)phenol could be achieved by methylation of 2-mercaptophenol. Similarly, *S*-methylation of other thiol-containing precursors (e.g., 2-mercapto-1,4-benzenediol) could be envisaged to form substituted (methylthio) phenols (e.g., 2-(methylthio)benzene-1,4-diol), although the origins of these precursors remain to be found in plants. Alternatively, substituted (methylthio)phenols could be formed from 2-(methylthio)phenol via general biosynthetic methylation, hydroxylation and oxidation pathways (Figure 15.3f).

15.4.5.3 Acetophenones

Wong et al. (2017b) proposed two alternative hypotheses for the biosynthesis of 2-hydroxy-6-methylacetophenone in *Caladenia plicata*: a polyketide biosynthetic route (Figure 15.3g) based on evidence from radiolabeling studies in an ant (Tecle et al., 1986), or production via coenzyme A-dependent β-oxidation of phenylpropanoid precursors (Negrel and Javelle, 2010). Initial evidence favors the polyketide pathway hypothesis, since strong tissue-specific differential expression of a putative polyketide synthase (*CpPKS1*) has already been found in the orchid (Xu et al., 2017).

15.5   SEMIOCHEMICALS, SEXUAL DECEPTION AND SPECIATION

15.5.1   INTRODUCTION TO POLLINATOR-DRIVEN ECOLOGICAL SPECIATION

Sexually deceptive orchids have long been considered prime candidates for pollinator-driven speciation. Indeed, with the accumulating evidence that floral volatile chemistry is of paramount importance for pollinator attraction and specificity, the plausibility of pollinator-driven speciation in sexually deceptive orchids has been reinforced, and hypotheses on the speciation process refined in both empirical studies (e.g., Schiestl and Ayasse, 2002, Peakall et al., 2010, Xu et al., 2011, Peakall and Whitehead, 2014) and in a series of reviews (e.g., Schlüter and Schiestl, 2008, Schiestl and Schlüter, 2009, Ayasse et al., 2011).

Central to the current hypotheses on the evolution of sexual deception are the following assumptions: (1) pollinator specificity is controlled by floral odor chemistry, (2) pollinator-mediated divergent selection for floral odor chemistry is highly likely, (3) reproductive isolation is predominantly achieved by the pollinator (as strong ethological floral isolation), (4) genic speciation is plausible. Below we briefly evaluate the empirical evidence supporting these assumptions.

15.5.2   IS POLLINATOR SPECIFICITY CONTROLLED BY FLORAL ODOR CHEMISTRY?

The often highly specialized pollination of orchids sets them apart from many plant groups with more generalized pollination, where a given plant species has many different species of pollinator. For example, Schiestl and Schlüter (2009) found the mean number of pollinator species per orchid species was just 2.3 (across 31 studies spanning diverse genera). The degree of pollinator specialization is even higher among sexually deceptive systems, with the mean number of pollinators per orchid species estimated to be 1.2 in *Ophrys*, 1.14 in *Caladenia*, and 1.0 in *Chiloglottis* (Schiestl and Schlüter, 2009), with more recent estimates of 1.1 for *Chiloglottis* (Peakall et al., 2010), 1.75 for *Drakaea* and 1.05 for *Caladenia* (estimated from supplementary data of Phillips et al., 2017).

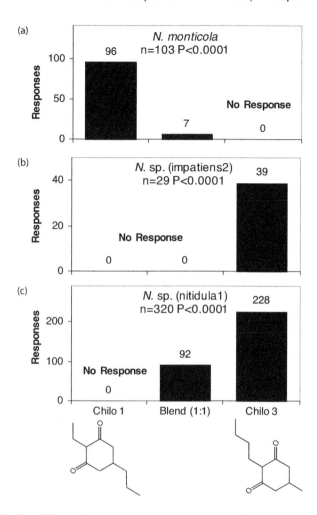

FIGURE 15.4 Male *Neozeleboria* thynnine wasp responses to choice tests revealing variation among *Chiloglottis* orchid pollinator species in semiochemical specificity and preference. The results show the total responses of three closely related wasp pollinators to choice tests offering chiloglottone 1 (Chilo 1), chiloglottone 3 (Chilo 3) and an equal blend of the two compounds (Blend 1:1). (a) Male *Neozeleboria monticola* the pollinator of *C. valida*, (b) *N.* sp. (impatiens2) the pollinator of *C.* aff. *jeanesii*, (c) *N.* sp. (nitidula1) the pollinator of *C. chlorantha*. (Based on a subset of the data shown in Figure 5 of Peakall, R. et al., *New Phytol.*, 188, 437–450, 2010.)

Presently, the strongest experimental evidence that pollinator specificity is controlled by chemistry is found in *Chiloglottis*, where the outcome of extensive bioassays have confirmed that pollinator specificity is either controlled by single compounds, or specific blends of two compounds (Peakall et al., 2010; Figures 15.2 and 15.4). In other sexually deceptive orchid genera, a strong role for chemistry in controlling pollinator specificity is indicated by the specific blends, and the often unusual compounds involved (Bohman et al. 2016, this chapter).

15.5.3 EVIDENCE FOR POLLINATOR-MEDIATED DIVERGENT SELECTION

To test for signatures of pollinator-mediated selection among three sympatric *Ophrys* species, Mant et al. (2005b) compared patterns of species divergence across active and nonactive volatiles, and neutral microsatellite genetic markers. Only the traits predicted to be under selection, the bioactive alkanes/alkenes, were strongly divergent. Xu et al. (2012) later reported the outcomes of additive bioassays. These showed that (Z)-7 alkenes strongly inhibited the attraction of the pollinator of *O. sphegodes* (which uses a (Z)-9 alkene as pollinator attractant). Conversely, (Z)-9 alkenes inhibited the pollinator of *O. exaltata* (which uses a (Z)-7 alkene as pollinator attractant). Thus, strong pollinator-mediated divergent selection for alkene double-bond position is highly likely between these species.

Choice tests with *Chiloglottis* orchid pollinators also support the hypothesis of pollinator-mediated selection (Peakall et al. 2010). For example, the pollinator of *C. valida*, responded very strongly to chiloglottone 1 (93%), not at all to chiloglottone 3, and rarely (7%) to a 1:1 blend (Figure 15.4a). Conversely, the pollinator of *C. aff. jeanesii*, only responded to chiloglottone 3 (Figure 15.4b). The pollinator of *C. chlorantha*, responded strongly to chiloglottone 3 (71%), not at all to chiloglottone 1, and partially (29%) to a 1:1 blend (Figure 15.4c). Thus, strong pollinator-driven selection for floral chemistry matching the sex pheromone of the wasp is highly likely. One theory is that blends arising from initially neutral mutations might provide a bridge for pollinator-driven selection to drive chemical shifts and potentially speciation (Peakall et al., 2010, Peakall and Whitehead, 2014).

15.5.4 The Strength of Floral Isolation

The strength of reproductive isolation (RI) barriers between two species can be estimated quantitatively as an RI index (range negative to 1), where a value of 1 indicates complete isolation. Across plants generally, relatively few studies have quantified RI indices across a full range of pre- and postzygotic reproductive isolating barriers (Martin and Willis, 2007, Lowry et al., 2008). Among sexually deceptive orchids, two parallel studies in European *Ophrys* (Xu et al., 2011) and Australian *Chiloglottis* (Whitehead and Peakall, 2014) have comprehensively quantified the strength of pollinator, post-mating prezygotic and postzygote isolation as well as genetically assessed the natural rate of hybridization. Both studies found pollinator RI values >0.98, while postzygote RI values were zero or negative. Furthermore, very limited evidence for hybrids was found in the *Ophrys* case (2 out of 146 genetic samples), and no evidence for hybridization in the *Chiloglottis* case (0 out of 571 genetic samples). Hence, both cases stand out as extreme examples of the pre-eminence of pollinator isolation (Xu et al., 2011, Whitehead and Peakall, 2014). Furthermore, given that the pollinator specificity that maintains species boundaries is controlled by chemistry in these species, genic speciation (sensu Wu, 2001) via speciation genes associated with floral volatiles seems highly plausible.

15.5.5 Speciation Genes

Nosil and Schlüter (2011) defined a speciation gene as a locus whose differentiation between populations makes a significant contribution to the evolution of reproductive isolation (RI). Thus, divergence selection at a speciation gene increases the total amount of RI between populations, but by definition must occur before speciation is complete. Schlüter and Schiestl (2008) predicted that simple genetic changes in enzymes acting in the biosynthetic pathway of a floral odor compound crucial for specific pollinator attraction, could readily impact RI. Likewise, changes in odor blends, due to changes in gene expression levels, could also be important. Consequently, the structural and/ or regulatory genes involved directly or indirectly in floral odor production are strong candidates for speciation genes in sexually deceptive orchids.

Alkene biosynthesis in *Ophrys* has long been a research target (Schlüter et al., 2011, Xu et al., 2012, Sedeek et al., 2013, 2014, Xu and Schlüter, 2015), with considerable progress now made in understanding the role of the stearoyl-acyl carrier protein desaturases (SAD) in the double-bond formation (see Section 15.4.1). Remarkably, the C16 specificity of SAD5, which is predicted to give rise to 7-alkenes (as opposed to 9-alkenes with C18 specificity) is linked to changes in just three amino acid residues within the putative substrate-binding cavity. Reconstruction of inferred ancestral state proteins at just these three amino acid sites restored C18 functionality of the protein *in vitro* (Sedeek et al., 2016). Thus, gene duplication to give *SAD5* from the housekeeping *SAD3*, followed by pollinator-mediated selection for alleles exhibiting reduced activity, or even complete loss of function for C18 activity, offers a plausible evolutionary pathway (Schlüter et al., 2011, Xu et al., 2011, 2012, Sedeek et al., 2016). To confirm *SAD5* as a speciation gene will now require confirmation that the changes occurred pre-speciation, and that such changes led to increased reproductive isolation (Nosil, 2012).

15.6   CONCLUSIONS

While most of the world's sexually deceptive plants are orchids, the several hundred known examples are drawn from across four continents and represent a wide cross section of the terrestrial orchid flora. The types of male insect pollinators exploited by these orchids are perhaps even more extraordinary, spanning Hymenoptera, Diptera and Coleoptera (Table 15.1).

As demonstrated by field bioassays (reviewed Bohman et al., 2016a; Chapter 3, and this chapter), precise semiochemical blends underpin the highly specific pollinator attraction of sexually deceptive orchids. Yet, while we have learned much about the chemical basis of pollinator attraction in European *Ophrys* and a diverse cross section of Australian sexually deceptive orchids, the semiochemicals involved still remain unknown for the vast majority of sexually deceptive plants (Table 15.1). Furthermore, to simply extrapolate the likely chemical basis of pollinator attraction to other cases, is fraught with danger. For example, while other *Andrena* bee pollinated *Ophrys* species may well use variants of alkane/alkene blends, it is already clear that this assumption does not extend to species pollinated by *Eucera* bees (Cuervo et al., 2017) or *Bombus* bees (Gögler et al., 2011). Similarly, among the sexually deceptive thynnine-pollinated Australian orchids, five unrelated chemical systems are already known (Schiestl et al., 2003, Bohman et al., 2014, 2017, Xu et al., 2017, Bohman et al., 2020), despite more than 200 more cases that are yet to be investigated.

While advances in our understanding of the biosynthesis of the semiochemicals that underpin sexual deception are being made, there is still a long way to go before the complete pathways are fully elucidated. A decisive case for genic speciation will depend on further work. Nonetheless, the emerging ecological, genetic and molecular evidence strongly point to pollinator-mediated ecological speciation as the major mode of evolution, and it will be no surprise if some of the genes involved in semiochemical production are confirmed as "speciation genes" in future research. Expanding work on all aspects of sexual deception, and across the full diversity of plants and insects involved, may also offer a wider and deeper understanding into the chemistry, biosynthesis and evolution of plant-animal interactions more generally.

ACKNOWLEDGMENTS

Björn Bohman was supported by an Australian Research Council (ARC) Discovery Early Career Researcher Award (DE 160101313). Darren Wong was supported by ARC Discovery project (DP150102762) to Rod Peakall and Eran Pichersky. We thank Jorun Tharaldsen and Manfred Ayasse for photographs of *Ophrys sphegodes* (JT) and *Ophrys leochroma* and *O. speculum* (MA), respectively. All other photographs by Rod Peakall.

REFERENCES

Arakaki N., K. Yasuda, S. Kanayama, S. Jitsuno, M. Oike and S. Wakamura. 2016. Attraction of males of the cupreous polished chafer *Protaetia pryeri pryeri* (Coleoptera: Scarabaeidae) for pollination by an epiphytic orchid *Luisia teres* (Asparagales: Orchidaceae). *Applied Entomology and Zoology* 51:241–246

Attieh J., R. Djiana, P. Koonjul, C. Étienne, S. A. Sparace and H. S. Saini. 2002. Cloning and functional expression of two plant thiol methyltransferases: A new class of enzymes involved in the biosynthesis of sulfur volatiles. *Plant Molecular Biology* 50:511–521.

Ayasse M., F. P. Schiestl, H. F. Paulus, F. Ibarra and W. Francke. 2003. Pollinator attraction in a sexually deceptive orchid by means of unconventional chemicals. *Proceedings of the Royal Society of London Series B-Biological Sciences* 270:517–522.

Ayasse M., J. Stökl and W. Francke. 2011. Chemical ecology and pollinator-driven speciation in sexually deceptive orchids. *Phytochemistry* 72:1667–1677.

Bino R. J., A. Dafni and A. D. J. Meeuse. 1982. The pollination ecology of *Orchis galilaea* (Bornm. et Schulze) Schltr. (Orchidaceae). *New Phytologist* 90:315–319.

Blanco M. A., and G. Barboza. 2005. Pseudocopulatory pollination in *Lepanthes* (Orchidaceae: Pleurothallidinae) by fungus gnats. *Annals of Botany* 95:763–772.

Bohman B., A. Karton, G. R. Flematti, A. Scaffidi and R. Peakall. 2018a. Structure-activity studies of semiochemicals from the spider orchid *Caladenia plicata* for sexual deception. *Journal of Chemical Ecology* 44:436–443.

Bohman B., A. Karton, R. C. M. Dixon, R. A. Barrow and R. Peakall. 2016b. Parapheromones for thynnine wasps. *Journal of Chemical Ecology* 42:17–23.

Bohman B., A. M. Weinstein, R. D. Phillips, R. Peakall and G. R. Flematti. 2019. 2-(Tetrahydrofuran-2-yl) acetic acid and ester derivatives as long-range pollinator attractants in the sexually deceptive orchid *Cryptostylis ovata*. *Journal of Natural Products* 82:1107–1113.

Bohman B., and R. Peakall. 2014. Pyrazines attract *Catocheilus* thynnine wasps. *Insects* 5:474–487.

Bohman B., G. R. Flematti, R. A. Barrow, E. Pichersky and R. Peakall. 2016a. Pollination by sexual deception: It takes chemistry to work. *Current Opinion in Plant Biology* 32:37–46.

Bohman B., L. Jeffares, G. R. Flematti et al. 2012a. Discovery of tetrasubstituted pyrazines as semiochemicals in a sexually deceptive orchid. *Journal of Natural Products* 75:1589–1594.

Bohman B., L. Jeffares, G. R. Flematti et al. 2012b. The discovery of 2-hydroxymethyl-3-(3-methylbutyl)-5-methylpyrazine: A semiochemical in orchid pollination. *Organic Letters* 14:2576–2578.

Bohman B., R. D. Phillips, G. R. Flematti and R. Peakall. 2018b. (Methylthio)phenol semiochemicals are exploited by deceptive orchids as sexual attractants for *Campylothynnus* thynnine wasps. *Fitoterapia* 126:78–82.

Bohman B., R. D. Phillips, G. R. Flematti, R. A. Barrow and R. Peakall. 2017. The spider orchid *Caladenia crebra* produces sulfurous pheromone mimics to attract its male wasp pollinator. *Angewandte Chemie International Edition* 56:8455–8458.

Bohman B., R. D. Phillips, M. H. M. Menz et al. 2014. Discovery of pyrazines as pollinator sex pheromones and orchid semiochemicals: Implications for the evolution of sexual deception. *New Phytologist* 203:939–952.

Bohman B., M. M. Y. Tan, R. D. Phillips et al. 2020. A specific blend of drakolide and hydroxymethylpyrazines: An unusual pollinator sexual attractant used by the endangered orchid *Drakaea micrantha*. *Angewandte Chemie International Edition* 132:1140–1144.

Borg Karlson A.-K. 1989. Attraction of *Phyllopertha horticola* (Coleoptera, Scarabaeidae) males to fragrance components of *Ophrys* flowers (Orchidaceae, section Fuciflorae). *Entomologisk Tidskrift* 109:105–109.

Borg-Karlson A.-K. 1990. Chemical and ethological studies of pollination in the genus *Ophrys* (Orchidaceae). *Phytochemistry* 29:1359–1387.

Borg-Karlson A.-K., J. Tengö, I. Valterova et al. 2003. (S)-(+)-linalool, a mate attractant pheromone component in the bee *Colletes cunicularius*. *Journal of Chemical Ecology* 29:1–14.

Brock N. L., M. Menke, T. A. Klapschinski and J. S. Dickschat. 2014. Marine bacteria from the *Roseobacter* clade produce sulfur volatiles via amino acid and dimethylsulfoniopropionate catabolism. *Organic & Biomolecular Chemistry* 12:4318–4323.

Cabello-Hurtado F., Y. Batard, J. P. Salaün, F. Durst, F. Pinot and D. Werck-Reichhart. 1998. Cloning, expression in yeast, and functional characterization of CYP81B1, a plant cytochrome P450 that catalyzes in-chain hydroxylation of fatty acids. *Journal of Biological Chemistry* 273:7260–7267.

Chase M. W., K. M. Cameron, J. V. Freudenstein et al. 2015. An updated classification of Orchidaceae. *Botanical Journal of the Linnean Society* 177:151–174.

Ciotek L., P. Giorgis, S. Benitez-Vieyra and A. A. Cocucci. 2006. First confirmed case of pseudocopulation in terrestrial orchids of South America: Pollination of *Geoblasta pennicillata* (Orchidaceae) by *Campsomeris bistrimacula* (Hymenoptera, Scoliidae). *Flora* 201:365–369.

Coiner H., G. Schröder, E. Wehinger et al. 2006. Methylation of sulfhydryl groups: A new function for a family of small molecule plant O-methyltransferases. *The Plant Journal* 46:193–205.

Cuervo M., D. Rakosy, C. Martel, S. Schulz and M. Ayasse. 2017. Sexual deception in the *Eucera*-pollinated *Ophrys leochroma*: A chemical intermediate between wasp- and *Andrena*-pollinated species. *Journal of Chemical Ecology* 43:469–479.

Dickschat J. S., S. Wickel, C. J. Bolten, T. Nawrath, S. Schulz and C. Wittmann. 2010. Pyrazine biosynthesis in *Corynebacterium glutamicum*. *European Journal of Organic Chemistry* 2010:2687–2695.

Dunlevy J. D., E. G. Dennis, K. L. Soole, M. V. Perkins, C. Davies and P. K. Boss. 2013. A methyltransferase essential for the methoxypyrazine-derived flavour of wine. *The Plant Journal* 75:606–617.

Ellis A. G., and S. D. Johnson. 2010. Floral mimicry enhances pollen export: The evolution of pollination by sexual deceit outside of the Orchidaceae. *American Naturalist* 176:E143–E151.

Falara V., R. Amarasinghe, J. Poldy, E. Pichersky, R. A. Barrow and R. Peakall. 2013. The production of a key floral volatile is dependent on UV light in a sexually deceptive orchid. *Annals of Botany* 111:21–30.

Flach A., A. J. Marsaioli, R. B. Singer et al. 2006. Pollination by sexual mimicry in *Mormolyca ringens*: A floral chemistry that remarkably matches the pheromones of virgin queens of *Scaptotrigona* sp. *Journal of Chemical Ecology* 32:59–70.

Franke S., F. Ibarra, C. M. Schulz et al. 2009. The discovery of 2,5-dialkylcyclohexan-1,3-diones as a new class of natural products. *Proceedings of the National Academy of Sciences of the USA* 106:8877–8882.

Gaskett A. C. 2011. Orchid pollination by sexual deception: Pollinator perspectives. *Biological Reviews* 86:33–75.

Gögler J., R. Twele, W. Francke and M. Ayasse. 2011. Two phylogenetically distinct species of sexually deceptive orchids mimic the sex pheromone of their single common pollinator, the cuckoo bumblebee *Bombus vestalis*. *Chemoecology* 21:243–252.

Heldt H.-W., and B. Piechulla. 2004. *Plant Biochemistry*. London, UK: Academic Press.

Lowry D. B., J. L. Modliszewski, K. M. Wright, C. A. Wu and J. H. Willis. 2008. The strength and genetic basis of reproductive isolating barriers in flowering plants. *Philosophical Transactions of the Royal Society B-Biological Sciences* 363:3009–3021.

Mant J., C. Brandli, N. J. Vereecken, C. M. Schulz, W. Francke and F. P. Schiestl. 2005a. Cuticular hydrocarbons as sex pheromone of the bee *Colletes cunicularius* and the key to its mimicry by the sexually deceptive orchid, *Ophrys exaltata*. *Journal of Chemical Ecology* 31:1765–1787.

Mant J., R. Peakall and F. P. Schiestl. 2005b. Does selection on floral odor promote differentiation among populations and species of the sexually deceptive orchid genus *Ophrys*? *Evolution* 59:1449–1463.

Martel C., W. Francke and M. Ayasse. 2019. The chemical and visual basis in the pollination of the neotropical sexually deceptive orchid *Telipogon peruvianus* (Orchidaceae). *New Phytologist* 223:1989–2001.

Martin N. H., and J. H. Willis. 2007. Ecological divergence associated with mating system causes nearly complete reproductive isolation between sympatric *Mimulus* species. *Evolution* 61:68–82.

Nawrath T., J. S. Dickschat, B. Kunze and S. Schulz. 2010. The biosynthesis of branched dialkylpyrazines in Myxobacteria. *Chemistry & Biodiversity* 7:2129–2144.

Negrel J., and F. Javelle. 2010. The biosynthesis of acetovanillone in tobacco cell-suspension cultures. *Phytochemistry* 71:751–759.

Nosil P. 2012. *Ecological Speciation*. Oxford, UK: Oxford University Press.

Nosil P., and D. Schluter. 2011. The genes underlying the process of speciation. *Trends in Ecology and Evolution* 26:160–167.

Paulus H. F. 2018. Pollinators as isolation mechanisms: Field observations and field experiments regarding specificity of pollinator attraction in the genus *Ophrys* (Orchidaceae und Insecta, Hymenoptera, Apoidea). *Entomologia Generalis* 37:261–316.

Paulus H. F., and C. Gack. 1990. Pollinators as prepollinating isolation factors: Evolution and speciation in *Ophyrs* (Orchidaceae). *Israel Journal of Botany* 39:43–79.

Peakall R. 1990. Responses of male *Zaspilothynnus trilobatus* Turner wasps to females and the sexually deceptive orchid it pollinates. *Functional Ecology* 4:159–167.

Peakall R., and M. R. Whitehead. 2014. Floral odour chemistry defines species boundaries and underpins strong reproductive isolation in sexually deceptive orchids. *Annals of Botany* 113:341–355.

Peakall R., C. C. Bower, A. E. Logan and H. I. Nicol. 1997. Confirmation of the hybrid origin of *Chiloglottis x pescottiana* (Orchidaceae: Diurideae). 1. Genetic and morphometric evidence. *Australian Journal of Botany* 45:839–855.

Peakall R., D. Ebert, J. Poldy et al. 2010. Pollinator specificity, floral odour chemistry and the phylogeny of Australian sexually deceptive *Chiloglottis* orchids: Implications for pollinator-driven speciation. *New Phytologist* 188:437–450.

Peters R. S., L. Krogmann, C. Mayer et al. 2017. Evolutionary history of the Hymenoptera. *Current Biology* 27:1013–1018.

Phillips R. D., D. Scaccabarozzi, B. A. Retter et al. 2014. Caught in the act: Pollination of sexually deceptive trap-flowers by fungus gnats in *Pterostylis* (Orchidaceae). *Annals of Botany* 113:629–641.

Phillips R. D., G. R. Brown, K. W. Dixon, C. Hayes, C. C. Linde and R. Peakall. 2017. Evolutionary relationships among pollinators and repeated pollinator sharing in sexually deceptive orchids. *Journal of Evolutionary Biology* 30:1674–1691.

Phillips R. D., R. Faast, C. C. Bower, G. R. Brown and R. Peakall. 2009. Implications of pollination by food and sexual deception for pollinator specificity, fruit set, population genetics and conservation of *Caladenia* (Orchidaceae). *Australian Journal of Botany* 57:287–306.

Pinot F., and F. Beisson. 2011. Cytochrome P450 metabolizing fatty acids in plants: Characterization and physiological roles. *The FEBS Journal* 278:195–205.

Poldy J., R. Peakall and R. A. Barrow. 2012. Identification of the first alkenyl chiloglottone congener. *The European Journal of Organic Chemistry* 2012:5818–5827.

Poldy J., R. Peakall and R. Barrow. 2009. Synthesis of chiloglottones: Semiochemicals from sexually deceptive orchids and their pollinators. *Organic and Biomolecular Chemistry* 7:4296–4300.

Pridgeon A. M., P. J. Cribb, M. W. Chase and F. N. Rasmussen. 2001. *Genera Orchidacearum. Volume 2. Orchidoideae (Part 1)*. Oxford, UK: Oxford University Press.

Reiter N., M. Freestone, G. Brown and R. Peakall. 2019. Pollination by sexual deception of fungus gnats (Keroplatidae and Mycetophilidae) in two clades of *Pterostylis* (Orchidaceae). *Botanical Journal of the Linnean Society* 190:101–116.

Samuels L., L. Kunst and R. Jetter. 2008. Sealing plant surfaces: Cuticular wax formation by epidermal cells. *Annual Reviews of Plant Biology* 59:683–707.

Schiestl F. P., and M. Ayasse. 2002. Do changes in floral odor cause speciation in sexually deceptive orchids? *Plant Systematics and Evolution* 234:111–119.

Schiestl F. P., and P. M. Schlüter. 2009. Floral isolation, specialized pollination, and pollinator behavior in orchids. *Annual Review of Entomology* 54:425–446.

Schiestl F. P., and R. Peakall. 2005. Two orchids attract different pollinators with the same floral odour compound: Ecological and evolutionary implications. *Functional Ecology* 19:674–680.

Schiestl F. P., M. Ayasse, H. F. Paulus et al. 1999. Orchid pollination by sexual swindle. *Nature* 399:421–422.

Schiestl F. P., M. Ayasse, H. F. Paulus et al. 2000. Sex pheromone mimicry in the early spider orchid (*Ophrys sphegodes*): Patterns of hydrocarbons as the key mechanism for pollination by sexual deception. *Journal of Comparative Physiology A Sensory Neural and Behavioral Physiology* 186:567–574.

Schiestl F. P., R. Peakall, J. G. Mant et al. 2003. The chemistry of sexual deception in an orchid-wasp pollination system. *Science* 302:437–438.

Schlüter P. M., and F. P. Schiestl. 2008. Molecular mechanisms of floral mimicry in orchids. *Trends in Plant Science* 13:228–235.

Schlüter P. M., S. Q. Xu, V. Gagliardini et al. 2011. Stearoyl-acyl carrier protein desaturases are associated with floral isolation in sexually deceptive orchids. *Proceedings of the National Academy of Sciences USA* 108:5696–5701.

Schwab W., and M. Wüst. 2015. Understanding the constitutive and induced biosynthesis of mono- and sesquiterpenes in grapes (*Vitis vinifera*): A key to unlocking the biochemical secrets of unique grape aroma profiles. *Journal of Agricultural and Food Chemistry* 63:10591–10603.

Sedeek K. E. M., E. Whittle, D. Guthörl, U. Grossniklaus, J. Shanklin and P. M. Schlüter. 2016. Amino acid change in an orchid desaturase enables mimicry of the pollinator's sex pheromone. *Current Biology* 26:1505–1511.

Sedeek K. E. M., G. Scopece, Y. M. Staedler et al. 2014. Genic rather than genome-wide differences between sexually deceptive *Ophrys* orchids with different pollinators. *Molecular Ecology* 23:6192–6205.

Sedeek K. E. M., W. Qi, M. A. Schauer et al. 2013. Transcriptome and proteome data reveal candidate genes for pollinator attraction in sexually deceptive orchids. *PLoS One* 8:e64621.

Ševčík J., D. Kaspřák, M. Mantič et al. 2016. Molecular phylogeny of the megadiverse insect infraorder Bibionomorpha sensu lato (Diptera). *PeerJ* 4:e2563.

Singer R. B. 2002. The pollination mechanism in *Trigonidium obtusum* Lindl (Orchidaceae: Maxillariinae): Sexual mimicry and trap-flowers. *Annals of Botany* 89:157–163.

Singer R. B., A. Flach, S. Koehler, A. J. Marsaioli and M. D. E. Amaral. 2004. Sexual mimicry in *Mormolyca ringens* (Lindl.) Schltr. (Orchidaceae: Maxillariinae). *Annals of Botany* 93:755–762.

Steiner K. E., V. B. Whitehead and S. D. Johnson. 1994. Floral and pollinator divergence in two sexually deceptive South African orchids. *American Journal of Botany* 81:185–194.

Stökl J., R. Twele, D. H. Erdmann, W. Francke and M. Ayasse. 2007. Comparison of the flower scent of the sexually deceptive orchid *Ophrys iricolor* and the female sex pheromone of its pollinator *Andrena morio*. *Chemoecology* 17:231–233.

Tecle B., J. J. Brophy and R. F. Toia. 1986. Biosynthesis of 2-hydroxy-6-methylacetophenone in an Australian ponerine ant, *Rhytidoponera aciculata* (Smith). *Insect Biochemistry* 16:333–336.

The Plant List. 2013. Version 1.1. Published on the Internet; http://www.theplantlist.org/ (Accessed June 1, 2019).

Tijet N., C. Helvig, F. Pinot et al. 1998. Functional expression in yeast and characterization of a clofibrate-inducible plant cytochrome P-450 (CYP94A1) involved in cutin monomers synthesis. *The Biochemical Journal* 332:583–589.

Vereecken N. J., C. A. Wilson, S. Hötling, S. Schulz, S. A. Banketov and P. Mardulyn. 2012. Pre-adaptations and the evolution of pollination by sexual deception: Cope's rule of specialization revisited. *Proceedings of the Royal Society B: Biological Sciences* 279:4786–4794.

Whitehead M. R., and R. Peakall. 2014. Pollinator specificity drives strong prepollination reproductive isolation in sympatric sexually deceptive orchids. *Evolution* 68:1561–1575.

Wong D. C. J., E. Pichersky and R. Peakall. 2017b. The biosynthesis of unusual floral volatiles and blends involved in orchid pollination by deception: Current progress and future prospects. *Frontiers in Plant Science* 8:1955.

Wong D. C. J., R. Amarasinghe, C. Rodriguez-Delgado, R. Eyles, E. Pichersky and R. Peakall. 2017a. Tissue-specific floral transcriptome analysis of the sexually deceptive orchid *Chiloglottis trapeziformis* provides insights into the biosynthesis and regulation of its unique UV-B dependent floral volatile, chiloglottone 1. *Frontiers in Plant Science* 8:1260.

Wong D. C. J., R. Amarasinghe, E. Pichersky and R. Peakall. 2018. Evidence for the involvement of fatty acid biosynthesis and degradation in the formation of insect sex pheromone-mimicking chiloglottones in sexually deceptive *Chiloglottis* orchids. *Frontiers in Plant Science* 9:839.

Wong D. C. J., R. Amarasinghe, V. Falara, E. Pichersky and R. Peakall. 2019. Duplication and selection in β-ketoacyl-ACP synthase gene lineages in the sexually deceptive *Chiloglottis* (Orchidaceace). *Annals of Botany* 123:1053–1066.

Wu C. I. 2001. The genic view of the process of speciation. *Journal of Evolutionary Biology* 14:851–865.

Xu H., B. Bohman, D. C. J. Wong et al. 2017. Complex sexual deception in an orchid is achieved by co-opting two independent biosynthetic pathways for pollinator attraction. *Current Biology* 27:1867–1877.

Xu S. Q., P. M. Schlüter, G. Scopece et al. 2011. Floral isolation is the main reproductive barrier among closely related sexually deceptive orchids. *Evolution* 65:2606–2620.

Xu S., and P. M. Schlüter. 2015. Modeling the two-locus architecture of divergent pollinator adaptation: How variation in SAD paralogs affects fitness and evolutionary divergence in sexually deceptive orchids. *Ecology and Evolution* 5:493–502.

Xu S., P. M. Schlüter, U. Grossniklaus and F. P. Schiestl. 2012. The genetic basis of pollinator adaptation in a sexually deceptive orchid. *PLOS Genetics* 8. e1002889.

Zhao N., J. L. Ferrer, H. S. Moon et al. 2012. A SABATH Methyltransferase from the moss *Physcomitrella patens* catalyzes S-methylation of thiols and has a role in detoxification. *Phytochemistry* 81:31–41.

16 Behavioral Responses to Floral Scent

Experimental Manipulations and Multimodal Plant–Pollinator Communication

Robert A. Raguso

CONTENTS

16.1 INTRODUCTION

The first edition of this book (Dudareva and Pichersky 2006) marked a foundational phase in the interdisciplinary study of floral scent. My original chapter (Raguso 2006) served as a call to integrate floral scent into the mainstream of floral ecology and evolution (Raguso 2008a, 2008b). This integration is now well underway in research on floral reproductive isolation (Byers et al. 2014; Bischoff et al. 2015), phenotypic selection (Schiestl et al. 2011; Parachnowitsch et al. 2012; Gross et al. 2016) and the structure of plant–pollinator community networks (Larue et al. 2016; Kantsa et al. 2018). Recent studies also have explored the reciprocal impacts of floral VOCs on microbes, herbivores, larcenists and predators as third-party agents of selection (Junker et al. 2011a, 2011b; Huang et al. 2012; Junker and Tholl 2013; Knauer et al. 2018).

Below I outline the experimental approaches now available, with the goal of understanding how floral scent impacts pollinator behavior at appropriate scales and contexts, with a focus on insights gained since 2006. Hebets and Papaj (2005) have provided the most useful conceptual framework by outlining hypotheses concerning multimodal signal evolution in animal behavior. Leonard, Dornhaus and Papaj (2011a, 2011b) re-phrased these hypotheses in the context of plant–pollinator communication, describing how floral scent interacts with color, shape, texture, taste and other sensory information to guide (or mislead) pollinator behavior (Table 16.1). Here, I review the insights gained from the experimental manipulation of flowers, drawing upon the multimodal

---

TABLE 16.1

Functional Hypotheses for Multicomponent Floral Signals as Related to Floral Scent

I. Content-Based Hypotheses

Multiple Messages

Distance attraction and floral landing/feeding mediated by different signals[5–12]

Different signals used as innate attractants vs. learned after experience[13–15]

Multiple signals reflect different functions for multiple receivers[16–20]

Redundant Signals

CO_2 behaves as a scent but does not alter the function of scent[21]

Discrimination learning results in higher visitation rates to rewarding flowers when signals are multimodal[22]

II. Efficacy-Based Hypotheses

Signal Transmission

Efficacy Backup

Multiple signals facilitate communication under variable ambient conditions[23–25]

Efficacy Tradeoff

One signal is used from a distance, another at closer range[26,27]

Signal Detection

Movement draws attention to a visual signal[30]; scent to an acoustic signal[9]

Signal Processing

Parallel Processing

Simultaneous tracking of different modalities improves flower handling[31,32]

Perceptual Variability

Differences in diet, sex, experience selectively alter responses to some signals but not to others[33]

Different signals and modalities attract different pollinators in generalized or bimodal systems[34,35]

III. Inter-Signal Interaction Hypotheses

Attention-Altering

Selective attention: the presence of one signal compels sharper discrimination of another[36]

A learned multimodal signal elicits greater constancy, suggesting working memory[23,37]

Context

One signal synergizes or contextualizes responses to another signal[27,31,38–41]

Presence of scent results in acquisition of more useful information about color (e.g., peak shift)[42,43]

Generalization

Learned olfactory spatial patterns can be transferred to visual modality[44]

Source: Content modified from references [1–4]. [1]Hebets and Papaj 2005; [2]Leonard et al. 2012a; [3]Leonard et al. 2012b; [4]Junker and Parachnowitsch 2015; [5]Bischoff et al. 2015; [6]Raguso and Weiss 2015; [7]Angioy et al. 2004; [8]Hansen et al. 2011; [9]Gonzalez-Terrazas et al. 2016; [10]von Arx et al. 2012; [11]Harrap et al. 2017; [12]Clarke et al. 2013. [13]Russell et al. 2018; [14]Milet-Pinheiro et al. 2013; [15]Dötterl et al. 2011; [16]Kessler et al. 2008; [17]Kessler et al. 2013; [18]Hoballah et al. 2005; [19]Klahre et al. 2011; [20]Kessler and Baldwin 2007; [21]Goyret et al. 2008; [22]Kulahci et al. 2008; [23]Kaczorowski et al. 2012a; [24]Lawson et al. 2017; [25]von Helversen and von Helversen 2003; [26]Hossaert-McKey et al. 1994; [27]Streinzer et al. 2010; [30]Sprayberry and Daniel 2007; [31]Goyret et al. 2007; [32]Roth et al. 2016; [33]Goyret et al. 2009; [34]Waelti et al. 2008; [35]Muchhala et al. 2009; [36]Kunze and Gumbert 2001; [37]Gegear and Laverty 2005; [38]Raguso and Willis 2005; [39]Pellmyr and Patt 1986; [40]Spaethe et al. 2010; [41]Raguso and Willis 2002; [42]Leonard et al. 2011c; [43]Wright et al. 2009; [44]Lawson et al. 2018.

signaling framework provided by Papaj and colleagues to explain how floral scent mediates pollinator behavior, in conjunction with other floral traits.

16.2   MULTIMODAL FLORAL COMMUNICATION

No floral trait exists in a vacuum. Floral color accompanies other elements of display (pattern, shape, texture, orientation and architecture), chemical traits (scent, surface chemistry, nectar, pollen, resins and oils) and timing (diel rhythms of flower opening, scent emission and reward presentation, blooming phenology), all of which impact pollinator attraction and fidelity. Despite the fact that most floral traits and their associated sensory physiology in animal pollinators have been studied in isolation, most flowers advertise themselves in complex, multimodal sensory channels (Raguso 2004, 2008b). The focus of this chapter is to understand when and how floral scent works in concert with other floral traits. The multimodal signaling concept structure is outlined in three classes of nonmutually exclusive hypotheses (Table 16.1): (1) *Content-related hypotheses* (what information is furnished by flowers?), (2) *Efficacy-related hypotheses* (how do floral signals function against variable biotic and abiotic backgrounds?), and (3) *Inter-signal interaction hypotheses* (how do cognitive aspects of signal perception impact signal function?). This framework has been utilized in recent reviews of concerted changes in floral color and scent (Raguso and Weiss 2015) and holistic evaluations of floral phenotypic integration (Junker and Parachnowitsch 2015). These hypotheses provide conceptual tools for interpreting the complex interactions between scent and other floral traits revealed through manipulative bioassays.

16.2.1   FLORAL MANIPULATION – A BRIEF PREHISTORY

Nearly a century ago, Clements and Long (1923) outlined ingenious methods for manipulating the color, orientation and symmetry of living flowers *in situ*, providing a wealth of tools for subsequent studies (Figure 16.1a). Such methods and their modern counterparts include resupination (rotating floral orientation (Fulton and Hodges 1999), painting flowers to alter their colors (Meléndez-Ackerman et al. 1997) and floral "mutilation" (removing petals, cutting nectar spurs or unzipping hoods to reveal other floral organs within (Pellegrino et al. 2017). Daumer (1958) used a similar approach, inverting the ray florets of *Helianthus rigidus* (Asteraceae) and other species with UV-absorbing centers, to which bees responded by probing for nectar at the flower's periphery rather than its center (Figure 16.1b–d).

Clements and Long (1923) decoupled the visual and olfactory displays of hawkmoth-pollinated flowers (*Oenothera cespitosa;* Onagraceae), using colored crêpe paper to conceal these fragrant flowers, whereas Knoll (1926a) sandwiched scented flowers of *Lonicera implexa* (Caprifoliaceae) between glass, compelling hawkmoths to choose between visual targets and a displaced scent plume. Knoll (1926b) extended this approach in his study of *Arum nigrum* (Araceae), a brood site-deceptive plant that traps its pollinators within a kettle-like inflorescence using fecal scent and heat. Knoll experimentally decoupled scent and heat by adding a scented (dissected) spadix to heated or unheated glass model spathes to construct floral "chimeras". In the century since Knoll's work, the "model spathe" approach has been adopted by botanists world-wide to decouple visual, olfactory and thermal floral stimuli from Arum family plants in behavioral assays with their pollinators (Patt et al. 1995; Miyake and Yafuso 2003).

Finally, the simplest (but not always straightforward) way to experimentally modify floral scent is to add single compounds or reconstituted blends to living flowers, typically using emitter devices or septa that control emission rates (Dobson et al. 2005). Manning (1956) pioneered this approach by adding essential oils to blooming and nonblooming *Cynoglossum officinale* (Boraginaceae) plants, to test their impact on bumblebee attention and foraging decisions. The novel odorants had no effect on approaches but reduced landings by half when flowers were present, either masking or disrupting the odor-color stimulus learned previously by the bees. Subsequent studies have established

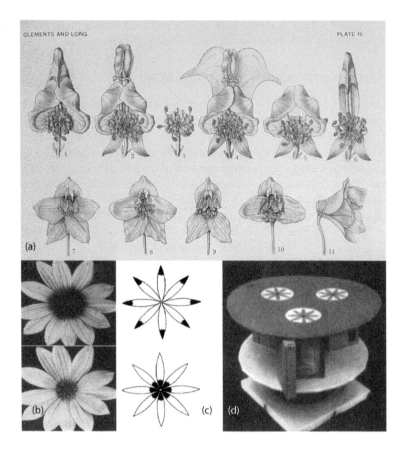

FIGURE 16.1 Early efforts at experimental floral manipulation. (a) Floral "mutilation" or targeted dissection of *Aconitum* and *Delphinium* flower parts (Ranunculaceae) by Clements and Long (1923), experimentally modifying floral symmetry, display size and reward accessibility in contrast to unmodified controls (left; drawings by Edith Clements). (From Clements, F.E. and F.L. Long. *Experimental Pollination: an Outline of the Ecology of Flowers and Insects.* Carnegie Institution of Washington, Washington, DC, 1923.) (b–d) Studies of ultraviolet (UV) absorbance patterns by Daumer (1958), showing differences in human visible vs. UV wavelengths for *Helianthus rigida* (b), a schematic drawing of the experimental inversion of ray florets to relocate the UV absorbance target (c), and an apparatus used to couple reconstructed floral displays with sugar rewards in bee behavioral choice assays (d). (From Daumer, K. *Z. Vergl. Physiol.*, 41, 49–100, 1958. Image in panel (a) is reprinted with permission from the Carnegie Institute of Science. Images in panels b–d are reprinted with permission from Springer-Nature, license 4605651447443.)

dose-dependent impacts of floral scent on the behavior of pollinators and natural enemies (Galen et al. 2011), highlighting the importance of measuring and matching natural emissions.

These foundational studies encompass three categories of experimental manipulation: (1) adding scent to living flowers (*floral augmentation*), (2) decoupling or subtracting specific cues from living flowers (*floral deconstruction*), and (3) reconstituting the essence of floral display (*floral reconstruction*) (Raguso 2006). Below, I review progress made using each of these three approaches, emphasizing more recent work (along with the specific hypotheses that were tested in each case) and highlighting emerging tools, ideas and technologies whenever possible.

16.2.2 Floral Augmentation

In floral augmentation, headspace samples, floral extracts, synthetic blends or single volatiles are added to living flowers to test behavioral, ecological or even evolutionary hypotheses (Table 16.2).

TABLE 16.2
Studies of Floral Augmentation, Organized by the Primary Question Tested by Manipulating Floral Scent

Study	Focal Plant Species and Animal Visitors	Scent(s) Added to Flowers	Response Variable	Role of Scent Revealed
Scent-mediated Behavior of Specialized Pollinators				
Hossaert-McKey et al. (1994)	*Ficus carica* figs *Blastophaga psenes* wasps	Pentane extracts of receptive figs added to ostiole of immature figs	Fig wasp landing, entry into fig via ostiole	Receptive fig scent guides upwind flight and entry
Schiestl and Ayasse (2001)	*Ophrys sphegodes* orchids *Andrena nigroaenea* bees	Farnesyl hexanoate, to mimic post-mating scent of female bees	Attempted copulation with flowers	Farnesyl hexanoate directs male bees away from pollinated flowers
Stensmyr et al. (2002) and Angioy et al. (2004)	*Helicodiceros muscivorus* Calliphorid blowflies	Oligosulfide blend to day-old spathe using heating coil on old appendix	Fly landings on spathe entries into floral chamber	Scent guides distance attraction heat enhances landing on appendix and entry into floral chamber
Scent-mediated Preference/Constancy of Generalized Pollinators				
Pellmyr (1986)	*Cimicifuga simplex* *Argynnis* butterflies	Isoeugenol & methyl anthranilate to populations lacking them	Butterfly approaches and landing on treated flowers	Either compound increases attraction but combination synergizes feeding
Dobson et al. (1999)	*Rosa rugosa* roses *Bombus terrestris* bees	Tetradecyl acetate, eugenol added to anther-less flowers	Bee landing, vibrating wing muscles (pollen collection)	Pollen volatiles increase bee landing and vibratory behavior
Cunningham et al. (2004)	*Nicotiana tabacum* *Helicoverpa armigera* moths	Forward pairing phenylacetaldehyde or α-pinene with sugar reward	Upwind flight in wind tunnel flower choice, preference	Scents learned as conditioned stimuli can be transferred across learning contexts
Ashman et al. (2005)	*Fragaria virginiana* Halictid bees	Pentane extracts of anthers, petals of hermaphrodites to female flowers	Approaches and landings by wild, experienced bees	Pollen scent (2-phenylethanol) increases visitation by bees to female flowers
Larue et al. (2016)	*Achillea millefolia* and *Cirsium arvense* flowers Diverse insect visitors	Reciprocal exchange of pentane extracts between species via evaporation	Plant–pollinator network links and their changes over time	Scent overwhelms visual differences guiding attraction and landing, reversibly restructuring visitation patterns & network structure
Effects of Scent Emission Rates on Seed Fitness				
Majetic et al. (2009)	*Hesperis matronalis* mustards Syrphid fly pollinators	Pentane extracts from night added to flowers during day	Pollinator visitation seed fitness	Selection shapes the timing and amount of floral scent emission
Galen et al. (2011)	*Polemonium viscosum* *Bombus balteatus* bees *Formica neorufibarbus* ants	Flush nectar and replace with sucrose solutions scented with different 2-phenylethanol doses	Bee visitation, seed fitness repellence and toxicity to ants, pollen limitation	Optimization of floral defense and pollen export links scent emissions with floral morphology (corolla flare)
Theis and Adler (2012)	*Cucurbita pepo* var. *texana* *Peponapis* and *Apis* bees *Acalyma vittatum* beetles	45-fold increase of major scent (1,4-dimethoxybenzene) to both staminate and pistallate flowers	Insect approaches, landings fruit mass, seed fitness	Augmented scent reduced bee visits to staminate flowers, attracted more beetles, reduced seed fitness

(Continued)

TABLE 16.2 (*Continued*)

Studies of Floral Augmentation, Organized by the Primary Question Tested by Manipulating Floral Scent

Study	Focal Plant Species and Animal Visitors	Scent(s) Added to Flowers	Response Variable	Role of Scent Revealed
Scent-mediated Floral Isolation and Pollinator Shifts				
Waelti et al. (2008)	*Silene dioica, S. latifolia* diurnal bees, butterflies nocturnal moths	Phenylacetaldehyde to reduce species-specific differences in scent composition	Track fluorescent powder movement as a surrogate for pollen transfer	Reproductive isolation is mediated by species-specific scents, breaks down when similar despite different colors
Shuttleworth and Johnson (2010)	*Eucomis* "pineapple lilies" Carrion flies, Pompilid wasps	Oligosulfide blend present in fly-pollinated *Eucomis* species added to wasp-pollinated *Eucomis* species	Fly visitation to novel species fluorescent powder movement as pollen surrogate	Oligosulfides sufficient to attract novel fly pollinators to a wasp-pollinated plant
Bischoff et al. (2015)	*Ipomopsis aggregata* and *Ipomopsis tenuituba, Hyles lineata* hawkmoths	Indole added to flowers with vs. without painting them red or white	Approaches and floral visits by wild and captive hawkmoths in mixed and unmixed arrays	Indole increases approaches, white color needed for probing: combination of traits mediates floral isolation
Campbell et al. (2016)	*Zaluzianskya natalensis* and *Z. microsiphon* plants *Hippotion, Basiothia* moths, *Prosoeca* long-tongued flies	Linalool and methyl benzoate present in moth-pollinated species added to fly-pollinated species, flower orientation also manipulated	Visits by wild hawkmoths in binary choice field assays	Flower orientation sufficient to mediate species-specific moth preferences in presence of scent; added scent did not alter visitation patterns
Gervasi et al. (2017)	*Ophrys insectifera, Ophrys aymoninii; Argogorytes* Digger wasps, *Andrena* bees	Four EAD-active compounds from *O. aymoninii* added to flowers of *O. insectifera*	Approaches and landings by *Andrena combinata* bees	Scent differences contribute significantly to prezygotic isolation via distance attraction
Moré et al. (2019)	*Jaborosa laciniata, Jaborosa integrifolia;* Calliphorid flies	Oligosulfide blend present in fly-pollinated species added to moth-pollinated species (*J. integrifolia*)	Fly visitation to novel species	Oligosulfides sufficient to attract novel fly pollinators to a moth-pollinated plant

For example, scent from receptive figs confers instant attraction of fig wasps to immature syconia (Hossaert-McKey et al. 1994), a post-pollination orchid volatile mimics the post-copulatory repellent of a mated female bee, enhancing outcrossing by reducing male copulatory attempts (Schiestl and Ayasse 2001), and oligosulfides restore the attractiveness of a spent, scentless Dead Horse Arum to female carrion flies, provided that the fetid inflorescence also is heated (Stensmyr et al. 2002; Angioy et al. 2004).

16.2.2.1 Preference and Constancy in Generalized Pollination

Pioneering work by Pellmyr (1986) and Dobson et al. (1999) showed that augmentation with specific volatiles could trigger nectar foraging by butterflies or pollen-collecting behavior by bees, respectively. Learned associations of volatiles with nectar or pollen rewards often reflect the pairing of profitable floral rewards with reliable (= low variance) stimuli (Wright and Schiestl 2009; Knauer and Schiestl 2015). Although salient olfactory learning in social bees is a core component of the literature on classical conditioning (Dötterl and Vereecken 2010), most models of floral constancy focus on color or limits to working memory. Thus, it was surprising that Larue et al. (2016) could reversibly rearrange generalized plant–pollinator network hubs in a German meadow community, including thistle (*Cirsium arvense*) and yarrow (*Achillea millefolium*; both Asteraceae), simply by augmenting these flowers with pentane extracts of each other's volatiles. Similarly, although plant–pollinator network structure is thought to reflect floral density (abundance), phenology (timing) and morphological (size) matching (Olesen et al. 2008; Stang et al. 2009), the manipulation of scent alone collapsed the network structure for *C. arvense, A. millefolium* and their respective floral visitors to complete connectance (a null network) within 5 minutes of scent augmentation, reverting to the original network structure when pentane extracts evaporated (Larue et al. 2016). When community-wide network models include floral scent along with color, density and phenology, it is found to mediate network links in generalized hub species (e.g., lavender and rock rose), which feed large bee communities, as well as beetle-pollinated poppies and sexually deceptive orchids peripheral to the core network (Kantsa et al. 2017, 2018).

16.2.2.2 Conflicting Selection between Pollinators and Enemies

Floral augmentation has been used to test the prediction that scent is under conflicting selection between pollinator attraction and herbivore avoidance in both emission rates and diel rhythms. Indeed, Theis and Adler (2012) found that spiking staminate and pistillate flowers of *Cucurbita texana* (Cucurbitaceae) with an excess of the dominant floral VOC (1,4-dimethoxybenzene) significantly increased attraction of (and floral damage by) *Acalyma vittatum* beetles without increased attraction of specialized (*Peponapis pruinosa*) or generalized (*Apis mellifera*) bee pollinators. Similarly, Galen et al. (2011) augmented the alpine flowers of *Polemonium viscosum* (Polemoniaceae) with different dosages of the most abundant floral VOC, 2-phenylethanol (2PE), solved in sucrose solution, after rinsing natural nectar from focal flowers. The highest (natural) dosages of augmented 2PE were repellent to bumblebee pollinators as well as nectar thieving ants, resulting in increased pollen limitation (and stronger pollinator-mediated selection on corolla morphology) by bees and reduced floral sterilization (through style removal) by ants, when compared with lower 2PE dosages. Interestingly, in *Hesperis matronalis* (Brassicaceae), the augmentation of daytime flowers with two-fold stronger pentane extracts of scent from nighttime flowers enhanced seed fitness through increased diurnal pollinator visitation, without attendant increases in herbivore attraction (Majetic et al. 2009). These results suggest that diurnal scent emission rates of *H. matronalis* should increase in North America, where it is an invasive plant with fewer natural enemies.

16.2.2.3 Pollinator Shifts and Reproductive Isolation

Waelti et al. (2008) provided proof-of-concept for scent-mediated reproductive isolation by using floral augmentation to abolish scent differences between two co-blooming species of *Silene* (Caryophyllaceae) pollinated by different insect functional groups. Floral augmentation with phenylacetaldehyde, a generic floral VOC resulted in the transfer of fluorescent powder (as a pollen surrogate) between the two *Silene* species, indicating the breakdown of prezygotic reproductive barriers. Shuttleworth and Johnson (2010) extended this approach to study *Eucomis* lilies (Asparagaceae) in South Africa, where species with similar floral color and morphology attract different pollinator guilds (Pompilid wasps, carrion flies, rodents) through divergent scent chemistry (Shuttleworth and Johnson 2008; Wester et al. 2019). Augmentation of wasp-pollinated flowers of *E. comosa* and *E. autumnalis* with a 1:1 mix of dimethyl disulfide and dimethyl trisulfide increased fly approaches and landings in the absence of morphological changes, with only a mild reduction of wasp visits. A similar outcome was observed in Argentina, when evening-blooming, hawkmoth-pollinated flowers of *Jaborosa integrifolia* (Solanaceae) augmented with the same volatile sulfide blend during daytime were visited instantly by Calliphorid flies (Moré et al. 2019). Carrion mimicry by angiosperms has evolved repeatedly world-wide, mediated in nearly all cases by the same volatile sulfides (Jürgens et al. 2013).

The last two examples relate to the role of scent in pollinator shifts involving short-tongued hawkmoths. In western North America, *Ipomopsis aggregata* and *I. tenuituba* (Polemoniaceae) show complete reproductive isolation at low elevation but produce hybrid swarms at some higher elevation sites (Aldridge and Campbell 2009). The hummingbird-pollinated flowers of *I. aggregata* are red, whereas the hawkmoth-pollinated flowers of *I. tenuituba* are white and scented with indole (Bischoff et al. 2013). Flowers of *I. aggregata* augmented with indole are more frequently approached by *Hyles lineata* moths but probing occurs only when indole-scented flowers are painted white, indicating that the scent and color of *I. tenuituba* mediate floral isolation through multiple messages (Bischoff et al. 2015; Table 16.1). However, in mixed species arrays augmented with indole, moths forage indiscriminately, suggesting how occasional introgression of floral scent and color might erode species boundaries within hybrid swarms. A parallel system in South Africa involves hawkmoth-pollinated *Zaluzianskya natalensis* and its long-tongued fly-pollinated relative, *Z. microsiphon* (Scrophulariaceae), which also form occasional hybrids (Campbell et al. 2016). These species differ in anthesis times (evening vs. morning), orientation (vertical- vs. horizontal-faced) and scent (linalool and methyl benzoate), but the long tubular corollas of both species are dark red, whereas the upper petal surfaces are bright white. Augmentation with linalool and methyl benzoate did not increase visits by wild *Hippotion celerio* and *Basiothia schenki* hawkmoths to the fly-pollinated flowers of *Z. microsiphon*. However, re-orientation to a vertical position elicited hawkmoth visits to *Z. microsiphon*, whereas hawkmoth visits to *Z. natalensis* were reduced when its flowers were positioned horizontally (Campbell et al. 2016). Ambient floral scent may have been sufficient to trigger hawkmoth foraging regardless of augmentation, but floral orientation is sufficient to mediate ethological isolation in *Zaluzianskya*, due to hawkmoth perception of visual contrast.

16.2.3 Floral Deconstruction

The purpose of floral deconstruction is to experimentally decouple contrasting elements of floral display, either to distinguish between components of the same modality (e.g., visual pattern and color (Weiss 1991; Hansen et al. 2011)) or to isolate different sensory channels. This is the fastest growing category of floral manipulation (Table 16.3). Historically, floral deconstruction was accomplished by enclosing flowers within transparent glass, plastic vials or oven bags to remove scent, or concealing them within dyed mesh or cheesecloth to remove visual display (Figure 16.2). A pair of recent studies model the care that should be taken to measure the impact of bags on deconstructed floral stimuli, including aspects of light, odor plume structure and sensory physiology (Riffell and Alarcón 2013; Russell et al. 2018; Table 16.3). Dötterl and colleagues have overcome these limitations by presenting floral stimuli to oligolectic bees in cylinders constructed from opaque or UV-transmitting clear quartz with a polycarbonate sleeve, in which perforations allow scent to escape the cylinder (Figure 16.2d).

16.2.3.1 Oligolectic Bees and Their Host Flowers

"Cylinder assays" on three specialized European bees provide valuable tests of the prediction that oligolectic bees use chemical cues in host selection (Table 16.3). Both naïve and experienced Megachilid bees (*Hoplitis adunca*) that specialize on pollen of *Echium vulgare* (Boraginaceae) use blue color alone to find these flowers, as well as related, nonhost plants with blue flowers (*Anchusa officinalis*) that provide nectar (Burger et al. 2010). However, recognition of *E. vulgare* as a pollen host is innate due to a combination of blue color and 1,4-benzoquinone. This requirement is relaxed as experienced bees learn more generic components of the *E. vulgare* volatile blend (Burger et al. 2012). Parallel studies of another Megachilid bee species (*Chelostoma rapunculi*) revealed a similar relationship with its pollen host plant, *Campanula trachelium* (Campanulaceae), in which both naïve and experienced bees can use blue color or scent to distinguish pollen hosts from nonhost flowers but prefer

TABLE 16.3
Studies of Floral Deconstruction; Decoupling Scent From Other Traits by Masking or Dissecting Flower Parts

Study	Focal Plant Species and Animal Visitors	Experimental Manipulation	Response Variable	Role of Scent Revealed
Using Transparent Bags and Dark Cloth to Decouple Visual and Olfactory Cues				
Raguso and Willis (2005)	*Datura wrightii* flowers *Manduca sexta* moths	Mask visual cues using cheesecloth mask scent using clear oven bags	Approaches and probing by wild hawkmoths in field assays	Moths require both scent and visual contrast for floral approach, feeding
Riffell and Alarcón (2013)	*D. wrightii, Agave palmeri Manduca sexta* moths	Shade cloth, oven bags, paper cone flowers to decouple and recombine visual and olfactory floral stimuli	Approaches and probing by captive hawkmoths in dual choice assays	Moths prefer scented flowers, make decisions more accurately and quickly when scent was present
Russell et al. (2018)	*Solanum houstonii* nightshades *Bombus impatiens* bees	Masking visual or olfactory floral traits with clear or opaque bags, with or without poricidal anthers	Approaches, landings and sonication behavior by bees in floral arrays	Naïve bees use corolla display to visit flowers but learn to use anther display to land after experience. Anther scent triggers sonication behavior by bees
Dissecting Innate vs. Learned Sensory Preferences of Host-Specialized or Solitary Bees				
Burger et al. (2010)	*Echium vulgare* (pollen host) *Anchusa officinalis* (nonhost) *Hoplitis adunca* bees	Clear vs. dark quartz cylinders with or without perforations, used to decouple floral scent and color	Approaches to different cylinders by bees	Naïve and experienced bees attracted to color of *Echium* flowers but distinguish between it and nonhost flowers using color + scent; recognition requires both.
Dötterl and Schäffler (2011)	*Lysimachia punctata* flowers *Macropis fulvipes* oil bees	Clear vs. dark quartz cylinders with or without perforations, used to decouple floral scent and color	Approaches to different cylinders by bees	Naïve females prefer scent but learn to generalize on yellow; naïve males also prefer scent, show stronger yellow bias
Milet-Pinheiro et al. (2012)	*Campanula trachelium* (host) *Echium vulgare* (nonhost) *Potentilla recta* (nonhost) *Chelostoma rapunculi* bees	Clear vs. dark quartz cylinders with or without perforations, used to decouple floral scent and color	Approaches to different cylinders by bees	Naïve bees prefer host to nonhost traits experience does not alter preference: combined > color > scent of host flower
Milet-Pinheiro, et al. (2013)	*Campanula trachelium Chelostoma rapunculi* bees	Present scent compounds or blends from dark quartz cylinders (no color) present scent of dissected flower parts from cloaked vials (no color)	Bee approaches to cylinder in dual choice assays, landings and feeding from hidden vials in cafeteria assays	Naïve bees prefer full blend or unique spiroacetals; experienced bees learn more generic scent compounds, scent of floral nectarines

(Continued)

TABLE 16.3 (*Continued*)

Studies of Floral Deconstruction; Decoupling Scent From Other Traits by Masking or Dissecting Flower Parts

Study	Focal Plant Species and Animal Visitors	Experimental Manipulation	Response Variable	Role of Scent Revealed
Wiemer et al. (2009)	*Cyclopogon elatus* orchids *Pseudoagapostemon jensenii* *Augochlora nausicaa* bees	Concealment of flowering scape within a paper chimney	Wild bee approaches in natural habitat	A simple floral volatile (trans-DMNT) is sufficient to attract bees to aperture of the paper chimney
Decoupling Visual and Olfactory Elements of Kettle Traps in Araceae				
Pellmyr and Patt (1986)	*Lysichiton americanum* aroids *Pelecomalius testaceum* beetles	Dissection of spadix and spathe, concealing yellow spathe under leaf concealing scent under glass	Approaches and landings by rove beetle pollinators in field settings	Beetles attracted scent of hidden spathe approaches to yellow spathe synergized by scent of spathe
Patt et al. (1995)	*Peltandra virginica* aroids *Elachiptera formosa* midges	Sham inflorescences constructed by adding scented spathes from different sexual phases to unscented spadices	Oviposition by midges into sham inflorescences in field settings	Oviposition varied with sexual stage of excised spathe, consistent with shifting ratios of unique volatile compounds
Miyake and Yafuso (2003)	*Alocasia odora* aroids *Calocasiomyia* flies	Partition spadices into female, male and sterile zones with bags, removal of lengths of scented appendix	Flies trapped using different exposed zones of spadix fruit set in field setting	Most flies attracted to distal half of the staminate spadix, 4-fold reduction in fruit set if scented appendix removed
Floral Dissection or "Mutilation" of Sexually Deceptive Orchids				
Spaethe et al. (2007)	*Ophrys heldreichiii* orchids *Eucera berlandi* bees	Remove pink perigon from attractive flowers, present flower photos above concealed (scented) orchid labella	Male bee approaches and copulatory attempts in a field setting	Pink perigon enhances sexual deception by simulating floral encounter sites for female bees (ecological context). Male bees prefer photos with labellum scent
Streinzer et al. (2010)	*Ophrys heldreichiii* (mimic) *O. dictynnae* (nonmimic) *Eucera berlandi* bees	Replace pink perigons from two related orchid spp. with photos; bees choose between labella augmented with scent	Male bee approaches and copulatory attempts	Pink perigon of *O. heldreichii* is more attractive than that of *O. dictynnae* to male bees; differences in labellum color pattern had no impact on bee approach
Rakosy et al. (2012)	*Ophrys heldreichii* orchids *Eucera berlandi* bees	Replace pink perigon with paper dummy, field tests of reproductive success (vs. control flowers)	Removal and deposit of pollinaria as a surrogate for male, female fitness	Low overall pollination success but greater for flowers with real or paper perigon than for those lacking pink

(Continued)

TABLE 16.3 (*Continued*)
Studies of Floral Deconstruction; Decoupling Scent From Other Traits by Masking or Dissecting Flower Parts

Study	Focal Plant Species and Animal Visitors	Experimental Manipulation	Response Variable	Role of Scent Revealed
Vereecken and Schiestl (2009)	*Ophrys arachnitiformis* orchids *Colletes cunicularius* bees	Add green or white orchid perigon to plastic beads with or without solvent extracts of labellum scent	Inspecting flights and contact with plastic bead in field settings	Colorful perigons have no effect on approaches or copulatory attempts by male Colletid bees
de Jager and Peakall (2016)	*Chiloglottis trapeziformis* & *C. valida* orchids *Neozeleborea cryptoides* & *N. monticola* wasps	Experimental reduction of (scented) labellum length in two sexually deceptive orchid species	Male wasp attraction, copulation attempts and orientation during visits in field settings	Shorter labella did not reduce male wasp attraction but altered duration of visits and orientation during copulatory effort
Rakosy et al. (2017)	*Ophrys leochroma* orchids *Eucera kullenbergi* bees	Removal of different components of labellum responsible for emission of scent and male bee orientation	Male bee landings and copulation attempts in field settings	Alterations did not reduce attraction to flower but impacted mechanical fit and effective pollen transfer
Pellegrino et al. (2017)	*Serapias lingua, S. vomeracea* bees seeking mates, shelter	Removal of (scented) labellum, paint callosity to increase contrast, open floral hood (source of shelter)	Removal and deposit of pollinaria, seed fitness in field settings	Loss of labellum eliminated seed set for both spp.; other manipulations enhance fitness in *S. lingua* (sexual deception)
Phillips and Peakall (2018)	*Caladenia pectinata* and *Drakaea livida* orchids *Zaspilothynnus nigripes* wasps	Excise osmophores of both orchids concealed beneath scentless labella of food deceptive *Caladenia* species.	Distance attraction and attempted copulations by male wasps	Sexual behavior of wasps did not depend on the floral tissue or appearance as long as mimetic sex pheromone was present
Metabolic Blockers and Gene Silencing to Delete Scent				
Kessler et al. (2008)	*Nicotiana attenuata* tobacco *Hyles lineata* hawkmoths *Archilochus* hummingbirds *Xylocopa* carpenter bees *Heliothis* moth larvae	RNAi silencing of genes governing nicotine (*pmt1/2*) and benzyl acetone (*chal1*) emission and the double mutant	Number and duration of floral visits, nectar robbing bud damage, seed fitness and siring success	Nicotine knockout increases feeding duration but reduces frequency of flower visits, increases nectar robbing and bud herbivory. Both nicotine and benzyl acetone needed to maximize fitness
Huang et al. (2012)	*Arabidopsis thaliana* plants *Pseudomonas syringae* bacteria	*tps21* knockout (β-caryophyllene) or ectopically expressed in leaves then inoculated with *P. syringae*	Pathogenic bacterial growth, seed fitness and stigmatic tissue necrotic damage	β-caryophyllene serves an antimicrobial function in self-pollinated *Arabidopsis* flower tissues

(*Continued*)

TABLE 16.3 (*Continued*)

Studies of Floral Deconstruction; Decoupling Scent From Other Traits by Masking or Dissecting Flower Parts

Study	Focal Plant Species and Animal Visitors	Experimental Manipulation	Response Variable	Role of Scent Revealed
Junker et al. (2011a)	*Phlox paniculata* plants *Lasius niger* ants *Episyrphus balteatus* flies	inhibition of monoterpenes using fosmidomycin, sesquiterpenes via mevinolin, metabolic inhibitors	ant choice in olfactometers hoverfly choice in field setting	ants repelled by control flowers were neutral toward terpene-inhibited flowers; no effect on hoverfly choices
Kessler et al. (2013)	*Petunia x hybrida* plants *Diabrotica undecimpunctata* *Oecanthus fultoni* crickets	RNAi silencing of genes governing some (*bpbt*, *cfat*) or all (*odo1*) scent emissions of benzenoids at night	Floral herbivore damage by beetles and wood crickets	Low floral damage in wild type (scented) and *odo1* mutants (no scent), damage greatest when isoeugenol or benzyl benzoate silenced, suggesting defensive roles for minor floral volatiles
Kessler et al. (2015)	*Nicotiana attenuata* tobacco *Hyles lineata* and *Manduca sexta* hawkmoths, *Archilochus alexandri* hummingbirds	RNAi silencing of benzyl acetone (*cha1*) and nectar production (*sweet9*) and double mutant to decouple nectar and scent production	Seed set in field settings with individual pollinators oviposition by female *M. sexta* moths	Reduced pollination by hummingbirds in nectar-, by *M. sexta* in nectar-, scent- or double mutant, by *H. lineata* in double mutant. Oviposition by *M. sexta* reduced when nectar silenced
Boachon et al. (2015)	*Arabidopsis thaliana* plants *Frankliniella occidentalis* thrips *Spodoptera* & *Plutella* moths	Bioassays with *cyp76c1* null mutants (emit (R) (–)linalool) vs. wild type (which convert to linalool oxides and lilac aldehydes, alcohols)	Choices by different pest insects in olfactometer bioassays	Linalool derivatives play defensive roles in self-pollinating *Arabidopsis* flowers; linalool itself is attractive to herbivores, florivores and pollen thieves
Trait Swapping with Recombinant Inbred Lines (RILS) and Near-Isogenic Lines (NILS) to Decouple Scent and Color				
Hoballah et al. (2007)	*Petunia x hybrida* plants, *P. axillaris*, *P. integrifolia Manduca sexta* moths *Bombus terrestris* bees	Create RILS by introgressing myb (*an2*) alleles from *P. axillaris* (white) or *P. integrifolia* (pink) into *P. hybrid* background; transgenic *P. axillaris* with pink *an2* allele (*P. integrifolia*)	Approaches and probing by hawkmoths and bees in flight cages, greenhouse	Moths and bees prefer white and pink, respectively, whether in parental spp. or in the RILS; in transgenic *P. axillaris*, moths prefer white over pink *an2* allele; nectar and scent are unchanged

(*Continued*)

TABLE 16.3 (Continued)
Studies of Floral Deconstruction; Decoupling Scent From Other Traits by Masking or Dissecting Flower Parts

Study	Focal Plant Species and Animal Visitors	Experimental Manipulation	Response Variable	Role of Scent Revealed
Klahre et al. (2011)	*Petunia axillaris, P. exserta; Manduca sexta* moths *Bombus terrestris* bees	Reciprocal backcrosses of QTL (*odo1* myb factor) for scent, *P. exserta* (red, scentless) and *P. axillaris* (white, scented) to create NILS with swapped traits (red-scented vs. white unscented)	Approaches and probing by hawkmoths in dual choice wind tunnel assays	Hawkmoths prefer scented > unscented when same flower color, but show no preference when floral trait combination is in conflict (white/ unscented vs. red/scented with methyl benzoate)
Dell'Olivo and Kuhlemeier (2013)	*Petunia axillaris, P. inflata* *Manduca sexta* moths *Bombus terrestris* bees	Reciprocal introgression of *an2* alleles from *P. axillaris* (white) or *P. inflata* (pink) to create NILS with swapped colors but species-specific flower form, nectar and scent	Approaches and probing by hawkmoths and bees	Moths prefer white *an2* morph in small, weakly-scented *P. inflata* background, but no preference in large, strong scent *P. axillaris* background. bees prefer pink *an2* morph in large, scented background of *P. axillaris* but no preference in small weakly-scented *P. inflata* background

the color–scent combination (Milet-Pinheiro et al. 2012). Subsequent bioassays using dissected flower parts and synthetic scent blends concealed within quartz cylinders or beneath dark cheesecloth revealed that naïve *C. rapunculi* bees are attracted to spiroacetals emitted by all floral organs, but experienced bees learn a blend of conventional VOCs associated with nectaries (Milet-Pinheiro et al. 2013). Finally, floral deconstruction revealed how specialized oil bees (*Macropis fulvipes*: Melittidae) recognize and handle flowers of *Lysimachia punctata* (Lythraceae), their primary source of floral oils. Female *M. fulvipes* bees collect floral oils to provision their young, whereas males aggregate at *L. punctata* flowers seeking mates. Naïve females prefer floral scent to color and the combination thereof to either trait in isolation, whereas foraging-experienced females generalize on the bright yellow display of *L. punctata* but still prefer the scent–color combination (Dötterl et al. 2011). Newly emerged males have a stronger color-bias than females but also show preference for scent, especially in combination with color. Coupled gas chromatography-electroantennography (GC-EAD) and behavioral assays revealed a compound (diacetin) that is structurally related to the floral oils, as a key behavioral attractant for this and other oil flower-bee associations world-wide (Schäffler et al. 2015). Given its likely derivation from the floral oils, diacetin represents a rare example of an index signal for the presence of floral rewards.

FIGURE 16.2 Methods in floral deconstruction. (a) Cut inflorescences of *Yucca glauca* in water, for use in field bioassays with yucca moths (*Tegeticula* spp.). From L to R: open (positive) control, oven-bagged (color, no scent), unopened buds (negative control) and cheesecloth-bagged (scent, no color). (b–d) Quartz cylinders for decoupling the floral traits of *Echium vulgare* inflorescences. Cylinders are connected to a membrane pump to allow diffusion of floral scent through a perforated ring. Opaque cylinder (b) removes visual cues while transparent cylinder (c) allows bee responses to UV and human-visible wavelengths of light. Panel (d) shows a female *Hoplitis adunca* bee attracted to multi-modal floral display. (e) Cut flowers of *Solanum houstonii* prepared for bee bioassays in which floral parts (corollas or stamens) are isolated or the flowers are bagged (From Russell, A.L., et al. *Animal Behav.*, 135, 223–236, 2018.), as in panel (a). (f) Sham and mosaic flowers assembled from dissected parts of *Solanum houstonii* and *Exacum affine* (From Russell, A.L., et al. *Behav. Ecol.*, 27, 731–740, 2016. (b–d) from Hannah Burger, with permission; images in (e) are reprinted with permission from Elsevier, license 4605941411875; (f) are reprinted with permission from Oxford University Press.)

16.2.3.2 Floral Dissection in Specialized Pollination Systems

Another approach to floral deconstruction involves translocating or removing dissected flower parts, with the obvious need to control for wound volatiles. One set of studies exploits the division of labor in pollinator attraction between the spadix (a column combining male and female florets and, at times, a sterile appendix) and spathe (a leafy or waxy bract enveloping the spadix) in the kettle-trap inflorescences of the Araceae (Table 16.3). Obligate "nursery pollination" mutualisms between Chloropid midges and *Peltandra virginica* (Patt et al. 1995) and Drosophilid flies and *Alocasia odora* (Miyake and Yafuso 2003) are mediated by volatiles exclusive to the spathe or spadix, respectively. Experimental exchange or removal of these floral organs revealed the timing and magnitude of female fly attraction, oviposition and consequences for seed fitness. Two other cases involve pollination by beetles that use Araceae plant inflorescences as mating sites. Deconstruction of *Lysichiton americanum* inflorescences showed that Staphylinid (rove) beetle pollinators are attracted to bright yellow spathes only when coupled with scented spadices (Pellmyr and Patt 1986), one of many cases in which floral scent and visual display contextualize each other (Table 16.1). A more recent study confirmed that yellow sticky traps placed in the habitat of *L. americanum* plants only capture rove beetles when baited with indole or two unique, EAG-active alkenes (Brodie et al. 2018), a simple case of floral reconstruction (Table 16.4).

Our current understanding of floral scent is indebted to long-term studies of sexual deception in the Mediterranean orchid genus *Ophrys*, for which GC-EAD-guided fractionation and field bioassays

TABLE 16.4
Studies of Floral Reconstruction; Experimental Recombination of Scent and Other Floral Traits

Study	Focal Plant Species and Animal Visitors	Experimental Manipulation	Response Variable	Role of Scent Revealed
Scent Augmentation of Model Flowers or Colored Traps				
Roy and Raguso (1997)	*Arabis (Boechera)* mustards *Puccinia monoica* rust fungi Alpine bees and flies	White, yellow paper with rust extracts or single volatiles vs. (+) and (−) controls	Insect approaches and landings on different treatments	Bees attracted to yellow color and most scents; flies only to (+) control rust-infected plants
Kessler and Baldwin (2007)	*Nicotiana attenuata* tobacco *Manduca sexta* moths *Archilocus* & *Selasphorus* Hummingbirds, *Solenopsis xyloni* ants	Use of artificial flowers, feeders to test combinations of nicotine & benzyl acetone solved within nectar or in headspace with visual cues necessary to attract moths and birds. Ant bio-assays adding scent to sugar solutions in small tubes	Visitation rate, feeding time and consumption by birds, moths. Total consumption of sugar solution (with/without) test volatiles by ants	Opposing effects of benzyl acetone and nicotine when present in nectar; nicotine compelled shorter feeding bouts by ants, moths and hummingbirds
Burger et al. (2012)	*Echium vulgare* plants; *Hoplitis adunca* bees	Added floral extracts and specific VOC (1,4-benzoquinone) to blue paper cone artificial flowers	Landings by captive bees	Naïve female bees attracted by floral scent blend or 1,4-benzoquinone equally experienced bees learned to respond to other scent compounds
Steenhuisen et al. (2013)	*Protea caffra, P. simplex Atrichelaphinis tigrina* beetles	Added EAG-active benzaldehyde, linalool & lin. oxides, methyl benzoate, methyl salicylate to green or yellow bucket traps	Trap capture of chafer beetles in field settings	Yellow traps captured 10-fold more than green over all treatments; all scents except benzaldehyde attracted more beetles than controls
Moré et al. (2013)	*Jaborosa rotacea* plants; Calliphorid blowflies	Added oligosulphides to paper, velvet artificial flowers; lures placed among flower models or separated spatially	Fly approaches and visits in field settings	Oligosulphides required to attract flies, increased approaches and landings to white when decoupled spatially
Byers et al. (2014)	*Mimulus lewisii, Mimulus cardinalis* monkeyflowers *Bombus vosnesenskii* bees	Artificial flowers scented with floral headspace, synthetic blends or single compounds from *Mimulus* flowers	Bee visitation in dual choice assays in captive settings	Bees trained on the scent of *M. lewisii* preferred it (frequency, duration) to the scent of *M. cardinalis* and to individual monoterpenes (limonene, β-myrcene)

(Continued)

TABLE 16.4 (*Continued*)

Studies of Floral Reconstruction; Experimental Recombination of Scent and Other Floral Traits

Study	Focal Plant Species and Animal Visitors	Experimental Manipulation	Response Variable	Role of Scent Revealed
Oelschlägel et al. (2015)	*Aristolochia rotunda* plants *Trachysiphonella rufipes* Choropid midges	Sticky traps created from colorless plastic tubes charged with EAG-active volatiles from Mirid bug haemolymph and *A. rotunda*	Number of attracted or trapped midge pollinators	Synthetic blend of EAG-active scents from *A. rotunda* was as attractive as dead Mirid bugs, used by midges as a food source
Chen et al. (2015)	*Amorphophallus konjak* aroids Calliphorid, Muscid and Sarcophagid carrion flies	Artificial inflorescence made with filter paper stained using pigments from *A. konjak*, charged with oligosulphides on cotton	Wild insects trapped by different flower models	Flies in a semi-natural setting were attracted by artificial flowers, with double the visits to scented + dyed models than to scent alone
Suinyuy et al. (2015)	*Encephalartos villosus* cycads *Porthetes pearsonii*, *Metacucujus goodei*, *Antliarhinus zamiae* weevils	Yellow vs. green bucket traps at different geographic locations within natural range of plants baited with EAG-active scents: (3*E*)-1,3-octadiene, 2-methoxy-3-isopropylpyrazine, 1,8-cineole	Trap capture of beetles (both) pollinators and seed predators	Beetles in northern range trapped by 1,3- octadiene (present in southern populations) beetles in southern range trapped by pyrazine (present in southern populations). 1,8-cineole did not trap any beetles across range
Brodie et al. (2018)	*Lysichiton americanum* aroids *Pelecomalium testaceum* rove beetle pollinators	Yellow sticky traps to simulate spathe, unscented or baited with EAG-active scent compounds	Trap capture of beetles in field settings	Indole, alone or with (*E*)-4-nonene and (*E*)-5-undecene, attracted more beetles than unscented yellow traps
Multimodal Choice Experiments with Bees and Bats				
Kaczorowski et al. (2012a)	Artificial flowers *Bombus impatiens* bees	Bees trained to scent OR color vs. multimodal cues, trained under different light levels	Bee accuracy and visitation rates to conditioned flower traits	Bees learned bimodal flowers more quickly than unimodal (color) ones accuracy suffered at low light for color but not for bimodal flowers
Lawson et al. (2017)	Artificial flowers *Bombus terrestris* bees	Artificial flowers with color and essential oils presented under ambient conditions compromising olfaction (wind, chemical noise)	Bee preference, learning speed, performance	Bees learned more quickly, performed better under windy or chemically noisy conditions when flowers were colored

(Continued)

TABLE 16.4 (*Continued*)

Studies of Floral Reconstruction; Experimental Recombination of Scent and Other Floral Traits

Study	Focal Plant Species and Animal Visitors	Experimental Manipulation	Response Variable	Role of Scent Revealed
Lawson et al. (2018)	Artificial flowers *Bombus terrestris* bees	Artificial flowers with spatial patterns of scent and visual contrast	Bee preference, learning speed and performance	Bees could transfer learned intra-floral patterns of scent into spatially similar color patterns to which they had not been trained
Muchhala and Serrano (2015)	artificial flowers, *Anaura caudifer, A appositus* bats	Bats trained to artificial flowers with shape-scent combinations, binary choice assays under open vs. cluttered settings in tropical forest	Bat learning and choices	Bats visited both flower shapes equally in open setting but preferred training scent in a cluttered setting. Dependence on trained scent increased during nectar foraging, especially dimethyldisulfide
Gonzalez-Terrazas et al. (2016)	*Pachycereus pringlei* cacti; *Leptonycteris yerbabuenae*	Sound-absorbing boxes with solo floral stimuli (scent = hidden flower; acoustic = ultrasonic microphone and acrylic parabola)	Bat approaches, first choices, feeding frequency, pulse rates (echolocation) in flight cage	Bats responded strongest to combined olfactory and acoustic cues over uni-modal display, but actual visits often involved probing parabolic model; mean responses higher to multimodal cues

Use of 3D-Printing Technology to Test Specific Floral Hypotheses

Study	Focal Plant Species and Animal Visitors	Experimental Manipulation	Response Variable	Role of Scent Revealed
Campos et al. (2015)	Artificial flowers *Manduca sexta* moths	3D-printed artificial flowers in funnel shapes, combined with sucrose and a 7-scent blend from *Datura wrightii*	Moth preference, constancy number of flowers emptied	In presence of scent, moths drink nectar most efficiently from funnel-shaped flowers but don't show preference or constancy to shape
Sponberg et al. (2015)	Artificial flowers *Manduca sexta* moths	3D-printed artificial flowers with sucrose and a 7-scent blend on a movable base for visual tracking in dim light	Moth visual tracking of artificial flowers	In presence of scent, moths take longer to visually track moving flowers in dim light, simulating windy conditions, dusk
Roth et al. (2016)	Artificial flowers *Manduca sexta* moths	3D-printed artificial flowers with sucrose and a 7-scent blend, but both flower and nectar tube can move independently, decoupling visual and mechanosensory inputs	Moth visual and mechanosensory tracking of flowers	In presence of scent, moths can use either visual or mechanosensory inputs to track flower movement, reinforcing feeding behavior through redundant sensory inputs

(Continued)

TABLE 16.4 (*Continued*)

Studies of Floral Reconstruction; Experimental Recombination of Scent and Other Floral Traits

Study	Focal Plant Species and Animal Visitors	Experimental Manipulation	Response Variable	Role of Scent Revealed
Policha et al. (2016)	*Dracula lafleuri* orchids *Zygothrica* fungus flies	3D-print molded silicone model flowers to decouple scent, visual display (color, pattern) of labellum and sepals recombined with flower parts as chimaeric flowers	Fly approaches and visits	Flies equally attracted to scented labellum, flower scent made visual display more attractive, chemistry and pattern of sepals necessary for fly movement to column, pollination
Nordström et al. (2018)	Floral communities in Sikkim, Bangalore and Uppsala *Eristalis tenax* & *Episyrphus balteatus* hoverflies	3D-printed artificial flowers reflecting unbiased consensus (color, shape, scent, humidity) of hoverfly-visited	Fly visits to artificial flowers in natural settings	Certain trait combinations universally attractive (without reward) including those not representing actual plant species. Other lure combinations were highly site- and context-dependent

with chemical blends have repeatedly "cracked the codes" of pheromone mimicry (Ayasse et al. 2011). However, *Ophrys* flowers also have complex shapes, color patterns and hair-like trichomes along with sometimes colorful perigon displays, inciting debate on nonchemical contributions to insect mimicry (Bradshaw et al. 2010; Vignolini et al. 2012). Accordingly, floral deconstruction was used to assess how visual traits impact the effectiveness of scent-mediated pheromone mimicry in *Ophrys* orchids. In Crete, flowers of *O. heldreichii* have a bright pink perigon and are pollinated by male longhorn bees (*Eucera berlandi*: Apidae) that patrol patches of pink flowers (e.g., *Salvia fruticosa, Malva sylvestris*) in search of nectar-feeding females (Spaethe et al. 2007). Removal of the perigon reduced the attractiveness of *O. heldreichii* flowers in choice assays. *Eucera berlandi* bees prefer the deeper pink of *O. heldreichii* to the pallid hues of a neighboring orchid, *O. dictynnae*, but do not distinguish between the ornate visual patterns on the labella of these species when they are presented to bees with cutout photographic dummy perigons and a scent plume provided by *O. heldreichii* (Streinzer et al. 2010). The removal and deposit of pollinia both suffer in field experiments when the pink perigon is removed but are rescued when labella are augmented with pink cardboard perigons (Rakosy et al. 2012).

When the same experiment is performed on flowers of *O. arachnitiformis* in southern France, the removal of yellowish green or white perigons has no impact on the attraction of male *Colletes cunicularius* bees as pollinators (Vereecken and Schiestl 2009). Unlike longhorn bees, male Andrenid and Colletid bees engage in scramble competition for mates in open sandy areas where virgin females emerge: flower color is not an important contextual signal in *Ophrys* species pollinated by such bees (Dötterl and Vereecken 2010). Similar studies on *Ophrys* and *Serapias* orchids in Europe (Pellegrino et al. 2017; Rakosy et al. 2017), along with other genera of sexually deceptive orchids in Australia (de Jager and Peakall 2016; Phillips and Peakall 2018), have evaluated the importance of scent vs. labellum size, color and morphology for pollinator attraction, placement and effectiveness (Table 16.3).

16.2.3.3   Metabolic Inhibitors and Gene Silencing

A missing tool in floral manipulation has been the ability to delete scent from floral display, as the converse of floral augmentation. Junker et al. (2011a) used fosmidomycin and mevidolin to block monoterpene and sesquiterpene VOC biosynthesis, respectively, in cut flowers of *Phlox paniculata*, resulting in reduced germacrene D, linalool and its derivatives (e.g., lilac aldehydes). The modified floral bouquet was more attractive to *Lasius niger* ants, presumably due to reduced amounts of linalool (a known ant repellent) but had no effect on *Episyrphus balteatus* hoverflies. However, metabolic inhibitors may have unintended pleiotropic consequences in plant metabolism. Targeted gene silencing provides an opportunity to surgically alter the expression VOCs at finer resolution. The creative use of gene silencing has exposed two fallacies in pollinator-centric thinking: (1) that pollinator attraction is sufficient for plant reproductive success, and (2) that the function of floral traits is primarily attractive.

In their study of a North American tobacco, *Nicotiana attenuata* (Solanaceae), Kessler et al. (2008) used inverted repeat RNAi constructs to silence the production of floral nicotine (solved in nectar), benzyl acetone (the major VOC in corolla tissues) or both (double mutant), tallied pollinator and natural enemy visitation and scored male (siring success) and female (seed production) fitness with anther-ectomized flowers, to avoid self-pollination. The absence of nicotine increased nectar-feeding duration and reduced visitation frequency by hummingbirds and hawkmoths but also increased nectar robbing by *Xylocopa* bees and florivory by *Heliothis* caterpillars. The benzyl acetone knockouts suffered lower pollinator visitation by deleting a compound attractive to moths as an odor and to birds as a flavor, whereas double mutants no longer attracted hawkmoths. Together with seed fitness and paternity results, these outcomes indicate that the push–pull combination of nicotine and benzyl acetone optimizes floral defense, pollinator attraction and pollen flow in this desert tobacco. In a subsequent experiment, these authors tested whether floral advertisements or rewards held stronger sway in the foraging decisions of birds and hawkmoths visiting *N. attenuata*, this time silencing genes responsible for benzyl acetone emission and/or nectar secretion (Kessler et al. 2015). Hummingbird visitation dropped in the absence of nectar, whereas large hawkmoths (*Manduca sexta*) were sensitive to all knockouts, requiring both scent and nectar for sustained feeding, while smaller hawkmoths (*Hyles lineata*) were willing to visit scentless or nectarless flowers, but not the double mutants. When bioassays focused on mated *M. sexta* moths (which use *N. attenuata* as a host plant as well as a floral nectar source), nectar knockouts had the strongest impact, by reducing oviposition, highlighting a task-specific shift in decision-making. Gene silencing studies targeting low-abundance floral volatiles have revealed additional defensive functions, including protection against facultative flower damage by beetles and wood crickets on *Petunia x hybrida* (Kessler et al. 2013) (due to isoeugenol and benzyl benzoate) and prophylaxis against bacterial pathogens (β-caryophyllene) (Huang et al. 2012) and thrips (lilac aldehydes and other linalool derivatives) in *Arabidopsis thaliana* (Brassicaceae) (Boachon et al. 2015).

16.2.3.4   Trait Swaps, Plant Breeding, RILS and NILS

The goal of remixing floral trait combinations against controlled genetic backgrounds can be achieved with living plants using the tools of plant breeding, as exemplified by Kuhlemeier and colleagues' research program using *Petunia* (Solanaceae) as a model system. In their first study, recombinant inbred lines (RILs) were prepared by introgressing different alleles of a myb-like transcription factor governing corolla color (*an2*), from wild species with different pollination syndromes into a common *P. x hybrida* genetic background (Hoballah et al. 2007). *Bombus terrestris* bees preferentially visited the small, pink, weakly-scented *P. integrifolia*, and their preference transferred to a *P. x hybrida* inbred line expressing the pink *an2* allele. Conversely, *M. sexta* moths preferentially visited the long-tubed, white, strongly scented *P. axillaris* and the *P. x hybrida* inbred line expressing the white *an2* allele, in a scent-saturated greenhouse. Next, the authors prepared a transgenic line of *P. axillaris* expressing the pink *an2* allele from *P. integrifolia*, in which all other traits (nectar, scent, morphology) bred true. Moths and bees again showed assortative attraction by flower color, this time in the absence of other allelic variation. Dell'Olivo and Kuhlemeier (2013) revisited this comparison between *P. axillaris* and *P. inflata*, another bee-pollinated species with small pink flowers. They constructed near-isogenic lines (NILs) by back-crossing interspecific hybrids to either parental species, thereby introgressing *an2* alleles from *P. axillaris* and *P. inflata* into each other's genetic backgrounds. NIL preparation resulted in a color swap between species without altering wild type morphology, nectar or scent production. Moths and bees showed no color preferences when visiting *an2*-swapped NILs of *P. axillaris* or *P. inflata*, respectively, from which they could feed profitably. Instead, they showed white-bias (moths) or pink-bias (bees) when visiting "against-syndrome" at flowers that were less rewarding or more difficult to handle. However, these pigment-focused studies presented *M. sexta* moths with recombinant floral displays immersed within a scent cloud in greenhouse or flight cage environments, at which point they and other hawk-moth species reliably show visually-guided flower choice (Raguso and Willis 2002, 2005; Bischoff et al. 2015). Using plant breeding to test the role of scent would require a different behavioral arena; a laminar flow wind tunnel.

Klahre et al. (2011) prepared NILs between *P. axillaris* and *P. exserta*, a rare, hummingbird-pollinated species with red, scentless flowers, now focusing on *odo1*, another myb-like transcription factor responsible for substrate flux through the benzenoid volatile biosynthetic pathway in *Petunia* (Verdonk et al. 2005). By exchanging *odo1* alleles, the authors added methyl benzoate to red-flowered *P. exserta* and deleted scent from white-flowered *P. axillaris*, which could then be competed against the parental species in dual choice wind tunnel assays. Moths always preferred the scented morph when choosing between lines with the same flower color but showed no preference when choosing between flowers with syndrome-conflicting traits (i.e., a white, scentless *P. axillaris* NIL and a red, scented *P. exserta* NIL). This result is consistent with previous studies of *M. sexta*, in which floral scent can "gate" probing at unscented flowers, either in a lab setting (Raguso and Willis 2002; Goyret et al. 2007) or in nature (Raguso and Willis 2005), but probing is unlikely in the absence of a visual target (Riffell and Alarcón 2013).

16.2.4   FLORAL RECONSTRUCTION

A third experimental approach is floral reconstruction, the preparation of artificial flowers models or dummies to which scents and/or floral rewards are added (Table 16.4). The sprawling literature on artificial flowers ranges from simple colored tubes or paper squares with scented or colored sugar rewards (Roy and Raguso 1997; Gegear and Laverty 2005; Hansen et al. 2006; Makino and Sakai 2007) to the use of photographed flowers or parts (Gegear et al. 2017; Streinzer et al. 2010) and detailed attempts to match complex floral morphology (Chen et al. 2015) or pollen presentation (Russell and Papaj 2016). Other examples of floral reconstruction outlined in Table 16.4 include choice bioassays using scented artificial flowers (Kessler and Baldwin 2007; Burger et al. 2012; Moré et al. 2013; Byers et al. 2014; Milet-Pinheiro et al. 2015) and field trapping experiments using scent–color combinations (Steenhuisen et al. 2013; Oelschlägel et al. 2015; Suinyuy et al. 2015; Brodie et al. 2018) to complement or confirm insights gained from floral deconstruction.

16.2.4.1   Bees, Bats and the Efficacy Backup Hypothesis

The efficacy backup hypothesis suggests that multimodal plant–pollinator communication is adaptive under variable ambient conditions, when a single sensory channel might be compromised by low light or high wind (Leonard et al. 2011a; Table 16.1). Kaczorowski et al. (2012a) trained *Bombus impatiens* bees to distinguish between artificial flowers with similar colors (shades of blue), odors (linalool and geraniol) or their combinations through associative learning. Foraging success was measured under three levels of illuminance, with the prediction that multimodal flowers would be easier to distinguish under low light because they were scented. Bees learned bimodal flower cues more rapidly and distinguished learned stimuli more accurately at low light levels. However, their improved accuracy did not increase visitation rates to rewarding flowers, suggesting that multimodal signals benefit flowers (pollen transfer via constancy) more than their pollinators (no increased profitability). Similarly, Lawson et al. (2017) tested whether multimodal floral learning by *Bombus terrestris* bees could buffer a converse set of challenges, in which wind speed or a noisy olfactory background nullify responses to scent as a conditioned stimulus. Learning speed, visitation rates and accurate choices were impaired by wind and chemical interference when they had trained only on scent, whereas their performance was comparable to control assays lacking olfactory noise or wind when bees trained on flowers with both scent and color.

Environmental challenges for nectar-foraging bats play out on different spatial and temporal scales, as they contend with differences in habitat complexity and (lunar) illumination over greater foraging areas and longer lifetimes than bees. Muchhala and Serrano (2015) explored the potential for multisensory nectar-foraging behavior in two related Glossophagine bats, *Anaura caudifer* and *A. geoffroyi*, in cloud forest field sites in Ecuador. Captive bats trapped in mist nets were trained to specific (rewarding) combinations of artificial flower shapes and scents (banana oil vs. dimethyl disulfide) while rotating position to avoid place learning. On the second night, bats chose between two artificial flowers; the conditioned flower shape paired with a novel scent vs. the conditioned scent paired with a novel shape, either in an empty flight cage or one cluttered with vegetation. In an empty cage, the bats generalized on flower shape irrespective of training scent, but in a cluttered cage they preferred novel-shaped flowers with the training scent, especially when that was dimethyl disulfide, a compound common to New World bat-pollinated flowers (Bestmann et al. 1997). Thus, in dense rainforest vegetation bats use scent as an efficacy backup strategy. In a desert setting, Gonzalez-Terrazas et al. (2016) examined whether another species of nectar-feeding Glossophagine bat (*Leptonycteris yerbabuenae*) shows sensory preferences while nectar foraging from columnar cacti (*Pachycereus pringlei*). Floral reconstruction resulted in single- and multimodal combinations of stimuli (acoustic: an ultrasonic microphone with an acrylic parabola-shaped flower dummy; olfactory: scent from hidden cactus flowers) with Petri dish nectaries at different stations within a flight cage. Initial bat responses often were triggered by scent, but most attempted feeding events were drawn by the parabolic shapes, and multimodal stations elicited significantly more approaches plus inspections than did either single-modality stimulus. Echolocation pulses increased in frequency when bats fed from Petri dishes or parabolic dummies, which never occurred at odor sources. These desert bats, like their rainforest cousins, can use multiple sensory channels as redundant signals but appear to favor multimodal foraging, at least when confronted with novel floral resources or settings.

16.2.4.2 Using High-Tech Flowers to Revisit Classical Questions

In recent years, behavioral experiments using floral reconstruction have broadened our collective understanding of floral advertisements and rewards. For example, simple artificial flowers were augmented to identify the importance of CO_2 (Goyret et al. 2008), relative humidity (von Arx et al. 2012), nectar salts (Afik et al. 2006), bacteria and yeast (Vannette et al. 2012). More sophisticated mechanical flowers were engineered to test the importance of petal surface texture (epoxy casts of conical cells (Whitney et al. 2009a) and iridescent patches (Whitney et al. 2009b), the "echo fingerprints" of bat-pollinated flower parts (embedded and cast using dental materials) (von Helversen and von Helversen 2003), floral temperature (insulated plastic cylinders with a resistance coil) (Harrap et al. 2017) and floral static electric fields (epoxy-steel disks with a shielded coaxial cable; Clarke et al. 2013) to discriminating pollinators.

Studies of artificial flowers inspired by the contours and morphometric relationships of natural flowers have revealed the importance of proboscis mechano-sensation in flower feeding by *Manduca sexta* moths (Goyret and Raguso 2006; Kaczorowski et al. 2012b). Daniel and associates have

FIGURE 16.3 Methods in floral reconstruction. (a) Series of still images from high-definition video of a 3D-printed funnel flower being visited by *Manduca sexta* (Campos, E.O., et al., *Funct. Ecol.*, 29, 462–468, 2015). (b) Silicone cast flower models and natural flower of *Dracula lafleuri* orchid in the Los Cedros Reserve cloud forest, Ecuador; clockwise from center: uncolored silicone flower (material control), colored (unscented) silicone flower, natural flower (positive control) and colored + scented silicone flower (Policha, T., et al., *New Phytol.* 210, 1058–1071, 2016). (c–e) An array of multi-modal 3D-printed flowers presented to syrphid flies in Bangalore, India (c; grey disk at bottom right is the negative control); *Eristalis tenax* hoverfly visiting a "hot" natural (*Brassica rapa*) flower (d) and a "hot" 3D-printed artificial flower (e; Nordström, K., et al., *Proc. Natl. Acad. Sci.*, 114, 13218–13223, 2017. Images in (a) from Tanvi Deora, Mahad Ahmed and Tom Daniel, in (b) from Bitty Roy and (c, d) from Shannon Olsson and (e) from Deepa Rajan, With permission.)

employed 3D-printed polystyrene flowers to better understand the roles of the hawkmoth proboscis in foraging behavior (Figure 16.3a). In each study, naïve moths were exposed to white 3D-printed robotic flowers or arrays thereof, in combination with sucrose solution and the presence of a 7-component scent blend derived from *Datura wrightii* flowers, which effectively releases feeding behavior by *M. sexta* (Riffell et al. 2009). These moths extract nectar most efficiently from funnel-shaped flowers with a specific curvature (Campos et al. 2015) and take longer to visually track mechanically oscillating flowers in low light (simulating windy conditions at dusk) (Sponberg et al. 2015) but can compensate for this through parallel visual and mechanosensory inputs (Roth et al. 2016), reinforcing feeding performance through parallel sensory processing (see Leonard et al. 2011b).

The final examples of floral reconstruction take 3D-printing into remote natural habitats. The ornate flowers of *Dracula* orchids harmonize a gilled, waxy mushroom-like labellum with pubescent, visually-patterned sepals, upon which fungus-breeding Drosophilid flies (*Zygothrica* spp.) aggregate and court, as they do on mushrooms (Endara et al. 2010). Gene silencing and introgressive breeding approaches would be impractical on any of the 100-plus described *Dracula* orchid species, and no paper flower could do justice to their multi-dimensional phenotypes. Policha et al. (2016) used a novel combination of floral deconstruction and reconstruction, with high resolution CT scans of plaster-positive casts to 3D-print a negative as a mold, with which the sepals and labella of *Dracula lafleuri* could be cast in surgical silicone and impregnated with scentless pigments to match (or alter) the flowers' visual display (Figure 16.3b). Floral deconstruction reduced fly visitation nearly to zero, indicating the importance of multimodal display and the crudeness of the plastic and cloth bags. Floral reconstruction was more effective, as colored silicone flower casts augmented with solvent extracts (9:1 hexane and acetone) of real flowers attracted more fly approaches and landings than did any single-modality flower model. The strongest insights were gained through the assembly of floral chimeras, combining labella or sepals dissected from living flowers with complementary flower parts cast in silicone. The presence of a real labellum with silicone sepals elicited as many landings per hour as the natural flower, but intra-floral movement by the flies was disrupted, leading fewer of them to the column, where pollen transfer occurs. The alternative chimera (real sepals, silicone labellum) attracted two-thirds fewer flies but restored intra-floral movement between sepals and column.

Although intriguing, the pollination of *Dracula* orchids is unusual for its reliance on Drosophilid flies that normally attend to fungi, whereas hoverflies (Syrphidae) are ubiquitous flower visitors and can be important pollinators globally (Larson et al. 2001). Nordström et al. (2017) asked whether widespread hoverfly species such as *Episyrphus balteatus* and *Eristalis tenax* show the same sensory preferences for flowers across their ranges, or do their patterns of visitation shift with the local availability of floral resources? Nordström and colleagues tracked the flowers that were visited ("hot") or ignored ("cold") by cosmopolitan syrphid flies that were observed foraging in alpine Sikkim and tropical Bangalore, India, and boreal Uppsala, Sweden, then measured their shapes, spectral reflectance, scent composition and other cues (CO_2, RH). Multivariate analyses were performed separately for "hot" and "cold" flowers in each habitat, producing consensus multimodal flower lures; platonic archetypes, as it were, of floral attractiveness to syrphid flies. These lures were reconstructed with requisite shape, color and scent blends (first from paper, more recently using 3D printing; Figure 16.3c, d) and placed into the field at low density in natural habitats in India and Sweden, with positive and negative controls accounting for suspected bias (yellow, based on *Potentilla fruticosa*) and a gray sham flower. The results were unexpected and fascinating: some "hot" lures (for Uppsala and the positive control) were universally attractive to syrphids, others were attractive only in their own environment (Bangalore). The approach modeled by this study is ideal for resolving the intellectual tension between sensory bias and density/optimality approaches to studying pollination, because the flowers can be printed *en masse*, grouped in different densities, with different neighbors (as pollinator magnets or competitors) and endowed with different reward values. The tools are now available to design experimental floral communities to address how extrinsic ecological factors and intrinsic behavioral responses by pollinators shape the evolution of multimodal floral displays.

16.3 CONCLUSIONS

I have outlined experimental approaches used to measure the behavioral responses of pollinators and other flower visitors to floral scent at different spatial and temporal scales. While there is still latitude for simple manipulations that cut to the heart of a plant–pollinator interaction, I have emphasized several emerging or mature experimental tools, including gene silencing, plant breeding and 3D-printing approaches to modifying floral scent in the context of other floral traits. The next round of innovation is bound to include gene editing approaches, both in floral signal production and in pollinator perception, with clear candidate genes already available in plant biosynthetic and insect olfactory pathways. Finally, I have evaluated many of the studies reviewed here (Tables 16.2 through 16.4) in the light of emerging hypotheses on the evolution of multimodal signals in floral communication (Table 16.1), where additional insights are likely to come from research on neural and cognitive aspects of pollinator learning and foraging behavior.

ACKNOWLEDGMENTS

Thanks to Eran Pichersky and Natalia Dudareva for the invitation to reprise my original chapter, to Dan Papaj for comments on the manuscript, to Tina McDowell (Carnegie Institute of Science) for permission to reprint Edith Clement's line drawings in Figure 16.1a, to Mahad Ahmed, Hannah Burger, Tom Daniel, Tanvi Deora, Stefan Dötterl, Shannon Olsson, Deepa Rajan, Bitty Roy and Avery Russell for permission to use their images, and to my many co-authors for shared experiences in floral manipulation. I gratefully acknowledge support from US National Science Foundation award DEB-1342792.

REFERENCES

Afik, O., A. Dag, Z. Kerem and S. Shafir. 2006. Analyses of avocado (*Persea americana*) nectar properties and their perception by honey bees (*Apis mellifera*). *Journal of Chemical Ecology* 32:1949–1963.

Aldridge, G. and D.R. Campbell. 2009. Genetic and morphological patterns show variation in frequency of hybrids between *Ipomopsis* (Polemoniaceae) zones of sympatry. *Heredity* 102:257–265.

Angioy, A.M., M.C. Stensmyr, I. Urru, M. Puliafito, I. Collu and B.S. Hansson. 2004. Function of the heater: The dead horse arum revisited. *Proceedings of the Royal Society of London. Series B: Biological Sciences* 271:S13–S15.

Ashman, T.L., M. Bradburn, D.H. Cole, B.H. Blaney and R.A. Raguso. 2005. The scent of a male: The role of floral volatiles in pollination of a gender dimorphic plant. *Ecology* 86:2099–2105.

Ayasse, M., J. Stökl and W. Francke. 2011. Chemical ecology and pollinator-driven speciation in sexually deceptive orchids. *Phytochemistry* 72:1667–1677.

Bestmann, H.J., L. Winkler and O. von Helversen. 1997. Headspace analysis of volatile flower scent constituents of bat-pollinated plants. *Phytochemistry* 46:1169–1172.

Bischoff, M., A. Jürgens and D.R. Campbell. 2013. Floral scent in natural hybrids of *Ipomopsis* (Polemoniaceae) and their parental species. *Annals of Botany* 113:533–544.

Bischoff, M., R.A. Raguso, A. Jürgens and D.R. Campbell. 2015. Context-dependent reproductive isolation mediated by floral scent and color. *Evolution* 69:1–13.

Boachon, B., R.R. Junker, L. Miesch et al. 2015. CYP76C1 (Cytochrome P450)-mediated linalool metabolism and the formation of volatile and soluble linalool oxides in *Arabidopsis* flowers: A strategy for defense against floral antagonists. *The Plant Cell* 27:2972–2990.

Bradshaw, E., P.J. Rudall, D.S. Devey, M.M. Thomas, B.J. Glover and R.M. Bateman. 2010. Comparative labellum micromorphology of the sexually deceptive temperate orchid genus *Ophrys*: Diverse epidermal cell types and multiple origins of structural colour. *Botanical Journal of the Linnean Society* 162: 504–540.

Brodie, B.S., A. Renyard, R. Gries et al. 2018. Identification and field testing of floral odorants that attract the rove beetle *Pelecomalium testaceum* (Mannerheim) to skunk cabbage, *Lysichiton americanus* (L.). *Arthropod-Plant Interactions* 12:591–599.

Burger, H., S. Dötterl and M. Ayasse. 2010. Host-plant finding and recognition by visual and olfactory floral cues in an oligolectic bee. *Functional Ecology* 24:1234–1240.

Burger, H., S. Dötterl, C.M. Häberlein, S. Schulz and M. Ayasse. 2012. An arthropod deterrent attracts specialised bees to their host plants. *Oecologia* 168:727–736.

Byers, K.J., J.P. Vela, F. Peng, J.A. Riffell and H.D. Bradshaw Jr. 2014. Floral volatile alleles can contribute to pollinator-mediated reproductive isolation in monkeyflowers (*Mimulus*). *The Plant Journal* 80:1031–1042.

Campbell, D.R., A. Jürgens and S.D. Johnson. 2016. Reproductive isolation between *Zaluzianskya* species: The influence of volatiles and flower orientation on hawkmoth foraging choices. *New Phytologist* 210:333–342.

Campos, E.O., H.D. Bradshaw, Jr. and T.L. Daniel. 2015. Shape matters: Corolla curvature improves nectar discovery in the hawkmoth *Manduca sexta*. *Functional Ecology* 29:462–468.

Chen, G., X.K. Ma, A. Jürgens et al. 2015. Mimicking livor mortis: A well-known but unsubstantiated color profile in sapromyiophily. *Journal of Chemical Ecology* 41:808–815.

Clarke, D., H. Whitney, G. Sutton and D. Robert. 2013. Detection and learning of floral electric fields by bumblebees. *Science* 340:66–69.

Clements, F.E. and F.L. Long. 1923. *Experimental Pollination: An Outline of the Ecology of Flowers and Insects* (No. 336). Washington, DC: Carnegie institution of Washington.

Cunningham, J.P., C.J. Moore, M.P. Zalucki and S.A. West. 2004. Learning, odour preference and flower foraging in moths. *Journal of Experimental Biology* 207:87–94.

Daumer, K. 1958. Blumenfarben wie sie die Bienen sehen. *Zeitschrift für vergleichende Physiologie* 41:49–100.

de Jager, M.L. and R. Peakall. 2016. Does morphology matter? An explicit assessment of floral morphology in sexual deception. *Functional Ecology* 30:537–546.

Dell'Olivo, A. and C. Kuhlemeier. 2013. Asymmetric effects of loss and gain of a floral trait on pollinator preference. *Evolution* 67:3023–3031.

Dobson, H.E.M., E.M. Danielson and I.D.V. Wesep. 1999. Pollen odor chemicals as modulators of bumble bee foraging on *Rosa rugosa* Thunb. (Rosaceae). *Plant Species Biology* 14:153–166.

Dobson, H.E.M., R.A. Raguso, J.T. Knudsen and M. Ayasse. 2005. Scent as an attractant. In *Practical Pollination Biology*, eds. A. Dafni, P.G. Kevan and. B.C. Husband, pp. 197–230. Cambridge, Canada: Enviroquest, Ltd.

Dötterl, S. and I. Schäffler. 2007. Flower scent of floral oil-producing *Lysimachia punctata* as attractant for the oil-bee *Macropis fulvipes*. *Journal of Chemical Ecology* 33:441–445.

Dötterl, S. and N.J. Vereecken. 2010. The chemical ecology and evolution of bee–flower interactions: A review and perspectives. *Canadian Journal of Zoology* 88:668–697.

Dötterl, S., K. Milchreit and I. Schäffler. 2011. Behavioural plasticity and sex differences in host finding of a specialized bee species. *Journal of Comparative Physiology A* 197:1119–1126.

Dudareva, N. and E. Pichersky. 2006. *Biology of Floral Scent*. Boca Raton, FL: CRC Press/Taylor & Francis Group.

Endara, L., D.A. Grimaldi and B.A. Roy. 2010. Lord of the flies: Pollination of Dracula orchids. *Lankesteriana International Journal on Orchidology* 10:1–11.

Fulton, M. and S.A. Hodges. 1999. Floral isolation between *Aquilegia formosa* and *Aquilegia pubescens*. *Proceedings of the Royal Society of London. Series B: Biological Sciences* 266:2247–2252.

Galen, C., R. Kaczorowski, S.L. Todd, J. Geib and R.A. Raguso. 2011. Dosage-dependent impacts of a floral volatile compound on pollinators, larcenists, and the potential for floral evolution in the alpine skypilot *Polemonium viscosum*. *The American Naturalist* 177:258–272.

Gegear, R.J. and T.M. Laverty. 2005. Flower constancy in bumblebees: A test of the trait variability hypothesis. *Animal Behaviour* 69:939–949.

Gegear, R.J., R. Burns and K.A. Swoboda-Bhattarai. 2017. "Hummingbird" floral traits interact synergistically to discourage visitation by bumble bee foragers. *Ecology* 98:489–499.

Gervasi, D.D., M.A. Selosse, M. Sauve, W. Francke, N.J. Vereecken, S. Cozzolino and F.P. Schiestl. 2017. Floral scent and species divergence in a pair of sexually deceptive orchids. *Ecology and Evolution* 7:6023–6034.

Gonzalez-Terrazas, T.P., C. Martel, P. Milet-Pinheiro, M. Ayasse, E.K. Kalko and M. Tschapka. 2016. Finding flowers in the dark: Nectar-feeding bats integrate olfaction and echolocation while foraging for nectar. *Royal Society Open Science* 3(8):160199. https://doi.org/10.1098/rsos.160199

Goyret, J. and R.A. Raguso. 2006. The role of mechanosensory input in flower handling efficiency and learning by *Manduca sexta*. *Journal of Experimental Biology* 209:1585–1593.

Goyret, J., A. Kelber, M. Pfaff and R.A. Raguso. 2009. Flexible responses to visual and olfactory stimuli by foraging *Manduca sexta*: Larval nutrition affects adult behaviour. *Proceedings of the Royal Society B: Biological Sciences* 276:2739–2745.

Goyret, J., P.M. Markwell and R.A. Raguso. 2007. The effect of decoupling olfactory and visual stimuli on the foraging behavior of *Manduca sexta*. *Journal of Experimental Biology* 210:1398–1405.

Goyret, J., P.M. Markwell and R.A. Raguso. 2008. Context-and scale-dependent effects of floral CO_2 on nectar foraging by *Manduca sexta*. *Proceedings of the National Academy of Sciences* 105:4565–4570.

Gross, K., M. Sun and F.P. Schiestl. 2016. Why do floral perfumes become different? Region-specific selection on floral scent in a terrestrial orchid. *PLoS One* 11(2):e0147975. https://doi.org/10.1371/journal.pone.0147975

Hansen, D.M., K. Beer and C.B. Müller. 2006. Mauritian coloured nectar no longer a mystery: A visual signal for lizard pollinators. *Biology Letters* 2:165–168.

Hansen, D.M., T. van der Niet and S.D. Johnson. 2011. Floral signposts: Testing the significance of visual "nectar guides" for pollinator behaviour and plant fitness. *Proceedings of the Royal Society B: Biological Sciences* 279:634–639.

Harrap, M.J., S.A. Rands, N.H. de Ibarra and H.M. Whitney. 2017. The diversity of floral temperature patterns, and their use by pollinators. *eLife* 6:e31262. https://doi.org/10.7554/eLife.31262.001

Hebets, E.A. and D.R. Papaj. 2005. Complex signal function: Developing a framework of testable hypotheses. *Behavioral Ecology and Sociobiology* 57:197–214.

Hoballah, M.E., J. Stuurman, T.C. Turlings, P.M. Guerin, S. Connetable and C. Kuhlemeier. 2005. The composition and timing of flower odour emission by wild *Petunia axillaris* coincide with the antennal perception and nocturnal activity of the pollinator *Manduca sexta*. *Planta* 222:141–150.

Hoballah, M.E., T. Gübitz, J. Stuurman et al. 2007. Single gene–mediated shift in pollinator attraction in *Petunia*. *The Plant Cell* 19:779–790.

Hossaert-McKey, M., M. Gibernau and J.E. Frey. 1994. Chemosensory attraction of fig wasps to substances produced by receptive figs. *Entomologia Experimentalis et Applicata* 70:185–191.

Huang, M., A.M. Sanchez-Moreiras, C. Abel, R. Sohrabi, S. Lee, J. Gershenzon and D. Tholl, 2012. The major volatile organic compound emitted from *Arabidopsis thaliana* flowers, the sesquiterpene (E)-β-caryophyllene, is a defense against a bacterial pathogen. *New Phytologist* 193:997–1008.

Junker, R.R. and A.L. Parachnowitsch. 2015. Working towards a holistic view on flower traits – How floral scents mediate plant–animal interactions in concert with other floral characters. *Journal of the Indian Institute of Science* 95:43–68.

Junker, R.R. and D. Tholl. 2013. Volatile organic compound mediated interactions at the plant-microbe inter-face. *Journal of Chemical Ecology* 39:810–825.

Junker, R.R., C. Loewel, R. Gross, S. Dötterl, A. Keller and N. Blüthgen. 2011b. Composition of epiphytic bacterial communities differs on petals and leaves. *Plant Biology* 13:918–924.

Junker, R.R., J. Gershenzon and S.B. Unsicker. 2011a. Floral odor bouquet loses its ant repellent properties after inhibition of terpene biosynthesis. *Journal of Chemical Ecology* 37:1323–1331.

Jürgens, A., S.L. Wee, A. Shuttleworth and S.D. Johnson. 2013. Chemical mimicry of insect oviposition sites: A global analysis of convergence in angiosperms. *Ecology Letters* 16:1157–1167.

Kaczorowski, R.L., A.R. Seliger, A.C. Gaskett, S.K. Wigsten and R.A. Raguso. 2012b. Corolla shape vs. size in flower choice by a nocturnal hawkmoth pollinator. *Functional Ecology* 26:577–587.

Kaczorowski, R.L., A.S. Leonard, A. Dornhaus and D.R. Papaj. 2012a. Floral signal complexity as a possible adaptation to environmental variability: A test using nectar-foraging bumblebees, *Bombus impatiens*. *Animal Behaviour* 83:905–913.

Kantsa, A., R.A. Raguso, A.G. Dyer, J.M. Olesen, T. Tscheulin and T. Petanidou. 2018. Disentangling the role of floral sensory stimuli in pollination networks. *Nature Communications* 9(1):1041. https://www.nature.com/articles/s41467-018-03448-w

Kantsa, A., R.A. Raguso, A.G. Dyer, S.P. Sgardelis, J.M. Olesen and T. Petanidou. 2017. Community-wide integration of floral colour and scent in a Mediterranean scrubland. *Nature Ecology & Evolution* 1(10):1502–1510. https://www.nature.com/articles/s41559-017-0298-0

Kessler, D. and I.T. Baldwin. 2007. Making sense of nectar scents: The effects of nectar secondary metabolites on floral visitors of *Nicotiana attenuata*. *The Plant Journal* 49:840–854.

Kessler, D., C. Diezel, D.G. Clark, T.A. Colquhoun and I.T. Baldwin. 2013. *Petunia* flowers solve the defence/apparency dilemma of pollinator attraction by deploying complex floral blends. *Ecology Letters* 16:299–306.

Kessler, D., K. Gase and I.T. Baldwin. 2008. Field experiments with transformed plants reveal the sense of floral scents. *Science* 321:1200–1202.

Kessler, D., M. Kallenbach, C. Diezel, E. Rothe, M. Murdock and I.T. Baldwin. 2015. How scent and nectar influence floral antagonists and mutualists. *eLife* 4:e07641. doi:10.7554/eLife.07641

Klahre, U., A. Gurba, K. Hermann et al. 2011. Pollinator choice in *Petunia* depends on two major genetic loci for floral scent production. *Current Biology* 21:730–739.

Knauer, A.C. and F.P. Schiestl. 2015. Bees use honest floral signals as indicators of reward when visiting flow-ers. *Ecology Letters* 18:135–143.

Knauer, A.C., M. Bakhtiari and F.P. Schiestl. 2018. Crab spiders impact floral-signal evolution indirectly through removal of florivores. *Nature Communications* 9(1):1367. https://www.nature.com/articles/s41467-018-03792-x

Knoll, F. 1926a. Lichtsinn und Blütenbesuch des Falters von *Deilephila livornica*. *Zeitschrift für verglei-chende Physiologie* 2:328–380.

Knoll, F. 1926b. Die Arum-blütenstände und ihre Besucher. *Abhandlungen der Zoologisch–Botanischen Gesellschaft Wien* 12:382–481.

Kulahci, I.G., A. Dornhaus and D.R. Papaj. 2008. Multimodal signals enhance decision making in foraging bumble-bees. *Proceedings of the Royal Society B: Biological Sciences* 275:797–802.

Kunze, J. and A. Gumbert. 2001. The combined effect of color and odor on flower choice behavior of bumble bees in flower mimicry systems. *Behavioral Ecology* 12:447–456.

Larson, B.M.H., P.G. Kevan and D.W. Inouye. 2001. Flies and flowers: Taxonomic diversity of anthophiles and pollinators. *The Canadian Entomologist* 133:439–465.

Larue, A.A.C., R.A. Raguso and R.R. Junker. 2016. Experimental manipulation of floral scent bouquets restructures flower–visitor interactions in the field. *Journal of Animal Ecology* 85:396–408.

Lawson, D.A., H.M. Whitney and S.A. Rands. 2017. Colour as a backup for scent in the presence of olfac-tory noise: Testing the efficacy backup hypothesis using bumblebees (*Bombus terrestris*). *Royal Society Open Science* 4(11):170996. https://doi.org/10.1098/rsos.170996

Lawson, D.A., L. Chittka, H.M. Whitney and S.A. Rands. 2018. Bumblebees distinguish floral scent patterns, and can transfer these to corresponding visual patterns. *Proceedings of the Royal Society B: Biological Sciences* 285(1880):20180661. https://doi.org/10.1098/rspb.2018.0661

Leonard, A.S., A. Dornhaus and D.R. Papaj. 2011a. Why are floral signals complex? An outline of func-tional hypotheses. *Evolution of Plant–Pollinator Relationships*, ed. S. Patiny, pp. 261–282. Cambridge: Cambridge University Press.

Leonard, A.S., A. Dornhaus and D.R. Papaj. 2011b. Forget-me-not: Complex floral displays, inter-signal inter-actions, and pollinator cognition. *Current Zoology* 57:215–224.

Leonard, A.S., A. Dornhaus and D.R. Papaj. 2011c. Flowers help bees cope with uncertainty: Signal detection and the function of floral complexity. *Journal of Experimental Biology* 214:113–121.

Majetic, C.J., R.A. Raguso and T.L. Ashman. 2009. The sweet smell of success: Floral scent affects pollinator attraction and seed fitness in *Hesperis matronalis*. *Functional Ecology* 23:480–487.

Makino, T.T. and S. Sakai. 2007. Experience changes pollinator responses to floral display size: From size-based to reward-based foraging. *Functional Ecology* 21:854–863.

Manning, A. 1956. Some aspects of the foraging behaviour of bumble-bees. *Behaviour* 9:164–200.

Meléndez-Ackerman, E., D.R. Campbell and N.M. Waser. 1997. Hummingbird behavior and mechanisms of selection on flower color in *Ipomopsis*. *Ecology* 78:2532–2541.

Milet-Pinheiro, P., M. Ayasse and S. Dötterl. 2015. Visual and olfactory floral cues of *Campanula* (Campanulaceae) and their significance for host recognition by an oligolectic bee pollinator. *PLoS One* 10(6):e0128577. https://doi.org/10.1371/journal.pone.0128577

Milet-Pinheiro, P., M. Ayasse, C. Schlindwein, H.E.M. Dobson and S. Dötterl. 2012. Host location by visual and olfactory floral cues in an oligolectic bee: Innate and learned behavior. *Behavioral Ecology* 23:531–538.

Milet-Pinheiro, P., M. Ayasse, H.E.M. Dobson, C. Schlindwein, W. Francke and S. Dötterl. 2013. The chemical basis of host-plant recognition in a specialized bee pollinator. *Journal of Chemical Ecology* 39:1347–1360.

Miyake, T. and M. Yafuso. 2003. Floral scents affect reproductive success in fly-pollinated *Alocasia odora* (Araceae). *American Journal of Botany* 90:370–376.

Moré, M., A.A. Cocucci and R.A. Raguso. 2013. The importance of oligosulfides in the attraction of fly pollinators to the brood-site deceptive species *Jaborosa rotacea* (Solanaceae). *International Journal of Plant Sciences* 174:863–876.

Moré, M., P. Mulieri, M. Battán-Horenstein, A.A. Cocucci and R.A. Raguso. 2019. The role of fetid olfactory signals in the shift to saprophilous fly pollination in *Jaborosa* (Solanaceae). *Arthropod-Plant Interactions* 13:375–386.

Muchhala, N. and D. Serrano. 2015. The complexity of background clutter affects nectar bat use of flower odor and shape cues. *PLoS One* 10(10):e0136657. https://doi.org/10.1371/journal.pone.0136657

Muchhala, N., A. Caiza, J.C. Vizuete and J.D. Thomson. 2009. A generalized pollination system in the tropics: Bats, birds and *Aphelandra acanthus*. *Annals of Botany* 103:1481–1487.

Nordström, K., J. Dahlbom, V.S. Pragadheesh et al. 2017. In situ modeling of multimodal floral cues attracting wild pollinators across environments. *Proceedings of the National Academy of Sciences* 114:13218–13223.

Oelschlägel, B., M. Nuss, M. von Tschirnhaus et al. 2015. The betrayed thief–the extraordinary strategy of *Aristolochia rotunda* to deceive its pollinators. *New Phytologist* 206:342–351.

Olesen, J.M., J. Bascompte, H. Elberling and P. Jordano. 2008. Temporal dynamics in a pollination network. *Ecology* 89:1573–1582.

Parachnowitsch, A.L., R.A. Raguso and A. Kessler. 2012. Phenotypic selection to increase floral scent emission, but not flower size or colour in bee-pollinated *Penstemon digitalis*. *New Phytologist* 195:667–675.

Patt, J.M., J.C. French, C. Schal, J. Lech and T.G. Hartman. 1995. The pollination biology of Tuckahoe, *Peltandra virginica* (Araceae). *American Journal of Botany* 82:1230–1240.

Pellegrino, G., F. Bellusci and A.M. Palermo. 2017. Functional differentiation in pollination processes among floral traits in *Serapias* species (Orchidaceae). *Ecology and Evolution* 7:7171–7177.

Pellmyr, O. and J.M. Patt. 1986. Function of olfactory and visual stimuli in pollination of *Lysichiton americanum* (Araceae) by a staphylinid beetle. *Madroño* 33:47–54.

Pellmyr, O. 1986. Three pollination morphs in *Cimicifuga simplex*; incipient speciation due to inferiority in competition. *Oecologia* 68:304–307.

Phillips, R.D. and R. Peakall. 2018. An experimental evaluation of traits that influence the sexual behaviour of pollinators in sexually deceptive orchids. *Journal of Evolutionary Biology* 31:1732–1742.

Policha, T., A. Davis, M. Barnadas, B.T. Dentinger, R.A. Raguso and B.A. Roy. 2016. Disentangling visual and olfactory signals in mushroom-mimicking *Dracula* orchids using realistic three-dimensional printed flowers. *New Phytologist* 210:1058–1071.

Raguso, R.A. and M.A. Willis. 2002. Synergy between visual and olfactory cues in nectar feeding by naïve hawkmoths, *Manduca sexta*. *Animal Behaviour* 64:685–695.

Raguso, R.A. and M.A. Willis. 2005. Synergy between visual and olfactory cues in nectar feeding by wild hawkmoths, *Manduca sexta*. *Animal Behaviour* 69:407–418.

Raguso, R.A. and M.R. Weiss. 2015. Concerted changes in floral colour and scent, and the importance of spatio-temporal variation in floral volatiles. *Journal of the Indian Institute of Science* 95:69–92.

Raguso, R.A. 2004. Flowers as sensory billboards: Progress towards an integrated understanding of floral advertisement. *Current Opinion in Plant Biology* 7:434–440.

Raguso, R.A. 2006. Behavioral responses to floral scent: Experimental manipulations and the interplay of sensory modalities. *Biology of Floral Scent*, eds. N. Dudareva and E. Pichersky, pp. 297–320. Boca Raton: CRC Press/Taylor & Francis Group.

Raguso, R.A. 2008a. Wake up and smell the roses: The ecology and evolution of floral scent. *Annual Review of Ecology, Evolution, and Systematics* 39:549–569.

Raguso, R.A. 2008b. Start making scents: The challenge of integrating chemistry into pollination ecology. *Entomologia Experimentalis et Applicata* 128:196–207.

Rakosy, D., M. Cuervo, H.F. Paulus and M. Ayasse. 2017. Looks matter: Changes in flower form affect pollination effectiveness in a sexually deceptive orchid. *Journal of Evolutionary Biology* 30:1978–1993.

Rakosy, D., M. Streinzer, H.F. Paulus and J. Spaethe. 2012. Floral visual signal increases reproductive success in a sexually deceptive orchid. *Arthropod-Plant Interactions* 6:671–681.

Riffell, J.A. and R. Alarcón. 2013. Multimodal floral signals and moth foraging decisions. *PLoS One* 8(8):e72809. https://doi.org/10.1371/journal.pone.0072809

Riffell, J.A., H. Lei, T.A. Christensen and J.G. Hildebrand. 2009. Characterization and coding of behaviorally significant odor mixtures. *Current Biology* 19:335–340.

Roth, E., R.W. Hall, T.L. Daniel and S. Sponberg. 2016. Integration of parallel mechanosensory and visual pathways resolved through sensory conflict. *Proceedings of the National Academy of Sciences* 113:12832–12837.

Roy, B.A. and R.A. Raguso. 1997. Olfactory versus visual cues in a floral mimicry system. *Oecologia* 109:414–426.

Russell, A.L. and D.R. Papaj. 2016. Artificial pollen dispensing flowers and feeders for bee behavior experiments. *Journal of Pollination Ecology* 18:13–22.

Russell, A.L., K.B. Mauerman, R.E. Golden and D.R. Papaj. 2018. Linking components of complex signals to morphological part: The role of anther and corolla in the complex floral display. *Animal Behaviour* 135:223–236.

Russell, A.L., R.E. Golden, A.S. Leonard and D.R. Papaj. 2016. Bees learn preferences for plant species that offer only pollen as a reward. *Behavioral Ecology* 27:731–740.

Schäffler, I., K.E. Steiner, M. Haid et al. 2015. Diacetin, a reliable cue and private communication channel in a specialized pollination system. *Scientific Reports* 5:12779. https://www.nature.com/articles/srep12779?origin=ppub

Schiestl, F.P. and M. Ayasse. 2001. Post-pollination emission of a repellent compound in a sexually deceptive orchid: A new mechanism for maximising reproductive success? *Oecologia* 126:531–534.

Schiestl, F.P., F.K. Huber and J.M. Gomez. 2011. Phenotypic selection on floral scent: Trade-off between attraction and deterrence? *Evolutionary Ecology* 25:237–248.

Shuttleworth, A. and S.D. Johnson. 2008. A key role for floral scent in a wasp-pollination system in *Eucomis* (Hyacinthaceae). *Annals of Botany* 103:715–725.

Shuttleworth, A. and S.D. Johnson. 2010. The missing stink: Sulphur compounds can mediate a shift between fly and wasp pollination systems. *Proceedings of the Royal Society B: Biological Sciences* 277:2811–2819.

Spaethe, J., W.H. Moser and H.F. Paulus. 2007. Increase of pollinator attraction by means of a visual signal in the sexually deceptive orchid, *Ophrys heldreichii* (Orchidaceae). *Plant Systematics and Evolution* 264:31–40.

Sponberg, S., J.P. Dyhr, R.W. Hall and T.L. Daniel. 2015. Luminance-dependent visual processing enables moth flight in low light. *Science* 348:1245–1248.

Sprayberry, J.D. and T.L. Daniel. 2007. Flower tracking in hawkmoths: Behavior and energetics. *Journal of Experimental Biology* 210:37–45.

Stang, M., P.G. Klinkhamer, N.M. Waser, I. Stang and E. van der Meijden. 2009. Size-specific interaction patterns and size matching in a plant–pollinator interaction web. *Annals of Botany* 103:1459–1469.

Steenhuisen, S.L., A. Jürgens and S.D. Johnson. 2013. Effects of volatile compounds emitted by *Protea* species (Proteaceae) on antennal electrophysiological responses and attraction of Cetoniine beetles. *Journal of Chemical Ecology* 39:438–446.

Stensmyr, M.C., I. Urru, I. Collu, M. Celander, B.S. Hansson and A.M. Angioy. 2002. Pollination: Rotting smell of dead-horse arum florets. *Nature* 420:625–626.

Streinzer, M., T. Ellis, H.F. Paulus and J. Spaethe. 2010. Visual discrimination between two sexually deceptive *Ophrys* species by a bee pollinator. *Arthropod-Plant Interactions* 4:141–148.

Suinyuy, T.N., J.S. Donaldson and S.D. Johnson. 2015. Geographical matching of volatile signals and pollinator olfactory responses in a cycad brood-site mutualism. *Proceedings of the Royal Society B: Biological Sciences* 282(1816):20152053. https://doi.org/10.1098/rspb.2015.2053

Theis, N. and L.S. Adler. 2012. Advertising to the enemy: Enhanced floral fragrance increases beetle attraction and reduces plant reproduction. *Ecology* 93:430–435.

Vannette, R.L., M.P. Gauthier and T. Fukami. 2012. Nectar bacteria, but not yeast, weaken a plant-pollinator mutualism. *Proceeding of the Royal Society B* 280(1752):20122601. https://doi.org/10.1098/rspb.2012.2601

Verdonk, J.C., M.A. Haring, A.J. van Tunen and R.C. Schuurink. 2005. ODORANT1 regulates fragrance biosynthesis in petunia flowers. *The Plant Cell* 17:16121624.

Vereecken, N.J. and F.P. Schiestl. 2009. On the roles of colour and scent in a specialized floral mimicry system. *Annals of Botany* 104:1077–1084.

Vignolini, S., M.P. Davey, R.M. Bateman et al. 2012. The mirror crack'd: Both pigment and structure contribute to the glossy blue appearance of the mirror orchid, *Ophrys speculum*. *New Phytologist* 196:1038–1047.

von Arx, M., J. Goyret, G. Davidowitz and R.A. Raguso. 2012. Floral humidity as a reliable sensory cue for profitability assessment by nectar-foraging hawkmoths. *Proceedings of the National Academy of Sciences* 109:9471–9476.

von Helversen, D. and O. von Helversen. 2003. Object recognition by echolocation: A nectar-feeding bat exploiting the flowers of a rain forest vine. *Journal of Comparative Physiology* 189:327–336.

Waelti, M.O., J.K. Muhlemann, A. Widmer and F.P. Schiestl. 2008. Floral odour and reproductive isolation in two species of *Silene*. *Journal of Evolutionary Biology* 21:111–121.

Weiss, M.R. 1991. Floral colour changes as cues for pollinators. *Nature* 354:227–229.

Wester, P., S.D. Johnson and A. Pauw. 2019. Scent chemistry is key in the evolutionary transition between insect and mammal pollination in African pineapple lilies. *New Phytologist* 222:1624–1633. https://doi.org/10.1111/nph.15671

Whitney, H.M., L. Chittka, T.J. Bruce and B.J. Glover. 2009a. Conical epidermal cells allow bees to grip flowers and increase foraging efficiency. *Current Biology* 19:948–953.

Whitney, H.M., M. Kolle, P. Andrew, L. Chittka, U. Steiner and B.J. Glover. 2009b. Floral iridescence, produced by diffractive optics, acts as a cue for animal pollinators. *Science* 323:130–133.

Wiemer, A.P., M. Moré, S. Benitez-Vieyra, A.A. Cocucci, R.A. Raguso and A.N. Sersic. 2009. A simple floral fragrance and unusual osmophore structure in *Cyclopogon elatus* (Orchidaceae). *Plant Biology* 11:506–514.

Wright, G.A. and F.P. Schiestl. 2009. The evolution of floral scent: The influence of olfactory learning by insect pollinators on the honest signaling of floral rewards. *Functional Ecology* 23:841–851.

Wright, G.A., A.F. Choudhary and M.A. Bentley. 2009. Reward quality influences the development of learned olfactory biases in honeybees. *Proceedings of the Royal Society B: Biological Sciences* 276:2597–2604.

17 Herbivore-Induced Plant Volatiles as a Source of Information in Plant–Insect Networks

Marcel Dicke and Dani Lucas-Barbosa

CONTENTS

17.1 INTRODUCTION

Plants face a rich diversity of herbivorous insects, most of them being specialists feeding on one or a few related plant species (Schoonhoven et al. 2005). A single plant species may face attack by individuals from several tens up to more than one hundred insect species (Schoonhoven et al. 2005). Plants have evolved various defenses against insect herbivores, which can be classified as constitutive and inducible defenses (Chapter 9). Constitutive defenses form the basal level or the first line of plant defense, keeping most herbivorous insects from attacking the plant. Many classes of secondary metabolites can be involved in constitutive defense and the secondary metabolic profile of a plant is species specific to a large extent (Dudareva et al. 2013). This provides an effective barrier, as most insect species cannot deal with these toxic compounds (Schoonhoven et al. 2005). However, specialist insects have usually evolved mechanisms to deal with specific secondary metabolites (Schoonhoven et al. 2005) and, thus, form a particular threat to plants. To deal with specialist insect herbivores that can cope with constitutive defenses, plants have a second line of defense, consisting of inducible defenses. Inducible defenses may be expressed locally at the site of attack as well as systemically in other tissues of the same plant.

Plant defenses can also be classified according to their ecological effects on the organisms with which they interact. Thus, plant defense can be direct, i.e., negatively influencing the performance of the attacker or indirect, i.e., enhancing the effectiveness of natural enemies of the attacker (Dicke and Sabelis 1988). Plants can respond to herbivory with the production of inducible volatiles that attract the natural enemies of herbivores to the damaged plant. Volatiles facilitate the localization of herbivores by these natural enemies and, consequently, enhance their attack of the herbivores (Schuman et al. 2012, Turlings and Erb 2018). Not only indirect defenses, but also direct defenses can be based on plant volatiles. Constitutive as well as inducible plant volatiles may repel herbivorous insects, and thus be classified as direct defenses (Bruce and Pickett 2011).

The discovery that plants produce a diversity of volatiles upon exposure to attackers has unveiled intriguing ways in which plants defend themselves against herbivorous arthropods (Dicke and Baldwin 2010, Dudareva et al. 2013, Turlings and Erb 2018). The induced volatile blend of a plant is specific to the plant species and can also be specific to the attacking species (Turlings and Erb 2018). Moreover, the blend can be modulated by attack by a second herbivore, interactions with microbes, or interactions with pollinators. This raises important questions regarding how plant chemical defenses vary throughout ontogeny, how plants balance or prioritize defenses against different attackers, and ultimately whether volatile-mediated defenses can enhance plant reproduction and, thus, Darwinian fitness. Here, we focus on how plant volatile blends are modulated by interactions with multiple organisms and what responses they induce in plant-associated arthropods. We will address the genetic variation in herbivore-induced plant volatiles (HIPVs). Plant volatile blends are complex and dynamic mixtures of very different compounds (Dudareva et al. 2013) (Figure 17.1) and they influence many members of the plant-associated community (Stam et al. 2014, Turlings and Erb 2018) (Figure 17.2). We address the methods to include this complexity in the study of plant–insect interactions. Finally, we will discuss future developments in the research on HIPVs.

17.2 INDUCTION OF PLANT VOLATILES BY ARTHROPOD HERBIVORY: CHEMICAL DIVERSITY AND SPECIFICITY

HIPVs are emitted as complex blends of up to several hundred compounds, including aldehydes, alcohols, esters, ketones, nitrogen-containing compounds such as nitriles and oximes, sulfur and nitrogen-containing compounds such as isothiocyanates, terpenoids, and aromatic compounds such as benzoates, indole and methyl salicylate (Mumm and Dicke 2010). Among the early compounds that are rapidly produced and emitted in response to herbivory are fatty-acid derived compounds (Allmann et al. 2013, Turlings and Wäckers 2004). Volatiles such as terpenoids and phenylpropanoids are usually emitted later (Turlings and Wäckers 2004). Due to the involvement of different biosynthetic pathways (Dudareva et al. 2013) and their dynamics, the composition of a plant's volatile blend is dynamic in time.

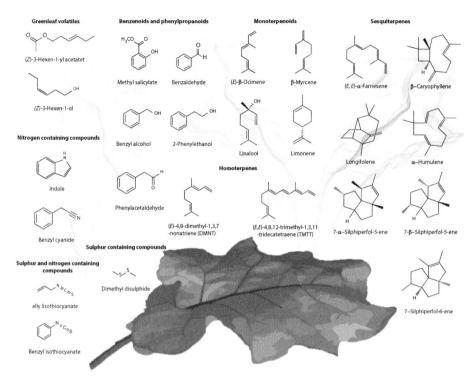

FIGURE 17.1   Major classes of herbivore-induced plant volatiles (HIPVs) are depicted with some examples, illustrating the diversity of compounds produced by plants. (Drawing of plant leaf; Courtesy of Camille Ponzio.)

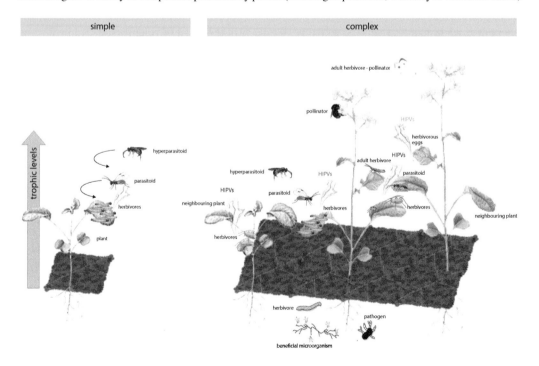

FIGURE 17.2   Schematic representation of how herbivore-induced plant volatiles can influence organisms at different trophic levels, considering a linearly simplified and a complex multivariate community. (Drawings of *Brassica nigra* plants, courtesy of Alison Shoorer and insect drawings, courtesy of Camille Ponzio.)

Yet, the most apparent characteristic of the blend is its high degree of specificity. Typically, HIPV blends emitted by plants of different species but damaged by the same herbivore species exhibit qualitative as well as quantitative differences. For instance, in a study of ten plant species, each infested by the two-spotted spider-mite *Tetranychus urticae*, the total volatile emission rate varied by a factor five. The composition of the HIPV blends differed considerably (Van den Boom et al. 2004) and predators of the two-spotted spider mites discriminate between the different blends (Van den Boom et al. 2002).

When different herbivore species feed on plants of the same species, this also results in different blends. When two caterpillar species feed from different individuals of the same plant species, the volatile blends emitted by the plants are distinct, but differences are smaller than the differences detected between plant species, and mostly represent quantitative differences. Moreover, HIPVs emitted by *Brassica rapa* plants each infested by one of ten different insect herbivore species shows that HIPV-blend composition not only varies with herbivore species, but HIPV blends had strong predictive power on the evolutionary history of the herbivore with the plant, diet breadth and feeding mode of the herbivore (Danner et al. 2018).

HIPVs are induced by feeding by taxonomically widely diverse arthropod species, including spider mites (Acari), rust mites (Acari), thrips (Thysanoptera), caterpillars (Lepidoptera), aphids (Hemiptera), planthoppers (Hemiptera), scale insects (Hemiptera), leafminer flies (Diptera), and beetles (Coleoptera) (Mumm and Dicke 2010). HIPVs are produced in response to feeding on leaves, stems, roots and flowers (Mumm and Dicke 2010, Turlings and Erb 2018) and, thus, their biosynthesis is not restricted to particular tissues. Moreover, not only feeding by arthropods, but already the deposition of eggs by adult herbivores can induce the production and emission of a blend of plant volatiles (Hilker and Fatouros 2015). For instance, egg deposition by the elm leaf beetle *Xanthogaleruca luteola* results in the induction of a blend of about 40 compounds in *Ulmus minor* plants. The majority of these compounds are terpenoids, such as the major blend components (*E, E*)-α-farnesene, β-caryophyllene, and (3*E*)-4,8-dimethyl-1,3,7-nonatriene (DMNT) (Wegener et al. 2001).

Deposition of eggs by insect herbivores may also influence the induction of plant volatiles by the feeding larvae that hatch from the eggs. For instance, egg deposition by *P. brassicae* butterflies on the leaves of *B. nigra* plants affects the quantitative aspects of the blend of volatiles induced by feeding neonate *P. brassicae* caterpillars and the most important compounds influencing the composition of the blends are the sesquiterpene silphiperfolenes 7-α-H-silphiperfol-5-ene, 7-β-H-silphiperfol-5-ene and silphiperfol-6-ene, as well as α-humulene, and β-caryophyllene (Pashalidou et al. 2015). Thus, the plants respond differentially to the feeding damage caused by herbivores, when they have been previously exposed to induction by butterfly eggs. In conclusion, feeding and/or oviposition by arthropods induce extensive changes in volatile emission by plants, thus altering the plant's phenotype.

17.3   RESPONSES OF PLANT-ASSOCIATED ARTHROPODS AT DIFFERENT TROPHIC LEVELS TO HIPVs

Plants are members of speciose communities and the change in a plant's phenotype as a result of HIPV emission influences its interactions with members of the associated community, such as herbivores, carnivores, and second-order carnivores (Figure 17.2). All members of this multitrophic community may respond to the HIPVs, provided that their chemosensory system can perceive these compounds.

17.3.1   Responses by Carnivores

Most information on responses to HIPVs comes from research on carnivorous enemies of the inducing arthropod herbivores, including insects, mites, entomopathogenic nematodes and birds (Amo et al. 2013, Mumm and Dicke 2010, Turlings and Erb 2018). Indeed, responses to HIPVs were first investigated for carnivorous enemies of the herbivorous arthropods. The question addressed in these first studies was how carnivorous arthropods could find their herbivorous prey, because natural selection on the prey would result in reduced provision of information that the carnivores could exploit to find them (Dicke and Sabelis 1988, Turlings et al. 1990). The predatory mite *Phytoseiulus persimilis*, a specialist predator of the spider-mite *T. urticae*, proved to be attracted by spider-mite induced plant volatiles, such as (*E*)-β-ocimene, DMNT, linalool, and methyl salicylate and the parasitoid *Cotesia marginiventris* was attracted to caterpillar-induced HIPV from corn seedlings (Turlings et al. 1990).

Extensive information is available on responses to HIPVs by parasitic wasps that parasitize different developmental stages of insect herbivores (Hilker and Fatouros 2015, Turlings and Wäckers 2004, Vet and Dicke 1992). Egg-induced plant volatiles attract parasitic wasps that lay their eggs in the eggs of the herbivorous insect, resulting in the death of the embryo (Hilker and Fatouros 2015). Egg-induced plant volatile blends partially overlap with the blends induced by insect herbivory on the same plant species (Fatouros et al. 2012, Wegener et al. 2001). Parasitic wasps that attack feeding herbivores are well-known to use HIPVs (Turlings and Erb 2018). This is true for wasps attacking juvenile herbivores such as caterpillars (Desurmont et al. 2016, Turlings et al. 1990) as well as wasps that attack adult herbivores, such as aphids (Blande et al. 2007, Pineda et al. 2013).

Moreover, also second-order parasitic wasps, that attack parasitic wasps that live in herbivorous insects, so-called hyperparasitoids, exploit HIPVs. For instance, when *P. brassicae* caterpillars are parasitized by the wasp *Cotesia glomerata*, the caterpillars induce the emission of an HIPV blend by the plant that differs quantitatively from the blend induced by healthy caterpillars (Zhu et al. 2015) and this allows hyperparasitoids to find caterpillars that contain their hosts, i.e., the primary parasitic wasp larvae feeding within the caterpillars (Zhu et al. 2015). It appears that a polydna-virus that is injected into the host caterpillar by the parasitoid wasp when she deposits here eggs, is responsible for the change in HIPVs that results in hyperparasitoid attraction (Zhu et al. 2018). In conclusion, there is ample information on carnivorous arthropods that exploit herbivore-induced plant volatiles to localize their host.

17.3.2 Responses by Herbivores

HIPVs represent public information and can be used by any organism associated with the emitting plant. For instance, herbivorous arthropods may eavesdrop on HIPVs. Plant volatiles can provide information on the quality of the tissues that herbivores consume.

HIPVs can elicit various responses from other herbivores (Knolhoff and Heckel 2014). Herbivores may use HIPVs as information on competition and avoid them to select other plants to feed on (Allmann et al. 2013, De Moraes et al. 2001, Sobhy et al. 2017). For instance, HIPVs emitted at night by tobacco plants infested with *Helicoverpa virescens* caterpillars, repel con-specific moths. The moths select undamaged plants as a substrate for oviposition, avoiding plants that were emitting HIPVs (De Moraes et al. 2001). Fatty-acid derived HIPVs, or GLVs, such as (*Z*)-3- and (*E*)-2 6-carbon-compounds represent a widespread group of HIPVs that convey herbivory-specific information via their isomeric composition. Feeding by *Manduca sexta* cater-pillars results in isomeric conversion of the (*Z*)-3- to the (*E*)-2-configuration. (*E*)-2-compounds and/or specific (*Z*)-3/ (*E*)-2-ratios provide information on caterpillar attack and this informa-tion is used by *M. sexta* moths to avoid oviposition on caterpillar-attacked plants (Allmann et al. 2013).

HIPVs may also make a plant more detectable to herbivores and attract conspecific or hetero-specific herbivores (Dicke and Baldwin 2010). When HIPVs attract herbivores to plants infested by conspecifics, this may be a component of aggregation behavior by the herbivores.

17.3.3 Responses by Pollinators

Pollinators help plants to reproduce and are rewarded with nectar and pollen. Complex odor bouquets and a wide spectrum of colors that flowers exhibit attract pollinators. It is remarkable that most studies on HIPVs have addressed plants in the vegetative stage (Mumm and Dicke 2010). However, when considering HIPVs in a plant defense context, it is important to address plant fitness in terms of seed production (Kessler and Halitschke 2009, Lucas-Barbosa 2016, Lucas-Barbosa et al. 2011). The limited studies on the effects of HIPVs on pollinators show that HIPVs can negatively or positively influence pollinator attraction (Kessler et al. 2011, Rusman et al. 2018, 2019a). For instance, infestation of *Brassica nigra* plants with the aphid *Brevicoryne brassicae* or the aphid *Lipaphis erysimi* differentially affect plant volatile emission and interactions with pollinators. Infestation with *B. brassicae* interferes with pollinator attraction whereas infestation with *L. erysimi* enhances pollinator attraction (Rusman et al. 2019b). Herbivore feeding on a plant's leaves may alter the volatiles that are systemically emitted by the flowers and this has been reported to influence pollinator attraction (Lucas-Barbosa et al. 2016, Schiestl et al. 2014). These effects may vary with plant genotype (Schiestl et al. 2014) and pollinator species (Lucas-Barbosa et al. 2016). Yet, such herbivory-mediated effects on pollinator attraction do not necessarily affect plant reproductive output (Lucas-Barbosa et al. 2016, Schiestl et al. 2014).

In *Nicotiana attenuata*, the terpenoid (*E*)-α-bergamotene that is emitted by flowers attracts the pollinating hawkmoth *Manduca sexta*, whereas its herbivorous larvae induce the same compound in the leaves of the plant, resulting in the attraction of natural enemies of the caterpillars (Zhou et al. 2017). Moreover, herbivores may also feed on the flowers. Uninfested flowers of the crucifer *Biscutella laevigata* emit the terpene (*E*)-β-ocimene that attracts bee pollinators as well as crab spiders that interfere with attraction of the bees. However, upon florivory, the plant induces the emission of (*E*)-β-ocimene and this is more pronounced in plant populations where crab spiders are present (Knauer et al. 2018). Thus, flowers have evolved to interact with pollinators while at the same time interacting with carnivorous enemies. However, if HIPVs result in colonization of flowers by carnivores, this may interfere with interactions with pollinators. This potential trade-off deserves to be further investigated (Rusman et al. 2019a).

17.4 MODIFICATION OF HIPV BLEND BY INTERACTION WITH OTHER MEMBERS OF THE PLANT-ASSOCIATED COMMUNITY

In nature, plants simultaneously or successively interact with multiple organisms, and a plant's response may be influenced by each of these players (Stam et al. 2014). A flowering plant, for instance, can be attacked by herbivores during the same period that it is interacting with pollinators, and responses to pollination can interfere with responses to herbivore attack (Lucas-Barbosa 2016, Lucas-Barbosa et al. 2016). Plants commonly face attack by several attackers at the same time (Stam et al. 2014). Plant responses to insect attack can facilitate or negatively affect the performance of a second attacker (Kaplan and Denno 2007). This suggests that plants may prioritize defense to one attacker over defense against another.

17.4.1 Effects of Dual Herbivore Infestation

Plant responses to multiple attack may vary with herbivore feeding mode or attack of different parts of the plant (Erb et al. 2010, Ponzio et al. 2014, 2016, Rasmann and Turlings 2007). The volatile blend of immature *B. nigra* plants can be distinguished based on whether they have been exposed to aphids or caterpillars (Ponzio et al. 2014). However, when plants are exposed to dual infestation by caterpillars and aphids, the blend resembles the volatile blend emitted by plants infested with caterpillars only. When *B. nigra* plants in the vegetative stage were exposed to herbivory and pathogen infestation, volatile blends of plants could be distinguished based on whether the plant had been exposed to caterpillars, irrespective of the presence and identity of a second attacker (Ponzio et al. 2014). Similarly, the blend of volatiles induced by simultaneous infestation of corn plants with the phloem feeder *Euscelidius variegatus* and caterpillars of *Spodoptera littoralis* was qualitatively and quantitatively similar to the blend induced by *S. littoralis* only, but different from that of plants induced exclusively by the phloem feeder (Erb et al. 2010). These examples suggest that plant responses to caterpillars in terms of volatile induction overrule responses to phloem feeders. This is further supported by untargeted metabolomics of *B. nigra* plants (Brassicaceae) that were exposed to caterpillars (*Pieris brassicae*), or caterpillars plus aphids (*Brevicoryne brassicae*): plants infested with caterpillars or with caterpillars and aphids were importantly characterized by glucosinolates whereas plants infested with aphids were especially characterized by sugars (Ponzio et al. 2017).

The volatiles emitted by plant leaves can also be influenced by an herbivore that feeds on the roots. For instance, *B. nigra* plants infested by larvae of the cabbage root fly that feed on the plant's roots characteristically emit dimethyl disulphide and dimethyl trisulfide, whereas the volatile blend of plants infested with leaf-feeding *P. brassicae* caterpillars is characterized by (*E*)-β-farnesene and DMNT. Plants infested by both the root feeder and the leaf feeder emit a blend that is characterized by dimethyl disulphide, dimethyl trisulfide, (*E*)-β-farnesene and DMNT. These changes in volatile emission concur with changes in the behavior of parasitic wasps. In a semi-field experiment, parasitic wasps parasitized a larger proportion of caterpillars on plants with undamaged roots than on plants infested by root-feeding fly larvae (Soler et al. 2007). Feeding by multiple herbivores can, therefore, differentially affect volatile emission by distinct parts of the plant, and this may be dependent on the feeding site of the herbivore and how resources and defensive traits are allocated to and activated in roots or shoots.

17.4.2 Effects of Simultaneous Infestation by a Microbial Pathogen

Simultaneous attack of the plant by a microbial pathogen may influence the induction of HIPVs by an insect herbivore. For instance, infestation of corn plants with the fungus *Setosphaeria turcica* reduces the induction of HIPVs in response to feeding *Spodoptera littoralis* caterpillars (Rostas et al. 2006). Infestation of pepper plants with the pathogenic bacterium *Xanthomonas campestris* pv. *vesicatoria* influences the induction of volatiles by *S. exigua* caterpillars with differential effects of a compatible versus an incompatible pathovar of the bacterium (Cardoza and Tumlinson 2006). The simultaneous infection of *Brassica rapa* plants with powdery mildew *(Erysiphe cruciferarum)* and *P. brassicae* caterpillars resulted in a reduced attraction of *C. glomerata* wasps (Desurmont et al. 2016). Such differences are likely associated with different signal-transduction pathways being induced by different pathogens. Infection of poplar leaves with the rust fungus *Melampsora larici-populina*, reduces herbivore-induced volatile emission and this results in the attraction of a generalist herbivore, *Lymantria dispar* caterpillars. Phytohormonal crosstalk likely underlies this effect also (Eberl et al. 2018). Thus, plant pathogens may have differential effects on HIPV emission and insect behavior.

17.4.3 Effects of Exposure to Rhizosphere Microorganisms

Plant roots are colonized by rhizobacteria and among these are plant growth promoting rhizobacteria (PGPR) that can prime plant defense (Pieterse et al. 2012, Pineda et al. 2013). Exposure to PGPR can also modulate the induction of HIPVs. Exposure of *Arabidopsis thaliana* roots to the rhizobacterium *Pseudomonas simiae*, modifies the composition of the blend of aphid-induced plant volatiles that is emitted by the plant's leaves and attenuates attraction of the parasitic wasp *Diaeretiella rapae*, a natural enemy of the aphids (Pineda et al. 2013). Exposure of *A. thaliana* roots to the rhizobacterium *P. simiae* also modifies the composition of caterpillar-induced plant volatiles, resulting in enhanced attraction of a parasitic wasp that attacks the caterpillars (Pangesti et al. 2015). Effects of PGPR on parasitism rates of aboveground herbivorous insects feeding on the PGPR-exposed plant have also been recorded under field conditions, resulting in effects on the population size of the insect herbivores (Gadhave et al. 2016).

Plants can also perceive and respond to volatiles emitted by soil-borne microbes, prior to colonization of roots by the microbes (Cordovez et al. 2017, Moisan et al. 2019, Song and Ryu 2013). Such plant responses influence plant development as well as plant volatile emission by belowground and aboveground tissues (Cordovez et al. 2017). Thus, such changes can potentially affect other organisms that interact with the plant, and, therefore, affect plant resistance and reproduction. These aspects deserve to be further investigated to assess their effects on interactions of the plant with its associated community members.

17.5   GENETIC VARIATION IN HIPVs

The emission of HIPVs and the behavioral responses of arthropods to HIPVs varies among plant genotypes. For instance, corn genotypes vary widely in the emission of HIPVs: the total amount, i.e., the sum of 23 selected compounds, emitted by seedlings of 31 corn inbred lines upon standardized elicitation varies by a factor of more than 75 (Degen et al. 2004). This variation proved to be largely genetically determined. Among 15 cucumber accessions, total HIPV emission varied by a factor of 28–77, depending on time since standardized elicitation (Kappers et al. 2010). Among 26 corn inbred lines, terpenoid emission rates varied by about 50 times (Richter et al. 2016) and among 27 *Arabidopsis* accessions the emission rate of the four major inducible compounds even varied by a factor of 572 (Huang et al. 2010).

In horsenettle (*Solanum carolinense*) plants that had been exposed to the natural herbivore community, outcrossed plants released approximately 1.5 times more volatiles than selfed plants and significant differences were recorded among some maternal families (Delphia et al. 2009). It is interesting to note that the evolutionary change from outbreeding to selfing has resulted in the loss of benzaldehyde as a major constituent of the floral scent in the brassicaceous genus *Capsella*. Inactivation of the gene *CNL1*, encoding for a cinnamate:CoA ligase-like protein is responsible for this evolutionary change in volatile production (Sas et al. 2016).

Variation among genotypes is also exhibited in terms of blend composition. The HIPV blend emitted by corn, cucumber and *Arabidopsis* plants displayed significant intraspecific variation (Degen et al. 2004, Kappers et al. 2010, Snoeren et al. 2010), resulting in variation in attraction of enemies of the herbivores (Degenhardt et al. 2009a, Gols et al. 2011, Rasmann et al. 2005). Large-scale phenotyping of a population of genetically well-characterized plant accessions may be used to unravel the genetic architecture underlying plant responses to environmental stresses such as insect herbivory (Thoen et al. 2017). So far, such approaches have not been made for induced plant volatiles as the phenotypic trait, likely because of the labor-intensive nature of full analysis of plant volatiles. Yet, through classical methods many genes underlying herbivore-induced volatile emission have been characterized.

17.5.1   GENES UNDERLYING VARIATION IN HIPV EMISSION AND ARTHROPOD ATTRACTION

For many HIPVs, the genes involved in their biosynthesis are known. For instance, a large family of terpene synthase genes is involved in the biosynthesis of the large diversity of terpenes (Figure 17.3). Tens of TPS genes are present in an individual plant species (Chen et al. 2011, Warren et al. 2015) and each TPS gene may yield from one to more than 50 terpenoid products (Degenhardt et al. 2009b). The products of a TPS gene may undergo further modification resulting in the volatiles that are emitted. For instance, the corn *TPS2* gene synthesizes linalool from the substrate geranyl diphosphate, (*E*)-nerolidol from farnesyl diphosphate and (*E,E*)-geranyllinalool from geranyl diphosphate. The latter two products yield DMNT and (3*E*,7*E*)-4,8,12-trimethyl-1,3,7,11-tridecatetraene (TMTT) as a result of the activity of two cytochrome P450 monooxygenases (Richter et al. 2016). This biosynthetic pathway as well as the two P450s are different from those yielding the production of

DMNT and TMTT in the model plant *Arabidopsis thaliana* (Sohrabi et al. 2015). This suggests that the evolution of these pathways has occurred independently in these two plant species.

For *Arabidopsis* there is ample detailed information on intraspecific genetic variation. In the SALK database (http://signal.salk.edu/atg1001/3.0/gebrowser.php), 855 different accessions are present that exhibit single nucleotide polymorphisms (SNPs) in the TPS02 gene, which encodes an ocimene/farnesene synthase. Many of these accessions have nonsynonymous SNPs resulting in a change in the protein. These nonsynonymous SNPs occur in at least 10 positions in the gene (Dicke, unpublished results, 2018).

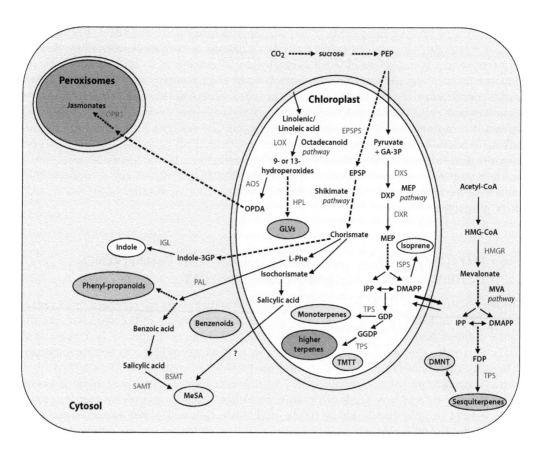

FIGURE 17.3 Schematic representation of the metabolic pathways underlying herbivore-induced plant volatiles (HIPVs) in plants that are known to have a role in indirect plant defense. Volatile compounds are shown inside ovals and names of crucial enzymes responsible for volatile biosynthesis are given in green text. Pathway names are written in italics. Solid arrows: one enzymatic step involved; dashed arrows: more than one enzymatic step is involved, only part is given. (Modified after Mumm, R. and Dicke, M., *Can. J. Zool.*, 88, 628–667, 2010.) Abbreviations: Acetyl-CoA, acetyl coenzyme-A; AOS, allene oxide synthase; BSMT, *S*-adenosyl-L-methionine:benzoic acid/salicylic acid carboxyl methyltransferase; DMAPP, dimethyl-allyl diphosphate; DMNT, (*E*)-4,8-dimethyl-1,3,7-nonatriene; DXP, 1-deoxy-D-xylulose 5-phosphate; DXR, DXP reductoisomerase; DXS, DXP synthase; EPSP, 5-enolpyruvylshikimate-3-phosphate; EPSPS, 5-enol-pyruvylshikimate-3-phosphate synthase; FDP, farnesyl diphosphate; GA-3P, glyceraldehyde-3-phosphate; GDP, geranyl diphosphate; GGDP, geranylgeranyl diphosphate; HMG-CoA, 3-hydroxy-3-methylglutaryl-CoA; HMGR, 3-hydroxy-3-methylglutaryl-CoA reductase; HPL, fatty acid hydroperoxide lyase; IGL, indole-3-glycerol phosphate lyase; Indole-3GP, indole 3-glycerol phosphate; IPP, isopentenyl diphosphate; ISPS, isoprene synthase; L-Phe, L-phenylalanine; LOX, lipoxygenase; MEP, 2-*C*-methyl-D-erythritol 4-phosphate; MeSA, methyl salicylate; MVA, mevalonate; OPDA, 12-oxophyto-dienoic acid; PAL, phenylalanine ammonia lyase; PEP, phosphoenolpyruvate; SAMT, *S*-adenosyl-L-methionine:salicylic acid carboxyl methyltransferase; TMTT, (*E,E*)-4,8,12-trimethyltrideca-1,3,7,11-tetraene; TPS, terpene synthase(s).

Genotypes differ in the degree of attraction of carnivorous arthropods upon HIPV emission. This has been recorded for, e.g., bean (Dicke et al. 1990a), corn (Hoballah et al. 2002), gerbera daisy (Krips et al. 2001), cabbage (Gols et al. 2011, Poelman et al. 2009), *Arabidopsis* (Snoeren et al. 2010), *Datura wrightii* (Hare 2007) and coyote tobacco (Schuman et al. 2009) plants. Inbreeding may result in genotypes with reduced attraction of carnivorous arthropods compared to outcrossed plants (Kariyat et al. 2012). Given that the HIPV blend consists of tens up to more than one hundred compounds and that the biosynthetic pathway underlying the production of each compound consists of various steps, it may seem a daunting task to unravel the genetic differences underlying differences in HIPV emission and carnivore attraction. Yet, for a limited number of compounds, a targeted approach has been successful. For instance, the major peak among a limited number of HIPVs emitted by corn roots infested by Western corn rootworm larvae consists of (E)-β-caryophyllene (Rasmann et al. 2005). This compound attracts entomopathogenic nematodes that attack the larvae and is responsible for the difference in attractiveness of European and North American corn accessions (Rasmann et al. 2005). North American corn accessions have a functional (E)-β-caryophyllene synthase gene, but they lack the gene's transcript. As a consequence, the production and emission of the sesquiterpene are lacking in North American lines. Complementation with an (E)-β-caryophyllene synthase gene from oregano restored the production of this sequiterpene in a North American accession as well as nematode attraction (Degenhardt et al. 2009a). Thus, knowledge of key genes controlling underlying the biosynthesis of individual HIPVs may be used to manipulate plants to alter their HIPV phenotype.

17.6 PHYTOHORMONAL MECHANISM UNDERLYING INTERFERENCE IN HIPV EMISSION BY HERBIVOROUS INSECTS

The biosynthetic pathways leading to HIPV production and emission are regulated by phytohormonal signaling pathways such as the octadecanoid pathway leading to jasmonic acid, the shikimate pathway leading to salicylic acid and the ethylene pathway (Figure 17.3) (Pieterse et al. 2012). Attack by chewing-biting insect herbivores, such as caterpillars and beetles, especially induces the octadecanoid pathway and the ethylene pathway, whereas phloem-feeding herbivores such as aphids and whiteflies especially induce the shikimate pathway (Stam et al. 2014). Exposing plants to jasmonic acid may result in a volatile blend that is similar, though not identical, to the blend induced by arthropod herbivory (Dicke et al. 1999, Gols et al. 1999, Ozawa et al. 2008). Silencing the *NaLOX3* gene that codes for a lipoxygenase mediating an early step in the biosynthesis of jasmonic acid in tobacco plants, results in a strong reduction in the herbivore-inducible volatile (E)-α-bergamotene and renders the plants much more acceptable to *Empoasca* sp. leafhoppers under field conditions (Kessler et al. 2004).

The two major phytohormones underlying plant responses to arthropod herbivory, i.e., jasmonic acid and salicylic acid, exhibit negative crosstalk, meaning that they interfere with each other's effects (Pieterse et al. 2012). For instance, co-infection of an SA-inducing herbivore, such as the whitefly *Bemisia tabaci*, or a pathogen, such as the powdery mildew *Erysiphe cruciferarum*, with a JA-inducing herbivorous arthropod results in interference with the JA-induced HIPV emission and leads to attenuation of the attraction of carnivorous enemies of the JA-inducing arthropod (Desurmont et al. 2016, Zhang et al. 2009). In Lima bean plants, co-infection of the JA-inducing two-spotted spider-mite *T. urticae* with the SA-inducer *B. tabaci* negatively affected the transcription of *PlLOX*, a gene which codes for an enzyme in the JA biosynthesis pathway as well as the gene *Ocimene Synthase* encoding an enzyme regulating the biosynthesis of the monoterpene (E)-β-ocimene. This effect could be mimicked by the application of synthetic SA. The resulting reduction in (E)-β-ocimene emission attenuated the attraction of the carnivorous predatory mite *P. persimilis*, a specialist natural enemy of the two-spotted spider-mite (Zhang et al. 2009).

Phytohormones may also positively interact. For instance, application of 1-aminocyclopropane-1-carboxylic acid (ACC), a precursor of ethylene, to Lima bean plants had a positive effect on the production of three JA-inducible volatiles, i.e., (*E*)-β-ocimene, (*Z*)-β-ocimene, and (*Z*)-3-hexenyl acetate. This consequently resulted in a positive effect on the attraction of the predatory mite *P. persimilis*, that is known to be attracted to (*E*)-β-ocimene (Horiuchi et al. 2001).

Other phytohormones involved in the induction of stress responses include, e.g., abscisic acid, cytokinines, auxin and gibberellins and these may modulate signaling by JA, SA and ethylene (Pieterse et al. 2012). Whether this modulation has consequences for HIPV induction remains to be investigated.

17.7 WHICH HIPVs INFLUENCE ARTHROPOD BEHAVIOR

Despite the more than 25 years of research on HIPVs, and the impressive increase in knowledge of the role of HIPVs in the interactions of plants and their associated insect community (Stam et al. 2014), information on which HIPVs are exploited by predators and parasitoids as information on prey and host presence has remained limited. Yet, one of the first studies on HIPVs identified some HIPVs that individually attract a predatory mite to Lima bean plants infested with its prey: (*E*)-β-ocimene, DMNT, linalool, and methyl salicylate (Dicke and Sabelis 1988, Dicke et al. 1990b). However, for most parasitoids such information on a behavioral effect of HIPV components is still lacking. An interesting observation is that the overexpression of a single TPS gene in corn can make a plant more attractive to a parasitoid (Schnee et al. 2006). The terpene synthase encoded by this gene (*ZmTPS10*) produces (*E*)-β-farnesene, (*E*)-α-bergamotene, and several other sesquiterpenes from the substrate farnesyl diphosphate. The parasitoids can learn to respond to (a selection of) these terpenes (Schnee et al. 2006). The capability of parasitic wasps to associatively learn to respond to HIPVs (Smid and Vet 2016) may be one component of the explanation why our knowledge of the exact blends of HIPVs that are used by especially parasitoids to locate their hosts has been limited to date.

Carnivorous arthropods commonly exploit a partial blend of the total blend of HIPVs available (Bruce and Pickett 2011, De Boer et al. 2004, van Wijk et al. 2011). Thus, in this context, manipulative experiments with multiple herbivores or comparative approaches involving multiple plant species that provide sufficient contrast as well as similarities between blends, may guide us to the elucidation of the partial volatile blend that mediates the attraction of carnivorous insects to plants infested with their herbivorous hosts or prey. Here, we will discuss the molecular and chemical approaches as well as the statistical analyses that can contribute in this direction.

The classical approach in the empirical studies of the chemical ecology of HIPVs involves five main steps: (1) infestation of plants with the target herbivore, (2) collection of volatiles from infested plants and from noninfested control plants, (3) qualitative and quantitative analysis of compounds emitted by plants followed by (4) statistical analyses, and (5) correlation with behavioral responses (Figure 17.4a). Each of these – apparently straightforward – steps comprises a laborious amount of work and a number of informed choices to be made.

Plant responses to herbivore exposure vary with herbivore identity, density, and the duration of exposure to the herbivore (Danner et al. 2018, Ponzio et al. 2016), and these will influence the decisions on the design of the experiment and the critical step 1. Plants can rapidly respond to herbivore induction with the production of volatiles – even within a few hours upon exposure to herbivores (Giacomuzzi et al. 2016, Turlings et al. 1998), and volatile emission varies with herbivore density (Ponzio et al. 2016). However, induction time is often confounded with herbivore density or amount of damage. Induction time can positively correlate with herbivore density when herbivores, such as aphids and spider mites, reproduce on the plant and the population grows with time, or when the amount of damage inflicted increases as the herbivore themselves grows, as is the case for caterpillars, for instance.

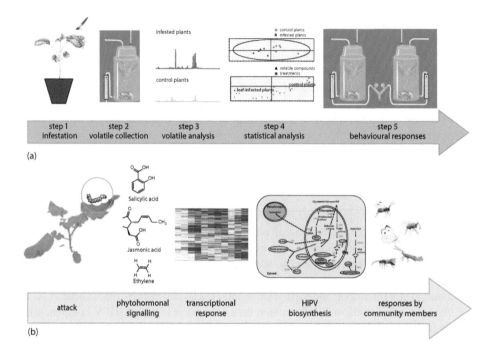

FIGURE 17.4 (a) Schematic representation of the five critical steps involved in research on the chemical ecology of herbivore-induced plants volatiles. (Drawing of the *Brassica nigra* plant, Courtesy of Alison Shoorer, drawing of the olfactometer modified from Fatouros, N.E. et al., *Plos One*, 7, e43607, 2012.) (b) Schematic representation of the temporal steps from the moment of insect attack, in terms of the emission of herbivore-induced plant volatiles and their effects on the responses by the community members. (Drawings of insects and the *Brassica nigra* plant, Courtesy of Camille Ponzio.)

17.7.1 MULTIVARIATE STATISTICAL ANALYSIS OF HIPV BLENDS

Although plants can produce volatiles *de novo* in response to herbivory, most differences between infested and noninfested plants are quantitative (Mumm and Dicke 2010). Statistical analysis of multicomponent blend composition requires a multivariate analysis (MVA), such as principal component analysis (PCA) and projection to latent structures–discriminant analysis (PLS-DA), and random forest have been used to infer differences between volatile blends emitted by plants with different treatments (Danner et al. 2018, McCormick et al. 2016, Pierre et al. 2011, Ponzio et al. 2014). The advantage of random forest analysis over the other multivariate methods is that this analysis can more effectively determine quantitative differences among minor volatile compounds (McCormick et al. 2014). However, although multivariate analysis of total blend composition provides important insight into chemical differences between blends, this does not necessarily reflect the differences in terms of biological activity. In other words, chemical differences determined by MVA do not necessary explain which volatile compounds are used as a source of information by carnivorous insects. Yet, MVA can provide valuable information in relation to the elucidation of compounds exploited by carnivorous insects if used to analyze data from manipulative comparative experiments with exposure of plants to multiple herbivores or multiple plant species that provide sufficient contrast, as well as similarities, between blends.

17.7.2 BIOASSAYS

When some insight is gained into which groups of compounds are possibly important in the attraction of parasitoids and predators to infested plants, then the ultimate proof is to bring the partial blend back to a bioassay where behavioral responses can be tested. The importance of a particular compound in the total HIPV blend can be demonstrated by careful manipulation of a complex blend. For instance, the importance of MeSA within the total blend of spider-mite induced Lima bean volatiles was revealed by exploiting JA as an elicitor of HIPV. JA induces most of the HIPVs that spider mites induce and JA application renders Lima bean plants attractive to a predatory mite. Yet, JA-induced Lima bean volatiles were less attractive than the natural HIPV blend induced by spider mites. In contrast to spider mites, JA does not induce MeSA and adding MeSA to the JA-induced volatile blend compensated for the lower attractiveness compared to the natural HIPV blend (De Boer and Dicke 2004). Preparative gas-chromatography can also be used to collect part of the volatile blend, or to exclude compounds of interest from the HIPV blend, and bring the partial volatile blend back to a bioassay to test whether the phenotypic trait or response is maintained or lost (van Beek et al. 2009). This approach is innovative because the original concentrations and ratios of complex HIPV blends are preserved and can be tested again in a behavioral bioassay. In one of the first studies on HIPVs all 11 major compounds identified in the blend of corn plants infested by *S. exigua* caterpillars were purified by preparative GC and then recombined into a synthetic blend and offered to a parasitoid of the caterpillars (Turlings et al. 1991). This blend was slightly less attractive to the parasitoids than the natural blend. However, when the parasitoids were given a host to oviposit in, in the presence of the synthetic blend, the attraction was much stronger than when the parasitoids had had an oviposition experience in the presence of the natural HIPV blend (Turlings et al. 1991). This indicates the importance of previous experiences of the parasitoids that has also been demonstrated for many other parasitoid species (Smid and Vet 2016, Turlings et al. 1993, Vet and Dicke 1992).

Other innovative approaches for assessing biological activity of volatile compounds are the use of metabolic inhibitors and/or transgenic plants. By treating plants with inhibitors of specific biosynthetic pathways, the emission of particular components of the blend can be reduced, and behavioral responses of parasitoids and predators tested (Mumm et al. 2008). Similarly, the use of mutant plants or lines overexpressing volatiles can be valuable tools to test the role of particular volatiles in attracting natural enemies of the herbivores (Degenhardt et al. 2009a, Erb and Robert 2016, Halitschke et al. 2008, Kappers et al. 2005, McCormick et al. 2014, Schnee et al. 2006).

17.8 DISCUSSION AND FUTURE PERSPECTIVES

The research on HIPVs was initiated in the 1980s (Dicke and Sabelis 1988, Sabelis and Van de Baan 1983, Turlings et al. 1990) and has rapidly expanded from behavioral and chemical research to include mechanistic studies on underlying biochemistry, biosynthetic pathways, physiological and transcriptional regulation (Howe and Jander 2008, Kessler and Baldwin 2002) as well as ecological consequences for species interactions, population dynamics and community ecology (Kessler and Heil 2011, Schuman and Baldwin 2016, Stam et al. 2014, Vet and Dicke 1992).

In this review, we have highlighted the chemical diversity, dynamic complexity of blend composition, and current methodological approaches to study the effects of HIPVs on plant-associated arthropods. With the development of more sensitive analytical equipment, more compounds have been recorded in HIPV mixtures, indicating the tremendous degree of variation in blend composition and, thus, specificity that plants may display. Simultaneously, electrophysiological and behavioral studies have shown that arthropod responses are elicited by mixtures of compounds (Bruce and Pickett 2011, De Boer et al. 2004, van Wijk et al. 2011). Thus, HIPV blends appear to be multivariate entities and investigating differences between blends has developed from comparison of emission rates of individual compounds to assessing differences in blend composition through multivariate analysis.

Some of the most prominent frontiers for research on HIPVs are highlighted below.

17.8.1 High-Throughput Blend Analysis

Now that molecular techniques for quantification of transcriptional responses have made revolutionary progress, so that high-resolution temporal patterns in transcriptional responses can be investigated in a high-throughput manner via RNASeq analysis (Hickman et al. 2017), detailed information comes available on especially the onset of HIPV-mediated phenotypic changes in plants. High-throughput phenotyping of insect behavior has been initiated (Jongsma et al. 2019) that will provide an important step forward toward a more detailed understanding of the ecological effects of HIPV blends. For high-resolution chemical analysis, first steps have been made, e.g., with the development of real-time methods such as the e-nose (Wilson 2013) or Proton-Transfer Reaction Mass Spectrometry (PTR-MS) (Steeghs et al. 2004), but so far these methods do not allow for analysis of HIPV blends to the extent that is needed for a thorough appreciation of their role as specific blends in ecological species interactions. Full analysis of multicomponent HIPV blends still relies to a large extent on labor-intensive handwork, especially because the blends are so complex that automated analysis still has many drawbacks.

17.8.2 CONNECTING DIFFERENT LEVELS OF BIOLOGICAL COMPLEXITY

The emission of HIPVs by a plant in response to herbivory is a process that starts with wounding of the plant, followed by the introduction of elicitors and effectors (Bonaventure 2012), early signaling events including rapid changes in the plasma membrane potential (V_m) involving changes in cytosolic Ca^{2+} concentrations and the generation of reactive oxygen species, gene transcriptional responses, activation of the phytohormonal signaling network and then the biosynthesis and emission of HIPVs (Figure 17.4b). Upon emission of the HIPVs, the information is available to any organism in the environment and each individual may respond behaviorally, leading to changes in spatial distribution and population responses, finally leading to changes in community composition and dynamics. Although this sequence of events seems rather straightforward, its temporal regulation is not yet well understood. Connecting the different steps is a major challenge. How do transcriptional changes translate into biosynthetic responses and how are the temporal processes at one level of biological organization connected to those at the next (see for example Wurtzel and Kutchan 2016)? To explain changes in HIPV emission in terms of the underlying mechanistic regulation, detailed knowledge is needed on temporal aspects of the regulatory pathways, including spatial aspects such as the involvement of different organelles or even organs (Brockmöller et al. 2017, Gulati et al. 2014). This further underlines the need of high-resolution temporal analyses, but now at different levels of biological organization. Because at each level the response is multivariate, such as a whole suite of genes whose transcription is altered, or a diverse set of metabolites that is being synthesized, the connection of levels of biological organization requires the linking of multivariate networks that change in time.

17.8.3 ECOLOGICAL NETWORK ANALYSIS

The dynamic emission of an HIPV blend can influence the behavior of a suite of associated organisms with different temporal dynamics (Thaler et al. 2001). The effects on colonization by additional plant attackers has consequences for the induction of HIPVs with subsequent effects on the behavior of associated organisms (Poelman and Kessler 2016). For instance, in replicated field plots the dynamics of the insect community has been recorded at weekly intervals. The data show patterns in colonization of the plants through time. Early-season infestation by a particular herbivore species has community-wide effects and determines the likelihood of future attack by a particular group of attackers (Poelman et al. 2010). The emerging pattern of specialist herbivores acting as strong drivers of the future composition of an herbivore community biased to a specialized community has strong implications for inducible plant defenses.

Thus, the multivariate response of a plant in terms of gene expression, followed by the multivariate response in terms of metabolite emission is followed by a multivariate ecological response in terms of the behavioral responses of individuals of various species that either interact directly through, e.g., predation or indirectly via HIPV-mediated interactions.

17.8.4 Plant Ontogeny and HIPVs

The diversity of metabolites produced by a plant may vary with plant ontogeny (Barton and Koricheva 2010). Within an individual plant it is common to observe that younger leaves have higher concentrations of secondary metabolites than older leaves, which has been commonly associated with defensive traits being allocated to the most valuable tissues (Schoonhoven et al. 2005), also known as optimal defense (Barton and Boege 2017). For instance, secondary metabolites such as glucosinolates are present in much higher (3–5 times) concentrations in the youngest leaves compared to the older leaves of mustard plants (Traw and Feeny 2008). Young and old leaves of a plant also emit different HIPV blends, resulting in differential effects on the attraction of herbivorous as well as carnivorous arthropods (Brilli et al. 2009, Radhika et al. 2008, Rostas and Eggert 2008). As the reproductive organs, flowers are highly valuable tissues. Therefore, it is not surprising that short-lived plants preferentially allocate resources and accumulate defensive traits in flower tissues, constitutively as well as in response to herbivory (Bruinsma et al. 2014). The metabolic composition of flower tissues is generally very different from that of leaves (Bruinsma et al. 2014, Lucas-Barbosa et al. 2016). Flower tissues usually contain higher contents of secondary metabolites that give rise to the vivid colors and rich odor bouquets, and can be of defensive nature as well as important for interactions with pollinators. Thus, it is important to investigate defenses of plants in the flowering stage, as well as defense mechanisms in flowers themselves. After all, these tissues are the most directly linked to Darwinian fitness of plants.

17.9 CONCLUSION

More than 25 years of research on HIPVs has yielded a wealth of information on chemical diversity of HIPVs, emission dynamics, underlying biosynthetic and regulatory pathways, as well as ecological effects. This impressive body of information has been collected by chemists, physiologists, molecular biologists, and ecologists. Progress has been especially possible as a result of multidisciplinary collaboration within new disciplines such as chemical ecology and molecular ecology, resulting in novel insights into networks at different levels of biological integration, such as transcriptomic networks, chemical networks and ecological networks. In recent years the research field has expanded by including community ecologists and statisticians and modelers and future challenges include the linking of the networks at different levels of biological integration (Gulati et al. 2014, Thoen et al. 2017). This process will require truly multidisciplinary collaboration to understand the complex interactions between species in an ecological network based on induced plant volatiles.

ACKNOWLEDGMENTS

The authors thank Camille Ponzio for providing us with most of the insect drawings and Alison Shoorer for the *Brassica nigra* drawing. This work benefited from funding through a Spinoza Award to MD by the Netherlands Organisation for Scientific Research (NWO).

REFERENCES

Allmann, S., Spathe, A., Bisch-Knaden, S., et al. 2013. Feeding-induced rearrangement of green leaf volatiles reduces moth oviposition. *eLife* 2:e00421.

Amo, L., Jansen, J.J., van Dam, N.M., Dicke, M. and Visser, M.E. 2013. Birds exploit herbivore-induced plant volatiles to locate herbivorous prey. *Ecology Letters* 16:1348–1355.

Barton, K.E. and Boege, K. 2017. Future directions in the ontogeny of plant defence: Understanding the evolutionary causes and consequences. *Ecology Letters* 20:403–411.

Barton, K.E. and Koricheva, J. 2010. The ontogeny of plant defense and herbivory: Characterizing general patterns using meta-analysis. *American Naturalist* 175:481–493.

Blande, J.D., Pickett, J.A. and Poppy, G.M. 2007. A comparison of semiochemically mediated interactions involving specialist and generalist *Brassica*-feeding aphids and the braconid parasitoid *Diaeretiella rapae. Journal of Chemical Ecology* 33:767–779.

Bonaventure, G. 2012. Perception of insect feeding by plants. *Plant Biology* 14:872–880.

Brilli, F., Ciccioli, P., Frattoni, M., Prestininzi, M., Spanedda, A.F. and Loreto, F. 2009. Constitutive and herbivore-induced monoterpenes emitted by *Populus x euroamericana* leaves are key volatiles that orient *Chrysomela populi* beetles. *Plant Cell and Environment* 32:542–552.

Brockmöller, T., Ling, Z., Li, D., Gaquerel, E., Baldwin, I.T. and Xu, S. 2017. Nicotiana attenuata Data Hub (NaDH): An integrative platform for exploring genomic, transcriptomic and metabolomic data in wild tobacco. *BMC Genomics* 18:79.

Bruce, T.J.A. and Pickett, J.A. 2011. Perception of plant volatile blends by herbivorous insects: Finding the right mix. *Phytochemistry* 72:1605–1611.

Bruinsma, M., Lucas-Barbosa, D., Ten Broeke, C.J.M., et al. 2014. Folivory affects composition of nectar, floral odor and modifies pollinator behavior. *Journal of Chemical Ecology* 40:39–49.

Cardoza, Y.J. and Tumlinson, J.H. 2006. Compatible and incompatible *Xanthomonas* infections differentially affect herbivore-induced volatile emission by pepper plants. *Journal of Chemical Ecology* 32:1755–1768.

Chen, F., Tholl, D., Bohlmann, J. and Pichersky, E. 2011. The family of terpene synthases in plants: A midsize family of genes for specialized metabolism that is highly diversified throughout the kingdom. *Plant Journal* 66:212–229.

Cordovez, V., Mommer, L., Moisan, K., et al. 2017. Plant phenotypic and transcriptional changes induced by volatiles from the fungal root pathogen *Rhizoctonia solani. Frontiers in Plant Science* 8:1262.

Danner, H., Desurmont, G.A., Cristescu, S.M. and van Dam, N.M.C. 2018. Herbivore-induced plant volatiles accurately predict history of coexistence, diet breadth, and feeding mode of herbivores. *New Phytologist* 220:726–738.

De Boer, J.G. and Dicke, M. 2004. The role of methyl salicylate in prey searching behavior of the predatory mite *Phytoseiulus persimilis. Journal of Chemical Ecology* 30:255–271.

De Boer, J.G., Posthumus, M.A. and Dicke, M. 2004. Identification of volatiles that are used in discrimination between plants infested with prey or nonprey herbivores by a predatory mite. *Journal of Chemical Ecology* 30:2215–2230.

De Moraes, C.M., Mescher, M.C. and Tumlinson, J.H. 2001. Caterpillar-induced nocturnal plant volatiles repel nonspecific females. *Nature* 410:577–580.

Degen, T., Dillmann, C., Marion-Poll, F. and Turlings, T.C.J. 2004. High genetic variability of herbivore-induced volatile emission within a broad range of maize inbred lines. *Plant Physiology* 135:1928–1938.

Degenhardt, J., Hiltpold, I., Kollner, T.G., et al. 2009a. Restoring a maize root signal that attracts insect-killing nematodes to control a major pest. *Proceedings of the National Academy of Sciences of the United States of America* 106:13213–13218.

Degenhardt, J., Kollner, T.G. and Gershenzon, J. 2009b. Monoterpene and sesquiterpene synthases and the origin of terpene skeletal diversity in plants. *Phytochemistry* 70:1621–1637.

Delphia, C.M., Rohr, J.R., Stephenson, A.G., De Moraes, C.M. and Mescher, M.C. 2009. Effects of genetic variation and inbreeding on volatile production in a field population of horsenettle. *International Journal of Plant Sciences* 170:12–20.

Desurmont, G.A., Xu, H. and Turlings, T.C.J. 2016. Powdery mildew suppresses herbivore-induced plant volatiles and interferes with parasitoid attraction in *Brassica rapa. Plant Cell and Environment* 39:1920–1927.

Dicke, M. and Baldwin, I.T. 2010. The evolutionary context for herbivore-induced plant volatiles: Beyond the "cry for help". *Trends in Plant Science* 15:167–175.

Dicke, M. and Sabelis, M.W. 1988. How plants obtain predatory mites as bodyguards. *Netherlands Journal of Zoology* 38:148–165.

Dicke, M., Gols, R., Ludeking, D. and Posthumus, M.A. 1999. Jasmonic acid and herbivory differentially induce carnivore-attracting plant volatiles in lima bean plants. *Journal of Chemical Ecology* 25:1907–1922.

Dicke, M., Sabelis, M.W., Takabayashi, J., Bruin, J. and Posthumus, M.A. 1990a. Plant strategies of manipulating predator-prey interactions through allelochemicals: Prospects for application in pest control. *Journal of Chemical Ecology* 16:3091–3118.

Dicke, M., Van Beek, T.A., Posthumus, M.A., Ben Dom, N., Van Bokhoven, H. and De Groot, A.E. 1990b. Isolation and identification of volatile kairomone that affects acarine predator-prey interactions: Involvement of host plant in its production. *Journal of Chemical Ecology* 16:381–396.

Dudareva, N., Klempien, A., Muhlemann, J.K. and Kaplan, I. 2013. Biosynthesis, function and metabolic engineering of plant volatile organic compounds. *New Phytologist* 198:16–32.

Eberl, F., Hammerbacher, A., Gershenzon, J. and Unsicker, S.B. 2018. Leaf rust infection reduces herbivore-induced volatile emission in black poplar and attracts a generalist herbivore. *New Phytologist* 220:760–772.

Erb, M. and Robert, C.A.M. 2016. Sequestration of plant secondary metabolites by insect herbivores: Molecular mechanisms and ecological consequences. *Current Opinion in Insect Science* 14:8–11.

Erb, M., Foresti, N. and Turlings, T.C.J. 2010. A tritrophic signal that attracts parasitoids to host-damaged plants withstands disruption by non-host herbivores. *BMC Plant Biology* 10:247.

Fatouros, N.E., Lucas-Barbosa, D., Weldegergis, B.T., et al. 2012. Plant volatiles induced by herbivore egg deposition affect insects of different trophic levels. *Plos One* 7:e43607.

Gadhave, K.R., Finch, P., Gibson, T.M. and Gange, A.C. 2016. Plant growth-promoting *Bacillus* suppress *Brevicoryne brassicae* field infestation and trigger density-dependent and density-independent natural enemy responses. *Journal of Pest Science* 89:985–992.

Giacomuzzi, V., Cappellin, L., Khomenko, I., et al. 2016. Emission of volatile compounds from apple plants infested with *Pandemis heparana* larvae, antennal response of conspecific adults, and preliminary field trial. *Journal of Chemical Ecology* 42:1265–1280.

Gols, R., Bullock, J.M., Dicke, M., Bukovinszky, T. and Harvey, J.A. 2011. Smelling the wood from the trees: Non-linear parasitoid responses to volatile attractants produced by wild and cultivated cabbage. *Journal of Chemical Ecology* 37:795–807.

Gols, R., Posthumus, M.A. and Dicke, M. 1999. Jasmonic acid induces the production of gerbera volatiles that attract the biological control agent *Phytoseiulus persimilis*. *Entomologia Experimentalis et Applicata* 93:77–86.

Gulati, J., Baldwin, I.T. and Gaquerel, E. 2014. The roots of plant defenses: Integrative multivariate analyses uncover dynamic behaviors of gene and metabolic networks of roots elicited by leaf herbivory. *Plant Journal* 77:880–892.

Halitschke, R., Stenberg, J.A., Kessler, D., Kessler, A. and Baldwin, I.T. 2008. Shared signals: "Alarm calls" from plants increase apparency to herbivores and their enemies in nature. *Ecology Letters* 11:24–34.

Hare, J.D. 2007. Variation in herbivore and methyl jasmonate-induced volatiles among genetic lines of *Datura wrightii*. *Journal of Chemical Ecology* 33:2028–2043.

Hickman, R.J., Van Verk, M.C., Van Dijken, A.J.H., et al. 2017. Architecture and dynamics of the jasmonic acid gene regulatory network. *Plant Cell* 29:2086–2105.

Hilker, M. and Fatouros, N.E. 2015. Plant responses to insect egg deposition. *Annual Review of Entomology* 60:493–515.

Hoballah, M.E.F., Tamo, C. and Turlings, T.C.J. 2002. Differential attractiveness of induced odors emitted by eight maize varieties for the parasitoid *Cotesia marginiventris*: Is quality or quantity important? *Journal of Chemical Ecology* 28:951–968.

Horiuchi, J., Arimura, G., Ozawa, R., Shimoda, T., Takabayashi, J. and Nishioka, T. 2001. Exogenous ACC enhances volatiles production mediated by jasmonic acid in lima bean leaves. *FEBS Letters* 509:332–336.

Howe, G.A. and Jander, G. 2008. Plant immunity to insect herbivores. *Annual Review of Plant Biology* 59:41–66.

Huang, M.S., Abel, C., Sohrabi, R., et al. 2010. Variation of herbivore-induced volatile terpenes among *Arabidopsis* ecotypes depends on allelic differences and subcellular targeting of two terpene synthases, TPS02 and TPS03. *Plant Physiology* 153:1293–1310.

Jongsma, M.A., Thoen, M.P.M., Poleij, L.M., et al. 2019. An integrated system for the automated recording and analysis of insect behavior in T-maze arrays. *Frontiers in Plant Science* 10:20.

Kaplan, I. and Denno, R.F. 2007. Interspecific interactions in phytophagous insects revisited: A quantitative assessment of competition theory. *Ecology Letters* 10:977–994.

Kappers, I.F., Aharoni, A., van Herpen, T.W.J.M., Luckerhoff, L.L.P., Dicke, M. and Bouwmeester, H.J. 2005. Genetic engineering of terpenoid metabolism attracts, bodyguards to *Arabidopsis*. *Science* 309:2070–2072.

Kappers, I.F., Verstappen, F.W.A., Luckerhoff, L.L.P., Bouwmeester, H.J. and Dicke, M. 2010. Genetic variation in jasmonic acid- and spider mite-induced plant volatile emission of cucumber accessions and attraction of the predator *Phytoseiulus persimilis*. *Journal of Chemical Ecology* 36:500–512.

Kariyat, R.R., Mauck, K.E., Moraes, C.M., Stephenson, A.G. and Mescher, M.C. 2012. Inbreeding alters volatile signalling phenotypes and influences tri-trophic interactions in horsenettle (*Solanum carolinense* L.). *Ecology Letters* 15:301–309.

Kessler, A. and Baldwin, I.T. 2002. Plant responses to insect herbivory: The emerging molecular analysis. *Annual Review of Plant Biology* 53:299–328.

Kessler, A. and Halitschke, R. 2009. Testing the potential for conflicting selection on floral chemical traits by pollinators and herbivores: Predictions and case study. *Functional Ecology* 23:901–912.

Kessler, A. and Heil, M. 2011. The multiple faces of indirect defences and their agents of natural selection. *Functional Ecology* 25:348–357.

Kessler, A., Halitschke, R. and Baldwin, I.T. 2004. Silencing the jasmonate cascade: Induced plant defenses and insect populations. *Science* 305:665–668.

Kessler, A., Halitschke, R. and Poveda, K. 2011. Herbivory-mediated pollinator limitation: Negative impacts of induced volatiles on plant-pollinator interactions. *Ecology* 92:1769–1780.

Knauer, A.C., Bakhtiari, M. and Schiestl, F.P. 2018. Crab spiders impact floral-signal evolution indirectly through removal of florivores. *Nature Communications* 9:1367.

Knolhoff, L.M. and Heckel, D.G. 2014. Behavioral assays for studies of host plant choice and adaptation in herbivorous insects. *Annual Review of Entomology* 59:263–278.

Krips, O.E., Willems, P.E.L., Gols, R., Posthumus, M.A., Gort, G. and Dicke, M. 2001. Comparison of cultivars of ornamental crop *Gerbera jamesonii* on production of spider mite-induced volatiles, and their attractiveness to the predator *Phytoseiulus persimilis*. *Journal of Chemical Ecology* 27:1355–1372.

Lucas-Barbosa, D. 2016. Integrating studies on plant–pollinator and plant–herbivore interactions. *Trends in Plant Science* 21:125–133.

Lucas-Barbosa, D., Sun, P., Hakman, A., van Beek, T.A., van Loon, J.J.A. and Dicke, M. 2016. Visual and odour cues: Plant responses to pollination and herbivory affect the behaviour of flower visitors. *Functional Ecology* 30:431–441.

Lucas-Barbosa, D., van Loon, J.J.A. and Dicke, M. 2011. The effects of herbivore-induced plant volatiles on interactions between plants and flower-visiting insects. *Phytochemistry* 72:1647–1654.

McCormick, A.C., Gershenzon, J. and Unsicker, S.B. 2014. Little peaks with big effects: Establishing the role of minor plant volatiles in plant-insect interactions. *Plant Cell and Environment* 37:1836–1844.

McCormick, A.C., Reinecke, A., Gershenzon, J. and Unsicker, S.B. 2016. Feeding experience affects the behavioral response of polyphagous gypsy moth caterpillars to herbivore-induced poplar volatiles. *Journal of Chemical Ecology* 42:382–393.

Moisan, K., Cordovez, V., van de Zande, E.M., Raaijmakers, J.M., Dicke. M. and Lucas-Barbosa, D. 2019. Volatiles of pathogenic and nonpathogenic soil-borne fungi affect plant development and resistance to insects. *Oecologia* 190(3):589–604.

Mumm, R. and Dicke, M. 2010. Variation in natural plant products and the attraction of bodyguards involved in indirect plant defense. *Canadian Journal of Zoology* 88:628–667.

Mumm, R., Posthumus, M.A. and Dicke, M. 2008. Significance of terpenoids in induced indirect plant defence against herbivorous arthropods. *Plant Cell and Environment* 31:575–585.

Ozawa, R., Shiojiri, K., Sabelis, M.W. and Takabayashi, J. 2008. Maize plants sprayed with either jasmonic acid or its precursor, methyl linolenate, attract armyworm parasitoids, but the composition of attractants differs. *Entomologia Experimentalis et Applicata* 129:189–199.

Pangesti, N., Weldegergis, B.T., Langendorf, B., van Loon, J.J.A., Dicke, M. and Pineda, A. 2015. Rhizobacterial colonization of roots modulates plant volatile emission and enhances the attraction of a parasitoid wasp to host-infested plants. *Oecologia* 178:1169–1180.

Pashalidou, F.G., Gols, R., Berkhout, B.W., et al. 2015. To be in time: Egg deposition enhances plant-mediated detection of young caterpillars by parasitoids. *Oecologia* 177:477–486.

Pierre, P.S., Jansen, J.J., Hordijk, C.A., van Dam, N.M., Cortesero, A.M. and Dugravot, S. 2011. Differences in volatile profiles of turnip plants subjected to single and dual herbivory above- and belowground. *Journal of Chemical Ecology* 37:368–377.

Pieterse, C.M.J., Van der Does, D., Zamioudis, C., Leon-Reyes, A. and Van Wees, S.C.M. 2012. Hormonal modulation of plant immunity. *Annual Review of Cell and Developmental Biology* 28:489–521.

Pineda, A., Soler, R., Weldegergis, B.T., Shimwela, M.M., Van Loon, J.J.A. and Dicke, M. 2013. Non-pathogenic rhizobacteria interfere with the attraction of parasitoids to aphid-induced plant volatiles via jasmonic acid signalling. *Plant Cell and Environment* 36:393–404.

Poelman, E.H. and Kessler, A. 2016. Keystone herbivores and the evolution of plant defenses. *Trends in Plant Science* 21:477–481.

Poelman, E.H., Oduor, A.M.O., Broekgaarden, C., et al. 2009. Field parasitism rates of caterpillars on *Brassica oleracea* plants are reliably predicted by differential attraction of *Cotesia* parasitoids. *Functional Ecology* 23:951–962.

Poelman, E.H., Van Loon, J.J.A., Van Dam, N.M., Vet, L.E.M. and Dicke, M. 2010. Herbivore-induced plant responses in *Brassica oleracea* prevail over effects of constitutive resistance and result in enhanced herbivore attack. *Ecological Entomology* 35:240–247.

Ponzio, C., Cascone, P., Cusumano, A., et al. 2016. Volatile-mediated foraging behaviour of three parasitoid species under conditions of dual insect herbivore attack. *Animal Behaviour* 111:197–206.

Ponzio, C., Gols, R., Weldegergis, B.T. and Dicke, M. 2014. Caterpillar-induced plant volatiles remain a reliable signal for foraging wasps during dual attack with a plant pathogen or non-host insect herbivore. *Plant Cell and Environment* 37:1924–1935.

Ponzio, C., Papazian, S., Albrectsen, B.R., Dicke, M. and Gols, R. 2017. Dual herbivore attack and herbivore density affect metabolic profiles of *Brassica nigra* leaves. *Plant Cell and Environment* 40:1356–1367.

Radhika, V., Kost, C., Bartram, S., Heil, M. and Boland, W. 2008. Testing the optimal defence hypothesis for two indirect defences: Extrafloral nectar and volatile organic compounds. *Planta* 228:449–457.

Rasmann, S. and Turlings, T.C.J. 2007. Simultaneous feeding by aboveground and belowground herbivores attenuates plant-mediated attraction of their respective natural enemies. *Ecology Letters* 10:926–936.

Rasmann, S., Kollner, T.G., Degenhardt, J., et al. 2005. Recruitment of entomopathogenic nematodes by insect-damaged maize roots. *Nature* 434:732–737.

Richter, A., Schaff, C., Zhang, Z.W., et al. 2016. Characterization of biosynthetic pathways for the production of the volatile homoterpenes DMNT and TMTT in Zea mays. *Plant Cell* 28:2651–2665.

Rostas, M. and Eggert, K. 2008. Ontogenetic and spatio-temporal patterns of induced volatiles in *Glycine max* in the light of the optimal defence hypothesis. *Chemoecology* 18:29–38.

Rostas, M., Ton, J., Mauch-Mani, B. and Turlings, T.C.J. 2006. Fungal infection reduces herbivore-induced plant volatiles of maize but does not affect naive parasitoids. *Journal of Chemical Ecology* 32:1897–1909.

Rusman, Q., Lucas-Barbosa, D. and Poelman, E.H. 2018. Dealing with mutualists and antagonists: Specificity of plant-mediated interactions between herbivores and flower visitors, and consequences for plant fitness. *Functional Ecology* 32:1022–1035.

Rusman, Q., Lucas-Barbosa, D., Poelman, E.H. and Dicke, M. 2019a. Ecology of plastic flowers. *Trends in Plant Science* 24:725–740.

Rusman, Q., Poelman, E.H., Nowrin, F., Polder, G. and Lucas-Barbosa, D. 2019b. Floral plasticity: Herbivore-species-specific induced changes in flower traits with contrasting effects on pollinator visitation. *Plant, Cell & Environment* 42:1882–1896.

Sabelis, M.W. and Van de Baan, H.E. 1983. Location of distant spider mite colonies by phytoseiid predators: Demonstration of specific kairomones emitted by *Tetranychus urticae* and *Panonychus ulmi*. *Entomologia Experimentalis et Applicata* 33:303–314.

Sas, C., Muller, F., Kappel, C., et al. 2016. Repeated inactivation of the first committed enzyme underlies the loss of benzaldehyde emission after the selfing transition in Capsella. *Current Biology* 26:3313–3319.

Schiestl, F.P., Kirk, H., Bigler, L., Cozzolino, S. and Desurmont, G.A. 2014. Herbivory and floral signaling: Phenotypic plasticity and tradeoffs between reproduction and indirect defense. *New Phytologist* 203:257–266.

Schoonhoven, L.M., van Loon, J.J.A. and Dicke, M. 2005. Insect-Plant Biology, 2nd edition. Oxford University Press, Oxford, UK, p. 400.

Schnee, C., Kollner, T.G., Held, M., Turlings, T.C.J., Gershenzon, J. and Degenhardt, J. 2006. The products of a single maize sesquiterpene synthase form a volatile defense signal that attracts natural enemies of maize herbivores. *Proceedings of the National Academy of Sciences of the United States of America* 103:1129–1134.

Schuman, M.C. and Baldwin, I.T. 2016. The layers of plant responses to insect herbivores. *Annual Review of Entomology* 61:373–394.

Schuman, M.C., Barthel, K. and Baldwin, I.T. 2012. Herbivory-induced volatiles function as defenses increasing fitness of the native plant *Nicotiana attenuata* in nature. *eLife* 1:e00007.

Schuman, M.C., Heinzel, N., Gaquerel, E., Svatos, A. and Baldwin, I.T. 2009. Polymorphism in jasmonate signaling partially accounts for the variety of volatiles produced by *Nicotiana attenuata* plants in a native population. *New Phytologist* 183:1134–1148.

Smid, H.M. and Vet, L.E.M. 2016. The complexity of learning, memory and neural processes in an evolutionary ecological context. *Current Opinion in Insect Science* 15:61–69.

Snoeren, T.A.L., Kappers, I.F., Broekgaarden, C., Mumm, R., Dicke, M. and Bouwmeester, H.J. 2010. Natural variation in herbivore-induced volatiles in *Arabidopsis thaliana*. *Journal of Experimental Botany* 61:3041–3056.

Sobhy, I.S., Woodcock, C.M., Powers, S.J., Caulfield, J.C., Pickett, J.A. and Birkett, M.A. 2017. cis-Jasmone elicits aphid-induced stress signalling in potatoes. *Journal of Chemical Ecology* 43:39–52.

Sohrabi, R., Huh, J.H., Badieyan, S., et al. 2015. In planta variation of volatile biosynthesis: An alternative biosynthetic route to the formation of the pathogen-induced volatile homoterpene DMNT via triterpene degradation in arabidopsis roots. *Plant Cell* 27:874–890.

Soler, R., Harvey, J.A., Kamp, A.F.D., et al. 2007. Root herbivores influence the behaviour of an aboveground parasitoid through changes in plant-volatile signals. *Oikos* 116:367–376.

Song, G.C. and Ryu, C.M. 2013. Two volatile organic compounds trigger plant self-defense against a bacterial pathogen and a sucking insect in cucumber under open field conditions. *International Journal of Molecular Sciences* 14:9803–9819.

Stam, J.M., Kroes, A., Li, Y.H., et al. 2014. Plant interactions with multiple insect herbivores: From community to genes. *Annual Review of Plant Biology* 65:689–713.

Steeghs, M., Bais, H.P., de Gouw, J., et al. 2004. Proton-transfer-reaction mass spectrometry as a new tool for real time analysis of root-secreted volatile organic compounds in *Arabidopsis*. *Plant Physiology* 135:47–58.

Thaler, J.S., Stout, M.J., Karban, R., and Duffey, S.S. 2001. Jasmonate-mediated induced plant resistance affects a community of herbivores. *Ecological Entomology* 26: 312–324.

Thoen, M.P.M., Olivas, N.H.D., Kloth, K.J., et al. 2017. Genetic architecture of plant stress resistance: Multitrait genome-wide association mapping. *New Phytologist* 213:1346–1362.

Traw, M.B. and Feeny, P. 2008. Glucosinolates and trichomes track tissue value in two sympatric mustards. *Ecology* 89:763–772.

Turlings, T.C.J. and Erb, M. 2018. Tritrophic interactions mediated by herbivore-induced plant volatiles: Mechanisms, ecological relevance, and application potential. *Annual Review of Entomology* 63:433–452.

Turlings, T.C.J. and Wäckers, F.L. 2004. Recruitment of predators and parasitoids by herbivore-injured plants. In *Advances in Insect Chemical Ecology*, eds. R. T. Cardé and J. G. Millar, pp. 21–75. Cambridge, MA: Advances in Insect Chemical Ecology.

Turlings, T.C.J., Lengwiler, U.B., Bernasconi, M.L. and Wechsler, D. 1998. Timing of induced volatile emissions in maize seedlings. *Planta* 207:146–152.

Turlings, T.C.J., Tumlinson, J.H. and Lewis, W.J. 1990. Exploitation of herbivore-induced plant odors by host-seeking parasitic wasps. *Science* 250:1251–1253.

Turlings, T.C.J., Tumlinson, J.H., Heath, R.R., Proveaux, A.T. and Doolittle, R.E. 1991. Isolation and identification of allelochemicals that attract the larval parasitoid, *Cotesia marginiventris* (Cresson), to the microhabitat of one of its hosts. *Journal of Chemical Ecology* 17:2235–2251.

Turlings, T.C.J., Wäckers, F.L., Vet, L.E.M., Lewis, W.J. and Tumlinson, J.H. 1993. Learning of host-finding cues by hymenopterous parasitoids. In *Insect Learning: Ecological and Evolutionary Perspectives*, eds. D.R. Papaj and A.C. Lewis, pp. 51–78. New York: Chapman & Hall.

van Beek, T.A., Tetala, K.K.R., Koleva, I.I., et al. 2009. Recent developments in the rapid analysis of plants and tracking their bioactive constituents. *Phytochemistry Reviews* 8:387–399.

Van den Boom, C.E.M., Van Beek, T.A. and Dicke, M. 2002. Attraction of *Phytoseiulus persimilis* (Acari: Phytoseiidae) towards volatiles from various *Tetranychus urticae*-infested plant species. *Bulletin of Entomological Research* 92:539–546.

Van den Boom, C.E.M., Van Beek, T.A., Posthumus, M.A., De Groot, A. and Dicke, M. 2004. Qualitative and quantitative variation among volatile profiles induced by *Tetranychus urticae* feeding on plants from various families. *Journal of Chemical Ecology* 30:69–89.

van Wijk, M., de Bruijn, P.J.A. and Sabelis, M.W. 2011. Complex odor from plants under attack: Herbivore's enemies react to the whole, not its parts. *Plos One* 6(7):e21742.

Vet, L.E.M. and Dicke, M. 1992. Ecology of infochemical use by natural enemies in a tritrophic context. *Annual Review of Entomology* 37:141–172.

Warren, R.L., Keeling, C.I., Yuen, M.M.S., et al. 2015. Improved white spruce (*Picea glauca*) genome assemblies and annotation of large gene families of conifer terpenoid and phenolic defense metabolism. *Plant Journal* 83:189–212.

Wegener, R., Schulz, S., Meiners, T., Hadwich, K. and Hilker, M. 2001. Analysis of volatiles induced by oviposition of elm leaf beetle *Xanthogaleruca luteola* on *Ulmus minor*. *Journal of Chemical Ecology* 27:499–515.

Wilson, A.D. 2013. Diverse applications of electronic-nose technologies in agriculture and forestry. *Sensors* 13:2295–2348.

Wurtzel, E.T. and Kutchan, T.M. 2016. Plant metabolism, the diverse chemistry set of the future. *Science* 353:1232–1236.

Zhang, P.J., Zheng, S.J., van Loon, J.J.A., et al. 2009. Whiteflies interfere with indirect plant defense against spider mites in Lima bean. *Proceedings of the National Academy of Sciences of the United States of America* 106:21202–21207.

Zhou, W.W., Kuegler, A., McGale, E., et al. 2017. Tissue-specific emission of (*E*)-alpha-bergamotene helps resolve the dilemma when pollinators are also herbivores. *Current Biology* 27:1336–1341.

Zhu, F., Broekgaarden, C., Weldegergis, B.T., et al. 2015. Parasitism overrides herbivore identity allowing hyperparasitoids to locate their parasitoid host using herbivore-induced plant volatiles. *Molecular Ecology* 24:2886–2899.

Zhu, F., Cusumano, A., Bloem, J., et al. 2018. Symbiotic polydnavirus and venom reveal parasitoid to its hyperparasitoids. *Proceedings of the National Academy of Sciences of the United States of America* 115:5205–5210.

18 Belowground Plant Volatiles
Plant–Plant, Plant–Herbivore and Plant–Microbial Interactions

Yifan Jiang, Dorothea Tholl, and Feng Chen

CONTENTS

18.1 INTRODUCTION

One distinction of land plants compared to their aquatic relatives is the presence of roots in vascular plants and rhizoids in bryophytes, which provide anchorage to the plant and enable the plant to absorb water and micronutrients from the soil. The rhizosphere, which is defined as the narrow region of soil that surrounds roots/rhizoids (Estabrook and Yoder, 1998), is enormously rich in nonplant species, including bacteria, fungi, nematodes, protozoa and arthropods (Wall and Moore, 1999). Deciphering the composition of the rhizosphere community associated with various plants is a challenge being taken up by many investigators. The interactions between roots/rhizoids and other soil organisms in the rhizosphere can influence a number of ecosystem processes, including net primary production (Setala and Huhta, 1991). Therefore, it is a question of central importance to ask how plant roots/rhizodes and other soil organisms interact.

The interactions between plant roots and soil organisms are complex. Unlike aboveground plant interactions, which can be relatively easily observed in natural settings, the hidden nature of roots makes it challenging to investigate their interactions with organisms in the belowground environment. Nonetheless, it has become increasingly evident that root exudates play critical roles in defining specific belowground interactions. Root exudates are broadly defined as metabolites that are produced by roots and rhizoids and secreted into the rhizosphere (Chapter 11; Bais et al., 2006). Diverse in chemical classes, root exudates include amino acids, sugars, organic acids, polysaccharides and proteins of high molecular mass (Roshchina and Roshchina, 1993; Bais et al., 2006). Some of these metabolites are part of plant specialized metabolism, unique to specific lineages of plants. The functions of some constituents of root exudates have been studied for mediating rhizosphere interactions, mainly in vascular plants (Schumann et al., 2018).

Only relatively recently has it been recognized that roots also release volatile compounds. Much progress has been made in identifying root volatiles and characterizing their biosynthesis,

FIGURE 18.1 Schematic of interactions between plant roots and other organisms mediated by root volatiles. Based on the mechanism of their biosynthesis, root volatiles are divided into three categories: those that are constitutively made (green dots), those that are induced by herbivory (red dots) and those induced by microbial association or inter-plant priming (yellow dots). Root volatiles may serve as attractants (arrow) or repellent/chemical defenses (T sign) to other organisms. Solid arrows and labeled arrows depict associations that are beneficial and detrimental to roots, respectively. Based on the organisms with which plant roots interact, three types of interactions mediated by root volatiles are depicted: root–root, root–herbivore and root–microbe. Volatile-mediated root–root interactions include self-recognition, allelopathy (inhibiting seed germination and seedling growth) and inter-plant priming. Root–herbivore interactions include direct defense, indirect defense and cues for herbivore foraging. Root–microbe interactions are divided into two types: interactions with beneficial microbes and interactions with pathogenic microbes. Question marks indicate the root volatiles-mediated interactions for which solid evidence is still needed.

detailed in Chapters 1 and 11. Some progress has also been made in our understanding of the functions of root volatiles in mediating various types of interactions between roots with other soil organisms (summarized in Figure 18.1 and described in detail in the next five sections). In this chapter, we will describe four types of such volatile-mediated interactions: (1) belowground plant–plant interactions; (2) belowground plant–insect interactions; (3) root-beneficial microbe interactions and (4) plant defense against microbial pathogens. We also briefly describe the functions of volatiles from belowground structures of geophytes (see Section 18.6), particularly those with modified stems.

18.2 BELOWGROUND PLANT–PLANT INTERACTIONS MEDIATED BY PLANT VOLATILES

Neighboring plants share resources both aboveground (e.g., sunlight) and belowground (e.g., water and micronutrients). When resources become limited, competition between plants arises. To avoid intraspecific competition, it is important for roots to have the ability to detect neighbors and differentiate self, kin, and nonkin roots (Gadagkar, 1985). Kin-recognition ability has been demonstrated with a number of plants such as pea (Falik et al., 2003), soybean (Gersani et al., 2001) and rice (Fang et al., 2013). Once recognition has occurred, the roots may exhibit different behaviors (Mahall and Callaway, 1991; Gersani et al., 2001). A plant may avoid self-competition or the roots of a superior competitor. On the other hand, a plant may enhance its own root development and inhibit the development of the roots of a nonself-competitor. Among the various mechanisms proposed to explain how plant roots identify neighbors and subsequently compete is the involvement of root exudates (Faget et al., 2013). Here, we focus our discussions on root volatiles.

Allelopathy is defined as the inhibition of growth of one plant by another plant due to the release of chemicals into the shared environment. There are many examples in which belowground volatiles function as allelopathic agents. Starting with the first discovery by Muller and Muller (1964), many studies have demonstrated the allelopathic effect of root-derived volatiles on seed germination (White, 1991; Viles and Reese, 1996; Paavolainen et al., 1998; Ens et al., 2009; Jassbi et al., 2010; Araniti et al., 2016). For example, a blend of unidentified root volatiles of *Echinacea angustifolia* showed allelopathic effects on seed germination of *Lactuca sativa*, switchgrass (*Panicum virgatum*), and Prairie Drop seed (*Sporobolus heterolepis*) (Viles and Reese, 1996). In another study, a blend of sesquiterpenes-dominant volatiles extracted from the roots of bitou bush (*Chrysanthemoides monilifera* ssp. *rotundata*) exhibited inhibitory effects on seed germination of native sedge (*Isolepis nodosa*) (Ens et al., 2009).

Besides inhibiting seed germination, some root-derived volatiles have been demonstrated to negatively affect seedling growth. For example, the blend of monoterpenes and sesquiterpenes from the roots of sagebrush (*Artemisia tridentate*) exhibited phytotoxic effects on the seedlings of naturally co-occurring tobacco (*Nicotiana attenuate*) (Jassbi et al., 2010). Many nonterpene root volatiles have also been shown to be allelopathic. As an example, 2,4-di-tert-butyl phenol released from *Eucalyptus grandis* roots showed strong inhibitory effects on seedling growth of *Vigna radiata*, *Raphanus sativus* and *Lactuca sativa* (Wang et al., 2015).

There have been some studies attempting to uncover the cellular mechanisms underlying volatile-caused allelopathy. The treatment of *Arabidopsis* seedling with an allelopathic agent, the monoterpene aldehyde citral, disrupted the development of microtubules (Chaimovitsh et al., 2010). Araniti et al. (2016) discovered that treatment of *Arabidopsis* roots with farnesene leads to a loss of gravitropism, which is related to microtubule disorganization. In another study, it was found that the inhibition of the growth of tomato seedlings by diallyl disulfide, a sulfur-containing volatile, is related to affecting cell division, phytohormone balance, and the expression of expansin genes (Cheng et al., 2016).

Some plants on the receiving end may use such chemical messengers to their advantage. The witchweeds (*Striga* spp.) are serious parasitic weeds and can cause severe yield losses of crops (Rodenburg and Bastiaans, 2011). The seeds of striga rely on chemicals released from the roots of host plants to regulate their germination (Yoder and Scholes, 2010). It has been demonstrated that parasitic seedlings use aboveground volatiles of host plants for host searching (Runyon et al., 2006). It will be interesting to determine whether root volatiles of host plants also serve as a chemical messenger in host plant recognition. Such function has been demonstrated for nonvolatile belowground compounds such as strigolactones (Yoneyama et al., 2010).

18.3 BELOWGROUND PLANT–HERBIVORE INTERACTIONS MEDIATED BY PLANT VOLATILES

Among soil arthropods are mites, springtails, and immature stages of insects including *Coleoptera* and *Diptera* (Wall and Moore, 1999). Some of these arthropods are herbivores while others are predators of herbivorous arthropods. Besides arthropods, an incredibly diverse array of nematodes live in the soil. Some species of nematodes feed on the roots, while others feed on other nematodes or herbivorous arthropods (Yeates, 1979). The roles of volatiles in mediating plant–herbivore interactions are well studied in aboveground parts of the plants, ranging from direct and/or indirect plant defenses against herbivores to chemical cues exploited by foraging herbivores (Chapter 17). Although much less studied, root volatiles have similar functions.

Some root volatiles are exploited by herbivores as foraging signals. It was found that different carrot cultivars have different levels of resistance to *Psila rosae*. While the mechanism of resistance was not fully determined, the roots of resistant cultivars emitted five times less volatiles than the susceptible cultivar (Guerin and Ryan, 2011). Since these volatile have been shown to positively influence host-searching behavior by the larvae, the defense strategy of carrot roots against *P. rosae* is to suppress the release of volatiles.

Some root volatiles are toxic to herbivores, thus functioning as direct defense against herbivory. In addition, some nonvolatile root compounds may be converted during herbivory to toxic volatile breakdown products (Hopkins et al., 2009). For example, sulfur-containing volatiles in *Brassicaceae* plants, which are the enzymatic breakdown products of glucosinolates (GSLs), showed direct defensive activities against belowground herbivores (Halkier and Gershenzon, 2006; Xu et al., 2016). In some studies, root volatiles were found to function as antifeedants. For example, studies in *Arabidopsis* demonstrated anti-feedant activities of constitutively produced semi-volatile diterpenes named rhizathalenes against opportunistic root herbivores such as *Bradysia* larvae (Vaughan et al., 2013).

To compensate for the ineffectiveness of direct defenses toward rapid chemical adaption of specialist herbivores, roots of some plants greatly increase the production and emission of volatiles upon herbivory. As in aboveground plant tissues, blends of volatiles from roots induced by the infestation of herbivorous arthropods or nematodes often consist of fatty acid derivatives, benzenoids and terpenoids (Fineschi and Loreto, 2012). Root volatiles induced by herbivores may serve as chemical cues to attract arthropod predators as natural enemies of herbivores, i.e., to function as indirect defense. As an example, dimethyl trisulfide (DMTS), a sulfur-containing volatile released from *Brassica nigra* roots attacked by cabbage root fly larvae (*Delia rdicum*), was found to be an attractant of soil-dwelling beetle predators of root fly larvae (Ferry et al., 2007) or sugarcane spittlebug (*Mahanarva fimbriolata*) at the third trophic level (Tonelli et al., 2016).

Some nematodes are natural enemies of belowground herbivores. They may also be attracted by herbivore-induced root volatiles for indirect defense (van Tol et al., 2001). It was shown that sesquiterpenes such as (E)-β-caryophyllene, α-humulene, α-copaene and caryophyllene-oxide emitted from maize roots in response to infestation by larvae of the herbivore Western corn root worm (*Diabrotica virgifera*) (Rasmann et al., 2005; Rasmann and Turlings, 2007; Hiltpold and Turlings, 2008; Hiltpold et al., 2011; Robert et al., 2012), were able to recruit nematodes (*Heterorhabditis megidis* and *H. bacteriophora*) of *D. virgifera* as an indirect defense (Hiltpold et al., 2010; Hiltpold and Turlings, 2012). Moreover, after the infestation by the weevil *Otiorhynchus sulcatu*, unidentified volatiles were induced from roots of *Thujaoccidentalis* and strawberry roots and served as attractants of the parasitic nematode *H. megidis* (Boff et al., 2001). In another example, the terpenoid pregeijerene (1,5-dimethylcyclodeca-1,5,7-triene) was identified as the main volatile released by the roots of *Citrus* "Swingle citrumelo" rootstock infested by *D. abbreviatus* larvae (Ali et al., 2010, 2011), and shown to attract entomopathogenic nematodes (*Steinernema diaprepesi*, *S. riobrave*, and *H. indica*), leading to higher mortality rates of insect larvae in the field (Ali et al., 2012, 2013).

Root volatiles have also recently been shown to positively affect neighboring plants. Herbivore-induced plant volatiles that benefit neighboring plants were first described aboveground (Engelberth et al., 2004). In early rhizosphere studies, volatiles from the roots of aphid-infested plants were found to cause adjacent intact plants to become more attractive to parasitoids (Chamberlain et al., 2001). Similarly, Dicke and Dijkman (2001) discovered that herbivore-damaged plants may prime the undamaged neighboring plants through root volatiles in the soil. On the other hand, root volatiles may affect the growth of neighboring plants to make them more susceptible to herbivores (Huang et al., 2018). In this study, *Centaurea stoebe*, the roots of which constitutively release high amounts of sesquiterpenes into the rhizosphere, was used as a volatile emitter and *Taraxacum officinale* plants were tested as the receivers. When *T. officinale* roots were analyzed after being exposed to root volatiles from *C. stoebe*, their defense compounds were not changed but levels of carbohydrates and total protein were increased. In another study, the exposure to root volatiles increased the growth of cockchafer larvae (*Melolontha melolontha*) on *T. officinale* plants, suggesting that *C. stoebe* plants achieve a defense against herbivores by making their neighbors more susceptible through root–root interactions.

Besides the role for the defense as discussed above, numerous studies have demonstrated that root-derived volatiles are exploited by herbivorous arthropods or nematodes for feeding and infestation (Wolfson, 1987; Wilschut et al., 2017). For example, an early report indicated that volatile blends of perennial ryegrass roots can attract the larvae of *Costelytra zealandica* (Sutherland and Hillier, 1974). To identify individual volatile compounds involved in attraction, Palma et al. (2012) found that females of the clover root borer, *Hylastinus obscurus*, are particularly attracted to the C_6 volatile (*E*)-2 hexenal, which is emitted among other volatile compounds from roots of red clover plants (*Trifolium pratense* L.). Terpenoids are another class of important and diverse attractants of belowground herbivores. For example, β-pinene has been recognized as a volatile attractant for root-feeding larvae of *Porphyrophora sophorae*, an oligophagous pest of the Chinese liquorice, *Glycyrrhiza uralensis* (Liu et al., 2016). In another example, 1,8-cineole emitted from damaged roots of oak can attract larvae of the cockchafer *Melolontha hippocastani* for feeding (Weissteiner and Schütz, 2006; Weissteiner et al., 2012).

Interestingly, some root volatiles exhibit concentration-dependent effects on root–herbivore interactions. For example, the volatile from red clover may function as either attractant or deterrent of the herbivore *H. obsucurus* depending on its concentration (Tapia et al., 2007). Similarly, the concentrations of volatiles from conifer roots were found to be crucial for the orienting behavior of *Hylobius abietis* (Nordlander et al., 1986).

18.4 BELOWGROUND ROOT VOLATILES IN ROOT-BENEFICIAL MICROBE INTERACTIONS

Microorganisms in the rhizosphere are important for nutrient cycling and consequently primary productivity of plants (Tsunoda and Van Dam, 2017). Root-associated microbial communities show less diversity than those in the soil (Edwards et al., 2015), suggesting that the composition of root-associated communities is largely influenced by root exudates. Beneficial microorganisms are those that have positive effects on plant performance by helping plants with water and nutrient uptake, promoting growth by affecting plant hormones balance and providing resistance against herbivores and pathogens. Three types of beneficial microorganisms are best known: mycorrhizal fungi, nitrogen-fixing rhizobia and growth-promoting rhizobacteria.

Chemical communications between roots and beneficial microorganisms are important for establishing such mutualistic interactions. A number of nonvolatile metabolites from root exudates have been demonstrated to function as chemical messengers, such as strigolactones in host-finding of mycorrhizal fungi (Bouwmeester et al., 2007) and flavonoids in rhizobia infection and nitrogen-fixing nodule formation (Peters et al., 1986). In contrast, it is not well understood whether root volatiles function as signaling molecules in the initiation and establishment of such beneficial interactions. Nonetheless, volatile secondary metabolites released from plant roots could provide carbon to the ambient microbial community and promote the growth of specific microbiomes in the soil, especially in the case of bacteria (Zak et al., 1994; Owen et al., 2007; Gramms and Bergmann, 2008). For example, bacterial endophytes of the *Pseudomonadaceae* and *Enterobacteriaceae* families were shown to use sesquiterpenes in Vetiver (*Vetiveria zizanioides*) roots as sole carbon source (Del Giudice et al., 2010). *Pseudomonas fluorescens* can also take advantage of the monoterpene α-pinene as carbon source for growth (Kleinheinz et al., 1999). On the other hand, association with terpene-utilizing bacteria appears to be important for root volatile production. For example, Vetiver grass roots, when grown under sterile conditions, do not accumulate terpenoids (Del Giudice et al., 2010).

Besides their possible role as direct carbon source, root-derived volatiles may play a role in long-distance plant-microbe interactions. Root volatiles can function as chemoattractants of beneficial bacteria. For example, the root-derived volatiles released upon fungal infection can attract specific bacteria with antifungal properties over a long distance (Schulz-Bohm et al., 2018). A recent study demonstrated that eudesmane-2,11-diol, a sesquiterpenoid, can be detected in root exudates of maize in response to infection with the pathogenic fungus *Fusarium verticillioides* (Liang et al., 2018). Besides functioning as a chemical defense, it remains to be determined whether this compound has other roles in the rhizosphere, such as attracting beneficial microbes. These ongoing studies suggest a mutualistic interaction in which the plant provides volatile-based carbon source and at the same time benefits from the stimulated production of these compounds by using them for defense against pathogens.

There is ample evidence that root inoculation with beneficial microorganisms can impact volatiles aboveground, such as in Medicago (Leitner et al., 2010) and *Artemisia annua* (Rapparini et al., 2008), indicating root-to-shoot signaling. The impact of beneficial microbes on root volatiles is just beginning to emerge. Inoculation of sorghum roots with mycorrhizal fungi dramatically changed the profiles of root volatiles (Sun and Tang, 2013). Several other studies explored the ecological impact of such changed root volatiles. For example, the inoculation of corn roots with the beneficial rhizobacterium *Azospirillum brasilense* was found to enhance the emission of the volatile sesquiterpene (*E*)-β-caryophyllene (Santos et al., 2014). Equally interesting, the root herbivore *Diabrotica speciosa* showed no preference for inoculated plants, probably due to the higher emission of (*E*)-β-caryophyllene. In another study with corn, it was found that roots inoculated with growth-promoting *Pseudomonas* bacteria released elevated levels of (*E*)-β-caryophyllene upon herbivore feeding (Chiriboga et al., 2018), which may enhance the attractiveness of the roots to entomopathogenic nematodes. The latter two studies provide mechanistic explanations for the enhanced resistance of roots to herbivores when inoculated with beneficial microorganisms.

18.5   BELOWGROUND PLANT VOLATILES IN DEFENSE AGAINST MICROBIAL PATHOGENS

Volatile compounds often accumulate in the form of essential oils in plant roots and rhizomes (Penuelas et al., 2014). Many of these oils and their constituents such as monoterpenes and sesquiterpenes, phenolic and benzenoid compounds have been associated with antibacterial and antifungal properties based on *in vitro* assays (e.g., Al-Ja'fari et al., 2011; Joshi, 2013; Fraternale et al., 2014; Singh et al., 2016). In other cases, volatile compounds are released from roots without significant storage. For instance, roots of *Arabidopsis* emit a variety of different semi-volatile diterpenes, some of which have been found to mitigate fungal growth *in vitro* (Wang et al., 2016). In particular, macrocyclic dolabellane-type diterpenes retard the growth of the root-rot pathogen *Pythium irregulare*. Although the concentrations used in these bioassays exceeded those in planta (Wang et al., 2016), it is possible that low levels of these metabolites contribute synergistically or in additive effects to defenses against pathogens in the rhizosphere.

Besides terpenes and other essential oil constituents, sulfur-containing volatile compounds have been shown to have potent antimicrobial activities. For example, benzyl isothiocyanate, a major glucosinolate breakdown product in roots of the mustard tree *Salvadora persica* exhibits strong bactericidal effects (Sofrata et al., 2011). In another example, exudates of Chinese leek (*Allium tuberosum*) were found to contain sulfides such as dimethyl disulfide and dimethyl trisulfide, which inhibit the germination and growth of the fungal pathogen *Fusarium oxysporum* (Zuo et al., 2015). Taking advantage of this antifungal activity, intercropping of *A. tuberosum* with banana allowed for an efficient treatment of *Fusarium* wilt in banana plantations (Zuo et al., 2015).

Volatile compounds are also released from roots in induced responses to pathogen infection. Roots of chickpea release 1-penten-3-ol upon infection with the ascochyta blight fungus *Didymella rabiei* (Cruz et al., 2012). This alcohol reduces *in vitro* growth of *F. graminearum*, a fungal pathogen of wheat, by over 50% and may, among other volatiles, help suppress *Fusarium* disease of wheat in chickpea-wheat rotation systems (Cruz et al., 2012). In *Arabidopsis*, emission of volatiles is elicited in root tissue by treatment with several pathogens. Inoculation of roots with *Pseudomonas syringae* DC3000 causes a rapid release of 1,8-cineole, a common monoterpene in plant roots and rhizomes that is known to have antimicrobial effects (Steeghs et al., 2004; Vilela et al., 2009). By contrast, treatment with the root-rot oomycete pathogen *Pythium irregulare* elicits the emission of the volatile homoterpene (*E*)-4,8-dimethyl-1,3,7-nonatriene (DMNT) at an early stage of infection (Sohrabi et al., 2015). DMNT is synthesized in an unusual cleavage reaction of a triterpene alcohol precursor, and it significantly reduces *Pythium* oospore germination *in vitro* at picomolar concentrations (Sohrabi et al., 2015). Mutants impaired in the formation of DMNT or its nonvolatile precursor were found to exhibit enhanced susceptibility upon soil inoculation with the pathogen (Sohrabi et al., 2015).

In maize, treatment with different fungal pathogens leads to the accumulation of the sesquiterpene volatile β-selinene (Ding et al., 2017). This compound is further converted to the corresponding nonvolatile phytoalexin derivative β-costic acid, which promotes fungal pathogen resistance (Ding et al., 2017). Similarly, upon infection with *Fusarium verticillioides*, maize roots accumulate phytoalexins called zealexins, which are derived from the volatile bicyclic sesquiterpene (*S*)-β-macrocarpene, as well as dolabralexins, which are derivatives of the diterpene hydrocarbon dolabradiene (Huffaker et al., 2011; Vaughan et al., 2015; Mafu et al., 2018). These findings indicate that nonoxygenated volatile or semi-volatile terpenes may often function as intermediates on the way to nonvolatile metabolites that exert more potent bioactivities against microbial pathogens than their volatile precursors.

Taken together, plant roots accumulate and release a variety of different volatile compounds that, in addition to downstream derivatives, are implicated in chemical defenses against invading pathogens. The modes of action of volatile metabolites occur at different levels as summarized by Junker and Tholl (2013). A primary mechanism of interference with microbial cells is to disturb membrane integrity and increase ion permeability (Radulovic et al., 2013). Other activities include alterations of membrane protein composition (e.g., Horvath et al., 2009), modifications in lipid metabolism (e.g., de Carvalho and da Fonseca, 2007), or disturbance of protein function (e.g., Luciano and Holley, 2009). Despite this knowledge, the bioactivity of many compounds is still largely evaluated by *in vitro* assays, and limited evidence is available *in situ* about the extent and the stage at which soil-borne pathogens are exposed to constitutive pools of volatiles and/or the release of volatile compounds upon elicitation and cell damage. More *in planta* studies using biosynthetic mutants paired with better analytical abilities to measure or monitor volatiles in the root environment will be necessary to gain better insight into the effect of these compounds on pathogens and other microbes inhabiting the rhizosphere and endosphere.

In this context, it should be noted that plant-derived volatiles most likely are complemented by volatile compounds derived from root-associated microbes. An increasing number of studies clearly indicate that volatiles derived from rhizobacteria or soil-borne fungi have antagonistic activities against plant pathogens and may contribute to disease suppression and plant health

(Mendez-Bravo et al., 2018; Werner et al., 2016). Other approaches relying on mutants of both the host plant and the microbes may allow disentangling the role of diverse volatile compounds in the root environment to combat soil-borne pathogens.

18.6   BELOWGROUND VOLATILES FROM MODIFIED STEMS OF GEOPHYTES

Some plants have evolved belowground storage structures from modified stems or roots as a part of their life cycle. Such plants are called geophytes and their storage structures include bulbs, tubers and rhizomes (Kamenetsky and Okubo, 2013). Because of the economic significance of some geophytes, we would like to briefly consider their volatiles and their role in mediating ecological interactions. Similarly to roots and rhizoids, geophyte-derived volatiles were also identified to be involved in alle-lopathy, indirect defense at the third trophic level, and foraging of herbivorous insects (Suttle et al., 2016; Laznik et al., 2016; Das et al., 2007; Karlsson et al., 2009). Diverse volatiles have been char-acterized from rhizomes and bulbs from a wide range of species (Iijima et al., 2014; Süzgeç-Selçuk et al., 2017; Plaingam et al., 2017; El-Hawaz et al., 2018; see also above), but potato tubers, which are derived from stems, are the best studied geophyte for the function of belowground volatiles.

Continuous exposure of potato tubers to cineole resulted in an allelopathic effect on sprout growth, suggesting an inhibitory effect of cineole on gibberellin synthesis or action (Suttle et al., 2016). Laznik et al. (2016) showed that four commercial entomopathogenic nematodes species (*Steinernema feltiae*, *S. carpocapsae*, *S. kraussei*, and *H. bacteriophora*) can be attracted by seven compounds (decanal, nonanal, octanal, undecane, 6-methyl-5-hepten-2-one, and 1,2,4-trimethylbenzene) released from insect-damaged (*Melolontha hippocastani*) potato tubers, suggesting that herbivore-induced volatiles play a role of indirect defense to attract entomopathogenic nematodes as natural enemies. Among these volatiles, decanal was identified as the most effective attractant for *H. bacteriophora* and *S. kraussei* (Laznik et al., 2016). Electroantennograms recorded that males and females of the potato tuber moth, (*Phthorimaea operculella*) have different prefer-ences to different classes of plant volatile compounds released from potato tuber, suggesting a role of volatiles as cues to directly attract the oligophagous pest (Das et al., 2007). Moreover, the head-space of potato tubers was studied with regard to volatiles that mediate host-finding and oviposition in the Guatemalan moth (*Tecia solanivora*). Fresh potato tubers were found to emit only trace amounts of a few sesquiterpenes. Female and male moths were attracted to methyl phenylacetate, which may accordingly contribute to female attraction to tuber-bearing potato plants in the field. Electroantennogram tests further revealed that the antennae of the moth can respond to methyl phenylacetate (Karlsson et al., 2009).

18.7   CONCLUSION

Plant roots synthesize and release mixtures of volatile organic compounds into the rhizosphere. In this chapter, we provided an overview on the various root-soil organismal interactions mediated by root volatiles. Some of the volatiles might be exploited by root-feeding herbivores, pathogenic microbes and parasitic plants for host searching. More importantly, these volatiles serve the benefits of the roots in various ways. They can function as toxic metabolites in direct defense. They may also function as signaling molecules to attract carnivorous arthropods/entomopathogenic nematodes as natural enemies of rood-feeding herbivores in indirect defense. Some volatiles function as signaling molecules to attract beneficial microbes for the establishment of mutualistic associations, which in turn may alter the production of root volatiles and their ability in defense (Figure 18.1).

Although significant progress has been made in our understanding of the ecological roles of root volatiles, much remains to be learned in this relatively young and rapidly developing field. Further advancements will be required in several areas. One would be the improvement of the detection of root volatiles as referred to in Chapter 1. Another area would be the improvement of bioassays. The development of new techniques, including high-throughput phenotyping of roots and accurate measurements of the diversity of soil organisms in the rhizosphere, will be important. As discussed in this chapter, much of our knowledge about belowground plant volatiles was obtained from vascular plants with true roots. With the continued advancement in bryophyte research, our understanding of the types and functions of volatiles from rhizoids will certainly grow (Chen et al., 2018), especially with some of the model bryophytes such as the liverwort *Marchantia polymorpha* (Bowman et al., 2017) and the moss *Physcomitrella patens* (Rensing et al., 2007).

ACKNOWLEDGMENTS

We thank Chi Zhang for his assistance with the preparation of Figure 18.1.

REFERENCES

Ali, J.G., Alborn, H.T., and Stelinski, L.L. 2010. Subterranean herbivore induced volatiles released by Citrus roots upon feeding by *Diaprepes abbreviatus* recruit entomopathogenic nematodes. *J Chem Ecol* 36:361–368.

Ali, J.G., Alborn, H.T., and Stelinski, L.L. 2011. Constitutive and induced subterranean plant volatiles attract both entomopathogenic and plant parasitic nematodes. *J Ecol* 99:26–35.

Ali, J.G., Alborn, H.T., Campos-Herrera, R., Kaplan, F., Duncan, L.W., Rodriguez-Saona, C., Koppenhofer, A.M., and Stelinski, L.L. 2012. Subterranean, herbivore-induced plant volatile increases biological control activity of multiple beneficial nematode species in distinct habitats. *PLoS One* 7:e38146.

Ali, J.G., Campos-Herrera, R., Alborn, H.T., Duncan, L.W., and Stelinski, L.L. 2013. Sending mixed messages: A trophic cascade produced by a belowground herbivore-induced cue. *J Chem Ecol* 3:1140–1147.

Al-Ja'fari, A.H., Vila, R., Freixa, B., Tomi, F., Casanova, J., Costa, J., and Canigueral, S. 2011. Composition and antifungal activity of the essential oil from the rhizome and roots of *Ferula hermonis*. *Phytochemistry* 72:1406–1413.

Araniti, F., Graña, E., Krasuska, U., Bogatek, R., Reigosa, M.J., Abenavoli, M.R., and Sánchez-Moreiras, A.M. 2016. Loss of gravitropism in farnesene-treated *Arabidopsis* is due to microtubule malformations related to hormonal and ROS unbalance. *PLoS One* 11:e0160202.

Bais, H.P., Weir, T.L., Perry, L.G., Gilroy, S., and Vivanco, J.M. 2006. The role of root exudates in rhizosphere interactions with plants and other organisms. *Annu Rev Plant Biol* 57:233–266.

Boff, M.I.C., Zoon, F.C., and Smits, P.H. 2001. Orientation of *Heterorhabditis megidis* to insect hosts and plant roots in a Y-tube sand olfactometer. *Entomol Exp Appl* 98:329–337.

Bouwmeester, H.J., Roux, C., Lopez-Raez, J.A., and Becard, G. 2007. Rhizosphere communication of plants, parasitic plants and AM fungi. *Trends Plant Sci* 12:224–230.

Bowman, J.L., Kohchi, T., Yamato, KT., Jenkins, J., Shu, S; Ishizaki, K., Yamaoka, S., Nishihama, R., Nakamura, Y., Berger, F et al. 2017. Insights into land plant evolution garnered from the *Marchantia polymorpha* genome. *Cell* 171:287–304.

Chaimovitsh, D., Abu-Abied, M., Belausov, E., Rubin, B., Dudai, N., and Sadot, E. 2010. Microtubules are an intracellular target of the plant terpene citral. *Plant J* 61:399–408.

Chamberlain, K., Guerrieri, E., Pennacchio, F., Petterssond, J., Picketta, J.A., Poppya, G.M., Powella, W., Wadhamsa, L.J., and Woodcocka, C.M. 2001. Can aphid induced plant signals be transmitted aerially and through the rhizosphere? *Biochem Syst Ecol* 29:1063–1074.

Chen, F., Ludwiczuk, A., Wei, G., Chen, X., Crandall-Stotler, B., and Bowman J.L. 2018. Terpenoid secondary metabolites in bryophytes: Chemical diversity, biosynthesis and biological functions. *Crit Rev Plant Sci* 37:210–231.

Cheng, F., Cheng, Z., and Meng, H. 2016. Transcriptomic insights into the allelopathic effects of the garlic allelochemical diallyl disulfide on tomato roots. *Sci Rep* 6:38902.

Chiriboga, X.M., Guo, H., Campos-Herrera, R., Röder, G., Imperiali, N., Keel, C., Maurhofer, M., and Turlings, T.C.J. 2018. Root-colonizing bacteria enhance the levels of (E)-β-caryophyllene produced by maize roots in response to rootworm feeding. *Oecologia* 187:459–468.

Cruz, A.F., Hamel, C., Yang, C., Matsubara, T., Gan, Y.T., Singh, A.K., Kuwada, K., and Ishii, T. 2012. Phytochemicals to suppress *Fusarium* head blight in wheat-chickpea rotation. *Phytochemistry* 78:72–80.

Das, P.D., Raina R., Prasad, A.R., and Sen, A. 2007. Electroantennogram responses of the potato tuber moth, *Phthorimaea operculella* (Lepidoptera; Gelichiidae) to plant volatiles. *J Biosci* 32:339–349.

de Carvalho, C.C.C.R. and da Fonseca, M.M.R. 2007. Preventing biofilm formation: Promoting cell separation with terpenes. *Fems Microbiol Ecol* 61:406–413.

Del Giudice, L., Massardo, D.R., Pontieri, P., Bertea, C.M., Mombello, D., Carata, E., Tredici, S.M., Talà, A., Mucciarelli, M., Groudeva, V.I., et al. 2008. The microbial community of Vetiver root and its involvement into essential oil biogenesis. *Environ Microbiol* 10:2824–2841.

Dicke, M. and Dijkman, H. 2001. Within-plant circulation of systemic elicitor of induced defence and release from roots of elicitor that affects neighbouring plants. *Biochem Syst Ecol* 29:1075–1087.

Ding, Y., Huffaker, A., Köllner, T.G., Weckwerth, P., Robert, C.A.M., Spencer, J.L., Lipka, A.E., and Schmelz, E.A. 2017. Selinene volatiles are essential precursors for maize defense promoting fungal pathogen persistence. *Plant Physiol* 175:1455–1468.

Edwards, K.R. 2015. Effect of nutrient additions and site hydrology on belowground production and root nutrient contents in two wet grasslands. *Ecol Eng* 84:325–335.

El-Hawaz, R.F., Grace, M.H., Janbey, A., Lila, M.A., and Adelberg, J.W. 2018. In vitro mineral nutrition of *Curcuma longa* L. affects production of volatile compounds in rhizomes after transfer to the greenhouse. *BMC Plant Biol* 18:122.

Engelberth, J., Alborn, H.T., Schmelz, E.A., and Tumlinson, J.H. 2004. Airborne signals prime plants against insect herbivore attack. *Proc Natl Acad Sci USA* 101:1781–1785.

Ens, E.J., Bremner, J.B., French, K., and Korth, J. 2009. Identification of volatile compounds released by roots of an invasive plant, bitou bush (*Chrysanthemoides monilifera* spp. *rotundata*), and their inhibition of native seedling growth. *Biol Invasions* 11:275–287.

Estabrook, E.M. and Yoder. J.I. 1998. Plant-Plant Communications: Rhizosphere signaling between parasitic angiosperms and their hosts. *Plant Physiol* 116:1–7.

Faget, M., Nagel, K.A., Walter, A., Herrera, J.M., Jahnke, S., Schurr, U., and Termperton, V.M. 2013. Root-root interactions: Extending our perspective to be more inclusive of the range of theories in ecology and agriculture using in-vivo analyses. *Ann Bot* 112:253–266.

Falik, O., Reides, P., Gersani, M., and Novoplansky, A. 2003. Self/non-self discrimination in roots. *J Ecol* 91:525–531.

Fang, S., Clark, R.T., Zheng, Y., Iyer-Pascuzzi, A., Weitz, J.S., and Kochian, L.V., Edelsbrunner. H, Liao H., and Benfey P.N. 2013. Genotypic recognition and spatial responses by rice roots. *Proc Natl Acad Sci USA* 110:2670–2675.

Ferry, A., Dugravot, S., Delattre, T., Christides, J.P., Auger, J., Bagneres, A.G., Poinsot D., and Cortesero, A.M. 2007. Identification of a widespread monomolecular odor differentially attractive to several *Delia radicum* ground-dwelling predators in the field. *J Chem Ecol* 33:2064–2077.

Fineschi, S. and Loreto, F. 2012. Leaf volatile isoprenoids: An important defensive armament in forest tree species. *iForest* 5:13–17.

Fraternale, D., Flamini, G., and Ricci, D. 2014. Essential oil composition and antimicrobial activity of *Angelica archangelica* L. (*Apiaceae*) roots. *J Med Food* 17:1043–1047.

Gadagkar, R. 1985. Kin recognition in social insects and other animals—A review of recent findings and a consideration of their relevance for the theory of kin selection. *Proc Anim Sci* 94: 587–621.

Gersani, M., Brown, J.S., O'Brien, E.E., and Abramsky, M.Z. 2001. Tragedy of the commons as a result of root competition. *J Ecol* 89:660–669.

Gramms, G. and Bergmann, H. 2008. Role of plants in the vegetative and reproductive growth of saprobic basidiomycetous ground fungi. *Microb Ecol* 56:660–670.

Guerin, P.M. and Ryan, M.F. 2011. Relationship between root volatiles of some carrot cultivars and their resistance to the carrot fly, *Psila rosae*. *Entomol Exp Appl* 36:217–224.

Halkier, B.A. and Gershenzon, J. 2006. Biology and biochemistry of glucosinolates. *Annu Rev Plant Biol* 57:303–333.

Hiltpold I., Toepfer S., Kuhlmann U., and Turlings T.C.J. 2010. How maize root volatiles affect the efficacy of entomopathogenic nematodes in controlling the western corn rootworm? *Chemoecology* 20:155–162.

Hiltpold, I. and Turlings, T.C.J. 2008. Belowground chemical signaling in maize: When simplicity rhymes with efficiency. *J Chem Ecol* 34:628–635.

Hiltpold, I. and Turlings, T.C.J. 2012. Manipulation of chemically mediated interactions in agricultural soils to enhance the control of crop pests and to improve crop yield. *J Chem Ecol* 38: 641–650.

Hiltpold, I., Erb, M., Robert, C.A.M., and Turlings, T.C.J. 2011. Systemic root signalling in a belowground, volatile-mediated tritrophic interaction. *Plant Cell Environ* 34:1267–1275.

Hopkins, R.J., van Dam, N.M., and van Loon, J.J.A. 2009. Role of glucosinolates in insect-plant relationships and multitrophic interactions. *Ann Rev Entomol* 54:57–83.

Horvath, G., Kovacs, K., Kocsis, B., and Kustos, I. 2009. Effect of thyme (*Thymus vulgaris* L.) essential oil and its main constituents on the outer membrane protein composition of erwinia strains studied with microfluid chip technology. *Chromatographia* 70:1645–1650.

Huang, W., Zwimpfer, E., Hervé, M.R., Bont, Z., and Erb, M. 2018. Neighborhood effects determine plant-herbivore interactions below ground. *J Ecol* 106:347–356

Huffaker, A., Kaplan, F., Vaughan, M.M., Dafoe, N.J., Ni, X., Rocca, J.R., Alborn, H.T., Teal, P.E.A., and Schmelz, E.A. 2011. Novel acidic sesquiterpenoids constitute a dominant class of pathogen-induced phytoalexins in maize. *Plant Physiol* 156:2082–2097.

Iijima, Y., Koeduka, T., Suzuki, H., and Kubota, K. 2014. Biosynthesis of geranial, a potent aroma compound in ginger rhizome (*Zingiber officinale*): Molecular cloning and characterization of geraniol dehydrogenase. *Plant Biotechnol* 31:525–534.

Jassbi, A.R., Zamanizadehnajari, S., and Baldwin, I.T. 2010. Phytotoxic volatiles in the roots and shoots of *Artemisia tridentata* as detected by headspace solid-phase microextraction and gas chromatographic-mass spectrometry analysis. *J Chem Ecol* 36:1398–1407.

Joshi, R.K. 2013. Chemical constituents and antibacterial property of the essential oil of the roots of *Cyathocline purpurea*. *J Ethnopharmacol* 145:621–625.

Junker, R.R. and Tholl, D. 2013. Volatile organic compound mediated interactions at the plant-microbe interface. *J Chem Ecol* 39:810–825.

Kamenetsky, R. and Okubo, H. 2013. Ornamental geophytes: From basic science to sustainable production; CRC Press: Boca Raton, FL, USA, p. 578.

Karlsson, M.F., Birgersson, G., Prado, A.M.C., Bosa, F., Bengtsson M., and Witzgall A.P. 2009. Plant odor analysis of potato: Response of guatemalan moth to above- and belowground potato volatiles. *J Agric Food Chem* 57:5903–5909.

Kleinheinz, G.T., Bagley, S.T., St John, W.P., Rughani J.R., and McGinnis, G.D. 1999. Characterization of alpha-pinene-degrading microorganisms and application to a bench-scale biofiltration system for VOC degradation. *Arch Environ Con Tox* 37:151–157.

Laznik, Ž. and Trdan, S. 2016. Attraction behaviors of entomopathogenic nematodes (*Steinernematidae* and *Heterorhabditidae*) to synthetic volatiles emitted by insect damaged potato tubers. *J Chem Ecol* 42:314–322.

Leitner, M., Kaiser, R., Hause, B., Boland, W., and Mithöfer A. 2010. Does mycorrhization influence herbivore-induced volatile emission in *Medicago truncatula*. *Mycorrhiza* 20:89–101.

Liang, J., Liu, J., Brown, R., Jia, M., Zhou, K., Peters, R.J, and Wang, Q. 2018. Direct production of di-hydroxylated sesquiterpenoids by a maize terpene synthase. *Plant J* 94:847–856.

Liu, X.F., Chen, H.H., Li, J.K., Zhang, R., Turlings, T.C.J., and Chen, L. 2016. Volatiles released by Chinese liquorice roots mediate host location behavior by neonate *Porphyrophora sophorae* (Hemiptera: Margarodidae). *Pest Manag Sci* 72:1959–1964.

Luciano, F.B. and Holley R.A. 2009. Enzymatic inhibition by allyl isothiocyanate and factors affecting its antimicrobial action against *Escherichia coli* O157:H7. *Int J Food Microbiol* 131: 240–245.

Mafu, S., Ding, Y, Murphy, K.M., Yaacoobi, O., Addison, J.B., Wang, Q., Shen, Z., Briggs, S.P., Bohlmann, J., Castro-Falcon, G et al. 2018. Discovery, biosynthesis and stress-related accumulation of dolabradiene-derived defenses in maize. *Plant Physiol* 176:2677.

Mahall, B.E. and Callaway, R.M. 1991. Root communication among desert shrubs. *Proc Natl Acad Sci USA* 88:874–876.

Mde, F. 2010. Do non-social insects get the (kin) recognition they deserve? *Ecol Entomol* 23:223–227.

Mendez-Bravo, A., Marian, C.M.E., Guevara-Avendano, E., Ceballos-Luna, O., Rodriguez-Haas, B., Kiel-Martinez, A.L., Hernandez-Cristobal, O., Guerrero-Analco, J.A., and Reverchon, F. 2018. Plant growth-promoting rhizobacteria associated with avocado display antagonistic activity against *Phytophthora cinnamomi* through volatile emissions. *Plos One* 13:e0194665.

Muller, W.H. and Muller, C.H. 1964. Volatile growth inhibitors produced by *Salvia* species. *Bull Torrey Bot Club* 91:327–330.

Nordlander, G., Eidmann, H.H., Jacobsson, U., Nordenhem, H., and Sjödin, K. 1986. Orientation of the pine weevil *Hylobius abietis* to underground sources of host volatiles. *Entomol Exp Appl* 41:91–100.

Owen, S.M., Clark, S., Pompe, M., and Semple, K.T. 2007. Biogenic volatile organic compounds as potential carbon sources for microbial communities in soil from the rhizosphere of *Populus tremula*. *FEMS Microbiol Lett* 268:34–39.

Paavolainen, L., Kitunen, V., and Smolander, A. 1998. Inhibition of nitrification in forest soil by monoterpenes. *Plant Soil* 205:147–154.

Palma, R., Mutis, A., Manosalva, L., Ceballos, R., and Quiroz, A. 2012. Behavioral and electrophysiological responses of *Hylastinus obscurus* to volatiles released from the roots of *Trifolium pratense* L. *J Plant Nutr Soil Sci* 12:183–193.

Penuelas, J., Asensio, D., Tholl. D., Wenke, K., Rosenkranz, M., Piechulla, B., and Schnitzler, J.P. 2014. Biogenic volatile emissions from the soil. *Plant Cell Environ* 37:1866–1891.

Peters, N.K., Frost, J.W., Long, S.R. 1986. A plant flavone, luteolin, induces expression of *Rhizobium meliloti* nodulation genes. *Science* 233:977–980.

Plaingam, W., Sangsuthum, S., Angkhasirisap, W., and Tencomnao, T. 2017. *Kaempferia parviflora* rhizome extract and *Myristica fragrans* volatile oil increase the levels of monoamine neurotransmitters and impact the proteomic profiles in the rat hippocampus: Mechanistic insights into their neuroprotective effects. *J Trad Complem Med* 7:538e552.

Radulovic, N.S., Blagojevic, P.D., Stojanovic-Radic, Z.Z., and Stojanovic, N.M. 2013. Antimicrobial plant metabolites: Structural diversity and mechanism of action. *Curr Med Chem* 20:932–952.

Rapparini, F., Llusià, J., and Peñuelas, J. 2008. Effects of arbuscular mycorrhizal (AM) colonization on terpene emission and content of *Artemisia annua* L. *Plant Biol* 10:108–122.

Rasmann, S. and Turlings, T.C.J. 2007. Simultaneous feeding by aboveground and belowground herbivores attenuates plant-mediated attraction of their respective natural enemies. *Ecol Lett* 10:926–936.

Rasmann, S., Köllner, T.G., Degenhardt, J., Hiltpold, I., Töpfer, S., Kuhlmann, U., and Turlings, T.C.J. 2005. Recruitment of entomopathogenic nematodes by insect-damaged maize roots. *Nature* 434:732–737.

Rensing, S.A., Ick, J., Fawcett, J.A., Lang, D., Zimmer, A., Peer, Y.V.D., and Reski, R. 2007. An ancient genome duplication contributed to the abundance of metabolic genes in the moss *Physcomitrella Patens*. *BMC Evol Biol* 7:130.

Robert, C.A.M., Erb, M., Duployer, M., Zwahlen, C., Doyen, G.R., and Turlings, T.C.J., 2012. Herbivore-induced plant volatiles mediate host selection by a root herbivore. *New Phytol* 194:1061–1069.

Rodenburg, J. and Bastiaans, L. 2011. Host-plant defence against *Striga* spp.: Reconsidering the role of tolerance. *Weed Res* 51:438–441.

Roshchina, V.V. and Roshchina, V.D. 1993. The excretory function of higher plants. Springer, Berlin, p. 25.

Runyon, J.B. 2006. Volatile chemical cues guide host location and host selection by parasitic plants. *Science* 313:1964–1967.

Santos, F., Peñaflor, M.F.G.V., Paré P.W., Sanches, P.A., Kamiya, A.C., Tonelli, M., Nardi, C., and Bento, J.M.S. 2014. A novel interaction between plant-beneficial rhizobacteria and roots: Colonization induces corn resistance against the root herbivore *Diabrotica speciose*. *PLoS One* 9: e113280.

Schulz-Bohm, K., Gerards, S., Hundscheid, M., Melenhorst, J., de Boer, W., and Garbeva, P. 2018. Calling from distance: Attraction of soil bacteria by plant root volatiles. *ISME J* 12:1252–1262.

Schumann, M., Ladin, Z.S., Beatens, J.M., Hiltpold, I. 2018. Navigating on a chemical radar: Usage of root exudates by foraging *Diabrotica virgifera* larvae. *J Appl Entomol* 142:911–920.

Setala, H. and Huhta, V. 1991. Soil Fauna Increase *Betula Pendula* Growth: Laboratory experiments with coniferous forest floor. *Ecology* 72:665–671.

Singh, Rajendra., Ahluwalia, Vivek., Singh, Pratap., Kumar, Naresh., Sati, OP., and Sati, N. 2016. Antifungal and phytotoxic activity of essential oil from root of *Senecio amplexicaulis* Kunth. (*Asteraceae*) growing wild in high altitude-Himalayan region. *Nat Prod Res* 30:1875.

Sofrata, A., Santangelo, E.M., Azeem, M., Borg-Karlson, A., Gustafsson, A., and Putsep, K. 2011. Benzyl isothiocyanate, a major component from the roots of *Salvadora persica* is highly active against gram-negative bacteria. *Plos One* 6:e23045.

Sohrabi, R., Huh, J.H., Badieyan, S., Rakotondraibe, L.H., Kliebenstein, D.J., Sobrado, P., and Tholl, D. 2015. In planta variation of volatile biosynthesis: An alternative biosynthetic route to the formation of the pathogen-induced volatile homoterpene DMNT via triterpene degradation in *Arabidopsis* roots. *Plant Cell* 27:874–890.

Steeghs, M., Bais, H.P., de Gouw, J., Goldan, P., Kuster, W., Northway, M., Fall, R., and Vivanco, J.M. 2004. Proton-transfer-reaction mass spectrometry as a new tool for real time analysis of root-secreted volatile organic compounds in *Arabidopsis*. *Plant Physiol* 135:47–58.

Sun, X.G. and Tang, M. 2013. Effect of arbuscular mycorrhizal fungi inoculation on root traits and root volatile organic compound emissions of *Sorghum bicolor*. *S Afr J Bot* 88:373–379.

Sutherland, O.R. and Hillier, J. 1974. Olfactory response of *Costelytra zealandica* (Coleoptera: Melolonthinae) to the roots of several pasture plants. *New Zeal J Zool* 1:365–369.

Suttle, J.C., Olson, L.L., and Lulai, E.C. 2016. The involvement of gibberellins in 1,8-cineole-mediated inhibition of sprout growth in russet burbank tubers. *Am J Potato Res* 93:72–79.

Süzgeç-Selçuk, S., Özek, G., Meriçli, AH., Baser, KHC., Haliloglu Y., and Özek T. 2017. Chemical and biological diversity of the leaf and rhizome volatiles of *Acorus calamus* L. from Turkey. *J Essent Oil Bear Plant* 20:1–16.

Tapia, T., Perich, F., Pardo, F., Palma, G., and Quiroz, A. 2007. Identification of volatiles from differently aged red clover (*Trifolium pratense*) root extracts and behavioural responses of clover root borer (*Hylastinus obscurus*) (Marsham) (*Coleoptera: Scolytidae*) to them. *Biochem Systemat Ecol* 35:61–67.

Tonelli, M., Peanflor, M.F.G.V., Leite, L.G., Silva, W.D., Martins, F. and Bento, J.M.S. 2016. Attraction of entomopathogenic nematodes to sugarcane root volatiles under herbivory by a sap-sucking insect. *Chemoecology* 26:59–66.

Tsunoda, T. and Van Dam, N.M. 2017. Root chemical traits and their roles in belowground biotic interactions. *Pedobiologia* 65:58–67.

van Tol, R.W.H.M., van der Sommen, A.T.C., Boff, M.I.C., van Bezooijen, J., Sabelis, M.W., and Smits, P.H. 2001. Plants protect their roots by alerting the enemies of grubs. *Ecol Lett* 4:292–294.

Vaughan, M.M., Christensen, S., Schmelz, E.A., Huffaker, A., McAuslane, H.J., Alborn, H.T., Romero, M., Allen, L.H., and Teal, P.E.A. 2015. Accumulation of terpenoid phytoalexins in maize roots is associated with drought tolerance. *Plant Cell Environ* 38:2195–2207.

Vaughan, M.M., Wang, Q.,Webster, F.X., Kiemle, D., Hong, Y.J., Tantillo, D.J., and Tholl, D. 2013. Formation of the unusual semivolatile diterpene rhizathalene by the *Arabidopsis* class I terpene synthase TPS08 in the root stele is involved in defense against belowground herbivory. *Plant Cell* 25:1108–1125.

Vilela, G.R., de Almeida, G.S., D'Arce, M., Moraes, M.H.D., Brito, J.O., da Silva, M., Silva, S.C., Piedade, S.M.D., Calori-Domingues, M.A., and da Gloria, E.M. 2009. Activity of essential oil and its major compound, 1,8-cineole, from *Eucalyptus globulus* Labill., against the storage fungi *Aspergillus flavus Link* and *Aspergillus parasiticus Speare*. *J Stored Prod Res* 45:108–111.

Viles, A.L. and Reese, R.N. 1996. Allelopathic potential of *Echinacea angustifolia* DC. *Environ Exp Bot* 36:39–43.

Wall, D. and Moore, J.C. 1999. Interactions underground. *BioScience* 49:109.

Wang, C.Z., Zhang, D.J., Zhang, J., Ji, T.W., Tang, Z.Q., and Zhao, Y.Y. 2015. Allelopathic effects of volatile compounds from *Eucalyptus grandis* on *Vigna radiata*, *Raphanus sativus* and *Lactuca sativa*. *Allelopathy J* 36:273–282.

Wang, Q., Jia, M.R., Huh, J.H., Muchlinski, A., Peters, R.J., and Tholl, D. 2016. Identification of a dolabellane type diterpene synthase and other root-expressed diterpene synthases in *Arabidopsis*. *Front Plant Sci* 7:1761.

Weissteiner, S. and Schütz, S. 2006. Are different volatile pattern influencing host plant choice of belowground living insects. *Mitt Dtsch Ges Allg Angew Ent* 15:51–55.

Weissteiner, S., Huetteroth, W., Kollmann, M., Weßbecker, B., Romani, R., Schachtner, J., and Schütz, S. 2012. Cockchafer larvae smell host root scents in soil. *PLoS One* 7:e45827.

Werner, S., Polle, A., and Brinkmann, N. 2016. Belowground communication: Impacts of volatile organic compounds (VOCs) from soil fungi on other soil-inhabiting organisms. *Appl Microbiol Biot* 100:8651–8665.

White, C.S. 1991. The role of monoterpenes in soil nitrogen cycling processes in ponderosa pine. *Biogeochem* 12:43–68.

Wilschut, R.A., Silva, J.C.P., Garbeva, P., and Putten, W.H.V.D. 2017. Belowground plant-herbivore interactions vary among climate-driven range-expanding plant species with different degrees of novel chemistry. *Front Plant Sci* 8:1861.

Wolfson, J.L. 1987. Impact of Rhizobium nodules on *Sitona hispidulus*, the clover root curculio. *Entomol Exp Appl* 43:237–243.

Xu, D., Hanschen, F.S., Witzel, K., Nintemann, S.J., Nour-Eldin, H.H., Schreiner, M., and Halkier, B.A. 2016. Rhizosecretion of stele-synthesized glucosinolates and their catabolites requires GTR-mediated import in *Arabidopsis*. *J Exp Bot* 68:3205–3214.

Yeates, G.W. 1979. Soil nematodes in terrestrial ecosystems, *J Nematol* 11:213–229.

Yoder, J.I. and Scholes, J.D. 2010. Host plant resistance to parasitic weeds; recent progress and bottlenecks. *Curr Opin Plant Biol* 13:478–484.

Yoneyama, K., Awad, A.A., Xie, X., Yoneyama, K., and Takeuchi, Y. 2010. Strigolactones as germination stimulants for root parasitic plants. *Plant Cell Physiol* 51:1095–1103.

Zak, J.C., Willig, M.R., Moorhead, D.L., and Wildmann, H.G. 1994. Functional diversity of microbial communities: A quantitative approach. *Soil Biol Biochem* 26:1101–1108.

Zuo, C., Li, C., Li, B., Wei, Y., Hu, C., Yang, Q., Yang, J., Sheng, O., Kuang, R., Deng, G., Biswas, M.K., and Yi, G. 2015. The toxic mechanism and bioactive components of Chinese leek root exudates acting against *Fusarium oxysporum* f.sp. cubense tropical race 4. *Eur J Plant Pathol* 143:447–460.

19 Tree Volatiles
Effects of Biotic and Abiotic Factors on Emission and Biological Roles

Erica Perreca, Jonathan Gershenzon, and Franziska Eberl

CONTENTS

19.1 INTRODUCTION

Most information on plant volatiles described in this book derives from herbaceous plants rather than trees and other woody plants. However, the fundamental physiological and ecological differences between trees and herbs may lead to significant difference in the patterns of volatiles emitted. Compared to herbs, trees have larger sizes and longer life spans and thus interact with a larger number and diversity of organisms that might be harmful or beneficial, and so induce the production of different volatile blends. With their larger size, trees also experience a greater range of environmental conditions than herbs, which might also influence volatile formation. Additionally, the complex architecture of trees and long distances between organs via vascular connections may lead to a greater reliance on volatiles for intra-plant signaling rather than vascular tissues. Further motivation to study tree volatiles comes from the influence of volatile compounds on atmospheric chemistry. Given the large biomass of trees in terrestrial ecosystems, they are responsible for emitting a much larger percentage of biological volatiles into the atmosphere than herbs.

The major classes of tree volatiles are the same as those emitted from herbaceous plants: terpenes, aromatics, fatty acid derivatives such as green leaf volatiles (GLVs), and nitrogen-containing compounds. We include all of these classes here, and focus on volatiles emitted from living tissue, especially foliage, and not those of litter or decaying wood.

19.2 TREE VOLATILES AND THE BIOTIC ENVIRONMENT

The sedentary lifestyles of plants limit the possibilities of communication with organisms at a distance. However, emission of volatile organic compounds (VOCs) provides a medium to send messages to animals, microbes and other plants many meters away. The role of VOCs in biotic interactions of plants has been especially well investigated in herbaceous plants, but has also been established for some woody plants (Chapter 5). Like herbs, trees and other woody plants emit VOCs upon biotic stresses such as herbivory or pathogen infection. The impact of these VOCs may be wide-ranging given the large biomass of trees and the many antagonistic, mutualistic and commensal organisms living in, on or around trees.

19.2.1 Effect of Herbivore and Pathogen Attack on Emission

Feeding by herbivores or infection with a pathogen can induce many chemical changes in trees, including alterations in the volatile profile, usually characterized by an increase in the quantity and diversity of VOCs. To our knowledge, all studies of herbivore-induced volatile emission in trees have found an increase in emission of at least some compounds upon herbivory, even though the amplitude of induction varies greatly. Some publications report an increase in all main volatile classes (terpenoids, GLVs, aromatic and nitrogenous compounds) upon herbivore feeding (Yoneya and Takabayashi, 2013; Clavijo McCormick et al., 2014b; Eberl et al., 2018a), while others observed the induction of only one or a few of these (Blande et al., 2010; Jiang et al., 2018; Fabisch et al., 2019).

The differences in herbivore-induced volatile profiles can be attributed to various factors, such as the emitting organ or the type of damage. For example, mechanical wounding of poplar foliage leads to increased volatile emission, but to a lesser extent than herbivory (Arimura et al., 2004), although different types of mechanical wounding induce different profiles of volatile emissions (Portillo-Estrada et al., 2015). Therefore for actual herbivory, it might not be surprising that different herbivore species with distinct feeding guilds cause different patterns of emission. So far, volatiles have been analyzed from trees infested with phloem-feeding aphids (Blande et al., 2010) or psyllids (Scutareanu et al., 2003), leaf-chewing caterpillars (Arimura et al., 2004; Danner et al., 2011; Clavijo McCormick et al., 2014b) or beetles (Brilli et al., 2009; Fabisch et al., 2019), or with stem-boring (Miller et al., 2005; Rodriguez-Saona et al., 2006; Heijari et al., 2011) or root-feeding insects (Abraham et al., 2015; Lackus et al., 2018). In general, leaf-chewers, which cause a high degree of tissue damage, induce a larger increase in volatiles than phloem-feeding herbivores. However, within the same feeding guild, differences in volatile emission can be observed among herbivores, even at the genus or species level (Jiang et al., 2018; Fabisch et al., 2019). Furthermore, the life stage (Yoneya and Takabayashi, 2013) and larval instar (Clavijo McCormick et al., 2014a) of the herbivore play an important role in volatile emission, as does the severity (Copolovici et al., 2017) and duration of damage (Clavijo McCormick et al., 2016).

Besides herbivores, microbial pathogens also influence volatile emission from tree foliage (Toome et al., 2010; Jiang et al., 2016; Martini et al., 2017; Eberl et al., 2018a; Patt et al., 2018) or stems (McLeod et al., 2005; Vezzola et al., 2018). However, fewer studies focus on pathogen-mediated changes in volatile emission compared to herbivore induction. When both pathogens and herbivores were tested in the same experiment, pathogen infection led to the release of lower amounts of volatiles and to a simpler blend. For example, rust-infected poplars increased only their emission of terpenoids and aromatic compounds instead of showing increases in all classes of volatiles induced in herbivore-infested poplars (Eberl et al., 2018a). Similarly, *Marssonina betulae* infection increased the emission of ocimene isomers but none of the other terpenoids known to be released from herbivore-damaged birch trees (Vuorinen et al., 2007).

Whereas the emission of most terpenoids increases upon pathogen infection, the emission of isoprene remains stable (Eberl et al., 2018b) or decreases (Brüggemann and Schnitzler, 2001; Toome et al., 2010; Jiang et al., 2016), depending on disease severity. A mild infection of a biotrophic pathogen, such as rust or mildew, does not influence isoprene emission (Brüggemann and Schnitzler, 2001; Eberl et al., 2018b), while a severe infection with necrosis leads to a decrease (Brüggemann and Schnitzler, 2001; Jiang et al., 2016). Accordingly, the time course of infection strongly influences the volatile emission from trees (Martini et al., 2017; Achotegui-Castells et al., 2015). However, both increasing and decreasing patterns of emission are observed for different VOCs. The number of studies performed is still too small to draw general conclusions about whether different types of pathogens (necrotroph *versus* biotroph) affect VOCs differently.

19.2.2 Molecular Regulation of Volatile Formation after Biotic Stress

The molecular regulation of induced volatile emission in trees has been most intensively investigated for terpenoids. As in herbaceous plants, trees produce mono- or sesquiterpenes from prenylated diphosphate precursors by catalysis of terpene synthases (Danner et al., 2011; Irmisch et al., 2014a). Nitrogen-containing VOCs are synthesized by cytochrome P450 enzymes (CYPs) from amino acids (Irmisch et al., 2013; Irmisch et al., 2014b). Both classes of enzymes, terpene synthases and CYPs, have been shown to be regulated at the transcriptional level, which is often induced by the phytohormone jasmonic acid (JA) (Danner et al., 2011; Irmisch et al., 2013; Irmisch et al., 2014b; Eberl et al., 2018a; Clavijo McCormick et al., 2019).

JA and its derivatives are crucial for inducing anti-herbivore defenses, including volatiles, and this role is also established in tree species (reviewed in Eberl et al., 2019). The exogenous application of JA or its volatile methyl ester, methyl jasmonate, induced the emission of volatiles in several trees such as poplar (Arimura et al., 2004; Clavijo McCormick et al., 2014a), ash (Rodriguez-Saona et al., 2006), citrus (Patt et al., 2018), pine (Lundborg et al., 2019) and spruce (Martin et al., 2003; Miller et al., 2005).

The role of another important stress-related phytohormone, salicylic acid (SA), is not well studied in trees. Depending on the species, induction of the SA pathway may have positive, negative or no effect on volatile emission (Eberl et al., 2018a; Patt et al., 2018). It has been hypothesized that SA contributes to defense responses in young trees or saplings, but not in mature trees (Germain and Séguin, 2011). In poplar, this hypothesis was supported by the observation that herbivory increases the levels of SA in young trees (Clavijo McCormick et al., 2014a) but not in old-growth trees (Clavijo McCormick et al., 2019). Furthermore, there is evidence for negative crosstalk between the two phytohormones, JA and SA (Eberl et al., 2018a; Patt et al., 2018), even though it is more equivocal than in herbaceous plants. More research at the molecular level is needed to better understand herbivore and pathogen-induced VOC emission in trees.

19.2.3 Role of Volatiles in Biotic Interactions

In order to function in biotic interactions, plant VOCs have to be perceived by other organisms. This requires that the compounds are present in the same place and time as the receiver organisms, and at sufficient concentration despite dilution and possible chemical modifications in the atmosphere (Atkinson and Arey, 2003).

The most well-studied function of herbivore-induced plant VOCs is the attraction of herbivore enemies, a form of "indirect defense" in plants (Arimura et al., 2005). VOCs derived from trees have been reported to attract different types of natural enemies, including parasitoids (Büchel et al., 2011; Clavijo McCormick et al., 2014b; Gossner et al., 2014), predatory insects (James, 2003), birds (Amo et al., 2003; Rubene et al., 2019) and nematodes (Ali et al., 2010). Whereas some studies demonstrated the attractiveness of a complex blend of herbivore-induced tree VOCs, others could identify active classes, such as sesquiterpenes attracting egg parasitoids (Büchel et al., 2011), or individual compounds, such as methyl salicylate that lures birds (Rubene et al., 2019) and green lacewings (James, 2003).

A well-characterized example of VOCs that function as herbivore enemy attractants is the nitrogenous compounds emitted from herbivore-infested poplar trees. A blend containing aldoximes and benzyl cyanide is released from damaged leaves after gypsy moth herbivory that is almost undetectable from nondamaged leaves (Irmisch et al., 2013; Clavijo McCormick et al., 2014a; Irmisch et al., 2014a), a pattern also observed under field conditions for old-growth trees (Clavijo McCormick et al., 2019). Furthermore, aldoximes are emitted exclusively during caterpillar feeding, stopping immediately after caterpillar removal, unlike many other compounds that continue to be emitted after removal of the herbivores (Clavijo McCormick et al., 2014b). These emission patterns give natural enemies that perceive aldoximes very precise information about when and where their prey or host is feeding. Indeed, a gypsy moth parasitoid is attracted to these volatiles and can detect them at low doses (Clavijo McCormick et al., 2014b). The same study also proved the activity of these compounds in attracting various parasitic Hymenopteran species under field conditions. Not all classes or individual volatile compounds have been so well studied as enemy attractants. In some cases, a general blend may be more important for attraction than the presence of specific individual VOCs (Clavijo McCormick et al., 2014c).

Apart from indirect defense, tree volatiles can also act directly against herbivores by being repellent or even toxic. For example, methyl salicylate, which is emitted by various tree species such as pear (James, 2003), oak (Copolovici et al., 2017), birch, alder (Blande et al., 2010) and poplar (Jiang et al., 2016), repels ambrosia beetles (Martini et al., 2017) as well as aphids (Petterson et al., 1994). Similarly, the nitrogen-containing compounds benzyl cyanide and methylbutyronitriles, emitted from herbivore-infested poplars, repel generalist caterpillars (Irmisch et al., 2014a). Another group of nitrogenous compounds in poplar leaves, the semivolatile phenylacetaldoximes, decreased the survival of caterpillars when supplemented in the diet (Irmisch et al., 2013).

On the other hand, tree volatiles can also benefit herbivores by helping them to identify their appropriate host. Bark beetles, for example, use volatile cues in order to identify trees that are infected with mutualistic fungi (McLeod et al., 2005; Kandasamy et al., 2019). Females of the emerald ash borer are attracted to host tree volatiles induced by beetle feeding or methyl jasmonate (Rodriguez-Saona et al., 2006). These herbivore behaviors can be exploited by using synthetic mixtures of VOCs identified from trees to trap pest species of beetles (Cossé et al., 2006,) and moths (Li et al., 2012). Additionally, tree VOCs can substantially enhance the attractiveness of pheromone lures employed to manage forest pests (Cossé et al., 2006; Li et al., 2012).

The effects of tree volatiles on microbes should not be neglected. VOCs emitted from trees can have bacteriostatic effects (Gao et al., 2005) or can inhibit or promote fungal growth (Mendgen et al., 2006). However, literature on this topic is scarce and volatile-mediated plant-microbe interactions need more research.

Plants can also employ volatiles to transfer information among their organs or to other plants. In fact, the first studies on volatile-mediated intra- and inter-plant signaling were conducted on trees (Baldwin and Schultz, 1983; Rhoades, 1983). The perception of VOCs by plants can cause physiological or biochemical changes directly (termed "induction"), as in the case of volatile hormones. Exposure to ethylene, for example, induces the formation of resin ducts in conifers (Hudgins and Franceschi, 2004). The effects of methyl jasmonate and methyl salicylate in activating defenses against herbivores or pathogens were described above. However, the perception of VOCs can also lead to a phenomenon termed "priming," the induction of a physiological state in which the plant is conditioned for a stronger and faster defense response upon actual herbivory (Martinez-Medina et al., 2016). Typically, no chemical or physiological changes can be observed during the primed state before herbivory stimulates defensive responses. The existence of volatile-mediated priming in woody plants has been demonstrated for various species, such as aspen (Li and Blande, 2017), hybrid poplar (Frost et al., 2007, 2008), willow (Yoneya et al., 2014) and birch (Girón-Calva et al., 2014). Priming of trees by volatiles negatively influences the preference and performance of herbivores (Rhoades, 1983; Dolch and Tscharntke, 2000; Yoneya et al., 2014). However, in such bioassay-based studies it is sometimes difficult to distinguish between priming and induction to determine if the plant was better defended already after VOC exposure or if the defensive response was only activated by herbivore feeding.

Regardless of the effect of VOCs on plants, the question arises of who is the intended receiver of these signals. Already in one of the early reviews on volatile-mediated communication in plants (Baldwin et al., 2002), it was pointed out that such inter-plant communication might be more in the interest of the receiver than the sender. In fact, it is difficult to see a benefit for a plant to send out signals to neighbors unless it is surrounded by close kin (Karban et al., 2003). It is currently believed that intra-plant signaling is likely to be much more important than inter-plant signaling, especially in trees because of their large size and lack of vascular connections between branches (Sprugel et al., 1991; Orians, 2005). However, when volatiles have been shown to function as intra-plant signals (Frost et al., 2007; Girón-Calva et al., 2014; Li and Blande, 2017), the existence of vascular connections between the studied branches mostly has not been investigated, for example by dye transport (Li and Blande, 2017). Interestingly, volatile signaling may have a specific ontogenetic context. Whereas young saplings showed systemic induction of volatile emission upon herbivory as mediated by vascular connections (Clavijo McCormick et al., 2014b), mature trees did not (Clavijo McCormick et al., 2019). Does this mean that young trees use vascular connections for information transmission, whereas mature trees rely on volatile signals? To answer this and related questions, we need more studies that investigate both vascular and volatile signaling together in trees of different ages. Moreover, much about the molecular aspects of volatile-mediated plant–plant communication is unknown, such as the mechanisms of volatile perception and signal transduction. A recent study describes a putative sesquiterpene receptor, which binds (E)-β-caryophyllene *in vitro* and in overexpressing tobacco lines, and acts as a transcriptional co-repressor (Nagashima et al., 2019). Nevertheless, the specificity of this receptor, downstream-signaling, transport of VOCs into the cells, as well as temporal patterns still have to be investigated.

19.3   TREE VOLATILES AND THE ABIOTIC ENVIRONMENT

The earliest studies on the emission of VOCs from trees already noted how the rate of emission is influenced by abiotic factors, such as light, temperature and water supply (Tingey et al., 1980). Since then, the effects of abiotic factors on volatile emission have gained increasing attention due to the question how global change could alter emission with impacts on atmospheric chemistry, air pollution and human health (Peñuelas and Staudt, 2010). The influence of VOCs on trees themselves has not been ignored, although more research has been carried out on herbaceous plants because of the comparative ease of experimentation.

19.3.1   Effect of Abiotic Factors on Emission

19.3.1.1   Light

The emission of VOCs from trees varies with light intensity and quality, with the C_5-terpene isoprene being most investigated in this context. Leaf isoprene emission is closely correlated with light intensity (Sharkey and Loreto, 1993) due at least in part to its close metabolic links to photosynthesis. Isoprene becomes rapidly labeled after switching from a $^{12}CO_2$ to a $^{13}CO_2$ atmosphere (Brilli et al., 2007; Karl et al., 2002b), consistent with its biosynthesis in the chloroplast from dimethylallyl diphosphate (DMAPP), a product of the methylerythritol 4-phosphate (MEP) pathway. The substrates of this pathway are pyruvate and glyceraldehyde-3-phosphate, which come directly from the reductive pentose phosphate pathway of photosynthesis.

While light regulates isoprene formation through control of substrate supply from the MEP pathway (Sharkey and Loreto, 1993), the activity of the final biosynthetic enzyme, isoprene synthase, is also light-dependent (Fall and Wildermuth, 1998, Sasaki et al., 2005). This enzyme has an alkaline pH optimum between 7.0 and 8.5, and a preference for bivalent cations as cofactors (Sasaki et al., 2005; Schnitzler et al., 2005). Under illumination, photosynthetic electron transport leads to the accumulation of Mg^{2+} in the stroma and increases the stromal pH (Schnitzler et al., 2010), thus contributing to isoprene synthase activity. Moreover, light effects on isoprene synthesis could also act at the transcriptional level. The expression of the isoprene synthase gene in *Populus canescens* correlates strongly with the light intensity and shows seasonal variations (Mayrhofer et al., 2005). The transcript levels of two other enzymes of the MEP pathway, 1-deoxy-D-xylulose-5-phosphate synthase (DXS) and hydroxy-3-methylbut-2-enyl-diphosphate reductase (HDR), also correlate with isoprene emission on a diurnal scale in *P. trichocarpa* (Wiberley et al., 2009).

Monoterpenes are also produced from products of the MEP pathway by condensation of one unit of DMAPP and one unit of isopentenyl diphosphate (IPP). However, the emission of monoterpenes responds to light variation differently in different tree species. In *Quercus* and other genera, monoterpene biosynthesis and emission, like that of isoprene, is light-dependent (Owen et al., 2002; Staudt and Lhoutellier, 2011), and accordingly monoterpenes become quickly labeled when trees are fed with $^{13}CO_2$ (Loreto et al., 1996b). Monoterpene synthase activity also increases in summer when more light is available (Fischbach et al., 2000). In contrast, for tree species that store monoterpenes in resin ducts and blisters (Pinaceae) and sub-epidermal cavities (Myrtaceae, Rutaceae) biosynthesis and emission are expected to be independent from light and the photosynthetic process. However, some conifers display a diurnal and seasonal pattern of monoterpene emission suggesting a certain dependency on light (Holzke et al., 2006). Further, emission of (*E*)-β-ocimene and linalool in *Pinus halepensis* showed a strong correlation with the light intensity and both compounds were biosynthesized *de novo* under illumination (Staudt et al., 2017). Ghirardo et al. (2010) showed that *de novo* biosynthesis accounts for up to 58% of monoterpene emission in illuminated *Pinus sylvestris*, 33% in *Picea abies*, and 9.8% in *Larix decidua*. These results suggest that monoterpene-storing species may also have separate transient pools that are synthesized and emitted immediately, although emission from stored pools may be occurring simultaneously (Staudt et al., 2017).

Sesquiterpenes are generally biosynthesized from products of the mevalonate pathway (MVA) found in the cytosol (Hemmerlin et al., 2012). In keeping with their extra-chloroplastidic origin, they show less dependence on light, but there is still some association (Duhl et al., 2008; Hansen and Seufert, 1999). For example, emission of (*E*)-β-caryophyllene from orange trees was best predicted by a logarithm integrating both temperature and light factors (Hansen and Seufert, 2003). Sesquiterpene emission was found to be independent of light in *Pinus halepensis* (Staudt et al., 2017; Staudt and Lhoutellier, 2011) and *Picea abies* (Bourtsoukidis et al., 2012), but had light dependency in *Quercus coccifera* (Staudt and Lhoutellier, 2011). Emission of GLVs does not show a strict light dependency. However, transition from light to dark could rapidly change the intracellular pH, and damages the stability of cellular membranes, which enhances the release of GLVs (Brilli et al., 2011; Karl et al., 2002a).

Additionally to intensity, the light quality may affect VOC emission from trees. Blue light seems to reduce the emission of monoterpenes and isoprene (Pallozzi et al., 2013). Under blue light chloroplasts are thought to assume a position that constrains the mesophyll conductance of CO_2 and so negatively affects photosynthesis (Pallozzi et al., 2013).

19.3.1.2 Temperature

Increased emission of VOCs at higher temperatures is expected because of increased vapor pressure, but the effects of temperature on emission are somewhat more complex. Isoprene emission is especially sensitive to temperature, increasing dramatically on heat stress (Loreto et al., 1996a; Sun et al., 2012). However, in some trees emission is maximal at 40°C but decreases at higher temperatures (Monson et al., 1992; Rasulov et al., 2015; Singsaas and Sharkey, 2000). The increase of isoprene emission rate with temperature is attributed in part to an increase in isoprene synthase activity (Kuzma and Fall, 1993; Sharkey et al., 2005), but the temperature optimum of this enzyme is actually above the temperature at which isoprene emission peaks (Lehning et al., 1999; Monson et al., 1992). This discrepancy might be explained by a limited supply of the substrate DMAPP at high temperatures (Rasulov et al., 2010), perhaps due to the inhibition of the photosynthetic electron transport and ribulose-1,5-bisphosphate carboxylase/oxygenase activity (Magel et al., 2006).

Monoterpene emission from nonstoring trees, such as species of *Quercus*, responds to rising temperatures with a maximum emission between 35°C and 40°C, followed by a drop at higher temperatures (Copolovici et al., 2005; Loreto et al., 1998). However, inhibition of monoterpene emission at higher temperature can be less evident than for isoprene emission because of a low release of monoterpenes from temporary storage sites in the lipid phase of the cell (Harley, 2013). Monoterpene emission also becomes acclimated to prolonged heat stress (Hanson and Sharkey, 2001; Staudt and Bertin, 1998). In monoterpene-storing species, rising temperatures could also increase emission depending on the resistance to diffusion specific to the storage structure (Harley, 2013). Sesquiterpene emission was also found to increase with rising temperature due to increased evaporation from storage structures, as in *Quercus ilex* (Staudt et al., 2017).

GLV emission is also reported to increase with temperature in species such as *Quercus coccifera* and *Picea abies* (Filella et al., 2007; Staudt and Lhoutellier, 2011). If the emission of GLVs is primarily associated with membrane damage to cells (Kleist et al., 2012), heat stress might act by increasing such damage.

19.3.1.3 Water Limitation

Water stress can be expected to reduce volatile emission by increasing stomatal closure or inhibiting volatile biosynthesis by negatively affecting photosynthesis (Flexas et al., 2016). However, moderate drought stress does not cause a drop of isoprene emission despite a reduction of photosynthesis (Pegoraro et al., 2004; Sharkey and Loreto, 1993). Alternative carbon sources have been found to supply the MEP pathway under moderate drought stress and sustain isoprene emission in poplar (Brilli et al., 2007). Candidates for these alternative carbon sources include xylem-transported glucose (Kreuzwieser et al., 2002), chloroplastidial starch degradation (Schnitzler et al., 2004), and CO_2 recycled by respiration (Kreuzwieser et al., 2002; Schnitzler et al., 2004). Labeling experiments with $^{13}CO_2$ in *Populus canescens* demonstrated that unlabeled carbon incorporated in isoprene under low CO_2 concentration is derived from cytosolic pyruvate (Trowbridge et al., 2012). Under severe drought stress, isoprene emission declines significantly (Brilli et al., 2007; Pegoraro et al., 2004; Sharkey and Loreto, 1993).

The effect of drought stress on monoterpene emission in nonstoring trees mirrors that of isoprene emission with little change under moderate stress and a strong decline under severe stress (Lavoir et al., 2009, Llusià and Peñuelas, 1998; Wu et al., 2015). The effect of drought on monoterpenes stored in special structures is less clear. *Pinus contorta* and *Pinus banksiana* seedlings emit monoterpenes at significantly lower rates under mild drought stress (Lusebrink et al., 2011), but *Picea abies* and *Pinus sylvestris* seedlings showed only slight reductions in emission (Wu et al., 2015). Furthermore, drought stress could change the composition of monoterpene emission via effects on stomatal closure. Reduced stomatal aperture causes steep declines in the emission of oxygenated monoterpenes, such as linalool and 1,8-cineole, due to their higher Henry's Law constants indicating greater potential to aqueous phase partitioning (Niinemets et al., 2002). In contrast, stomatal aperture has less effect on the emission of nonoxygenated monoterpenes, such as (E)-β-ocimene and limonene, since stomatal closure is balanced by an increase in the diffusion gradient to the surrounding atmosphere.

The effects of drought on sesquiterpene emission show a similar pattern. *Citrus sinensis* reduced its (E)-β-caryophyllene emission under severe drought stress (Hansen and Seufert, 2003). A reduction of sesquiterpene emission also occurred in *Pinus halepensis* and *Quercus coccifera* (Ormeño et al., 2007) under drought, but sesquiterpene emissions in *Quercus robur* and *Prunus serotina* showed only a slight reduction (Bourtsoukidis et al., 2014).

19.3.2 ROLE OF TREE VOLATILES IN ABIOTIC STRESS RESISTANCE

19.3.2.1 Protective Function of Isoprene against Oxidative Stress

Isoprene has long been thought to help plants resist oxidative stress arising from the accumulation of reactive oxygen species (ROS) formed during photosynthesis, respiration and other metabolic processes (Vickers et al., 2009). ROS are formed at elevated rates during high light, low temperature, drought and other abiotic stresses, such as ozone pollution. Plants possess a complex anti-oxidant system consisting of many components, including carotenoid pigments, ascorbate, glutathione and several enzymes (Pandhair and Sekhon, 2006). However, as a volatile, lipophilic anti-oxidant, isoprene may have a distinctive role.

Among the major discoveries, fumigation of birch and tobacco leaves with isoprene was found to decrease the extent of ozone damage, with protection occurring either inside the leaf or at the boundary layer (Loreto et al., 2001). In addition, inhibition of isoprene formation in *Platanus orientalis* leaves resulted in a greater accumulation of hydrogen peroxide, which increased lipid peroxidation (Velikova et al., 2006). The anti-oxidant properties of isoprene have been attributed to its ability to react directly with ROS (Vickers et al., 2009). This reaction produces methylvinyl ketone and methacrolein (Fuentes et al., 2000; Jardine et al., 2012).

19.3.2.2 Protective Function of Isoprene against High Temperature

A second protective function of isoprene in plants is its role in tolerating high temperature episodes. When leaves of *Platanus orientalis*, *Quercus alba* or various herbaceous plants were fumigated with isoprene, negative effects of heat were avoided and photosynthetic rates recovered much more rapidly (Singsaas et al., 1997; Velikova et al., 2011). Meanwhile, blocking of isoprene formation in oak leaves reduced thermotolerance of photosynthesis (Sharkey et al., 2001).

Isoprene has been suggested to confer heat tolerance because of its ability to partition into the lipid phases of the thylakoid membranes and stabilize the membranes through hydrophobic interactions (Singsaas et al., 1997; Velikova et al., 2011). This mechanism has gained support from theoretical studies (Siwko et al., 2007), but more recent evidence argues that isoprene is not abundant enough to significantly influence membrane stability (Harvey et al., 2015). Another intriguing mechanism proposed for isoprene's contribution to heat resistance is that volatilization could reduce the temperature of the phospholipid bilayer of the chloroplast membranes in a similar manner to the cooling effect of water released by the leaf (Pollastri et al., 2014).

Alternatively, isoprene may act as a signal. A recent study conducted on herbaceous species documented a long list of genes upregulated by ectopic expression of isoprene synthase in *Arabidopsis* and tobacco (Zuo et al., 2019). Included are genes upstream of many hormone signaling pathways, such as those for jasmonic acid, ethylene, gibberellins and cytokinins, and genes involved in biotic and abiotic stress tolerance. Besides the transcriptome, isoprene causes significant changes in the proteome (Vanzo et al., 2016) and metabolome (Ghirardo et al., 2014; Way et al., 2013), such as increases in chlorophyll and carotenoid pigments (Zuo et al., 2019).

19.3.2.3 Other Roles of Isoprene

Rosenstiel et al. (2004) hypothesized that isoprene emission is a mechanism for balancing the demand of DMAPP by higher isoprenoid biosynthesis (Rosenstiel et al., 2004). Isoprene biosynthesis may act as a safety valve by preventing the accumulation of DMAPP and an excessive sequestration of phosphate, when this substrate is not diverted to the downstream isoprenoids. It has been argued that this strategy is an efficient mechanism to control diversion of DMAPP and IPP to essential and nonessential isoprenoids (Owen and Peñuelas, 2005).

19.3.2.4 Other Volatiles in Abiotic Stress Resistance

A number of other volatiles of trees may also serve to provide protection against abiotic stresses. Besides isoprene, monoterpenes and sesquiterpenes have been studied in this regard because emission increases under drought, high light, and high temperature, all conditions causing elevated ROS levels (Holopainen and Gershenzon, 2010). In addition, some trees emit monoterpenes and sesquiterpenes directly without storage, and reaction occurs readily with ROS (Calogirou et al., 1999). In fact, many monoterpenes and sesquiterpenes, such as myrcene, α-terpinene, and (*E*)-β-caryophyllene, react more rapidly with ROS than with isoprene (Atkinson and Arey, 2003).

GLVs, as already discussed, are released during various abiotic stresses that damage the phospholipid composition of membranes (Jardine et al., 2015). Plants might conceivably use these compounds as signals to detect membrane damage and alleviate negative effects. Experiments with *Arabidopsis* showed that GLVs induce the expression of genes involved in abiotic stress responses (Alméras et al., 2003), but this finding still needs to be demonstrated in woody plants.

19.4 CONCLUSION

Trees and other woody plants can be expected to have different volatile emission profiles than herbs. Based on their larger sizes and longer lives, trees might have greater defensive needs, and there could be increased reliance on volatiles for communication within tree canopies. Unfortunately, not enough data have been collected yet to support or refute these hypotheses. We know that trees release the same major groups of volatile compounds as herbs, but do not know if they emit more diverse blends or at higher rates. Broader surveys of tree species are required, and it would be valuable to compare the effects of various biotic and abiotic factors in the same species. Further investigation on the importance of volatile vs. vascular signaling in trees would be especially welcome. The availability of robust, sensitive protocols for volatile collection and analysis should facilitate future progress in this area.

REFERENCES

Abraham, J, Giacomuzzi, V, Angeli, S. 2015. Root damage to apple plants by cockchafer larvae induces a change in volatile signals below and above-ground. *Entomol Exp Appl*, 156:279–289.

Achotegui-Castells, A, Danti, R, Llusià, J, Della Rocca, G, Barberini, S, Peñuelas, J. 2015. Strong induction of minor terpenes in Italian Cypress, *Cupressus sempervirens*, in response to infection by the fungus *Seiridium cardinale*. *J Chem Ecol*, 41:224–243.

Ali, JG, Alborn, HT, Stelinski, LL. 2010. Subterranean herbivore-induced volatiles released by citrus roots upon feeding by *Diaprepes abbreviatus* recruit entomopathogenic nematodes. *J Chem Ecol*, 36:361–368.

Alméras, E, Stolz, S, Vollenweider, S, Reymond, P, Mène-Saffrané, L, Farmer, EE. 2003. Reactive electrophile species activate defense gene expression in Arabidopsis. *Plant J*, 34:205–216.

Amo, L, Jansen, JJ, Van Dam, NM, Dicke, M, Visser, ME. 2013. Birds exploit herbivore-induced plant volatiles to locate herbivorous prey. *Ecol Lett*, 16:1348–1355.

Arimura, G-I, Kost, C, Boland, W. 2005. Herbivore-induced, indirect plant defences. *Biochim Biophys Acta*, 1734:91–111.

Arimura, GI, Huber, DP, Bohlmann, J. 2004. Forest tent caterpillars (*Malacosoma disstria*) induce local and systemic diurnal emissions of terpenoid volatiles in hybrid poplar (*Populus trichocarpa × deltoides*): cDNA cloning, functional characterization, and patterns of gene expression of (−)-germacrene D synthase, PtdTPS1. *Plant J*, 37:603–616.

Atkinson, R, Arey, J. 2003. Gas-phase tropospheric chemistry of biogenic volatile organic compounds: A review. *Atmos Environ*, 37:197–219.

Baldwin, IT, Kessler, A, Halitschke, R. 2002. Volatile signaling in plant–plant–herbivore interactions: What is real? *Curr Opin Plant Biol*, 5:351–354.

Baldwin, IT, Schultz, JC. 1983. Rapid changes in tree leaf chemistry induced by damage: Evidence for communication between plants. *Science*, 221:277–279.

Blande, JD, Korjus, M, Holopainen, JK. 2010. Foliar methyl salicylate emissions indicate prolonged aphid infestation on silver birch and black alder. *Tree Phys*, 30:404–416.

Bourtsoukidis, E, Kawaletz, H, Radacki, D, Schütz, S, Hakola, H, Hellén, H, … Bonn, B. 2014. Impact of flooding and drought conditions on the emission of volatile organic compounds of *Quercus robur* and *Prunus serotina*. *Trees*, 28:193–204.

Brilli, F, Barta, C, Fortunati, A, Lerdau, M, Loreto, F, Centritto, M. 2007. Response of isoprene emission and carbon metabolism to drought in white poplar (*Populus alba*) saplings. *New Phytol*, 175:244–254.

Brilli, F, Ciccioli, P, Frattoni, M, Prestininzi, M, Spanedda, AF, Loreto, F. 2009. Constitutive and herbivore-induced monoterpenes emitted by *Populus × euroamericana* leaves are key volatiles that orient *Chrysomela populi* beetles. *Plant Cell Environ*, 32:542–552.

Brilli, F, Ruuskanen, TM, Schnitzhofer, R, Müller, M, Breitenlechner, M, Bittner, V, ... Hansel, A. 2011. Detection of plant volatiles after leaf wounding and darkening by proton transfer reaction "time-of-flight" mass spectrometry (PTR-TOF). *PLoS ONE*, 6:e20419.

Brüggemann, N, Schnitzler, J-P. 2001. Influence of powdery mildew (*Microsphaera alphitoides*) on isoprene biosynthesis and emission of pedunculate oak (*Quercus robur* L.) leaves. *J Appl Botany*, 75:91–96.

Büchel, K, Malskies, S, Mayer, M, Fenning, TM, Gershenzon, J, Hilker, M, Meiners, T. 2011. How plants give early herbivore alert: Volatile terpenoids attract parasitoids to egg-infested elms. *Basic Appl Ecol*, 12:403–412.

Calogirou, A, Larsen, B, Kotzias, D. 1999. Gas-phase terpene oxidation products: A review. *Atmos Environ*, 33:1423–1439.

Clavijo McCormick, A, Boeckler, GA, Köllner, TG, Gershenzon, J, Unsicker, SB. 2014a. The timing of herbivore-induced volatile emission in black poplar (*Populus nigra*) and the influence of herbivore age and identity affect the value of individual volatiles as cues for herbivore enemies. *BMC Plant Biology*, 14:304.

Clavijo McCormick, A, Gershenzon, J, Unsicker, SB. 2014b. Little peaks with big effects: Establishing the role of minor plant volatiles in plant–insect interactions. *Plant Cell Environ*, 37:1836–1844.

Clavijo McCormick, A, Irmisch, S, Reinecke, A, Boeckler, GA, Veit, D, Reichelt, M, ... Unsicker, SB. 2014c. Herbivore-induced volatile emission in black poplar: Regulation and role in attracting herbivore enemies. *Plant Cell Environ*, 37:1909–1923.

Clavijo McCormick, A, Irmisch, S, Boeckler, GA, Gershenzon, J, Köllner, TG, Unsicker, SB. 2019. Herbivore-induced volatile emission from old-growth black poplar trees under field conditions. *Sci Rep*, 9:7714.

Clavijo McCormick, A, Reinecke, A, Gershenzon, J, Unsicker, SB. 2016. Feeding experience affects the behavioral response of polyphagous gypsy moth caterpillars to herbivore-induced poplar volatiles. *J Chem Ecol*, 42:382–393.

Copolovici, L, Pag, A, Kännaste, A, Bodescu, A, Tomescu, D, Copolovici, D, ... Niinemets, Ü. 2017. Disproportionate photosynthetic decline and inverse relationship between constitutive and induced volatile emissions upon feeding of *Quercus robur* leaves by large larvae of gypsy moth (*Lymantria dispar*). *Environ Exp Bot*, 138:184–192.

Copolovici, LO, Filella, I, Llusia, J, Niinemets, Ü, Penuelas, J. 2005. The capacity for thermal protection of photosynthetic electron transport varies for different monoterpenes in *Quercus ilex*. *Plant Physiol*, 139:485–496.

Cossé, AA, Bartelt, RJ, Zilkowski, BW, Bean, DW, Andress, ER. 2006. Behaviorally active green leaf volatiles for monitoring the leaf beetle, *Diorhabda elongata*, a biocontrol agent of saltcedar, *Tamarix* spp. *J Chem Ecol*, 32:2695–2708.

Danner, H, Boeckler, GA, Irmisch, S, Yuan, JS, Chen, F, Gershenzon, J, ... Köllner, TG. 2011. Four terpene synthases produce major compounds of the gypsy moth feeding-induced volatile blend of *Populus trichocarpa*. *Phytochemistry*, 72:897–908.

Dolch, R, Tscharntke, T. 2000. Defoliation of alders (*Alnus glutinosa*) affects herbivory by leaf beetles on undamaged neighbours. *Oecologia*, 125:504–511.

Duhl, T, Helmig, D, Guenther, A. 2008. Sesquiterpene emissions from vegetation: A review. *Biogeosciences*, 5:761–777.

Eberl, F, Hammerbacher, A, Gershenzon, J, Unsicker, SB. 2018a. Leaf rust infection reduces herbivore-induced volatile emission in black poplar and attracts a generalist herbivore. *New Phytol*, 220:760–772.

Eberl, F, Perreca, E, Vogel, H, Wright, LP, Hammerbacher, A, Veit, D, ... Unsicker, SB. 2018b. Rust infection of black poplar trees reduces photosynthesis but does not affect isoprene biosynthesis or emission. *Front Plant Sci*, 9.

Eberl, F, Uhe, C, Unsicker, SB. 2019. Friend or foe? The role of leaf-inhabiting fungal pathogens and endophytes in tree-insect interactions. *Fungal Ecol*, 38:104–112.

Fabisch, T, Gershenzon, J, Unsicker, SB. 2019. Specificity of herbivore defense responses in a woody plant, black poplar (*Populus nigra*). *J Chem Ecol*, 45:162–177.

Fall, R, Wildermuth, MC. 1998. Isoprene synthase: From biochemical mechanism to emission algorithm. *J Geophys Res: Atmos*, 103:25599–25609.

Filella, I, Wilkinson, MJ, Llusià, J, Hewitt, CN, Peñuelas, J. 2007. Volatile organic compounds emissions in Norway spruce (*Picea abies*) in response to temperature changes. *Physiol Plant*, 130:58–66.

Fischbach, RJ, Zimmer, I, Steinbrecher, R, Pfichner, A, Schnitzler, J-P. 2000. Monoterpene synthase activities in leaves of *Picea abies* (L.) Karst. and *Quercus ilex* (L.). *Phytochemistry*, 54:257–265.

Flexas, J, Díaz-Espejo, A, Conesa, M, Coopman, R, Douthe, C, Gago, J, ... Ribas-Carbo, M. 2016. Mesophyll conductance to CO_2 and Rubisco as targets for improving intrinsic water use efficiency in C3 plants. *Plant Cell Environ*, 39:965–982.

Frost, CJ, Appel, HM, Carlson, JE, De Moraes, CM, Mescher, MC, Schultz, JC. 2007. Within-plant signalling via volatiles overcomes vascular constraints on systemic signalling and primes responses against herbivores. *Ecol Lett*, 10:490–498.

Frost, CJ, Mescher, MC, Dervinis, C, Davis, JM, Carlson, JE, De Moraes, CM. 2008. Priming defense genes and metabolites in hybrid poplar by the green leaf volatile cis-3-hexenyl acetate. *New Phytol*, 180:722–734.

Fuentes, JD, Lerdau, M, Atkinson, R, Baldocchi, D, Bottenheim, J, Ciccioli, P, ... Guenther, A. 2000. Biogenic hydrocarbons in the atmospheric boundary layer: A review. *BAm Meteorol Soc*, 81:1537–1576.

Gao, Y, Jin, YJ, Li, HD, Chen, HJ. 2005. Volatile organic compounds and their roles in bacteriostasis in five conifer species. *J Integr Plant Biol*, 47:499–507.

Germain, H, Séguin, A. 2011. Innate immunity: Has poplar made its BED? *New Phytol*, 189:678–687.

Ghirardo, A, Koch, K, Taipale, R, Zimmer, I, Schitzler, J-P, Rinne, J. 2010. Determination of de novo and pool emissions of terpenes from four common boreal/alpine trees by 13CO$_2$ labelling and PTR-MS analysis. *Plant Cell Environ*, 33:781–792.

Ghirardo, A, Wright, LP, Bi, Z, Rosenkranz, M, Pulido, P, Rodríguez-Concepción, M, ... Schnitzler, J-P. 2014. Metabolic flux analysis of plastidic isoprenoid biosynthesis in poplar leaves emitting and nonemitting isoprene. *Plant Physiol*, 165:37–51.

Girón-Calva, PS, Li, T, Koski, T-M, Klemola, T, Laaksonen, T, Huttunen, L, Blande, JD. 2014. A role for volatiles in intra-and inter-plant interactions in birch. *J Chem Ecol*, 40:1203–1211.

Gossner, MM, Weisser, WW, Gershenzon, J, Unsicker, SB. 2014. Insect attraction to herbivore-induced beech volatiles under different forest management regimes. *Oecologia*, 176:569–580.

Hansen, U, Seufert, G. 1999. Terpenoid emission from *Citrus sinensis* (L.) OSBECK under drought stress. *Phys Chem Earth B*, 24:681–687.

Hansen, U, Seufert, G. 2003. Temperature and light dependence of β-caryophyllene emission rates. *J Geophys Res: Atmos*, 108.

Hanson, DT, Sharkey, TD. 2001. Rate of acclimation of the capacity for isoprene emission in response to light and temperature. *Plant Cell Environ*, 24:937–946.

Harley, PC. 2013. The roles of stomatal conductance and compound volatility in controlling the emission of volatile organic compounds from leaves. *Biology, Controls and Models of Tree Volatile Organic Compound Emissions*. Springer, the Netherlands.

Harvey, CM, Li, Z, Tjellström, H, Blanchard, GJ, Sharkey, TD. 2015. Concentration of isoprene in artificial and thylakoid membranes. *J Bioenerg Biomembr*, 47:419–429.

Heijari, J, Blande, JD, Holopainen, JK. 2011. Feeding of large pine weevil on Scots pine stem triggers localised bark and systemic shoot emission of volatile organic compounds. *Environ Exp Bot*, 71:390–398.

Hemmerlin, A, Harwood, JL, Bach, TJ. 2012. A raison d'être for two distinct pathways in the early steps of plant isoprenoid biosynthesis? *Prog Lipid Res*, 51:95–148.

Holopainen, JK, Gershenzon, J. 2010. Multiple stress factors and the emission of plant VOCs. *Trends Plant Sci*, 15:176–184.

Holzke, C, Hoffmann, T, Jaeger, L, Koppmann, R, Zimmer, W. 2006. Diurnal and seasonal variation of monoterpene and sesquiterpene emissions from Scots pine (*Pinus sylvestris* L.). *Atmos Environ*, 40:3174–3185.

Hudgins, J, Franceschi, VR. 2004. Methyl jasmonate-induced ethylene production is responsible for conifer phloem defense responses and reprogramming of stem cambial zone for traumatic resin duct formation. *Plant Physiol*, 135:2134–2149.

Irmisch, S, Jiang, Y, Chen, F, Gershenzon, J, Köllner, TG. 2014a. Terpene synthases and their contribution to herbivore-induced volatile emission in western balsam poplar (*Populus trichocarpa*). *BMC Plant Biology*, 14:270.

Irmisch, S, Clavijo McCormick, A, Günther, J, Schmidt, A, Boeckler, GA. 2014b. Herbivore-induced poplar cytochrome P450 enzymes of the CYP 71 family convert aldoximes to nitriles which repel a generalist caterpillar. *Plant J*, 80:1095–1107.

Irmisch, S, Mccormick, AC, Boeckler, GA, Schmidt, A, Reichelt, M, Schneider, B, ... Unsicker, SB. 2013. Two herbivore-induced cytochrome P450 enzymes CYP79D6 and CYP79D7 catalyze the formation of volatile aldoximes involved in poplar defense. *Plant Cell*, 25:4737–4754.

James, DG. 2003. Field evaluation of herbivore-induced plant volatiles as attractants for beneficial insects: Methyl salicylate and the green lacewing, *Chrysopa nigricornis*. *J Chem Ecol*, 29:1601–1609.

Jardine, KJ, Chambers, JQ, Holm, J, Jardine, AB, Fontes, CG, Zorzanelli, RF, ... Manzi, AO. 2015. Green leaf volatile emissions during high temperature and drought stress in a central Amazon rainforest. *Plants*, 4:678–690.

Jardine, KJ, Monson, RK, Abrell, L, Saleska, SR, Arneth, A, Jardine, A, … Huxman, T. 2012. Within-plant isoprene oxidation confirmed by direct emissions of oxidation products methyl vinyl ketone and methacrolein. *Global Change Biol*, 18:973–984.

Jiang, Y, Veromann-Jürgenson, LL, Ye, J, Niinemets, Ü. 2018. Oak gall wasp infections of *Quercus robur* leaves lead to profound modifications in foliage photosynthetic and volatile emission characteristics. *Plant, Cell & Environ*, 41:160–175.

Jiang, Y, Ye, J, Veromann, L-L, Niinemets, Ü. 2016. Scaling of photosynthesis and constitutive and induced volatile emissions with severity of leaf infection by rust fungus (*Melampsora larici-populina*) in *Populus balsamifera var. suaveolens. Tree Phys*, 36:856–872.

Kandasamy, D, Gershenzon, J, Andersson, MN, Hammerbacher, A. 2019. Volatile organic compounds influence the interaction of the Eurasian spruce bark beetle (*Ips typographus*) with its fungal symbionts. *ISME J*:1.

Karban, R, Maron, J, Felton, GW, Ervin, G, Eichenseer, H. 2003. Herbivore damage to sagebrush induces resistance in wild tobacco: Evidence for eavesdropping between plants. *Oikos*, 100:325–332.

Karl, T, Curtis, AJ, Rosenstiel, TN, Monson, RK, Fall, R. 2002a. Transient releases of acetaldehyde from tree leaves–products of a pyruvate overflow mechanism? *Plant Cell Environ*, 25:1121–1131.

Karl, T, Fall, R, Rosenstiel, T, Prazeller, P, Larsen, B, Seufert, G, Lindinger, W. 2002b. On-line analysis of the $13CO_2$ labeling of leaf isoprene suggests multiple subcellular origins of isoprene precursors. *Planta*, 215:894–905.

Kleist, E, Mentel, TF, Andres, S, Bohne, A, Folkers, A, Kiendler-Scharr, A, … Wildt, J. 2012. Irreversible impacts of heat on the emissions of monoterpenes, sesquiterpenes, phenolic BVOC and green leaf volatiles from several tree species. *Biogeosciences*, 9:5111–5123.

Kreuzwieser, J, Graus, M, Wisthaler, A, Hansel, A, Rennenberg, H, Schnitzler, JP. 2002. Xylem-transported glucose as an additional carbon source for leaf isoprene formation in *Quercus robur*. *New Phytol*, 156:171–178.

Kuzma, J, Fall, R. 1993. Leaf isoprene emission rate is dependent on leaf development and the level of isoprene synthase. *Plant Physiol*, 101:435–440.

Lackus, ND, Lackner, S, Gershenzon, J, Unsicker, SB, Köllner, TG. 2018. The occurrence and formation of monoterpenes in herbivore-damaged poplar roots. *Sci Rep*, 8:17936.

Lavoir, A, Staudt, M, Schnitzler, J, Landais, D, Massol, F, Rocheteau, A, … Rambal, S. 2009. Drought reduced monoterpene emissions from the evergreen Mediterranean oak *Quercus ilex*: Results from a throughfall displacement experiment. *Biogeosciences*, 6:1167, 1180.

Lehning, A, Zimmer, I, Steinbrecher, R, Brüggemann, N, Schnitzler, JP. 1999. Isoprene synthase activity and its relation to isoprene emission in *Quercus robur* L. leaves. *Plant Cell Environ*, 22:495–504.

Li, J, Valimaki, S, Shi, J, Zong, S, Luo, Y. 2012. Attraction of the gypsy moth to volatile organic compounds (VOCs) of damaged Dahurian larch. *Z Naturforsch Sect C*, 67:437–444.

Li, T, Blande, JD. 2017. Volatile-mediated within-plant signaling in hybrid aspen: Required for systemic responses. *J Chem Ecol*, 43:327–338.

Llusià, J, Peñuelas, J. 1998. Changes in terpene content and emission in potted Mediterranean woody plants under severe drought. *Can J Bot*, 76:1366–1373.

Loreto, F, Ciccioli, P, Brancaleoni, E, Cecinato, A, Frattoni, M, Sharkey, TD. 1996a. Different sources of reduced carbon contribute to form three classes of terpenoid emitted by *Quercus ilex* L. leaves. *PNAS*, 93:9966–9969.

Loreto, F, Ciccioli, P, Cecinato, A, Brancaleoni, E, Frattoni, M, Fabozzi, C, Tricoli, D. 1996b. Evidence of the photosynthetic origin of monoterpenes emitted by *Quercus ilex* L. leaves by 13C labeling. *Plant Physiol*, 110:1317–1322.

Loreto, F, Förster, A, Dürr, M, Csiky, O, Seufert, G. 1998. On the monoterpene emission under heat stress and on the increased thermotolerance of leaves of *Quercus ilex* L. fumigated with selected monoterpenes. *Plant Cell Environ*, 21:101–107.

Loreto, F, Mannozzi, M, Maris, C, Nascetti, P, Ferranti, F, Pasqualini, S. 2001. Ozone quenching properties of isoprene and its antioxidant role in leaves. *Plant Physiol*, 126:993–1000.

Lundborg, L, Sampedro, L, Borg-Karlson, A-K, Zas, R. 2019. Effects of methyl jasmonate on the concentration of volatile terpenes in tissues of Maritime pine and Monterey pine and its relation to pine weevil feeding. *Trees*, 33:53–62.

Lusebrink, I, Evenden, ML, Blanchet, FG, Cooke, JE, Erbilgin, N. 2011. Effect of water stress and fungal inoculation on monoterpene emission from an historical and a new pine host of the mountain pine beetle. *J Chem Ecol*, 37:1013.

Magel, E, Mayrhofer, S, Muller, A, Zimmer, I, Hampp, R, Schnitzler, JP. 2006. Photosynthesis and substrate supply for isoprene biosynthesis in poplar leaves. *Atmos Environ*, 40:S138–S151.

Martin, DM, Gershenzon, J, Bohlmann, J. 2003. Induction of volatile terpene biosynthesis and diurnal emission by methyl jasmonate in foliage of Norway spruce. *Plant Physiol*, 132:1586–1599.

Martinez-Medina, A, Flors, V, Heil, M, Mauch-Mani, B, Pieterse, CM, Pozo, MJ, … Conrath, U. 2016. Recognizing plant defense priming. *Trends Plant Sci*, 21:818–822.

Martini, X, Hughes, MA, Killiny, N, George, J, Lapointe, SL, Smith, JA, Stelinski, LL. 2017. The fungus *Raffaelea lauricola* modifies behavior of its symbiont and vector, the redbay ambrosia beetle (*Xyleborus glabratus*), by altering host plant volatile production. *J Chem Ecol*, 43:519–531.

Mayrhofer, S, Teuber, M, Zimmer, I, Louis, S, Fischbach, RJ, Schnitzler, J-P. 2005. Diurnal and seasonal variation of isoprene biosynthesis-related genes in grey poplar leaves. *Plant Physiol*, 139:474–484.

Mcleod, G, Gries, R, Von Reuss, SH, Rahe, JE, Mcintosh, R, König, WA, Gries, G. 2005. The pathogen causing Dutch elm disease makes host trees attract insect vectors. *Proc R Soc Lond, Ser B: Biol Sci*, 272:2499–2503.

Mendgen, K, Wirsel, SGR, Jux, A, Hoffmann, J, Boland, W. 2006. Volatiles modulate the development of plant pathogenic rust fungi. *Planta*, 224:1353–1361.

Miller, B, Madilao, LL, Ralph, S, Bohlmann, J. 2005. Insect-induced conifer defense. White pine weevil and methyl jasmonate induce traumatic resinosis, de novo formed volatile emissions, and accumulation of terpenoid synthase and putative octadecanoid pathway transcripts in Sitka spruce. *Plant Physiol*, 137:369–382.

Monson, RK, Jaeger, CH, Adams, WW, Driggers, EM, Silver, GM, Fall, R. 1992. Relationships among isoprene emission rate, photosynthesis, and isoprene synthase activity as influenced by temperature. *Plant Physiol*, 98:1175–1180.

Nagashima, A, Higaki, T, Koeduka, T, Ishigami, K, Hosokawa, S, Watanabe, H, … Touhara, K. 2019. Transcriptional regulators involved in responses to volatile organic compounds in plants. *J Biol Chem*, 294:2256–2266.

Niinemets, Ü, Reichstein, M, Staudt, M, Seufert, G, Tenhunen, JD. 2002. Stomatal constraints may affect emission of oxygenated monoterpenoids from the foliage of *Pinus pinea*. *Plant Physiol*, 130:1371–1385.

Orians, C. 2005. Herbivores, vascular pathways, and systemic induction: Facts and artifacts. *J Chem Ecol*, 31:2231–2242.

Ormeño, E, Mevy, J, Vila, B, Bousquet-Mélou, A, Greff, S, Bonin, G, Fernandez, C. 2007. Water deficit stress induces different monoterpene and sesquiterpene emission changes in Mediterranean species. Relationship between terpene emissions and plant water potential. *Chemosphere*, 67:276–284.

Owen, SM, Harley, P, Guenther, A, Hewitt, CN. 2002. Light dependency of VOC emissions from selected Mediterranean plant species. *Atmos Environ*, 36:3147–3159.

Owen, SM, Peñuelas, J. 2005. Opportunistic emissions of volatile isoprenoids. *Trends Plant Sci*, 10:420–426.

Pallozzi, E, Tsonev, T, Marino, G, Copolovici, L, Niinemets, Ü, Loreto, F, Centritto, M. 2013. Isoprenoid emissions, photosynthesis and mesophyll diffusion conductance in response to blue light. *Environ Exp Bot*, 95:50–58.

Pandhair, V, Sekhon, B. 2006. Reactive oxygen species and antioxidants in plants: An overview. *J Plant Biochem Biot*, 15:71–78.

Patt, JM, Robbins, PS, Niedz, R, Mccollum, G, Alessandro, R. 2018. Exogenous application of the plant signalers methyl jasmonate and salicylic acid induces changes in volatile emissions from citrus foliage and influences the aggregation behavior of Asian citrus psyllid (*Diaphorina citri*), vector of Huanglongbing. *PloS One*, 13:e0193724.

Pegoraro, E, Rey, A, Bobich, EG, Barron-Gafford, G, Grieve, KA, Malhi, Y, Murthy, R. 2004. Effect of elevated CO_2 concentration and vapour pressure deficit on isoprene emission from leaves of *Populus deltoides* during drought. *Funct Plant Biol*, 31:1137–1147.

Peñuelas, J, Staudt, M. 2010. BVOCs and global change. *Trends Plant Sci*, 15:133–144.

Pollastri, S, Tsonev, T, Loreto, F. 2014. Isoprene improves photochemical efficiency and enhances heat dissipation in plants at physiological temperatures. *J Exp Bot*, 65:1565–1570.

Portillo-Estrada, M, Kazantsev, T, Talts, E, Tosens, T, Niinemets, Ü. 2015. Emission timetable and quantitative patterns of wound-induced volatiles across different leaf damage treatments in aspen (*Populus tremula*). *J Chem Ecol*, 41:1105–1117.

Rasulov, B, Bichele, I, Hüve, K, Vislap, V, Niinemets, Ü. 2015. Acclimation of isoprene emission and photosynthesis to growth temperature in hybrid aspen: Resolving structural and physiological controls. *Plant, Cell Environ*, 38:751–766.

Rasulov, B, Hüve, K, Bichele, I, Laisk, A, Niinemets, Ü. 2010. Temperature response of isoprene emission *in vivo* reflects a combined effect of substrate limitations and isoprene synthase activity: A kinetic analysis. *Plant Physiol*, 154:1558–1570.

Rhoades, DF. 1983. Responses of alder and willow to attack by tent caterpillars and webworms: Evidence for pheromonal sensitivity of willows. ACS Publications.

Rodriguez-Saona, C, Poland, TM, Miller, JR, Stelinski, LL, Grant, GG, De Groot, P, ... Macdonald, L. 2006. Behavioral and electrophysiological responses of the emerald ash borer, *Agrilus planipennis*, to induced volatiles of Manchurian ash, *Fraxinus mandshurica*. *Chemoecology*, 16:75–86.

Rosenstiel, T, Ebbets, A, Khatri, W, Fall, R, Monson, R. 2004. Induction of poplar leaf nitrate reductase: A test of extrachloroplastic control of isoprene emission rate. *Plant Biol*, 7:12–21.

Rubene, D, Leidefors, M, Ninkovic, V, Eggers, S, Low, M. 2019. Disentangling olfactory and visual information used by field foraging birds. *Ecol Evol*, 9:545–552.

Sasaki, K, Ohara, K, Yazaki, K. 2005. Gene expression and characterization of isoprene synthase from *Populus alba*. *FEBS Lett*, 579:2514–2518.

Schnitzler, J-P, Graus, M, Kreuzwieser, J, Heizmann, U, Rennenberg, H, Wisthaler, A, Hansel, A. 2004. Contribution of different carbon sources to isoprene biosynthesis in poplar leaves. *Plant Physiol*, 135:152–160.

Schnitzler, J-P, Louis, S, Behnke, K, Loivamäki, M. 2010. Poplar volatiles – Biosynthesis, regulation and (eco) physiology of isoprene and stress-induced isoprenoids. *Plant Biol*, 12:302–316.

Schnitzler, J-P, Zimmer, I, Bachl, A, Arend, M, Fromm, J, Fischbach, R. 2005. Biochemical properties of isoprene synthase in poplar (*Populus* × *canescens*). *Planta*, 222:777–786.

Scutareanu, P, Bruin, J, Posthumus, MA, Drukker, B. 2003. Constitutive and herbivore-induced volatiles in pear, alder and hawthorn trees. *Chemoecology*, 13:63–74.

Sharkey, TD, Chen, X, Yeh, S. 2001. Isoprene increases thermotolerance of fosmidomycin-fed leaves. *Plant Physiol*, 125.

Sharkey, TD, Loreto, F. 1993. Water stress, temperature, and light effects on the capacity for isoprene emission and photosynthesis of kudzu leaves. *Oecologia*, 95:328–333.

Sharkey, TD, Yeh, S, Wiberley, AE, Falbel, TG, Gong, D, Fernandez, DE. 2005. Evolution of the isoprene biosynthetic pathway in kudzu. *Plant Physiol*, 137:700–712.

Singsaas, EL, Lerdau, M, Winter, K, Sharkey, TD. 1997. Isoprene increases thermotolerance of isoprene-emitting species. *Plant Physiol*, 115:1413–1420.

Singsaas, EL, Sharkey, TD. 2000. The effects of high temperature on isoprene synthesis in oak leaves. *Plant, Cell Environ*, 23:751–757.

Siwko, ME, Marrink, SJ, De Vries, AH, Kozubek, A, Uiterkamp, AJS, Mark, AE. 2007. Does isoprene protect plant membranes from thermal shock? A molecular dynamics study. *BiochimBiophys Acta*, 1768:198–206.

Sprugel, D, Hinckley, T, Schaap, W. 1991. The theory and practice of branch autonomy. *Annu Rev Ecol Syst*, 22:309–334.

Staudt, M, Bertin, N. 1998. Light and temperature dependence of the emission of cyclic and acyclic monoterpenes from holm oak (*Quercus ilex* L.) leaves. *Plant, Cell Environ*, 21:385–395.

Staudt, M, Bourgeois, I, Al Halabi, R, Song, W, Williams, J. 2017. New insights into the parametrization of temperature and light responses of mono—and sesquiterpene emissions from Aleppo pine and rosemary. *Atmos Environ*, 152:212–221.

Staudt, M, Lhoutellier, L. 2011. Monoterpene and sesquiterpene emissions from *Quercus coccifera* exhibit interacting responses to light and temperature. *Biogeosciences*, 8:2757–2771.

Sun, Z, Copolovici, L, Niinemets, Ü. 2012. Can the capacity for isoprene emission acclimate to environmental modifications during autumn senescence in temperate deciduous tree species *Populus tremula*? *J Plant Res*, 125:263–274.

Tingey, DT, Manning, M, Grothaus, LC, Burns, WF. 1980. Influence of light and temperature on monoterpene emission rates from slash pine. *Plant Physiol*, 65:797–801.

Toome, M, Randjärv, P, Copolovici, L, Niinemets, U, Heinsoo, K, Luik, A, Noe, SM. 2010. Leaf rust induced volatile organic compounds signalling in willow during the infection. *Planta*, 232:235–243.

Trowbridge, AM, Asensio, D, Eller, AS, Way, DA, Wilkinson, MJ, Schnitzler, J-P, ... Monson, RK. 2012. Contribution of various carbon sources toward isoprene biosynthesis in poplar leaves mediated by altered atmospheric CO_2 concentrations. *PLoS One*, 7:e32387.

Vanzo, E, Merl-Pham, J, Velikova, V, Ghirardo, A, Lindermayr, C, Hauck, SM, ... Schnitzler, J-P. 2016. Modulation of protein S-nitrosylation by isoprene emission in poplar. *Plant Physiol*, 170:1945–1961.

Velikova, V, Loreto, F, Tsonev, T, Brilli, F, Edreva, A. 2006. Isoprene prevents the negative consequences of high temperature stress in *Platanus orientalis* leaves. *Funct Plant Biol*, 33:931–940.

Velikova, V, Várkonyi, Z, Szabó, M, Maslenkova, L, Nogues, I, Kovács, L, … Loreto, F. 2011. Increased thermostability of thylakoid membranes in isoprene-emitting leaves probed with three biophysical techniques. *Plant Physiol*, 157:905–916.

Vezzola, LC, Michelozzi, M, Calamai, L, Gonthier, P, Giordano, L, Cherubini, P, Pelfini, M. 2018. Tree-ring volatile terpenes show potential to indicate fungal infection in asymptomatic mature Norway spruce trees in the Alps. *Forestry*, 92:149–156.

Vickers, CE, Gershenzon, J, Lerdau, MT, Loreto, F. 2009. A unified mechanism of action for volatile isoprenoids in plant abiotic stress. *Nat Chem Biol*, 5:283.

Vuorinen, T, Nerg, A-M, Syrjälä, L, Peltonen, P, Holopainen, JK. 2007. *Epirrita autumnata* induced VOC emission of silver birch differ from emission induced by leaf fungal pathogen. *Arthropod-Plant Inte*, 1:159.

Way, DA, Ghirardo, A, Kanawati, B, Esperschütz, J, Monson, RK, Jackson, RB, … Schnitzler, JP. 2013. Increasing atmospheric CO_2 reduces metabolic and physiological differences between isoprene-and non-isoprene-emitting poplars. *New Phytol*, 200:534–546.

Wiberley, AE, Donohue, AR, Westphal, MM, Sharkey, TD. 2009. Regulation of isoprene emission from poplar leaves throughout a day. *Plant, Cell Environ*, 32:939–947.

Wu, C, Pullinen, I, Andres, S, Carriero, G, Fares, S, Goldbach, H, … Kleist, E. 2015. Impacts of soil moisture on de novo monoterpene emissions from European beech, Holm oak, Scots pine, and Norway spruce. *Biogeosciences*, 12:177–191.

Yoneya, K, Kugimiya, S, Takabayashi, J. 2014. Leaf beetle larvae, *Plagiodera versicolora* (Coleoptera: Chrysomelidae), show decreased performance on uninfested host plants exposed to airborne factors from plants infested by conspecific larvae. *Appl Entomol Zool*, 49:249–253.

Yoneya, K, Takabayashi, J. 2013. Interaction–information networks mediated by plant volatiles: A case study on willow trees. *J Plant Interact*, 8:197–202.

Zuo, Z, Weraduwage, SM, Lantz, AT, Sanchez, LM, Weise, SE, Wang, J, … Sharkey, TD. 2019. Isoprene acts as a signaling molecule in gene networks important for stress responses and plant growth. *Plant Physiol*, 180:124–152.

Section IV

Genetic Improvements of Plant Volatiles

20 Metabolic Engineering of Plant Volatiles

Floral Scent, Flavors, Defense

*Milan Plasmeijer, Pan Liao, Michel A. Haring,
Natalia Dudareva, and Robert C. Schuurink*

CONTENTS

20.1 INTRODUCTION

For thousands of years, humans have domesticated wild plants to increase favorable traits for food production or, more recently, decorative purposes. In the process of plant breeding, ancient farmers simply selected plants with bigger fruits, brighter colors and higher yield without any knowledge of plant genetics, biochemistry or molecular biology. As a result, fruits have lost flavors, flowers have lost scent and plants have lost volatiles that are important signaling molecules for communication with its environment. Reintroduction of lost traits can be achieved by classical breeding, as was done in tomato by crossing it with its wild relatives *Solanum peruvianum* and *S. habrochaites* (Kamal et al. 2008; Bleeker et al. 2012). However, this is a laborious and time-consuming process that requires the monitoring of very complex traits. In addition, the extent of improvement can be

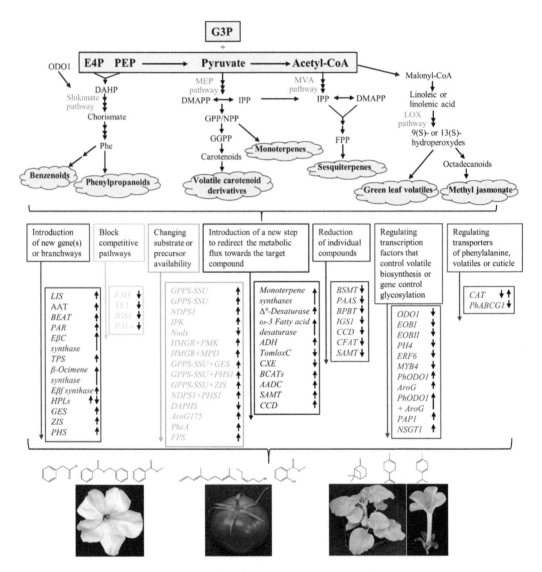

FIGURE 20.1 Overview of strategies for metabolic engineering of plant volatile organic compounds (VOCs). Precursors for plant VOCs derived from primary metabolism are shown in green box. Four major VOC biosynthetic pathways including shikimate/phenylalanine, the methylerythritol phosphate (MEP), the mevalonic acid (MVA), and lipoxygenase (LOX) pathways responsible for the biosynthesis of benzenoids/ phenylpropanoids, monoterpenes, volatile carotenoid derivatives, sesquiterpenes, green leaf volatiles and methyl jasmonate. Stacked arrows represent multiple enzymatic reaction involved. VOCs are marked with a blue cloud as background. Petunia flower, tomato fruit, *Nicotiana benthamiana* plants and its flower that have been used for metabolic engineering of VOCs are presented. Different strategies of metabolic engineering were highlighted with various boxes in one specific color. Genes that have been used for overexpression or downregulation to modify flower, fruit and leaf VOCs are shown with one specific color. ↑ represents that gene is over-expressed; ↓ represents that gene is downregulated. Abbreviations: AADC, amino acid decarboxylase; AAT, alcohol acetyltransferase; ADH, alcohol dehydrogenase; AroG, 3-deoxy-ᴅ-arabinoheptulosonate 7-phosphaste synthase; BCATs, branched-chain amino acid transaminases; BEAT, benzyl alcohol acetyltransferase; BPBT, benzyl alcohol/phenyl ethanol benzoyl-CoA transferase; BSMT, benzoic acid/salicylic acid carboxyl methyltransferase; CAT, cationic amino-acid transporter; CCD, carotenoid cleavage dioxygenase; CFAT, coniferyl alcohol acetyltransferase; CXE, carboxylesterase; DAHP, 3-deoxy-ᴅ-arabinoheptulosonate-7 phosphate; DAHPS, DAHP-synthase; DMAPP, dimethylallyl diphosphate; EβC, (*E*)-β-caryophyllene; Eβf, (*E*)-β-farnesene; EOB, EMISSION OF BENZENOIDS; ERF6, ETHYLENE RESPONSE FACTOR6;

limited by the genetic diversity available within sexually compatible species of any given crop. A more rapid strategy to achieve elevated levels of specific metabolites is metabolic engineering using genetic modification of a target crop.

Genetic engineering approaches rely heavily on a comprehensive understanding of the structure and regulatory properties of plant volatile biosynthetic machinery. In the last decades, biochemical pathways leading to volatile metabolites such as terpenes, benzenoids/phenylpropanoids and green leaf volatiles (GLVs) have become increasingly well documented (Figure 20.1). Not only have many enzymes been characterized, but also the function of their metabolic output has been extensively studied (see Sections II and III in this book). In general, the bioengineering of volatiles can be achieved by the introduction of new genes in the host plant, and the modification of carbon flux through existing pathways via up-/downregulation of biochemical step(s) or by blocking the competing pathways (Figure 20.1). Here, the most promising recent results of metabolic engineering of plant volatiles are summarized (Figure 20.1), and future perspectives for metabolic engineering via gene-editing through CRISPR-technologies and its potential in agriculture are discussed.

20.2 METABOLIC ENGINEERING OF FLORAL VOLATILES

Floral scent is an important trait for crop plants as floral volatiles provide chemical cues to pollinators. In case of ornamentals, floral scent enhances their aesthetic properties that play an important role in human life. Unfortunately, traditional breeding of ornamental crops has focused on producing cultivars with improved vase life, shipping characteristics, and visual aesthetic values (i.e., color, shape) while sacrificing floral scent. These ornamentals are an important target for the genetic manipulation of flower fragrance level and/or profile.

Floral scent bouquet has already been modified in several plant species including petunia (Underwood et al. 2005; Kaminaga et al. 2006; Orlova et al. 2006; Guterman et al. 2006; Dexter et al. 2007; Ben Zvi et al. 2008; Koeduka et al. 2008), carnations (Lavy et al. 2002; Zuker et al. 2002), tobacco (Lücker et al. 2004; Orlova et al. 2009) and, recently, roses (Ben Zvi et al. 2012). While in ornamentals introduced changes have to be detected by humans, more precise fine-tuned emissions synchronized with the activities of potential pollinators have to be achieved in crop plants. However, to date very little is known about the modification of floral volatiles in insect-pollinated crops. Metabolic engineering, in general, relies on successful transformation, which has been developed for some varieties of roses, chrysanthemums, carnations and gerbera, but remains in its infancy for most species. Another critical issue is the availability and selection of suitable promoters, to regulate the introduced transgene and restrict its expression to flowers thereby avoiding potential negative effects on plant development and coordinating emission of engineered volatiles with pollinators' activity.

FIGURE 20.1 (Continued)   F3H, flavanone 3-hydroxylase; FPP, farnesyl pyrophosphate; FPS, farnesyl diphosphate synthase; G3P, glyceraldehyde 3-phosphate; GES, geraniol synthase; GGPP, geranylgeranyl diphosphate; GPP, geranyl diphosphate; GPPS-SSU, geranyl diphosphate synthase small subunit; HMGR, 3-hydroxy-3-methyl-glutaryl-CoA reductase; HPL, hydroperoxide lyase; IGS1, isoeugenol synthase1; IPK, isopentenyl phosphate kinase; IPP, isopentenyl diphosphate; LIS, (S)-linalool synthase; MPD, phosphomevalonate decarboxylase; NDPS, neryl diphosphate synthase; NPP, neryl diphosphate; NSGT1, non-smoky glycosyltransferase1; Nudix, Nudix superfamily hydrolase; ODO1, ODORANT1; PAAS, phenylacetaldehyde synthase; PALs, phenylalanine ammonia lyases; PAP1, production of anthocyanin pigment 1; PAR, phenylacetaldehyde reductase; PEP, phosphoenolpyruvate; PH4, H^+-ATPase proton pump (an R2R3-MYB transcription factor); PhABCG1, *Petunia hybrida* adenosine triphosphate-binding cassette (ABC) transporter1; Phe, phenylalanine; PheA, chorismate mutase/prephenate dehydratase; PHS1, phellandrene synthase1; PhODO1, *Petunia hybrida* odorant1; PMK, phosphomevalonate kinase; SAMT, salicylic acid carboxyl methyltransferase; TE1, thioesterase1; TomloxC, tomato lipoxygenaseC; TPS, terpene synthase; ZIS, α-zingiberene synthase.

20.2.1 Introduction of New Gene(s) or Branchways to Redirect the Metabolic Flux to the Target Compound

The ectopic expression of the *Clarkia breweri (S)-linalool synthase* (LIS), which catalyzes the conversion of geranyl diphosphate (GPP) to (*S*)-linalool (Dudareva 1996), in *Petunia hybrida* and *Dianthus caryophyllus* (carnations), both of which lack this monoterpene, represents the first attempts to modify the floral scent bouquet via the introduction of a new gene (Lücker et al. 2001; Lavy et al. 2002) (Table 20.1). Linalool production was achieved in these transgenic plants, but its emission was undetectable by humans in both cases. While in petunia most of the linalool was sequestered as a nonvolatile linalool glycoside by the action of endogenous glucosyltransferase (Lücker et al. 2001), in transgenic carnations much of the linalool was further oxidized by an endogenous enzyme to *cis*- and *trans*-linalool oxide. Despite the fact that linalool and linalool oxide constituted almost 10% of the total volatiles emitted from the transgenic flowers, human mostly failed to detect a change in floral aroma in smell tests (Lavy et al. 2002). These pioneering studies revealed unexpected problems that can be encountered: (i) modification of the scent compound(s) by endogenous, nonspecific enzymes and (ii) masking of introduced compound(s) by other volatiles.

Another important issue is that production of desirable products in transgenic plants depends on the availability of precursors for the introduced enzyme. Indeed, the ectopic expression of tomato phenylacetaldehyde reductases, which convert phenylacetaldehyde to 2-phenylethanol, in petunia flowers with high levels of phenylacetaldehyde led to the production of higher levels of 2-phenylethanol with a subsequent decrease in phenylacetaldehyde levels (Tieman et al. 2007). In contrast, the ectopic expression of both the strawberry (SAAT) and rose alcohol acetyltransferases (RhAAT) in petunia (Beekwilder 2004; Guterman et al. 2006) as well as *C. breweri* benzyl alcohol acetyltransferase (BEAT) in lisianthus (Aranovich et al. 2007) resulted in no production of the expected compounds due to the lack of available substrates. However, rose RhAAT, although showing the highest preference to geraniol and citronellol, used phenylethyl alcohol and benzyl alcohol and produced the corresponding acetate esters *in vivo* (Guterman et al. 2006), suggesting that enzymes with broad substrate specificity can use other potential substrates in the absence of the preferred substrate (Aranovich et al. 2007).

The simultaneous introduction of three *Citrus limon* monoterpene synthases presents the first example of multigene engineering to produce new volatiles in tobacco plants (Lücker et al. 2004). Transgenic tobacco expressing three lemon monoterpene synthases, γ-terpinene cyclase, (+)-limonene cyclase, and (-)-β-pinene cyclase, produced 10- to 25-fold more monoterpenes than control plants. The magnitude of monoterpene emission in leaves was close to that predicted based on the K_m values of the enzymes for GPP, while the emission levels in flowers were comparable. Moreover, the level of endogenously produced linalool (in flowers), was hardly affected by introduced enzymes suggesting that the GPP pool in flowers was large enough not to limit monoterpene production (Lücker et al. 2004).

Overall, these examples show that the newly synthesized compound(s) may be affected by endogenous enzymes with broad substrate specificity, such as dehydrogenases, phosphatases and glucosyl transferases. Moreover, a detailed understanding of the regulation of the flux through the entire biochemical pathways is essential for the successful metabolic engineering of floral volatiles.

20.2.2 Modification of Existing Endogenous Metabolic Pathways

20.2.2.1 Modification of Floral Scent via Changing Substrate/Precursor Availability

Success in production of target compound relies on the availability and levels of substrates. The dependency of benzenoid/phenylpropanoid emission on the level of phenylalanine (Phe) was clearly demonstrated in petunia flowers. The reduction in Phe levels achieved by RNAi down-regulation of genes involved in the shikimate and Phe biosynthetic pathways resulted in decreased

TABLE 20.1

Reports of Metabolic Engineering of Flower Volatile Compounds

Gene(s)	Origin	Engineered Species	Strategy	Regulation	Changes in Volatile Spectrum[a]	Reference(s)
LIS	Clarkia breweri	Petunia	Ox	Constitutive	Linalool glycoside ↑	Lücker et al. (2001)
LIS	C. breweri	Carnation	Ox	Constitutive	Linalool ↑, linalool oxides ↑	Lavy et al. (2002)
SAAT	Strawberry	Petunia	Ox	Constitutive	No changes	Beekwilder et al. (2004)
RhAAT	Rose	Petunia	Ox	Constitutive	Benzyl acetate ↑, phenylethyl acetate ↑	Guterman et al. (2006)
BEAT	C. breweri	Lisianthus	Ox	Constitutive	No benzyl acetate detected	Aranovich et al. (2007)
BEAT	C. breweri	Lisianthus	Ox and substrate feeding	Constitutive	Benzyl acetate ↑	Aranovich et al. (2007)
PARs	Tomato	Petunia	Ox	Constitutive	2-phenylethanol ↑, 2-phenylacetaldehyde ↓	Tieman et al. (2007)
ADT1	Petunia	Petunia	Downregulation	Petal-specific	Phe ↓, volatiles ↓, shikimate ↓, Trp ↓, arogenate unaltered	Maeda et al. (2010)
ADT1	Petunia	Petunia	Downregulation and shikimate feeding	Petal-specific	Arogenate ↑, Phe ↑	Maeda et al. (2010)
DAHPS1	Petunia	Petunia	Downregulation	Constitutive	Total volatiles ↓, phenylacetaldehyde ↓, 2-phenylethanol ↓, benzyl alcohol ↓, benzaldehyde ↓	Langer et al. (2014)
DAHPS2	Petunia	Petunia	Downregulation	Constitutive	Total volatiles unaltered	Langer et al. (2014)
AroG175	Escherichia coli	Petunia	Ox	Constitutive	Phe-derived volatiles ↑	Oliva et al. (2015)
PheA	E. coli	Petunia	Ox	Constitutive	Phe-derived volatiles ↑	Oliva et al. (2017)
CM1	Petunia	Petunia	Downregulation	Constitutive	Total volatiles ↓,	Colquhoun et al. (2010)
CM2	Petunia	Petunia	Downregulation	Petal-specific	Phe ↓, Phe-derived volatiles ↓, tyrosine ↑, tryptophan ↓	Qian et al. (2019)
F3H	Carnation	Carnation	Downregulation	Constitutive	Methyl benzoate ↑, original orange/reddish color ↓	Zuker et al. (2002)

(Continued)

TABLE 20.1 (Continued)
Reports of Metabolic Engineering of Flower Volatile Compounds

Gene(s)	Origin	Engineered Species	Strategy	Regulation	Changes in Volatile Spectrum[a]	Reference(s)
TE1	Petunia	Petunia	Downregulation	Petal-specific	Benzylbenzoate ↑, phenylethylbenzoate ↑, volatile phenylpropenes↑, isoeugenol ↑, eugenol ↑, vanillin ↑, anthocyanins ↑	Adebesin et al. (2018)
IGS1	Petunia	Petunia	Downregulation	Petal-specific	Isoeugenol ↓, eugenol ↑	Koeduka et al. (2009)
Three PALs	Petunia	Petunia	Downregulation	Petal-specific	Phe ↑, shikimate ↑, tyrosine ↑, tryptophan ↑, benzaldehyde ↓, benzoic acid ↓, methyl benzoate ↓, benzyl alcohol ↓, benzyl benzoate ↓, isoeugenol ↓, eugenol ↓, phenylethylbenzoate ↓	Lynch et al. (2017)
GPPS.SSU	Snapdragon	Tobacco	Ox	Petal-specific	Total monoterpenes ↑, new compound (E)-β-ocimene produced in transgenic flowers	Orlova et al. (2009)
TER, PIN, LIM	Citrus limon	Tobacco	Ox	Constitutive	γ-Terpinene ↑, limonene ↑, β-pinene ↑, linalool ↓	Lücker et al. (2004)
BSMT	Petunia	Petunia	Downregulation	Constitutive	Methylbenzoate ↓	Underwood et al. (2005)
PAAS	Petunia	Petunia	Downregulation	Constitutive	Phenylacetaldehyde ↓, 2-phenylethanol ↓	Kaminaga et al. (2006)
BPBT	Petunia	Petunia	Downregulation	Petal-specific	Benzylbenzoate ↓, phenylethylbenzoate ↓, benzyl alcohol ↑, benzaldehyde ↑	Orlova et al. (2006)
CFAT	Petunia	Petunia	Downregulation	Constitutive	Isoeugenol ↓	Dexter et al. (2007)
CCD	Petunia	Petunia	Cosuppression	Constitutive	β-Ionone ↓	Simkin et al. (2004)
ODO1	Petunia	Petunia	Downregulation	Constitutive	Phenylpropanoid volatiles ↓	Verdonk et al. (2005)
EOB1	Petunia	Petunia	Downregulation	Constitutive	Phenylpropanoid volatiles ↓	Spitzer-Rimon et al. (2012)
EOBII	Petunia	Petunia	Downregulation	Constitutive	Phenylpropanoid volatiles ↓	Spitzer-Rimon et al. (2010)
PH4	Petunia	Petunia	Downregulation	Constitutive	Phenylpropanoid volatiles ↓	Cna'ani et al. (2015)

(Continued)

TABLE 20.1 (Continued)

Reports of Metabolic Engineering of Flower Volatile Compounds

Gene(s)	Origin	Engineered Species	Strategy	Regulation	Changes in Volatile Spectrum[a]	Reference(s)
ERF6	Petunia	Petunia	Downregulation	Constitutive	Phenylpropanoid volatiles ↑	Liu et al. (2017)
MYB4	Petunia	Petunia	Downregulation	Constitutive	Phenylpropanoid volatiles ↑	Colquhoun et al. (2011)
PAP1	Arabidopsis thaliana	Petunia	Ox	Constitutive	Anthocyanins ↑, phenylpropanoid volatiles ↑	Ben Zvi et al. (2008)
PAP1	A. thaliana	Rosa hybrida	Ox	Constitutive	Anthocyanins ↑, phenylpropanoid volatiles ↑, terpenoid volatiles ↑	Ben Zvi et al. (2012)
CAT	Petunia	Petunia	Downregulation	Petal-specific	Phe-derived volatiles ↓	Widhalm et al. (2015)
CAT	Petunia	Petunia	Ox	Petal-specific	Phe-derived volatiles ↑	Widhalm et al. (2015)
PhABCG1	Petunia	Petunia	Downregulation	Petal-specific	Phenylpropanoid volatiles ↓	Adebesin et al. (2017)

Abbreviations: ADT, arogenate dehydratase; AroG, 3-deoxy-D-arabinoheptulosonate 7-phosphate synthase; BEAT, benzyl alcohol acetyltransferase; BPBT, benzyl alcohol/phenyl ethanol benzoyl-CoA transferase; BSMT, benzoic acid/salicylic acid carboxyl methyltransferase; CAT, cationic amino-acid transporter; CCD, carotenoid cleavage dioxygenase; CFAT, coniferyl alcohol acetyltransferase; CM, chorismate mutase; DAHPS, 3-deoxy-D-arabinoheptulosonate 7-phosphate synthase; EOB, emission of benzenoids; ERF6, ethylene response factor6; F3H, flavanone 3-hydroxylase; GPPS.SSU, geranyl diphosphate synthase small subunit; IGS, isoeugenol synthase; LIS, (S)-linalool synthase; LIM, limonene synthase; MYB4, an R2R3-MYB transcription factor; ODO1, odorant1; Ox, overexpression; PAAS, phenylacetaldehyde synthase; PAL, phenylalanine ammonia lyase; PAR, phenylacetaldehyde reductase; PH4, H+-ATPase proton pump (an R2R3-MYB transcription factor); PhABCG1, *P. hybrida* adenosine triphosphate-binding cassette (ABC) transporter1; Phe, phenylalanine; PheA, chorismate mutase/ prephenate dehydratase; PAP1, production of anthocyanin pigment 1; PIN, β-pinene synthase; RhAAT, rose alcohol acetyltransferase; SAAT, strawberry alcohol acetyltransferase; TE, thioesterase; TER, (γ-terpinene synthase).

[a] *Flower volatiles that have been increased* (↑) *or reduced* (↓) *in amount.*

production of Phe-derived volatiles (Maeda et al. 2010; Langer et al. 2014; Colquhoun et al. 2010; Qian et al. 2019). Conversely, higher Phe levels led to enhanced emission of Phe-derived volatiles. Such increased Phe levels were obtained by expressing a bacterial gene encoding a feedback-insensitive 3-deoxy-D-arabino-heptulosonate-7-phosphate synthase (DAHPS), which catalyzes the first committed step of the shikimate pathway, or another bacterial gene, PheA, which encodes a bifunctional chorismate mutase/prephenate dehydratase (Oliva et al. 2015, 2017).

Another example involves carbon flux distribution between eugenol and isoeugenol in petunia flowers. As eugenol synthase (PhEGS1) and isoeugenol synthase (PhIGS1) compete for coniferyl acetate substrate, an increase in eugenol production was obtained upon RNAi downregulation of *PhIGS1* in petunia flowers (Koeduka et al. 2008, 2009). Interestingly, the combined emission output of eugenol and isoeugenol per flower remained very similar to that in wild-type, suggesting that the level of substrate in the cell determines the total production of these two compounds (Koeduka et al. 2009).

One approach to increase substrate pool includes blocking the competitive pathways. Both anthocyanins and methyl benzoate originate from the phenylpropanoid pathway, thus an increase in benzoic acid, a precursor of methyl benzoate, was achieved via redirecting the metabolic flux from the anthocyanin pathway by the antisense suppression of the *flavanone 3-hydroxylase* in transgenic carnations. Transgenic flowers lost their original orange/reddish color but had increased methyl benzoate emission that could be detected by humans (Zuker et al. 2002).

An alternative approach to increase the level of available substrate was demonstrated in tobacco. Ectopic expression of snapdragon (*Antirrhinum majus*) GPPS small subunit (*GPPS.SSU*) increased monoterpene emission from both leaves and flowers, the latter also emitted the monoterpene *(E)*-β-ocimene, which was absent in floral scent of nontransgenic flowers (Orlova et al. 2009). These results provide evidence that the GPP level for endogenous tobacco monoterpene synthases was limiting in control plants. Although the petal-specific *Clarkia breweri* linalool synthase promoter was used, the transgene was expressed to some extent also in leaves. Transgenic tobacco plants showed strong chlorosis and a reduction in stature suggesting that the flux redirection toward monoterpene formation led to a reduction in the GGPP-derived metabolites essential for plant growth and development (Orlova et al. 2009). These results also highlight the importance of tissue- and cell-specific promoters for successful metabolic engineering of floral volatiles.

Recently, a new metabolite reactivation strategy was proposed for increasing terpenoid formation in plants (Henry et al. 2015, 2018). A cytosolically localized pool of isopentenyl phosphate (IP) was discovered in plants, which could be returned to terpenoid biosynthetic network via overexpression of IP kinase (IPK), another IPP-generating enzyme that transforms IP to IPP (Henry et al. 2015). IPK functions together with dedicated Nudix hydrolases, which catalyze dephosphorylation of IPP to IP (Henry et al. 2018), to coordinately regulate the endogenous metabolite recycling and thus the concentration of IPP. Indeed, flowers of *Arabidopsis nudx1* and *nudx3* mutants emit up to 60% and 500% more of the sesquiterpene β-caryophyllene and the monoterpene linalool, respectively, relatively to wild-type flowers (Henry et al. 2018), suggesting that ratio of IPP to IP contributes to regulating the formation of both FPP- and GPP-derived terpenoids.

Henry et al. (2018) also showed that phosphomevalonate kinase (PMK), which phosphorylates phosphomevalonate (MVAP) to mevalonate diphosphate in peroxisomes, shares control of flux through the MVA pathway together with 3-hydroxy-3-methyl-glutaryl-CoA reductase (HMGR). Indeed, coexpression of *AtHMGR* and *AtPMK* in tobacco leaves enhanced not only emission of sesquiterpenes by up to 63-fold, but also monoterpenes on average by 4.6-fold. When *AtHMGR* was coexpressed with bacterial phosphomevalonate decarboxylase (MPD), which provides a bypass to the PMK-catalyzed reaction by decarboxylating MVAP to IP, emission of both sesquiterpenes and monoterpenes increased over 130-fold and 20-fold, respectively, relative to their levels in wild-type tobacco leaves (Henry et al. 2018). These approaches can now also be used to modify floral volatiles.

20.2.2.2 Modification of Floral Scent via Reduction of Individual Compounds

The elimination of individual volatile compounds from the floral bouquet is another approach to modify scent. Transgenic petunias lacking methyl benzoate (Underwood et al. 2005), phenylacetaldehyde (Kaminaga et al. 2006), benzylbenzoate and phenylethylbenzoate (Orlova et al. 2006), isoeugenol (Dexter et al. 2007) or β-ionone (Simkin et al. 2004) were obtained via RNA interference or by cosuppression in the case of β-ionone. It should be noted that RNAi downregulation of a single gene might have broader effect on the produced volatile spectrum. For example, downregulation of the coniferyl alcohol acyltransferase (CFAT) that catalyzes the formation of coniferyl acetate, the precursor of isoeugenol and eugenol, led to lower isoeugenol emission, but also affected emission of five other volatiles in petunia flowers (Dexter et al. 2007). This could be the result of inhibition of several enzymes by elevated levels of coniferyl aldehyde and homovanillic acid in transgenic plants (Dexter et al. 2007), once again highlighting the complexity of volatile metabolic network and its regulation.

20.2.2.3 Modification of Floral Scent Biosynthesis by Transcription Factors

The availability of transcription factors (TFs) specifically regulating the biosynthesis of volatiles provides another way to modify floral scent (Spitzer-Rimon et al. 2010, 2012). Several TFs have already been identified. These include R2R3-type MYBs *ODORANT1* (*ODO1*), *EMISSION OF BENZENOIDS* (*EOBI*), *EOBII*, and *PH4* from petunia that positively regulate the biosynthesis of floral volatiles, as well as *ETHYLENE RESPONSE FACTOR6* (*ERF6*) and *MYB4* TFs which exert a negative control (Verdonk et al. 2005; Spitzer-Rimon et al. 2010, 2012; Colquhoun et al. 2011; Liu et al. 2017; Cna'ani et al. 2015). Recently more TFs regulating floral scent have been reported (Ben Zvi et al. 2008, 2012; Yon et al. 2016); however, some of them regulate not only floral volatile production. Indeed, ectopic expression of *Arabidopsis Production of Anthocyanin Pigment 1* (PAP1) TF in petunia and rose plants led to a several-fold increase in both anthocyanin accumulation and scent production, although the two pathways are temporally separated during flower development (Ben Zvi et al. 2008, 2012). Although PAP1 TF is known to transcriptionally activate the phenylpropanoid pathway, its overexpression in roses also increased emission of terpenoids up to 6.5-fold (Ben Zvi et al. 2012). These experiments show that availability of TFs that specifically regulate the biosynthesis of volatiles and have no adverse effects on plant growth and color formation should be better targets for metabolic engineering of floral volatiles under tissue- and cell-specific promoters.

20.2.2.4 Modification of Floral Scent via Regulating Transporters

Biochemical steps of a single pathway can take place in multiple subcellular locations. As most metabolites are not membrane permeable, transporter(s) facilitates directional passage of metabolic intermediates across organellar membranes. Transport capacity may become a major constraint for efficient genetic manipulation and the regulation of transporters for Phe and volatiles emerged recently as another approach to modify floral scent. Overexpression of a plastidial cationic amino-acid transporter, which is involved in plastidial Phe export, increased emission of Phe-derived volatiles up to 18% relative to control petunia flowers, further suggesting that enhancement of transport capacities may promote emission of volatiles (Widhalm et al. 2015). Transporters also control emission of volatiles from intact cells into the environment. Recently, it has been shown that an adenosine triphosphate-binding cassette (ABC) transporter (PhABCG1) is directly involved in transport of phenylpropanoid and benzenoid volatiles across the plasma membrane in *P. hybrida* flowers (Adebesin et al. 2017). RNAi downregulation of *PhABCG1* expression resulted in decreased emission of VOCs, which accumulated in the plasma membrane to toxic levels (Adebesin et al. 2017). Given the detrimental effects of accumulated volatiles on membrane integrity, transport capacity across membrane could be one of the limiting obstacles in engineering of floral volatiles.

20.2.3 EFFECT OF METABOLIC ENGINEERING OF FLORAL VOLATILES ON INSECT BEHAVIOR

The first investigation of the consequences of metabolic engineering of floral VOCs on pollinator visitation was performed in transgenic tobacco *Nicotiana attenuata* plants with diminished levels of benzyl acetone (Kessler et al. 2008). Monitoring the activity of floral visitors in their native habitat revealed that plants lacking benzyl acetone attracted fewer hawkmoths and hummingbirds than flowers emitting this volatile. It was also shown that changes in more than one volatile compound can influence pollinator behavior. Indeed, upon remodeling of the scent bouquet in rose flowers overexpressing the *Arabidopsis PAP1* TF, the flowers were easily distinguished by honeybees (Ben Zvi et al. 2012), the native pollinators of some wild rose species (Shalit et al. 2004).

The studies on metabolic engineering of floral volatiles, the targeted genes and their metabolic output are summarized in Table 20.1.

20.3 METABOLIC ENGINEERING OF FRUIT VOLATILES

To date, tomato has predominantly been used as a model system for metabolic engineering of fruit aromas (Table 20.2). Tomato fruit aroma has already been changed through upregulating or downregulating the expression of key enzymes in fatty acid degradation (Wang et al. 1996; Prestage et al. 1999; Chen et al. 2004; Dominguez et al. 2010; Goulet et al. 2012; Shen et al. 2014), amino-acid production and catabolism (Tieman et al. 2006, 2010; Kochevenko et al. 2012; Tzin et al. 2013), carotenoid metabolism (Simkin et al. 2004; Gao et al. 2019) and terpenoid pathways (Lewinsohn et al. 2001; Simkin et al. 2004; Davidovich-Rikanati et al. 2007, 2008; Gutensohn et al. 2013, 2014) as well as via overexpression of TFs that control gene expression in one or more metabolic pathways (Xie et al. 2016; Zhang et al. 2016) (Table 20.2). However, in most cases, the effects of changes in fruit volatiles on the perception by humans were either not formally tested or minimal. In contrast, when *Ocimum basilicum* geraniol synthase (*GES*) was overexpressed in tomatoes, the untrained panelists recognized their novel aroma (Davidovich-Rikanati et al. 2007). While exhibiting reduced carotenoid levels, *GES* transgenic fruits accumulated not only geraniol (whose scent is similar to rose) but also other geraniol-derived monoterpenes like geranial and neral, which give a lemon flavor and are the result of action of endogenous enzymes (Davidovich-Rikanati et al. 2007).

Cultivated tomato fruits contain only minute amounts of mono- and sesquiterpenes due to the limited number of expressed endogenous terpene synthases (Falara et al. 2011). Introduction of enzymes with broad substrate specificity could allow access to a precursor pool available in host plants, an approach that was used for fruit metabolic engineering by the expression of a α-zingiberene synthase (*ZIS*), a sesquiterpene synthase from *Ocimum basilicum* which can also accept GPP. This manipulation resulted in the accumulation of both sesquiterpenes and monoterpenes in tomato fruits (Davidovich-Rikanati et al. 2008). Since ZIS has no plastidic transit peptide and is localized in the cytosol, monoterpene production in these transgenic tomatoes suggests that there is an available GPP pool in this compartment. Interestingly, carotenoids levels were slightly decreased in *ZIS* transgenic fruits, while no reduction was observed in the cytosolically synthesized sterols, suggesting that some terpene precursors were shuttled from plastids into the cytosol to make up for the increased precursor demand (Davidovich-Rikanati et al. 2008).

An alternative strategy to access a new precursor pool includes utilization of a single gene involved in more than one pathway. *TomLoxC*, which encodes a 13-lipoxygenase and previously known to be involved in the biosynthesis of fatty acid-derived C5 and C6 volatiles (Chen et al. 2004; Shen et al. 2014), was recently found to affect the production of carotenoid-derived volatile apocarotenoids as well (Gao et al. 2019). Downregulation of *TomLoxC* reduced not only fatty acid-derived volatiles, but also volatile apocarotenoids in ripe tomato fruits. While the exact role of *TomLoxC* in apocarotenoid production remains to be determined (Gao et al. 2019), these experiments provide a

TABLE 20.2

Reports of Metabolic Engineering of Tomato Fruit Volatile Compounds

Gene(s)	Origin	Engineered Species	Strategy	Regulation	Changes in Volatile Spectrum[a]	Reference(s)
Δ9-desaturase	Yeast	Tomato	Ox	Constitutive	Cis-3-hexenol ↑, 1-hexanol ↑, hexanal ↑, cis-3-hexenal ↑, 6-methyl-5-hepten-2-one ↑, 2-isobutylthiazole ↑	Wang et al. (1996)
ω-3 Fatty acid desaturases (FAD3 and FAD7)	FAD3 from *Brassica napus*, FAD7 from *Solanum tuberosum*	Tomato	Ox	Constitutive	(Z)-hex-3-enal ↑, (E)-hex-2-enal ↑, (Z)-hex-3-enol ↑, hexanal ↓	Dominguez et al. (2010)
ADH	Tomato	Tomato	Downregulation	Constitutive or fruit specific	(Z)-3-hexenol ↓, 3-methylbutanal ↑	Prestage et al. (1999)
TomloxC	Tomato	Tomato	Downregulation	Constitutive	Hexanal ↓, hexenal ↓, henenol ↓	Chen et al. (2004)
TomloxC	Tomato	Tomato	Downregulation	Constitutive	Fatty acid–derived volatiles ↓, apocarotenoids ↓	Gao et al. (2019)
CXE1	Tomato	Tomato	Downregulation	Constitutive	Acetate esters ↑	Goulet et al. (2012)
HPL	Tomato	Tomato	Downregulation	Constitutive	1-Penten-3-ol ↑, 1-penten-3-one ↑, pentanal ↑, 1-pentanol ↑, (Z)-2-penten-1-ol ↑, (Z)-3-hexenal ↓, hexanal ↓, (Z)-3-hexen-1-ol ↓, hexyl alcohol ↓	Shen et al., 2014
BCAT1 or BCAT3	Tomato	Tomato	Ox	Constitutive	3-Methylbutanol ↑, 3-methylbutanal ↑, 2-methylbutanol ↑, 2-methylbutanal ↑, isobutyl acetate ↑, isovaleronitrile ↑, 2-isobutylthiazole ↑	Kochevenko et al. (2012)
AADC1A or AADC2	Tomato	Tomato	Ox	Constitutive	2-Phenylacetaldehyde ↑, 2-phenylethanol ↑, 1-nitro-2-phenylethane ↑	Tieman et al. (2006)
AADC2	Tomato	Tomato	Downregulation	Constitutive	2-Phenylacetaldehyde ↓, 2-phenylacetonitrile ↓, 2-phenylethanol ↓, 1-nitro-2-phenylethane ↓	Tieman et al. (2006)
SAMT	Tomato	Tomato	Ox	Constitutive	Methyl salicylate ↑	Tieman et al. (2010)
CCD1	Tomato	Tomato	Downregulation	Constitutive	β-Ionone ↓, geranylacetone ↓	Simkin et al. (2004)

(*Continued*)

TABLE 20.2 (Continued)

Reports of Metabolic Engineering of Tomato Fruit Volatile Compounds

Gene(s)	Origin	Engineered Species	Strategy	Regulation	Changes in Volatile Spectrum[a]	Reference(s)
GPPS-SSU	Snapdragon	Tomato	Ox	Fruit specific	Geraniol ↑, geranial ↑, neral ↑, citronellol ↑, citronellal ↑, trans-lycopene ↓	Gutensohn et al. (2013)
GES	O. basilicum	Tomato	Ox	Fruit specific	Geraniol ↑, geranial ↑, geranyl acetate ↑, citronellol ↑, citronellal ↑, rose oxide ↑, lycopene ↓	Davidovich-Rikanati et al. (2007)
GPPS-SSU + GES	GPPS-SSU from Snapdragon, GES from Ocimum basilicum	Tomato	Ox	Fruit specific	Geraniol ↑, geranial ↑, neral ↑, citronellol ↑, citronellal ↑, citronellyl acetate ↑	Gutensohn et al. (2013)
ZIS	O. basilicum	Tomato	Ox	Fruit specific	α-Zingiberene ↑, α-bergamotene ↑, 7-epi-sesquithujene ↑, β-bisabolene ↑, β-curcumene ↑, α-thujene ↑, α-pinene ↑, β-phellandrene ↑ γ-terpinene ↑	Davidovich-Rikanati et al. (2008)
GPPS-SSU + ZIS	GPPS-SSU from Snapdragon, ZIS from O. basilicum	Tomato	Ox	Fruit specific	α-Thujene ↑, α-pinene ↑, sabinene ↑, γ-terpinene ↑, geraniol ↑, sesquiterpenes ↑	Gutensohn et al. (2013)
LIS	C. breweri	Tomato	Ox	Fruit specific	S-Linalool ↑, 8-hydroxylinalool ↑,	Lewinsohn et al. (2001)
PHS1	Tomato	Tomato	Ox	Fruit specific	Ocimene ↑, myrcene ↑	Gutensohn et al. (2014)
PHS1 + GPPS-SSU	Tomato	Tomato	Ox	Fruit specific	Ocimene ↑, myrcene ↑, citronellal ↑, citronellol ↑, neral ↑, geraniol ↑, geranial ↑	Gutensohn et al. (2014)
NDPS1	Tomato	Tomato	Ox	Fruit specific	Nerol ↑, neral ↑, geranial ↑, geraniol ↑, linalool ↑, α-terpineol ↑, terpinen-4-ol ↑, limonene ↑, lycopene ↓	Gutensohn et al. (2014)
PHS1 + NDPS1	Tomato	Tomato	Ox	Fruit specific	β-Phellandrene ↑, 2-carene ↑	Gutensohn et al. (2014)

Abbreviations: AADC, amino acid decarboxylase; ADH, alcohol dehydrogenase; BCAT, branched-chain amino acid transaminase; CCD, carotenoid cleavage deoxygenase; CXE, carboxylesterase; FAD, fatty acid desaturase; GES, geraniol synthase; GPPS-SSU, geranyl diphosphate synthase small subunit; HPL, hydroperoxide lyase; NDPS, neryl diphosphate synthase; Ox, overexpression; PHS, phellandrene synthase; SAMT, salicylic acid carboxyl methyltransferase; TomloxC, tomato lipoxygenaseC; ZIS, α-zingiberene synthase.

[a] Fruit volatiles that have been increased (↑) or reduced (↓) in amount.

new strategy to improve both the fatty acid-derived and carotenoid-derived fruit volatiles by over-expression of the bifunctional *TomLoxC* in tomato fruits.

Many volatiles accumulate in tomato fruit as glycoconjugates, representing a significant locked-in flavor resources (Ortiz-Serrano and Gil 2007). Thus, releasing volatiles from their glycosylated forms, represents another approach to modify fruit flavor. However, this strategy will not work for volatiles that negatively affect aroma properties, and arresting their release, rather than enhancing it, need to be the goal. Examples include guaiacol and other smoky-related phenylpropanoid volatiles, which are normally released from their glycosylated precursors upon fruit disruption thus contributing to "smoky" aroma that is generally disliked by consumers (Tikunov et al. 2013). The over-expression of the *non-smoky glycosyltransferase1* (*NSGT1*) gene, which encodes an enzyme that catalyzes the conversion of the cleavable diglycosides to noncleavable triglycosides, prevented the deglycosylation and release of the undesirable volatiles from tomato fruit in response to tissue disruption. A 2.8-fold reduction in "smoky" aroma was recognized in blind test by a sensory panel of 13 trained judges (Tikunov et al. 2013).

20.4 METABOLIC ENGINEERING OF PLANT VOLATILES TO INCREASE PLANT DEFENSE

An overview of reports on metabolic engineering to increase defense responses is described in Table 20.3.

20.4.1 Modification of Plant Volatiles via Constitutive Expression of Terpene Synthase Genes to Enhance Defense Responses

A very elegant demonstration of the importance of a volatile sesquiterpene in defense came from work using the natural biodiversity of maize (*Zea mays*). North American maize lines appeared not to have the ability to produce the sesquiterpene *(E)*-β-caryophyllene, while the European maize varieties and its common ancestor teosinte do and are resistant to corn rootworm (*Diabortica virgifera virgifera*) (Kollner et al. 2008). The production of this sesquiterpene was restored in North American lines by constitutive expression of an *(E)*-β-caryophyllene-synthase from *Origanum vulgare*. Lines with restored *(E)*-β-caryophyllene emission were now also effective in attracting below-ground entomopathogenic nematodes that parasitize on corn rootworm larvae and thus suffered less damage from the pest (Degenhardt et al. 2009), demonstrating the importance of *(E)*-β-caryophyllene in the plant's indirect defense. Remarkably, transgenic maize seedlings producing *(E)*-β-caryophyllene exhibited an increase in susceptibility to a pathogenic fungus (*Colletotrichum graminicola*) (Fantaye et al. 2015), indicating that *(E)*-β-caryophyllene is involved in several defense processes.

A role for *(E)*-β-caryophyllene in defenses was also established in rice (*Oryza sativa*). Overexpression of sesquiterpene synthase *OsTPS3* increased the amount of *(E)*-β-caryophyllene and these transgenic plants were preferred by *Anagrus nilaparvatae*, an egg parasitoid of the rice plant hopper *Nilaparvata lugens* (Cheng et al. 2007). Several other studies indicated that overexpression and downregulation of *OsTPS3* in rice led to an increase or decrease of rice plant hopper incidence, respectively, indicating that both parasitoids and herbivores use the same volatile signal to find their host (Xiao et al. 2012; Wang et al. 2015). These studies indicate a role for *(E)*-β-caryophyllene in defense responses and emphasize that metabolic engineering of volatiles to increase plant defenses could also have unexpected or unwanted effects.

Detection of volatile β-ocimene by neighboring plants can induce hormonal signaling pathways (e.g., jasmonic acid; JA, salicylic acid; SA or ethylene; ET) and thus prime their defense responses. For example, exogenous application of allo-ocimene, a structural isomer of β-ocimene, to *Arabidopsis thaliana* leaves induced a subset of defense-related genes

TABLE 20.3

Reports of Metabolic Engineering of Plant Volatiles to Increase Plant Defense

Gene(s)	Origin	Engineered Species	Strategy	Regulation	Changes in Volatile Spectrum[a]	Effect on Plant Defense[b]	Reference(s)
OS	Lima bean	Tobacco	Ox	Constitutive	methyl salicylate ↑, (E)-DMNT ↑, (Z)-DMNT↑, TMTT ↑, when exposed to transgenic plants emitting (E)-β-ocimene	Spider mites (*Mythimna separate*) ↓ Parasitic wasps (*Cotesia kariyai*) ↑	Muroi et al. (2011)
OS	Lima bean	Tomato	Ox	Constitutive	Methyl salicate ↑, (E)-3-hexenol ↑, when exposed to transgenic plants emitting (E)-β-ocimene	Aphids (*Macrosiphum euphorbiae*) ↓ Parasitic wasps (*Aphidius ervi*) ↑	Cascone et al. (2015)
TPS6	Oregano	Maize	Ox	Constitutive	(E)-β-caryophyllene ↑, α-humulene ↑	Corn rootworm (*Diabrotica virgifera*) ↓ Pathogenic fungi (*Colletotrichum graminicola, Fusarium graminearum*) ↓	Degenhardt et al. (2009), Fantaye et al. (2015), Robert et al. (2013)
HPL	*Capsicum annuum*	*A. thaliana*	Ox	Constitutive	(Z)-hexenal ↑, GLV biosynthesis ↑	Parasitic wasps (*Cotesia glomerata*) ↑ Pathogenic fungi (*Botrytis cinerea*) ↓	Shiojiri et al. (2006)
ZIS, FPS	ZIS from *Solanum habrochaites*; FPS from Tomato	Tomato	Ox	Trichome-specific	New compound 7-epizingiberene	White fly (*Bermisia tabaci*) ↓ Spider mites (*Tetranychus evansi, Tetranychus urticae*) ↓	Bleeker et al. (2012)
FPS	Tomato	Tomato	Ox	Trichome-specific	No changes	No changes	Kortbeek et al. (2016)

(Continued)

TABLE 20.3 (Continued)
Reports of Metabolic Engineering of Plant Volatiles to Increase Plant Defense

Gene(s)	Origin	Engineered Species	Strategy	Regulation	Changes in Volatile Spectrum[a]	Effect on Plant Defense[b]	Reference(s)
FPS	Chicken	Tomato	Ox	Trichome-specific	(E)-nerolidol ↑, α-farnesene ↑, monoterpenes ↓	White fly (*Bemisia tabaci*) ↓	Kortbeek et al. (2016)
HPL	*A. thaliana*	*A. thaliana*	Ox	Constitutive	(Z)-3-hexenal ↑, other GLVs ↑	Pathogenic fungi (*Botrytis cinereal*) ↓ Parasitic wasps (*Cotesia glomerata*) ↑	Shiojiri et al. (2006)
HPL	*A. thaliana*	*A. thaliana*	Down-regulation	Constitutive	(Z)-3-hexenal ↓, other GLVs ↓	Pathogenic fungi (*Botrytis cinereal*) ↑ Parasitic wasps (*Cotesia glomerata*) ↓	Shiojiri et al. (2006)
HPL3	Rice	Rice	Down-regulation	Constitutive	(Z)-3-hexenal ↓, other GLVs ↓	Planthoppers (*Nilaparvata lugens*) ↑ Moths (*Chilo suppressalis*) ↓ Planthoppers (*Sogatella furcifera*) ↓ Pathogenic fungi (*Xanthomonas oryzae*) ↓ Parasitic wasps (*Anagrus nilaparvatae*) ↑	Tong et al. (2012)
HPL2	Rice	Rice	Ox	Constitutive	(E)-2-hexenal ↑, other GLVs ↑	Pathogenic fungi (*Xanthomonas oryzae*) ↓	Gomi et al. (2010)

Abbreviations: CHS, chrysanthemol synthase; FPS, farnesyl diphosphate synthase; HPL, hydroperoxide lyase; NES1, linalool/nerolidol synthase; OS, (E)-β-ocimene synthase; Ox, overexpression; TPS, terpene synthase; ZIS, 7-epizingiberene synthases.

[a] Plant volatiles that have been increased (↑) or reduced (↓) in amount.
[b] Incidence and performance of herbivores, pathogens or parasitic organisms that have been increased (↑) or reduced (↓).

and increased resistance against the fungus *Botrytis cinerea* (Kishimoto et al. 2005). Also, application of synthetic β-ocimene on a cotton swab placed near Chinese cabbage (*Brassica pekinensis*) plants resulted in an increased accumulation of glucosinolates, increased attraction of parasitic wasps, reduced aphid performance and induction of JA and SA marker genes (Kang et al. 2018). A follow-up study used transgenic tobacco (*Nicotiana tabacum*) plants ectopically expressing *(E)*-β-ocimene synthase from lima bean (*Phaseolus lunatus*) as emitter. Upon perception of β-ocimene and infestation by herbivores, maize and lima bean receiver plants emitted higher amounts of monoterpenes, sequiterpenes, homoterpenes, methyl salicylate and hexyl acetate. Oviposition by the armyworm *Mythimna separata* on maize and the two-spotted spider mite *Tetranychus urticae* on lima bean plants was reduced. Moreover, the increased volatiles resulted in higher presence of parasitoid wasps (*Cotesia kariyai*) on maize and predatory mites (*Phytoseiulus persimillis*) on lima bean plants (Muroi et al. 2011). Interestingly, lima bean plants grown near lima bean plants infested with *T. urticae* emitted higher levels of the *(E)*-4,8-dimethyl-1,3,7-nonatriene (DMNT) and *(E,E)*-4,8,12-trimethyltrideca-1,3,7,11-tetraene (TMTT) and showed increased defense responses (Arimura et al. 2012). In a comparable study, tomato plants exposed to the mentioned above transgenic β-ocimene-emitter tobacco plants produced higher amounts of terpenes (β-ocimene and limonene), GLVs (*(E)*-3-hexenol and *(Z)*-3-hexenol) and benzenoids (methyl salicylate). Both indirect and direct defense responses against the potato aphid *Macrosiphum euphorbiae* were significantly enhanced, as more parasitoid wasps (*Aphidius ervi*) were attracted and thus impairing aphids reproduction (Cascone et al. 2015). These studies indicate that metabolic engineering to increase emission of airborne signals, such as terpenes, influenced plant–plant communication and improved direct and indirect plant defense responses.

Plant volatiles also play an important role in the so-called push-and-pull insect pest management system. The aim is to "push" herbivores from the field and "pull" them toward more attractive food sources outside the crop field by visual or chemical cues (Cook et al. 2007). It has been long known that aphids produce *(E)*-β-farnesene as an alarm signal to other conspecifics. By doing so, aphids betray themselves to predators and parasitoids who use *(E)*-β-farnesene for the localization of their food source. In transgenic *A. thaliana*, expression of a *(E)*-β-farnesene from peppermint (*Mentha piperita*) or maize (*Zea mays*), resulted in emission of *(E)*-β-farnesene and both repulsion of aphids and attraction of parasitoids was observed (Beale et al. 2006; Schnee et al. 2006). In a follow-up study, the peppermint *(E)*-β-farnesene-synthase was targeted to plastids of wheat (*Triticum aestivum*) to increase emission levels (Bruce et al. 2015). Under laboratory conditions, repulsion of different aphid species (*Sitobion avenae, Rhopalosiphum padi, and Metopolophium dirhodum*) was observed, and *Aphidius ervi* parasitoids spent more time on transgenic plants compared to wild type. To test whether this system could be a viable system for pest management in agriculture, a large-scale field trial was performed. During this trial neither enhanced direct nor indirect defense responses against aphids were observed (Bruce et al. 2015), suggesting that climate conditions, natural occurrence of both aphids and parasitoids and other ecological interactions might be important parameters affecting tritrophic interactions in the field.

Another field trial was performed with maize plants overexpressing an oregano *(E)*-β-caryophyllene synthase (TPS6). Maize plants emitted more *(E)*-β-caryophyllene and α-humulene, but suffered severe costs in terms of seed germination and biomass. Moreover, transgenic plants were more attractive to above-ground herbivores, while root damage by below-ground root nematodes was reduced (Robert et al. 2013). It appears that much needs to be learned about the ecological interactions in agricultural sites before such strategies can be applied in the field. Also, the need for tissue-specific terpene biosynthesis seems to be a desirable alternative to constitutive expression to reduce unwanted developmental and fitness costs.

20.4.2   Modification of Plant Volatiles via Tissue- and Cellular-Specific Targeting of Terpene Synthesis as Viable Tool to Increase Plant Defense

One promising way to reduce unwanted effects is to specifically engineer metabolite pathways in glandular trichomes. Trichomes are hair-like structures on plant epidermal surfaces (see Chapter 13) that produce, store and secrete upon mechanical disruption large amounts of secondary metabolites, such as terpenoids and phenylpropanoids. By restricting metabolic modifications to trichomes using trichome-specific promoters, unwanted effects on crop yield and plant fitness can be prevented. A comprehensive list of trichome-specific promoters is provided in Kortbeek et al. (2016) and references therein.

In *Solanum lycopersicum*, 14 out of 45 known terpene synthases are expressed in stem trichomes and two of these terpene synthases (SlTPS5 and SlTPS9) have been shown to be exclusively expressed in type VI glandular trichomes (Spyropoulou, Haring, and Schuurink 2014; Kortbeek et al. 2016). The induction of these TPSs by JA indicates a role for terpene synthesis in plant defense. The introduction of a 7-epizingiberene synthase from *Solanum habrochaites* (ShZIS) into *S. lycopersicum* under control of the SlTPS5 promoter led to the production of a novel sesquiterpene 7-epizingiberene when coexpressed with the precursor synthase *cis*-prenyltransferase (zFPS) (Bleeker et al. 2012). The resulting transgenic plants displayed severe reduction in fecundity of *Tetranychus urticae* (spider mites) with no associated plant growth retardation. Still, levels of 7-epizingiberene were very low in the transgenic plants compared to *S. habrochaites* (Bleeker et al. 2012). Apparently, the amount of substrate available to ShZIS was limiting, similar as was observed for *Z, Z*-FPP engineering in tobacco (Sallaud et al. 2009). When endogenous genes or closely related genes are overexpressed, such as zFPS, post-transcriptional regulation and feedback inhibition might attenuate the expected output.

Post-transcriptional regulation or feedback inhibition was observed when farnesyl diphosphate synthase (SlFPS) was expressed in cultivated tomato under the trichome-specific promoter of methyl ketone synthase from *S. habrochaites* (ShMKS1), as only moderately increased levels of FPS transcripts were obtained. However, transformation of the homolog of FPS from chicken (*Gallus gallus*) resulted in high expression under the same ShMKS1 promoter. The internal levels of FPP were much higher compared to wild type and to transgenic lines expressing SlFPS, resulting in higher levels of *(E)*-nerolidol and α-farnesene and a remarkable depletion of monoterpenes. Either the loss of monoterpenes or the gain of sesquiterpenes made the transgenic plants less attractive to *B. tabaci* (Kortbeek et al. 2016). Both transgenes were engineered to contain a signal peptide for targeting FPS to the plastids, as it was hypothesized that biosynthesis of terpenes can be improved by targeting the metabolism to non-native cellular compartments. These studies show the flexibility of terpene metabolism and the potential of metabolic engineering of terpenes in specific tissues.

20.4.3   Modification of Green Leaf Volatiles via the Lipoxygenase/ Hydroperoxide Lyase Pathway to Increase Plant Defense

Although the role of GLVs in herbivore- and pathogen-induced defense responses is well-studied (Ameye et al. 2018), few attempts have been made to specifically engineer the lipoxygenase/hydroperoxide lyase pathway for GLV biosynthesis as a viable approach for pest management in plants. In 2006, Shiojiri et al. (2006) investigated the effect of modified GLV emission by overexpression and downregulation of hydroperoxide lyase (HPL) in *A. thaliana* on enzyme activity, GLV emission, indirect defense against herbivores and fungal pathogen resistance. Compared to wild-type (WT), increased HPL enzyme activity in intact transgenic plants did not result in increased *(Z)*-3-hexenal emission, the first C6-aldehyde in the GLV biosynthetic pathway. But upon damage by herbivores (*Pieris rapae*) or infestation with pathogens (*Botrytis cinerea*), *(Z)*-3-hexenal levels increased more strongly in transgenic lines overexpressing *HPL* than in WT plants. The enhanced production of *(Z)*-3-hexenal led to a higher attractiveness to parasitic wasps (*Cotesia glomerata*) and smaller necrotic lesions after *B. cinerea* infestation. In contrast, silenced *HPL* lines produced less *(Z)*-3-hexenal, with lower incidence of parasitic wasps and more susceptibility to *B. cinerea* (Shiojiri et al. 2006). In rice, there are three *HPL* genes encoding enzymes with substrate specificity for 9-/13-hydroperoxides (*HPL1* and *HPL2*) or 13-hydroperoxy linolenic acid (*HPL3*). Interestingly, silencing of *HPL3* in rice resulted in decreased *(Z)*-3-hexenal emission, higher susceptibility to the rice plant hopper *Nilaparvata lugens*, but more resistance to other herbivores (*Chilo suppressalis* and *Sogatella furcifera*) and pathogens (*Xanthomonas oryzae*) (Tong et al. 2012). These rice plants also contain higher levels of JA due to the increased flux to JA upon *HPL3* silencing as JA biosynthesis and GLV production rely on the same precursor. Overexpression of *HPL2* in rice led to increased GLV production, more specifically *(E)*-2-hexenal, and an increase in resistance to bacterial blight, but the effect on JA production was not investigated (Gomi et al. 2010). These studies (Gomi et al. 2010; Tong et al. 2012) indicate that the exact chemical composition after herbivore damage needs to be studied in detail before targeted metabolic engineering experiments commence.

20.5   FUTURE PERSPECTIVE: CRISPR-TECHNOLOGIES

As the above studies will continue to direct breeders in the right direction, the new generation gene-editing techniques will facilitate a more elegant, direct and cost-efficient approach to alter volatile production.

CRISPR-based gene-editing techniques offer the possibility to introduce specific mutations at almost any location in the genome of the organism of interest. The induction of mutations relies on the expression of an endonuclease enzyme (Cas9) and a specific single guide RNA-molecule to guide the endonuclease to the desired sequence in the genome. This expression is usually achieved by integration of the transgene in the genome by *Agrobacterium*-mediated T-DNA transfer. Upon recognition of the target sequence, the endonuclease will form a double-stranded break (DSB) in the target DNA. The DSB can be repaired by any one of two endogenous repair mechanisms, nonhomologous end joining (NHEJ) or homology-directed repair (HDR). NHEJ is prone to errors and can induce small insertions/deletions, while HDR can facilitate the integration of donor template DNA, flanked by a sequence homologous to the sequence flanking the DSB, at the site of the DSB. After site-directed mutagenesis occurs, the T-DNA insertion can be removed by backcrossing, allowing the progeny to meet the USDA guidelines for CRISPR/Cas9-directed mutagenesis. Transgene-free mutagenesis can also occur via introduction of the CRISPR/Cas9-protein complex supplemented with gRNAs in protoplasts (Woo et al. 2015; Malnoy et al. 2016; Zhang et al. 2016). By doing so, scientist can generate homozygous mutants without the need to backcross, even in the T_0 generation.

What can we expect in this new biotechnology era when it comes to metabolic engineering of plant volatiles? First, CRISPR-directed editing can target multiple alleles and multiple (homologous) genes in one experiment. This allows for facile editing of polyploid genomes, a significant challenge in common gene-editing techniques. Moreover, in floral volatile biosynthesis, several biosynthetic genes have multiple homologs with similar functionalities. For example, in *Petunia x hybrida* caffeoyl-coenzyme A *O*-methyltransferase CCoAOMT1 is known to catalyze the conversion of caffeoyl-CoA to feruloyl-CoA in the production of eugenol and isoeugenol, but two more homologs are present. Simultaneous targeting of all homologs can thus further elucidate the role of CCoAOMTs in volatile biosynthesis.

With homology-directed repair, it is possible to swap a specific promoter with other promoters. With two sgRNAs flanking the promoter Cas9 can excise the promoter from the genomic DNA. When the new promoter is present as double-stranded DNA during DNA repair, it can serve as template and be integrated at the location of the double-stranded break by homologous recombination.

By doing so, overexpression or tissue/organelle specific expression of a gene of interest under an (endogenous) promoter can be achieved. Other possible outcomes using this same method are the reparation of defective genes, insertion of novel genes at a predetermined location in the genome or silencing of genes by DNA insertion.

Site-directed mutagenesis depending on NHEJ is a random process. Endonucleases like Cas9 can lose their catalytic domain by protein engineering (called dead Cas9, or dCas9), without losing the ability to be guided to specific DNA sequences. DNA modifying enzymes (e.g., DNA methyltransferases and deaminases) can be fused to dCas9 and induce specific (epigenetic) mutations without cleaving the DNA. By adding or removing methyl groups by DNA methyltransferases in promoters, genes can be silenced or activated respectively (Gallego-Bartolomé et al. 2018). Adenosine and cytosine deaminases lead to the conversion of A-T to G-C and C-G to T-A conversions respectively (Cohen 2017; Zong et al. 2017; Kang et al. 2018). The expected mutations can be predicted with precision, and these tools can therefore be used to induce specific mutations in *cis*-regulatory sequences in promoters or to modify proteins, for example in important domain sequences, without compromising expression of the target genes.

These tools are only at the dawn of a new era of innovation that will lead to a well-equipped toolbox for site-directed mutagenesis and facilitate scientists to study plant volatiles by metabolic pathway engineering. With increasing allowance of CRISPR-based editing techniques for commercial purposes, metabolic engineered plants might find their way to farmers to increase crop performance and floral scent.

ACKNOWLEDGMENTS

This work was supported by grant IOS-1655438 from the National Science Foundation, and by the USDA National Institute of Food and Agriculture Hatch project 177845 to ND and by grant 2014-01-04PRO from the Taskforce of Applied Research (SIA-RAAK) by the Netherlands Organization for Scientific Research (NWO). We thank Nelleke Kreike for fruitful discussions and critical comments on the manuscript.

REFERENCES

Adebesin, Funmilayo, Joshua R. Widhalm, Benoît Boachon, François Lefèvre, Baptiste Pierman, Joseph H. Lynch, Iftekhar Alam, et al. 2017. "Emission of Volatile Organic Compounds from Petunia Flowers Is Facilitated by an ABC Transporter." *Science* 356 (6345): 1386–88. doi:10.1126/science.aan0826.

Adebesin, Funmilayo, Joshua R. Widhalm, Joseph H. Lynch, Rachel M. McCoy, and Natalia Dudareva. 2018. "A Peroxisomal Thioesterase Plays Auxiliary Roles in Plant (β)-Oxidative Benzoic Acid Metabolism." *Plant Journal* 93 (5): 905–16. doi:10.1111/tpj.13818.

Ameye, Maarten, Silke Allmann, Jan Verwaeren, Guy Smagghe, Geert Haesaert, Robert C. Schuurink, and Kris Audenaert. 2018. "Green Leaf Volatile Production by Plants: A Meta-Analysis." *New Phytologist* 220 (3): 666–83. doi:10.1111/nph.14671.

Aranovich, Dina, Efraim Lewinsohn, and Michele Zaccai. 2007. "Post-Harvest Enhancement of Aroma in Transgenic Lisianthus (*Eustoma grandiflorum*) Using the *Clarkia breweri* Benzyl Alcohol Acetyltransferase (*BEAT*) Gene." *Postharvest Biology and Technology* 43 (2): 255–60. doi:10.1016/j.postharvbio.2006.09.001.

Arimura, Gen ichiro, Atsushi Muroi, and Masahiro Nishihara. 2012. "Plant-Plant-Plant Communications, Mediated by *(E)*-β-Ocimene Emitted from Transgenic Tobacco Plants, Prime Indirect Defense Responses of Lima Beans." *Journal of Plant Interactions* 7 (3): 193–96. doi:10.1080/17429145.2011.650714.

Beale, Michael H., Michael A. Birkett, Toby J. A. Bruce, Keith Chamberlain, Linda M. Field, Alison K. Huttly, Janet L. Martin, et al. 2006. "Aphid Alarm Pheromone Produced by Transgenic Plants Affects Aphid and Parasitoid Behavior." *Proceedings of the National Academy of Sciences* 103 (27): 10509–13. doi:10.1073/pnas.0603998103.

Beekwilder, Jules. 2004. "Functional Characterization of Enzymes Forming Volatile Esters from Strawberry and Banana." *Plant Physiology* 135 (4): 1865–78. doi:10.1104/pp.104.042580.

Ben Zvi, Michal M., Elena Shklarman, Tania Masci, Haim Kalev, Thomas Debener, Sharoni Shafir, Marianna Ovadis, and Alexander Vainstein. 2012. "PAP1 Transcription Factor Enhances Production of Phenylpropanoid and Terpenoid Scent Compounds in Rose Flowers." *New Phytologist* 195 (2): 335–45. doi:10.1111/j.1469-8137.2012.04161.x.

Ben Zvi, Michal M., Florence Negre-Zakharov, Tania Masci, Marianna Ovadis, Elena Shklarman, Hagit Ben-Meir, Tzvi Tzfira, Natalia Dudareva, and Alexander Vainstein. 2008. "Interlinking Showy Traits: Co-Engineering of Scent and Colour Biosynthesis in Flowers." *Plant Biotechnology Journal* 6 (4): 403–15. doi:10.1111/j.1467-7652.2008.00329.x.

Bleeker, Petra M., Rossana Mirabella, Paul J. Diergaarde, Arjen VanDoorn, Alain Tissier, Merijn R. Kant, Marcel Prins, Martin de Vos, Michel A. Haring, and Robert C. Schuurink. 2012. "Improved Herbivore Resistance in Cultivated Tomato with the Sesquiterpene Biosynthetic Pathway from a Wild Relative." *Proceedings of the National Academy of Sciences* 109 (49): 20124–29. doi:10.1073/pnas.1208756109.

Bruce, Toby J.A., Gudbjorg I. Aradottir, Lesley E. Smart, Janet L. Martin, John C. Caulfield, Angela Doherty, Caroline A. Sparks, et al. 2015. "The First Crop Plant Genetically Engineered to Release an Insect Pheromone for Defence." *Nature Scientific Reports* 5: 11183. doi:10.1038/srep11183.

Cascone, Pasquale, Luigi Iodice, Massimo E. Maffei, Simone Bossi, Gen ichiro Arimura, and Emilio Guerrieri. 2015. "Tobacco Overexpressing β-Ocimene Induces Direct and Indirect Responses against Aphids in Receiver Tomato Plants." *Journal of Plant Physiology* 173: 28–32. doi:10.1016/j.jplph.2014.08.011.

Chen, Guoping, Rachel Hackett, David Walker, Andy Taylor, Zhefeng Lin, and Donald Grierson. 2004. "Identification of a Specific Isoform of Tomato Lipoxygenase (TomloxC) Involved in the Generation of Fatty Acid-Derived Flavor Compounds." *Plant Physiology* 136 (1): 2641–51. doi:10.1104/pp.104.041608.

Cheng, Ai X., Cai Y. Xiang, Jian X. Li, Chang Q. Yang, Wen L. Hu, Ling J. Wang, Yong G. Lou, and Xiao Y. Chen. 2007. "The Rice *(E)*-β-Caryophyllene Synthase (OsTPS3) Accounts for the Major Inducible Volatile Sesquiterpenes." *Phytochemistry* 68 (12): 1632–41. doi:10.1016/j.phytochem.2007.04.008.

Cna'ani, Alon, Ben Spitzer-Rimon, Jasmin Ravid, Moran Farhi, Tania Masci, Javiera Aravena-Calvo, Marianna Ovadis, and Alexander Vainstein. 2015. "Two Showy Traits, Scent Emission and Pigmentation, Are Finely Coregulated by the MYB Transcription Factor PH4 in Petunia Flowers." *New Phytologist* 208 (3): 708–14. doi:10.1111/nph.13534.

Cohen, Jon. 2017. "'Base Editors' Open New Way to Fix Mutations." *Science* 358 (6362): 432–33. doi:10.1126/science.358.6362.432.

Colquhoun, Thomas A., Bernardus C. J. Schimmel, Joo Y. Kim, Didier Reinhardt, Kenneth Cline, and David G. Clark. 2010. "A Petunia Chorismate Mutase Specialized for the Production of Floral Volatiles." *Plant Journal* 61 (1): 145–55. doi:10.1111/j.1365-313X.2009.04042.x.

Colquhoun, Thomas A., Joo Young Kim, Ashlyn E. Wedde, Laura A. Levin, Kyle C. Schmitt, Robert C. Schuurink, and David G. Clark. 2011. "*PhMYB4* Fine-Tunes the Floral Volatile Signature of *Petunia×hybrida* through *PhC4H*." *Journal of Experimental Botany* 62 (3): 1133–43. doi:10.1093/jxb/erq342.

Cook, Samantha M., Zeyaur R. Khan, and John A. Pickett. 2007. "The Use of Push-Pull Strategies in Integrated Pest Management." *Annual Review of Entomology* 52 (1): 375–400. doi:10.1146/annurev.ento.52.110405.091407.

Davidovich-Rikanati, Rachel, Efraim Lewinsohn, Einat Bar, Yoko Lijima, Eran Pichersky, and Yaron Sitrit. 2008. "Overexpression of the Lemon Basil α-Zingiberene Synthase Gene Increases Both Mono- and Sesquiterpene Contents in Tomato Fruit." *The Plant Journal* 228–38. doi:10.1111/j.1365-313X.2008.03599.x.

Davidovich-Rikanati, Rachel, Yaron Sitrit, Yaakov Tadmor, Yoko Iijima, Natalya Bilenko, Einat Bar, Bentsi Carmona, et al. 2007. "Enrichment of Tomato Flavor by Diversion of the Early Plastidial Terpenoid Pathway." *Nature Biotechnology* 25 (8): 899–901. doi:10.1038/nbt1312.

Degenhardt, Jörg, Ivan Hiltpold, Tobias G. Kollner, Monika Frey, Alfons Gierl, Jonathan Gershenzon, Bruce E. Hibbard, Mark R. Ellersieck, and Ted C. J. Turlings. 2009. "Restoring a Maize Root Signal That Attracts Insect-Killing Nematodes to Control a Major Pest." *Proceedings of the National Academy of Sciences* 106 (32): 13213–18. doi:10.1073/pnas.0906365106.

Dexter, Richard, Anthony Qualley, Christine M. Kish, Choong Je Ma, Takao Koeduka, Dinesh A. Nagegowda, Natalia Dudareva, Eran Pichersky, and David Clark. 2007. "Characterization of a Petunia Acetyltransferase Involved in the Biosynthesis of the Floral Volatile Isoeugenol." *Plant Journal* 49 (2): 265–75. doi:10.1111/j.1365-313X.2006.02954.x.

Dominguez, Teresa, María L. Hernandez, Joyce C. Pennycooke, Pedro Jimenez, José M. Martinez-Rivas, Carlos Sanz, Eric J. Stockinger, José J. Sanchez-Serrano, and Maite Sanmartin. 2010. "Increasing ω-3 Desaturase Expression in Tomato Results in Altered Aroma Profile and Enhanced Resistance to Cold Stress." *Plant Physiology* 153 (2): 655–65. doi:10.1104/pp.110.154815.

Dudareva, Natalia. 1996. "Evolution of Floral Scent in *Clarkia*: Novel Patterns of S-Linalool Synthase Gene Expression in the *C. breweri* Flower." *Plant Cell* 8 (7): 1137–48. doi:10.1105/tpc.8.7.1137.

Falara, Vasiliki, Tariq A. Akhtar, Thuong T. H. Nguyen, Eleni A. Spyropoulou, Petra M. Bleeker, Ines Schauvinhold, Yuki Matsuba, et al. 2011. "The Tomato Terpene Synthase Gene Family." *Plant Physiology* 157 (2): 770–89. doi:10.1104/pp.111.179648.

Fantaye, Chalie A., Diana Köpke, Jonathan Gershenzon, and Jörg Degenhardt. 2015. "Restoring *(E)*-β-Caryophyllene Production in a Non-Producing Maize Line Compromises Its Resistance against the Fungus *Colletotrichum graminicola*." *Journal of Chemical Ecology* 41 (3): 213–23. doi:10.1007/s10886-015-0556-z.

Gallego-Bartolomé, Javier, Jason Gardiner, Wanlu Liu, Ashot Papikian, Basudev Ghoshal, Hsuan Yu Kuo, Jenny Miao-Chi Zhao, David J. Segal, and Steven E. Jacobsen. 2018. "Targeted DNA Demethylation of the *Arabidopsis* Genome Using the Human TET1 Catalytic Domain." *Proceedings of the National Academy of Sciences* 115 (9): E2125–34. doi:10.1073/pnas.1716945115.

Gao, Lei, Itay Gonda, Honghe Sun, Kan Bao, Denise M. Tieman, Tara L. Fish, Kaitlin A. Stromberg, et al. 2019. "The Tomato Pan-Genome Uncovers New Genes and a Rare Allele Regulating Fruit Flavor." *Nature Genetics* 51 (5): 1044-51. doi:10.1038/s41588-019-0410-2.

Gomi, Kenji, Masaru Satoh, Rika Ozawa, Yumi Shinonaga, Sachiyo Sanada, Katsutomo Sasaki, Masaya Matsumura, et al. 2010. "Role of Hydroperoxide Lyase in White-Backed Planthopper (*Sogatella furcifera Horváth*)-Induced Resistance to Bacterial Blight in Rice, *Oryza Sativa* L." *Plant Journal* 61 (1): 46–57. doi:10.1111/j.1365-313X.2009.04031.x.

Goulet, Charles, Melissa H. Mageroy, Nghi B. Lam, Abbye Floystad, Denise M. Tieman, and Harry J. Klee. 2012. "Role of an Esterase in Flavor Volatile Variation within the Tomato Clade." *Proceedings of the National Academy of Sciences* 109 (46): 19009–14. doi:10.1073/pnas.1216515109.

Gutensohn, Michael, Irina Orlova, Thuong T. H. Nguyen, Rachel Davidovich-Rikanati, Mario G. Ferruzzi, Yaron Sitrit, Efraim Lewinsohn, Eran Pichersky, and Natalia Dudareva. 2013. "Cytosolic Monoterpene Biosynthesis Is Supported by Plastid-Generated Geranyl Diphosphate Substrate in Transgenic Tomato Fruits." *Plant Journal* 75 (3): 351–63. doi:10.1111/tpj.12212.

Gutensohn, Michael, Thuong T. H. Nguyen, Richard D. Mcmahon, Ian Kaplan, Eran Pichersky, and Natalia Dudareva. 2014. "Metabolic Engineering of Monoterpene Biosynthesis in Tomato Fruits via Introduction of the Non-Canonical Substrate Neryl Diphosphate." *Metabolic Engineering* 24 (7): 107–16. doi:10.1016/j.ymben.2014.05.008.

Guterman, Inna, Tania Masci, Xinlu Chen, Florence Negre, Eran Pichersky, Natalia Dudareva, David Weiss, and Alexander Vainstein. 2006. "Generation of Phenylpropanoid Pathway-Derived Volatiles in Transgenic Plants: Rose Alcohol Acetyltransferase Produces Phenylethyl Acetate and Benzyl Acetate in Petunia Flowers." *Plant Molecular Biology* 60 (4): 555–63. doi:10.1007/s11103-005-4924-x.

Henry, Laura K., Michael Gutensohn, Suzanne T. Thomas, Joseph P. Noel, and Natalia Dudareva. 2015. "Orthologs of the Archaeal Isopentenyl Phosphate Kinase Regulate Terpenoid Production in Plants." *Proceedings of the National Academy of Sciences* 112 (32): 10050–55. doi:10.1073/pnas.1504798112.

Henry, Laura K., Suzanne T. Thomas, Joshua R. Widhalm, Joseph H. Lynch, Thomas C. Davis, Sharon A. Kessler, Jörg Bohlmann, Joseph P. Noel, and Natalia Dudareva. 2018. "Contribution of Isopentenyl Phosphate to Plant Terpenoid Metabolism." *Nature Plants* 4 (9): 721–29. doi:10.1038/s41477-018-0220-z.

Kamal, Hena M., Tadashi Takashina, Hiroaki Egashira, Hideki Satoh, and Shigeru Imanishi. 2008. "Introduction of Aromatic Fragrance into Cultivated Tomato from the 'Peruvianum Complex.'" *Plant Breeding* 120 (2): 179–81. doi:10.1046/j.1439-0523.2001.00581.x.

Kaminaga, Yasuhisa, Jennifer Schnepp, Greg Peel, Christine M. Kish, Gili Ben-Nissan, David Weiss, Irina Orlova, et al. 2006. "Plant Phenylacetaldehyde Synthase Is a Bifunctional Homotetrameric Enzyme That Catalyzes Phenylalanine Decarboxylation and Oxidation." *Journal of Biological Chemistry* 281 (33): 23357–66. doi:10.1074/jbc. M602708200.

Kang, Beum Chang, Jae Young Yun, Sang Tae Kim, You Jin Shin, Jahee Ryu, Minkyung Choi, Je Wook Woo, and Jin Soo Kim. 2018. "Precision Genome Engineering through Adenine Base Editing in Plants." *Nature Plants* 4 (7): 427–31. doi:10.1038/s41477-018-0178-x.

Kang, Zhi-Wei, Fang-Hua Liu, Zhan-Feng Zhang, Hong-Gang Tian, and Tong-Xian Liu. 2018. "Volatile β-Ocimene Can Regulate Developmental Performance of Peach Aphid *Myzus Persicae* through Activation of Defense Responses in Chinese Cabbage *Brassica pekinensis.*" *Frontiers in Plant Science* 9 (5): 1–12. doi:10.3389/fpls.2018.00708.

Kessler, Danny, Klaus Gase, and Ian T. Baldwin. 2008. "Field Experiments with Transformed Plants Reveal the Sense of Floral Scents." *Science* 321 (5893): 1200–1202. doi:10.1126/science.1160072.

Kishimoto, Kyutaro, Kenji Matsui, Rika Ozawa, and Junji Takabayashi. 2005. "Volatile C6-Aldehydes and Allo-Ocimene Activate Defense Genes and Induce Resistance against *Botrytis cinerea* in *Arabidopsis thaliana.*" *Plant and Cell Physiology* 46 (7): 1093–1102. doi:10.1093/pcp/pci122.

Kochevenko, Andrej, Wagner L. Araújo, Gregory S. Maloney, Denise M. Tieman, Phuc Thi Do, Mark G. Taylor, Harry J. Klee, and Alisdair R. Fernie. 2012. "Catabolism of Branched Chain Amino Acids Supports Respiration but Not Volatile Synthesis in Tomato Fruits." *Molecular Plant* 5 (2): 366–75. doi:10.1093/mp/ssr108.

Koeduka, Takao, Gordon V. Louie, Irina Orlova, Christine M. Kish, Mwafaq Ibdah, Curtis G. Wilkerson, Marianne E. Bowman, et al. 2008. "The Multiple Phenylpropene Synthases in Both *Clarkia breweri* and *Petunia hybrida* Represent Two Distinct Protein Lineages." *Plant Journal* 54 (3): 362–74. doi:10.1111/j.1365-313X.2008.03412.x.

Koeduka, Takao, Irina Orlova, Thomas J. Baiga, Joseph P. Noel, Natalia Dudareva, and Eran Pichersky. 2009. "The Lack of Floral Synthesis and Emission of Isoeugenol in *Petunia axillaris* Subsp. *parodii* Is Due to a Mutation in the Isoeugenol Synthase Gene." *Plant Journal* 58 (6): 961–69. doi:10.1111/j.1365-313X.2009.03834.x.

Kollner, Tobias G., Matthias Held, Claudia Lenk, Ivan Hiltpold, Ted C. J. Turlings, Jonathan Gershenzon, and Jörg Degenhardt. 2008. "A Maize *(E)*-β-Caryophyllene Synthase Implicated in Indirect Defense Responses against Herbivores Is Not Expressed in Most American Maize Varieties." *The Plant Cell* 20 (2): 482–94. doi:10.1105/tpc.107.051672.

Kortbeek, Ruy W.J., Jiesen Xu, Aldana Ramirez, Eleni Spyropoulou, Paul Diergaarde, Ilona Otten-Bruggeman, Michiel de Both, et al. 2016. "Engineering of Tomato Glandular Trichomes for the Production of Specialized Metabolites." *Methods in Enzymology* 576: 305–31. doi:10.1016/bs.mie.2016.02.014.

Langer, Kelly M., Correy R. Jones, Elizabeth A. Jaworski, Gabrielle V. Rushing, Joo Young Kim, David G. Clark, and Thomas A. Colquhoun. 2014. "PhDAHP1 Is Required for Floral Volatile Benzenoid/ Phenylpropanoid Biosynthesis in *Petunia × Hybrida* Cv 'Mitchell Diploid.'" *Phytochemistry* 103 (7): 22–31. doi:10.1016/j.phytochem.2014.04.004.

Lavy, Michal, Amir Zuker, Efraim Lewinsohn, Olga Larkov, Uzi Ravid, Alexander Vainstein, and David Weiss. 2002. "Linalool and Linalool Oxide Production in Transgenic Carnation Flowers Expressing the *Clarkia breweri* Linalool Synthase Gene." *Molecular Breeding* 9 (2): 103–11. doi:10.1023/A:1026755414773.

Lewinsohn, Efraim, Fernond Schalechet, Jack Wilkinson, Kenji Matsui, Yaakov Tadmor, Kyoung-Hee Nam, Orit Amar, et al. 2001. "Enhanced Levels of the Aroma and Flavor Compound." *Plant Physiology* 127: 1256–65. doi:10.1104/pp.010293.1256.

Liu, Fei, Zhina Xiao, Li Yang, Qian Chen, Lu Shao, Juanxu Liu, and Yixun Yu. 2017. "PhERF6, Interacting with EOBI, Negatively Regulates Fragrance Biosynthesis in Petunia Flowers." *New Phytologist* 215 (4): 1490–1502. doi:10.1111/nph.14675.

Lücker, Joost, Harro J. Bouwmeester, Wilfried Schwab, Jan Blaas, Linus H. W. van der Plas, and Harrie A. Verhoeven. 2001. "Expression of *Clarkia S*-Linalool Synthase in Transgenic Petunia Plants Results in the Accumulation of *S*-Linalyl-Beta-D-Glucopyranoside." *Plant Journal* 27 (4): 315–24. doi:10.1046/j.1365-313X.2001.01097.x.

Lücker, Joost, Wilfried Schwab, Bianca van Hautum, Jan Blaas, Linus H. W. van der Plas, and Harro J. Bouwmeester. 2004. "Increased and Altered Fragrance of Tobacco Plants after Metabolic Engineering Using Three Monoterpene Synthases from Lemon." *Plant Physiology* 134 (1): 510–19. doi:10.1104/pp.103.030189.

Lynch, Joseph H., Irina Orlova, Chengsong Zhao, Longyun Guo, Rohit Jaini, Hiroshi Maeda, Tariq Akhtar, et al. 2017. "Multifaceted Plant Responses to Circumvent Phe Hyperaccumulation by Downregulation of Flux through the Shikimate Pathway and by Vacuolar Phe Sequestration." *Plant Journal* 92 (5): 939–50. doi:10.1111/tpj.13730.

Maeda, Hiroshi, Ajit K. Shasany, Jennifer Schnepp, Irina Orlova, Goro Taguchi, Bruce R. Cooper, David Rhodes, Eran Pichersky, and Natalia Dudareva. 2010. "RNAi Suppression of Arogenate Dehydratase1 Reveals That Phenylalanine Is Synthesized Predominantly via the Arogenate Pathway in Petunia Petals." *Plant Cell* 22 (3): 832–49. doi:10.1105/tpc.109.073247.

Malnoy, Mickael, Roberto Viola, Min-Hee Jung, Ok-Jae Koo, Seokjoong Kim, Jin-Soo Kim, Riccardo Velasco, and Chidananda Nagamangala Kanchiswamy. 2016. "DNA-Free Genetically Edited Grapevine and Apple Protoplast Using CRISPR/Cas9 Ribonucleoproteins." *Frontiers in Plant Science* 7 (12): 1–9. doi:10.3389/fpls.2016.01904.

Muroi, Atsushi, Abdelaziz Ramadam, Masahiro Nishihara, Masaki Yamamoto, Rika Ozawa, Junji Takabayashi, and Gen-Ichiro Arimura. 2011. "The Composite Effect of Transgenic Plant Volatiles for Acquired Immunity to Herbivory Caused by Inter-Plant Communications." *Plos One* 6 (10): 25–29. doi:10.1371/journal.pone.0024594.

Oliva, Moran, Einat Bar, Rinat Ovadia, Avichai Perl, Gad Galili, Efraim Lewinsohn, and Michal Oren-Shamir. 2017. "Phenylpyruvate Contributes to the Synthesis of Fragrant Benzenoid–Phenylpropanoids in *Petunia × Hybrida* Flowers." *Frontiers in Plant Science* 8 (5): 1–13. doi:10.3389/fpls.2017.00769.

Oliva, Moran, Rinat Ovadia, Avichai Perl, Einat Bar, Efraim Lewinsohn, Gad Galili, and Michal Oren-Shamir. 2015. "Enhanced Formation of Aromatic Amino Acids Increases Fragrance without Affecting Flower Longevity or Pigmentation in *Petunia × Hybrida*." *Plant Biotechnology Journal* 13 (1): 125–36. doi:10.1111/pbi.12253.

Orlova, Irina, Amy Marshall-Colon, Jennifer Schnepp, Barbara Wood, Marina Varbanova, Eyal Fridman, Joshua J. Blakeslee, et al. 2006. "Reduction of Benzenoid Synthesis in Petunia Flowers Reveals Multiple Pathways to Benzoic Acid and Enhancement in Auxin Transport." *Plant Cell* 18 (12): 3458–75. doi:10.1105/tpc.106.046227.

Orlova, Irina, Dinesh A. Nagegowda, Christine M. Kish, Michael Gutensohn, Hiroshi Maeda, Marina Varbanova, Eyal Fridman, et al. 2009. "The Small Subunit of Snapdragon Geranyl Diphosphate Synthase Modifies the Chain Length Specificity of Tobacco Geranylgeranyl Diphosphate Synthase in Planta." *Plant Cell* 21 (12): 4002–17. doi:10.1105/tpc.109.071282.

Ortiz-Serrano, Pepa, and José Vicente Gil. 2007. "Quantitation of Free and Glycosidically Bound Volatiles in and Effect of Glycosidase Addition on Three Tomato Varieties (*Solanum lycopersicum* L.)." *Journal of Agricultural and Food Chemistry* 55 (22): 9170–76. doi:10.1021/jf0715673.

Prestage, Samantha, Rob S. T. Linforth, Andrew J. Taylor, Elizabeth Lee, Jim Speirs, and Wolfgang Schuch. 1999. "Volatile Production in Tomato Fruit with Modified Alcohol Dehydrogenase Activity." *Journal of the Science of Food and Agriculture* 79 (1): 131–36. doi:10.1002/(SICI)1097-0010(199901)79:1<131::AID-JSFA196>3.0.CO;2-Z.

Qian, Yichun, Joseph H. Lynch, Longyun Guo, David Rhodes, John A. Morgan, and Natalia Dudareva. 2019. "Completion of the Cytosolic Post-Chorismate Phenylalanine Biosynthetic Pathway in Plants." *Nature Communications* 10: 15. doi:10.1038/s41467-018-07969-2.

Robert, Christelle Aurélie Maud, Matthias Erb, Ivan Hiltpold, Bruce Elliott Hibbard, Mickaël David Philippe Gaillard, Julia Bilat, et al. 2013. "Genetically Engineered Maize Plants Reveal Distinct Costs and Benefits of Constitutive Volatile Emissions in the Field." *Plant Biotechnology Journal* 11 (5): 628–39. doi:10.1111/pbi.12053.

Sallaud, Christophe, Denis Rontein, Sandrine Onillon, Francoise Jabes, Philippe Duffé, Cécile Giacalone, Samuel Thoraval, et al. 2009. "A Novel Pathway for Sesquiterpene Biosynthesis from Z, Z-Farnesyl Pyrophosphate in the Wild Tomato *Solanum habroichaites*." *Plant Cell* 21: 301-17. doi: 10.1105/tpc.107.057885

Schnee, Christiane, Tobias G. Kollner, Matthias Held, Ted C. J. Turlings, Jonathan Gershenzon, and Jörg Degenhardt. 2006. "The Products of a Single Maize Sesquiterpene Synthase Form a Volatile Defense Signal That Attracts Natural Enemies of Maize Herbivores." *Proceedings of the National Academy of Sciences* 103 (4): 1129–34. doi:10.1073/pnas.0508027103.

Shalit, Moshe, Sharoni Shafir, Olga Larkov, Einat Bar, Dalit Kaslassi, Zach Adam, Dani Zamir, et al. 2004. "Volatile Compounds Emitted by Rose Cultivars: Fragrance Perception by Man and Honeybees." *Israel Journal of Plant Sciences* 52 (3): 245–55. doi:10.1560/p7g3-ft41-xjcp-1xfm.

Shen, Jiyuan, Denise Tieman, Jeffrey B. Jones, Mark G. Taylor, Eric Schmelz, Alisa Huffaker, Dawn Bies, Kunsong Chen, and Harry J. Klee. 2014. "A 13-Lipoxygenase, TomloxC, Is Essential for Synthesis of C5 Flavour Volatiles in Tomato." *Journal of Experimental Botany* 65 (2): 419–28. doi:10.1093/jxb/ert382.

Shiojiri, Kaori, Kyutaro Kishimoto, Rika Ozawa, Soichi Kugimiya, Soichi Urashimo, Genichiro Arimura, Junichiro Horiuchi, Takaaki Nishioka, Kenji Matsui, and Junji Takabayashi. 2006. "Changing Green Leaf Volatile Biosynthesis in Plants: An Approach for Improving Plant Resistance against Both Herbivores and Pathogens." *Proceedings of the National Academy of Sciences* 103 (45): 16672–76. doi:10.1073/pnas.0607780103.

Simkin, Andrew J.. 2004. "Circadian Regulation of the PhCCD1 Carotenoid Cleavage Dioxygenase Controls Emission of β-Ionone, a Fragrance Volatile of Petunia Flowers." *Plant Physiology* 136 (3): 3504–14. doi:10.1104/pp.104.049718.

Spitzer-Rimon, Ben, Elena Marhevka, Oren Barkai, Ira Marton, Orit Edelbaum, Tania Masci, Naveen-Kumar Prathapani, Elena Shklarman, Marianna Ovadis, and Alexander Vainstein. 2010. "*EOBII*, a Gene Encoding a Flower-Specific Regulator of Phenylpropanoid Volatiles' Biosynthesis in Petunia." *Plant Cell* 22 (6): 1961–76. doi:10.1105/tpc.109.067280.

Spitzer-Rimon, Ben, Moran Farhi, Boaz Albo, Alon Cna'ani, Michal Moyal Ben Zvi, Tania Masci, Orit Edelbaum, et al. 2012. "The R2R3-MYB-Like Regulatory Factor EOBI, Acting Downstream of EOBII, Regulates Scent Production by Activating *ODO1* and Structural Scent-Related Genes in Petunia." *Plant Cell* 24 (12): 5089–5105. doi:10.1105/tpc.112.105247.

Spyropoulou, Eleni A., Michel A. Haring, and Robert C. Schuurink. 2014. "RNA Sequencing on *Solanum lycopersicum* Trichomes Identifies Transcription Factors That Activate Terpene Synthase Promoters." *BMC Genomics* 15 (1): 1–16. doi:10.1186/1471-2164-15-402.

Tieman, Denise, Holly M. Loucas, Joo Young Kim, David G. Clark, and Harry J. Klee. 2007. "Tomato Phenylacetaldehyde Reductases Catalyze the Last Step in the Synthesis of the Aroma Volatile 2-Phenylethanol." *Phytochemistry* 68 (21): 2660–69. doi:10.1016/j.phytochem.2007.06.005.

Tieman, Denise, Mark Taylor, Nicolas Schauer, Alisdair R. Fernie, Andrew D. Hanson, and Harry J. Klee. 2006. "Tomato Aromatic Amino Acid Decarboxylases Participate in Synthesis of the Flavor Volatiles 2-Phenylethanol and 2-Phenylacetaldehyde." *Proceedings of the National Academy of Sciences* 103 (21): 8287–92. doi:10.1073/pnas.0602469103.

Tieman, Denise, Michelle Zeigler, Eric Schmelz, Mark G. Taylor, Sarah Rushing, Jeffrey B. Jones, and Harry J. Klee. 2010. "Functional Analysis of a Tomato Salicylic Acid Methyl Transferase and Its Role in Synthesis of the Flavor Volatile Methyl Salicylate." *Plant Journal* 62 (1): 113–23. doi:10.1111/j.1365-313X.2010.04128.x.

Tikunov, Yury M., Jos Molthoff, Ric C. H. de Vos, Jules Beekwilder, Adele van Houwelingen, Justin J. J. van der Hooft, Mariska Nijenhuis-de Vries, et al. 2013. "NON-SMOKY GLYCOSYLTRANSFERASE1 Prevents the Release of Smoky Aroma from Tomato Fruit." *Plant Cell* 25 (8): 3067–78. doi:10.1105/tpc.113.114231.

Tong, Xiaohong, Jinfeng Qi, Xudong Zhu, Bizeng Mao, Longjun Zeng, Baohui Wang, Qun Li, Guoxin Zhou, Xiaojing Xu, Yonggen Lou, and Zuhua He. 2012. "The Rice Hydroperoxide Lyase OsHPL3 Functions in Defense Responses by Modulating the Oxylipin Pathway." *Plant Journal* 71 (5): 763–75. doi:10.1111/j.1365-313X.2012.05027.x.

Tzin, Vered, Ilana Rogachev, Sagit Meir, Michal Moyal Ben Zvi, Tania Masci, Alexander Vainstein, Asaph Aharoni, and Gad Galili. 2013. "Tomato Fruits Expressing a Bacterial Feedback-Insensitive 3-Deoxy-D-Arabino-Heptulosonate 7-Phosphate Synthase of the Shikimate Pathway Possess Enhanced Levels of Multiple Specialized Metabolites and Upgraded Aroma." *Journal of Experimental Botany* 64 (14): 4441–52. doi:10.1093/jxb/ert250.

Underwood, Beverly A., Denise M. Tieman, Kenichi Shibuya, Richard J. Dexter, Holly M. Loucas, Andrew J. Simkin, Charles A. Sims, Eric A. Schmelz, Harry J. Klee, and David G. Clark. 2005. "Ethylene-Regulated Floral Volatile Synthesis in *Petunia corollas.*" *Plant Physiology* 138 (1): 255–66. doi:10.1104/pp.104.051144.

Verdonk, Julian C., Michel A. Haring, Arjen J. van Tunen, and Robert C. Schuurink. 2005. "*ODORANT1* Regulated Fragrance Biosynthesis in Petunia Flowers." *Plant Cell* 17 (5): 1612–1624. doi10.1105/tpc.104.028837.

Wang, Chunlin, Chee Kok Chin, Chi Tang Ho, Chin Fa Hwang, James J. Polashock, and Charles E. Martin. 1996. "Changes of Fatty Acids and Fatty Acid-Derived Flavor Compounds by Expressing the Yeast Δ-9 Desaturase Gene in Tomato." *Journal of Agricultural and Food Chemistry* 44 (10): 3399–3402. doi:10.1021/jf960174t.

Wang, Qi, Zhaojun Xin, Jiancai Li, Lingfei Hu, Yonggen Lou, and Jing Lu. 2015. "*(E)*-β-Caryophyllene Functions as a Host Location Signal for the Rice White-Backed Planthopper *Sogatella furcifera.*" *Physiological and Molecular Plant Pathology* 91 (7): 106–12. doi:10.1016/j.pmpp.2015.07.002.

Widhalm, Joshua R., Michael Gutensohn, Heejin Yoo, Funmilayo Adebesin, Yichun Qian, Longyun Guo, Rohit Jaini, et al. 2015. "Identification of a Plastidial Phenylalanine Exporter That Influences Flux Distribution through the Phenylalanine Biosynthetic Network." *Nature Communications* 6: 8142. doi:10.1038/ncomms9142.

Woo, Je Wook, Jungeun Kim, Soon Il Kwon, Claudia Corvalán, Seung Woo Cho, Hyeran Kim, Sang-gyu Kim, Sang-tae Kim, Sunghwa Choe, and Jin-soo Kim. 2015. "DNA-Free Genome Editing in Plants with Preassembled CRISPR-Cas9 Ribonucleoproteins." *Nature Biotechnology* 33 (11): 1162–64. doi:10.1038/nbt.3389.

Xiao, Yunhua, Qianjin Wang, Matthias Erb, Ted C.J. Turlings, Linmei Ge, Lingfei Hu, James Xinzhi Li, et al. 2012. "Specific Herbivore-Induced Volatiles Defend Plants and Determine Insect Community Composition in the Field." *Ecology Letters* 15 (10): 1130–39. doi:10.1111/j.1461-0248.2012.01835.x.

Xie, Qingjun, Zhongyuan Liu, Sagit Meir, Ilana Rogachev, Asaph Aharoni, Harry J. Klee, and Gad Galili. 2016. "Altered Metabolite Accumulation in Tomato Fruits by Coexpressing a Feedback-Insensitive *AroG* and the *PhODO1 MYB*-Type Transcription Factor." *Plant Biotechnology Journal* 14 (12): 2300–2309. doi:10.1111/pbi.12583.

Yon, Felipe, Youngsung Joo, Lucas Cortés Llorca, Eva Rothe, Ian T. Baldwin, and Sang Gyu Kim. 2016. "Silencing *Nicotiana attenuata* LHY and ZTL Alters Circadian Rhythms in Flowers." *New Phytologist* 209 (3): 1058–66. doi:10.1111/nph.13681.

Zhang, Bo, Denise M. Tieman, Chen Jiao, Yimin Xu, Kunsong Chen, Zhangjun Fe, James J. Giovannoni, and Harry J. Klee. 2016. "Chilling-Induced Tomato Flavor Loss Is Associated with Altered Volatile Synthesis and Transient Changes in DNA Methylation." *Proceedings of the National Academy of Sciences* 113 (44): 12580–85. doi:10.1073/pnas.1613910113.

Zhang, Yi, Zhen Liang, Yuan Zong, Yanpeng Wang, Jinxing Liu, Kunling Chen, Jin Long Qiu, and Caixia Gao. 2016. "Efficient and Transgene-Free Genome Editing in Wheat through Transient Expression of CRISPR/Cas9 DNA or RNA." *Nature Communications* 7: 12617. doi:10.1038/ncomms12617.

Zong, Yuan, Yanpeng Wang, Chao Li, Rui Zhang, Kunling Chen, Yidong Ran, Jin Long Qiu, Daowen Wang, and Caixia Gao. 2017. "Precise Base Editing in Rice, Wheat and Maize with a Cas9-Cytidine Deaminase Fusion." *Nature Biotechnology* 35 (5): 438–40. doi:10.1038/nbt.3811.

Zuker, Amir, Tzvi Tzfira, Hagit Ben-Meir, Marianna Ovadis, Elena Shklarman, Hanan Itzhaki, Gert Forkmann, et al. 2002. "Modification of Flower Color and Fragrance by Antisense Suppression of the Flavanone 3-Hydroxylase Gene." *Molecular Breeding* 9 (1): 33–41. doi:10.1023/A:1019204531262.

Index

Note: Page numbers in italic and bold refer to figures and tables, respectively.

Printed and bound by CPI Group (UK) Ltd, Croydon, CR0 4YY

17/10/2024

01775663-0007